INTERNATIONAL UNION OF CRYSTALLOGRAPHY
BOOK SERIES

IUCr Monographs on Crystallography

1 *Accurate molecular structures*
   A. Domenicano, I. Hargittai, editors

2 *P. P. Ewald and his dynamical theory of X-ray diffraction*
   D. W. J. Cruickshank, H. J. Juretschke, N. Kato, editors

3 *Electron diffraction techniques, Vol. 1*
   J. M. Cowley, editor

4 *Electron diffraction techniques, Vol. 2*
   J. M. Cowley, editor

5 *The Rietveld method*
   R. A. Young, editor

6 *Introduction to crystallographic statistics*
   U. Shmueli, G. H. Weiss

7 *Crystallographic instrumentation*
   L. A. Aslanov, G. V. Fetisov, J. A. K. Howard

8 *Direct phasing in crystallography*
   C. Giacovazzo

9 *The weak hydrogen bond*
   G. R. Desiraju, T. Steiner

10 *Defect and microstructure analysis by diffraction*
   R. L. Snyder, J. Fiala and H. J. Bunge

11 *Dynamical theory of X-ray diffraction*
   A. Authier

12 *The chemical bond in inorganic chemistry*
   I. D. Brown

13 *Structure determination from powder diffraction data*
   W. I. F. David, K. Shankland, L. B. McCusker, Ch. Baerlocher, editors

14 *Polymorphism in molecular crystals*
   J. Bernstein

15 *Crystallography of modular materials*
  G. Ferraris, E. Makovicky, S. Merlino
16 *Diffuse X-ray scattering and models of disorder*
  T. R. Welberry
17 *Crystallography of the polymethylene chain: an inquiry into the structure of waxes*
  D. L. Dorset
18 *Crystalline molecular complexes and compounds: structure and principles*
  F. H. Herbstein
19 *Molecular aggregation: structure analysis and molecular simulation of crystals and liquids*
  A. Gavezzotti
20 *Aperiodic crystals: from modulated phases to quasicrystals*
  T. Janssen, G. Chapuis, M. de Boissieu
21 *Incommensurate crystallography*
  S. van Smaalen
22 *Structural crystallography of inorganic oxysalts*
  S.V. Krivovichev
23 *The nature of the hydrogen bond: outline of a comprehensive hydrogen bond theory*
  G. Gilli, P. Gilli

IUCr Texts on Crystallography

 1 *The solid state*
  A. Guinier, R. Julien
 4 *X-ray charge densities and chemical bonding*
  P. Coppens
 7 *Fundamentals of crystallography, second edition*
  C. Giacovazzo, editor
 8 *Crystal structure refinement: a crystallographer's guide to SHELXL*
  P. Müller, editor
 9 *Theories and techniques of crystal structure determination*
  U. Shmueli
10 *Advanced structural inorganic chemistry*
  Wai-Kee Li, Gong-Du Zhou, Thomas Mak
11 *Diffuse scattering and defect structure simulations: a cook book using the program DISCUS*
  R. B. Neder, T. Proffen
12 *The basics of crystallography and diffraction, third edition*
  C. Hammond
13 *Crystal structure analysis: principles and practice, second edition*
  W. Clegg, editor

Structural Crystallography of Inorganic Oxysalts

SERGEY V. KRIVOVICHEV

Professor and Chairman
Department of Crystallography
St. Petersburg State University

OXFORD
UNIVERSITY PRESS

OXFORD

UNIVERSITY PRESS

Great Clarendon Street, Oxford OX2 6DP

Oxford University Press is a department of the University of Oxford.
It furthers the University's objective of excellence in research, scholarship,
and education by publishing worldwide in

Oxford New York

Auckland Cape Town Dar es Salaam Hong Kong Karachi
Kuala Lumpur Madrid Melbourne Mexico City Nairobi
New Delhi Shanghai Taipei Toronto

With offices in

Argentina Austria Brazil Chile Czech Republic France Greece
Guatemala Hungary Italy Japan Poland Portugal Singapore
South Korea Switzerland Thailand Turkey Ukraine Vietnam

Oxford is a registered trade mark of Oxford University Press
in the UK and in certain other countries

Published in the United States
by Oxford University Press Inc., New York

© Sergey Krivovichev 2009

The moral rights of the author have been asserted
Database right Oxford University Press (maker)

First published 2009

British Library Cataloguing in Publication Data

Data available

Library of Congress Cataloging in Publication Data

Krivovichev, Sergey V.
Structural crystallography of inorganic oxysalts/Sergey V. Krivovichev.
p. cm.—(International Union of Crystallography monographs on
crystallography; no. 22)
ISBN 978–0–19–921320–7
1. Oxysalts.   2. Inorganic compounds.   3. Crystal growth.
4. Crystallography.   I. Title.
QD194.K75 2008
546'.34—dc22   2008048455

Typeset by Newgen Imaging Systems (P) Ltd., Chennai, India
Printed in the UK
on acid-free paper by the MPG Books Group

ISBN 978–0–19–921320–7 (Hbk.)

1 3 5 7 9 10 8 6 4 2

To Irina, Ivan, Nikolay, Evfrasiya,
Vasilisa, Aleksey and Platon

Preface

Inorganic oxysalts represent one of the most important and widespread groups of inorganic compounds. They constitute more than half of the mineral species known today, they form in the environment around us, their structure and properties determine a wide range of natural and technological processes. This book deals with the crystallography of inorganic oxysalts, i.e. with a description of the organization of atoms in their crystal structures. The world of inorganic oxysalts is so diverse that a systematic and adequate description of this diversity seems to be almost impossible. The imagination and knowledge of structural crystallographers are very often restricted to their own objects of study. They work out the structure solution (sometimes easy, sometimes difficult), produce crystallographic information files (CIF) and leave the structure aside. The broad view and structure correlations are often neglected, leaving an important gap in our understanding of nature. The essential part of the book is devoted to various approaches and techniques used in the description of the structural architecture and classification of inorganic oxysalts. For this purpose, the author will use various topological theories such as graph theory, tilings, nets, space partitions, etc. Of course, this part of the book is rather subjective since it represents sometimes personally elaborated (and thus favorite) author's techniques. However, the author has tried to include into consideration all useful and interesting ideas that he has met in the literature. Chapter 1 contains a very short listing of basic facts concerning geometrical parameters and the history of structural classification of inorganic oxysalts. Many important concepts are only slightly touched on here, taking into account that several excellent books have appeared recently that tackle these problems in more detail (for instance, I.D. Brown's book on bond-valence theory published in the IUCr series in 2002). Chapter 2 deals with topological classification of low-dimensional structural units in inorganic oxysalts using graph theory. The structural diversity of such units is amazing as is the fact that most of these topologies can be derived from simple and highly symmetrical archetype graphs. In Chapter 3, framework structures are considered and some principles of their description and classification are outlined. In contrast to 2D graphs, it is shown that 3-dimensional nets that describe topologies of polyhedral linkages in inorganic oxysalts are often extremely complex and cannot be reduced to simple networks. Dense 2-dimensional structures are analyzed in Chapter 4 using the concept of anion topology developed by P.C. Burns and co-authors. Alternative approaches to structure description, such as anion-centered polyhedra and cation arrays, are discussed in Chapter 5. Chapter 6 is devoted to the analysis of correlations between the structure and composition of inorganic oxysalts using the principle of dimensional reduction.

I am quite aware that many aspects of crystallography of inorganic oxysalts have escaped my attention and expertise. In particular: no information is given on phase transitions in inorganic oxysalts and their changes under changing temperature and pressure; less attention is paid to oxysalts with large and low-valent cations than to those containing relatively small and highly charged cations, etc. I hope that these gaps will be covered in the future by those who are more experienced and qualified than me to discuss these topics.

Finally, I would like to express my hope that this book will be useful to the broad audience of scientists working with crystal structures of inorganic oxysalts, including crystallographers, mineralogists, material scientists, and specialists in solid-state chemistry and physics.

Sergey V. Krivovichev

Acknowledgements

First, my deep gratitude goes to Stanislav K. Filatov who introduced me into the field of inorganic crystal chemistry and carefully guided my first steps of scientific development during the economically difficult and sometimes morally unbearable days of "perestroika". International support was especially important at that time and I am grateful to Friedrich Liebau who helped us to incorporate our ideas into the current framework of modern structural and crystal chemistry. Over the years, I enjoyed collaboration with my friends and colleagues Thomas Armbruster, Peter Burns, Christopher Cahill, Wulf Depmeier, Andrew Locock, Viktor Yakovenchuk, Sergey Britvin, Dmitrii Chernyshov, and Joel Brugger. My former and current students provided many interesting and new insights (in chronological order): Evgenii Nazarchuk, Oleg Siidra, Andrey Zolotarev, Vladislav Gurzhiy, Darya Spiridonova, and Anastasiya Chernyatieva. I am grateful to Giovanni Ferraris for his continuous support and encourangement to write this book. I am indebted to my Russian colleagues who keep our national scientific institutes running and whose efforts guarantee that Russian scientific traditions will not vanish forever: Vadim S. Urusov, Vladimir Ya. Shevchenko, Dmitrii Yu. Pushcharovskii, Igor V. Pekov, Natalia I. Organova, Andrei G. Bulakh, and many others with whom I had the privilege and honor to collaborate. Special thanks are due to my parents, Vladimir G. Krivovichev and Galina L. Starova, for all they have done for me. I also gratefully acknowledge the support of the Russian Ministry of Science and Education and Russian Foundation for Basic Research that provided funding for our research in crystal chemistry of inorganic oxysalts. International collaboration was possible only with the support of the US National Science Foundation, the Alexander von Humboldt Foundation, the Swiss National Science Foundation, Deutsche Forschungsgemeinschaft, and the Austrian Science Foundation.

Contents

1	Basic concepts		1
	1.1	Structural classification of inorganic oxysalts	1
	1.2	Basic geometrical parameters	3
	1.3	OH and H_2O in inorganic oxysalts	3
2	Graph theory applied to low-dimensional structural units in inorganic oxysalts		6
	2.1	Symbolic description of topologies of heteropolyhedral structural units	6
	2.2	2D topologies: graphs with M–T links only	8
		2.2.1 Basic graph {**3.6.3.6**} and its derivatives	8
		2.2.1.1 M:T = 1:2	10
		2.2.1.2 Topological isomerism	15
		2.2.1.3 M:T = 2:3	18
		2.2.1.4 M:T = 3:4	18
		2.2.1.5 M:T = 13:18, 3:8, 5:8, 3:5, 1:1 and 1:3	22
		2.2.1.6 Geometrical isomerism and orientation matrix	24
		2.2.1.7 *Cis–trans* geometrical isomerism and connectivity diagrams	34
		2.2.2 Autunite topology and its derivatives	35
		2.2.3 Other basic graphs and their derivatives	40
		2.2.4 Modular approach to complex 2D topologies	42
		2.2.4.1 Topologies derived from the same basic graph	42
		2.2.4.2 Topologies derived from more than one basic graph	46
	2.3	2D topologies: graphs with M=T links and without M–M links	46
	2.4	2D topologies: graphs with M–M, M=M, or M≡M links	52
		2.4.1 Graphs with M–M links	52
		2.4.1.1 Graphs with finite subgraphs of black vertices	54
		2.4.1.2 Graphs with 1D subgraphs of black vertices	54
		2.4.1.3 Graphs with 2D subgraphs of black vertices	59
		2.4.1.4 Graphs with modular structure	60
		2.4.1.5 Other graphs	64
		2.4.2 Graphs with M=M links	65
		2.4.3 Graphs with M≡M links	68

	2.5	1D topologies: chains	69
	2.5.1	Some notes	69
	2.5.2	Orientational geometrical isomerism of chains	72
	2.5.3	Lone-electron-pair-induced geometrical isomerism	77
	2.5.4	*Cis–trans* isomerism	79
	2.6	0D topologies: finite clusters	81
	2.7	Nanoscale low-dimensional units in inorganic oxysalts: some examples	86

3	Topology of framework structures in inorganic oxysalts		94
	3.1	Regular and quasiregular nets	94
	3.2	Heteropolyhedral frameworks: classification principles	96
	3.3	Frameworks based upon fundamental building blocks (FBBs)	98
	3.3.1	Some definitions	98
	3.3.2	Leucophosphite-type frameworks	98
	3.3.3	Frameworks with oxocentered tetrahedral cores	100
	3.3.4	Pharmacosiderite-related frameworks	102
	3.3.5	Nasicon, langbeinite and related frameworks	103
	3.4	Frameworks based upon polyhedral units	105
	3.4.1	Polyhedra	106
	3.4.2	Tilings	106
	3.4.3	Example 1: minerals of the labuntsovite group	107
	3.4.4	Example 2: shcherbakovite–batisite series	108
	3.4.5	Combinatorial topology of polyhedral units	109
	3.4.6	Topological complexity of polyhedral units: petarasite net	112
	3.5	Frameworks based upon infinite chains	113
	3.5.1	Fundamental chains as bases of complex frameworks: an example	113
	3.5.2	Frameworks with non-parallel orientations of fundamental chains	115
	3.5.3	Frameworks with no M–M and T–T linkages	115
	3.5.3.1	Frameworks consisting of kroehnkite chains	115
	3.5.3.2	Other examples	120
	3.5.4	Frameworks with M–M and no T–T linkages	124
	3.5.4.1	Frameworks based upon chains of corner-sharing M octahedra	124
	3.5.4.2	Frameworks consisting of finite clusters of corner-sharing octahedra	128
	3.5.5	Frameworks with M=M and no T–T linkages	129
	3.5.6	Some frameworks based upon T_2O_7 double tetrahedra	131
	3.5.7	Frameworks with both T–T and M–M linkages	134
	3.5.7.1	Zorite and ETS-4	134
	3.5.7.2	Benitoite net as based upon arrangement of polyhedral units and tubular units	135

	3.5.7.3	Tubular units, their topology, symmetry and classification	137
3.6	Frameworks based upon 2D units		140
	3.6.1	Frameworks based upon sheets with no T–T linkages	140
	3.6.2	Frameworks based upon sheets with T–T linkages	143
		3.6.2.1 Octahedral–tetrahedral frameworks with 2D tetrahedral anions	143
		3.6.2.2 Umbite-related frameworks	150
		3.6.2.3 The use of 2D nets to recognize structural relationships	153
	3.6.3	Structure description versus intuition	154

4	Anion-topology approach		163
	4.1	The concept of anion topology	163
	4.2	Classification of anion topologies	164
	4.3	Anion topologies and isomerism	169
	4.4	1D derivatives of anion topologies	171

5	Alternative approaches to structure description		173
	5.1	Introductory remarks	173
	5.2	Anion-centered tetrahedra in inorganic oxysalts	173
		5.2.1 Historical notes	173
		5.2.2 0D units of oxocentered tetrahedra	175
		5.2.3 1D units of oxocentered tetrahedra	179
		5.2.4 2D units of anion-centered tetrahedra	187
		5.2.5 3D units of anion-centered tetrahedra	198
		5.2.6 Which cations may form anion-centered tetrahedra?	201
	5.3	Anion-centered octahedra in inorganic oxysalts	203
		5.3.1 Some notes on antiperovskites	203
		5.3.2 Chemical considerations	204
		5.3.3 Structural diversity	204
	5.4	Cation arrays in inorganic oxysalts	209

6	Dimensional reduction in inorganic oxysalts		215
	6.1	Introduction	215
	6.2	Alkali-metal uranyl molybdates	215
	6.3	Inorganic oxysalts in the $Ca_3(TO_4)_2 - H_3(TO_4) - (H_2O)$ system	217
	6.4	Inorganic oxysalts in the $M_2(TO_4)_3 - H_2(TO_4) - (H_2O)$ system	221
	6.5	Inorganic oxysalts in the $M(TO_4) - H_2(TO_4) - (H_2O)$ system	224
	6.6	Concluding remarks	225

| References | | | 227 |
| Index | | | 305 |

1

Basic concepts

1.1 Structural classification of inorganic oxysalts

Inorganic oxysalts are inorganic compounds that can be considered as salts of simple inorganic oxyacids, $H_n T^{q+} O_m$, where $q \geq 3$ and T has either triangular or tetrahedral coordination. The structure of an inorganic oxysalt consists of $(T^{q+} O_m)^{n-}$ anionic subunits interlinked by some additional charge-compensating agents. As a rule, description of the structure of an inorganic oxysalt is based upon subdivision in it of the strongest structural unit.

The $(T^{q+} O_m)^{n-}$ subunits may be polymerized into units of higher dimensionality (finite clusters, chains, sheets or frameworks) that are obviously the strongest units in the structure. Classification of such structures with condensed anions is usually based upon their topological and dimensional properties. The classical example of polymerized tetrahedral units is silicates. Bragg (1930) classified silicates according to the features of polymerization of $(Si,Al)O_4$ tetrahedra, and this scheme was developed in detail by Zoltai (1960) and Liebau (1985). Condensed phosphates containing anionic units of linked (PO_4) tetrahedra (polyphosphates) were classified and reviewed by Durif (1995). Borates may contain structural units of polymerized (BO_4) tetrahedral and (BO_3) triangles. Classification of triangular–tetrahedral anions in borates is much more intricate than that of silicates and polyphosphates and has been reviewed by Christ and Clark (1977), Burns *et al.* (1995), Grice *et al.* (1998), and Yuan and Xue (2007).

In the case of isolated $(T^{q+} O_m)^{n-}$ subunits, structures of inorganic oxysalts are described in terms of the next strongest structural subunits that unite the $(T^{q+} O_m)^{n-}$ subunits into extended complexes.

The principle of *structural hierarchy* plays an important role in the structural crystallography of inorganic oxysalts. The first and lowest level of hierarchy consists of atoms that combine to form structural subunits (second level). Structural subunits link together to form structural units. The structural unit is the basis of structural organization and thus its dimensionality and topology serves as a major criterion for the classification.

For oxysalts with isolated $(T^{q+} O_m)^{n-}$ subunits, subdivision of a structural unit as the strongest unit that forms the basis of a crystal structure has long been recognized. Moore (1970a, b, 1975, etc.) investigated structural hierarchies in phosphates and arsenates with octahedrally coordinated M cations (Fe, Al, Mn, Mg, etc.) and provided many important insights and instruments for the topological description of structural units (graph theory, Schlegel diagrams, structural and geometrical isomerism, combinatorial polymorphism, etc.). Structural classification of sulphates was proposed by Bokii and Gorogotskaya (1969), Sabelli and Trosti-Ferroni (1985), Rastsvetaeva and Pushcharovskii (1989). Hawthorne (1983, 1985, 1986, 1990, 1994, 1998) developed a systematical approach to the description of inorganic oxysalts using bond-valence theory (Brown 2002). According to Hawthorne (1983), structures can be classified according to the polymerization of cation-centered polyhedra with higher bond valences. These may form homo- or heteropolyhedral structural units, usually with negative formal charge. This charge is compensated by interstitial species, which are usually low-charge large cations (e.g. alkali-metal cations). This approach was developed for a large number of inorganic oxysalts, including sulphates (Hawthorne *et al.* 2000), phosphates (Huminicki and Hawthorne 2002), and Be-containing minerals (Hawthorne and Huminicki 2002). The same general principle was applied to U^{6+} oxides and oxysalts by Burns *et al.* (1996) and Burns (1999, 2005).

In the Russian school of crystal chemistry, a similar approach has been developed to the description of complex inorganic oxysalts. This theory, known as the theory of *mixed anionic radicals*, parallels the theory of heteropolyhedral structural units proposed by Hawthorne (1983). It was first proposed by Voronkov *et al.* (1975) and developed systematically by Sandomirskii and Belov (1985). According to this approach, a mixed anionic radical is a fragment of a structure consisting of linked coordination polyhedra of cations (~ heteropolyhedral unit). The topological and dimensional properties of mixed anionic radicals are used to construct structural classifications, to elucidate structural stability limits, etc. This theory was used as a basis for the classification of sulphates by Rastsvetaeva and Pushcharovskii (1989) (see also Pushcharovskii *et al.* 1998).

There are also numerous papers that deal with separate classes of inorganic oxysalts. Recently, a fundamental survey of rare-earth oxysalts was provided by Wickleder (2002). Reviews on the structural chemistry of actinide oxysalt compounds can be found in (Krivovichev *et al.* 2007f). Modular aspects of the crystallography of inorganic oxysalts have been considered in the monograph by Ferraris *et al.* (2004).

In this book, we shall basically use the principles of structural classifications of inorganic oxysalts similar to those developed by Hawthorne (1983) as the theory of heteropolyhedral structural units and by Sandomirskii and Belov (1985) as the theory of mixed anionic radicals. For the matter of terminology, we follow the recommendations of the Subcommittee on the Nomenclature of the Inorganic Structure Types of the International Union of Crystallography Commission on the Crystallographic Nomenclature (Lima-de-Faria *et al.* 1990). The structure is considered as based upon structural units that represent the linked array of the strongest coordination polyhedra. The charge (if any) of the structural unit is compensated by interstitial ions of low charge and high coordination numbers. As a rule, coordination polyhedra forming the structural unit (= subunits) are centered by cations. The exceptions are structures that can be described in terms of anion-centered polyhedra and that are considered in Chapter 6.

1.2   Basic geometrical parameters

In many cases, it is useful to have a reference list of typical T–O bond lengths of the $(T^{q+}O_m)^{n-}$ anionic subunits ($q \geq 3$; T has either triangular or tetrahedral coordination). Such a list is provided in Table 1.1. Polymerization of the $(T^{q+}O_m)^{n-}$ subunits by sharing bridging O_{br} atom results in elongation of the T–O_{br} bond in comparison to other T–O bonds. Detailed discussions of distortion of tetrahedral oxyanions and predictive relationships can be found in (Louisnathan *et al.* 1977; Griffen and Ribbe 1979; Baur 1974, 1981; Liebau 1985).

Protonation of the $(T^{q+}O_m)^{n-}$ subunits results in formation of acid $(H_kT^{q+}O_m)^{(n-k)-}$ subunits. The T–OH bonds are essentially elongated compared to normal T–O bonds that can be used as a sign of a protonated complex during structure refinement. Table 1.2 provides statistical data on the T–OH bonds for most common $(T^{q+}O_m)^{n-}$ subunits.

1.3   OH and H$_2$O in inorganic oxysalts

Hydrogen bonding plays an important role in the structural organization of inorganic oxysalts. The basic principles of hydrogen bonding can be found in (Jeffrey 1997). Hawthorne (1992) considered the role of OH and H$_2$O in oxysalt minerals in great detail. According to Hawthorne (1992), H$_2$O can play four different roles. 1. As a component of a structural unit, H$_2$O dictates the dimensional polymerization of the

Table 1.1   Average <T–O> bond lengths in $(T^{q+}O_m)^{n-}$ subunits (m = 3 and 4; $q \geq 3$)

m	T	q	<T–O> [Å]	Reference
4	B	3+	1.478	Baur 1981
	Si	4+	1.623	Baur 1981
	Ge	4+	1.756	Baur 1981
	P	5+	1.537	Huminicki and Hawthorne 2002
	As	5+	1.682	Baur 1981
	V	5+	1.721	Gopal and Calvo 1973
	S	6+	1.473	Hawthorne *et al.* 2000
	Se	6+	1.638	This work
	Mo	6+	1.762	Krivovichev 2004
	Cr	6+	1.647	Krivovichev 2004
	Cl	7+	1.435	This work
	Re	7+	1.718	This work
3	As	3+	1.788	This work
	B	3+	1.367	Baur 1981
	C	4+	1.285	This work
	Se	4+	1.709	Hawthorne *et al.* 1987
	N	5+	1.250	Baur 1981

Table 1.2 Average $<T–OH>$ bond lengths in protonated $(H_kT^{q+}O_m)^{n-}$ subunits ($m = 3$ and 4; $q \geq 3$)

T	m	Subunit	$<T–OH>$ [Å]	Reference
B^{3+}	3	HBO_3	1.372	Ferraris and Ivaldi 1984
		H_3BO_3	1.364	Ferraris and Ivaldi 1984
	4	HBO_4	1.465	Ferraris and Ivaldi 1984
		H_2BO_4	1.481	Ferraris and Ivaldi 1984
		H_3BO_4	1.485	Ferraris and Ivaldi 1984
C^{4+}	3	HCO_3	1.331	Ferraris and Ivaldi 1984
Se^{4+}	3	$HSeO_3$	1.759	Ferraris and Ivaldi 1984
		H_2SeO_3	1.736	Ferraris and Ivaldi 1984
Si^{4+}	4	$HSiO_4$	1.643	Nyfeler and Armbruster 1998
		H_2SiO_4	1.660	This work
N^{5+}	3	HNO_3	1.339	Ferraris and Ivaldi 1984
As^{5+}	4	$HAsO_4$	1.731	Ferraris and Ivaldi 1984
		H_2AsO_4	1.709	Ferraris and Ivaldi 1984
		H_3AsO_4	1.697	Ferraris and Ivaldi 1984
P^{5+}	4	HPO_4	1.581	Ferraris and Ivaldi 1984
		H_2PO_4	1.564	Ferraris and Ivaldi 1984
		H_3PO_4	1.544	Ferraris and Ivaldi 1984
S^{6+}	4	HSO_4	1.555	Ferraris and Ivaldi 1984
		H_2SO_4	1.530	Ferraris and Ivaldi 1984
Se^{6+}	4	$HSeO_4$	1.710	Ferraris and Ivaldi 1984
		H_2SeO_4	1.730	Ferraris and Ivaldi 1984

structural unit (see Chapter 4 for more discussion). 2. Interstitial H_2O bonded to an interstitial cation to form complex cations. The H_2O molecules of this type act as bond-valence transformers, moderating the Lewis acidity of interstitial cations. 3. Interstitial H_2O not bonded to an interstitial cation and occupying a well-defined position participates in a stable hydrogen-bonded network. 4. Occluded H_2O that is not bonded to an interstitial cation and not involved in a stable hydrogen-bonded network.

The H_2O molecules can be protonated that results in formation of hydronium ions, H_3O^+. Two types of H_3O^+ cations can be distinguished in inorganic oxysalts. H_3O^+ cations of the *first type* are strongly hydrogen-bonded to three adjacent anions. In this case, these can easily be recognized during structure refinement, even if the H positions could not be determined. For instance, the structure of $(H_3O)_2[C_{12}H_{30}N_2]_3[(UO_2)_4(SeO_4)_8](H_2O)_5$ (Krivovichev *et al.* 2005d) contains two symmetrically independent H_3O^+ ions that can easily be identified by the analysis of their coordination environment. The two corresponding O are coordinated by three oxygen atoms located at the O···O distances of of 2.47–2.64 Å. The coordination is a rather flat pyramid that can be inferred from the O···H_3O···O angles of 99.0–123.8°. The H_3O^+ molecules of the *second type* are not so well defined and usually disordered in the interstitial space of the structure. Their presence cannot be unambiguously identified from the structure data and is inferred on the basis of charge compensation. The typical example is hydronium jarosite, $(H_3O)Fe_3(SO_4)_2(OH)_6$ (Majzlan *et al.* 2004).

The structure of $H_5O_2^+$ cation involves formation of a symmetrical O–H–O bond between adjacent H_2O molecules. The O⋯O distance is in the range 2.40–2.50 Å with a bridging H^+ in between the O atoms (O–H = 1.10–1.30 Å). The data on more complex hydrogen-bonded H_2O clusters can be found in Jeffrey (1997). However, they are not so common in inorganic oxysalts.

Graph theory applied to low-dimensional structural units in inorganic oxysalts

2.1 Symbolic description of topologies of heteropolyhedral structural units

Symbolic representation of structural units using the graph approach is a useful and efficient method of description of complex structural topologies based upon cation-centered coordination polyhedra. Within this approach, the structure (or a part of it) is represented as a graph with nodes symbolizing coordination polyhedra. Nodes of different colors correspond to geometrically and/or chemically different polyhedra. Two nodes of the graph are linked only if corresponding polyhedra share at least one common ligand. Figures 2.1(a) and (b) show $[(UO_2)(MoO_4)_4]^{6-}$ finite clusters from the structure of $Rb_6[(UO_2)(MoO_4)_4]$ (Krivovichev and Burns 2002a). This unit consists of a UO_6 octahedron that shares four equatorial O atoms with four adjacent MoO_4 tetrahedra. Using a graphical representation, the topological structure of this unit (connectivity of its polyhedra) may be described by the black-and-white graph shown in Fig. 2.1(c). Here, the black vertex symbolizes U coordination polyhedron, and four white nodes correspond to four Mo tetrahedra. The central black vertex is linked to white nodes by four single lines that means that the corresponding U polyhedron shares four of its corners with four adjacent Mo polyhedra, one corner for each MoO_4 tetrahedron.

Thus, every structural unit based upon different coordination polyhedra is associated with a graph of corresponding complexity. If only two types of polyhedra are present (e.g. octahedra and tetrahedra), one needs two colors (black and white) to distinguish between two different polyhedra types. In this chapter, we shall use black and white colors to denote $M^{q+}O_n$ and $T^{r+}O_m$ polyhedra, respectively.

Following Liebau (1985), we define the connectedness, s, of a polyhedron as the number of adjacent polyhedra with which it shares common corners. The finite cluster shown in Fig. 2.1(b) consists of four-connected U polyhedron ($s_U = 4$) surrounded by one-connected Mo tetrahedra ($s_{Mo} = 1$). The corresponding graph shown in Fig. 2.1(c) is therefore (4;1)-connected.

Figure 2.2(a) shows a $[(UO_2)_3(CrO_4)_5]^{4-}$ sheet observed in the structure of $Mg_2[(UO_2)_3(CrO_4)_5](H_2O)_{17}$ (Krivovichev and Burns 2003e). Its black-and-white graph (Fig. 2.2(b)) is (5;3)-connected (black and white nodes are linked to five and

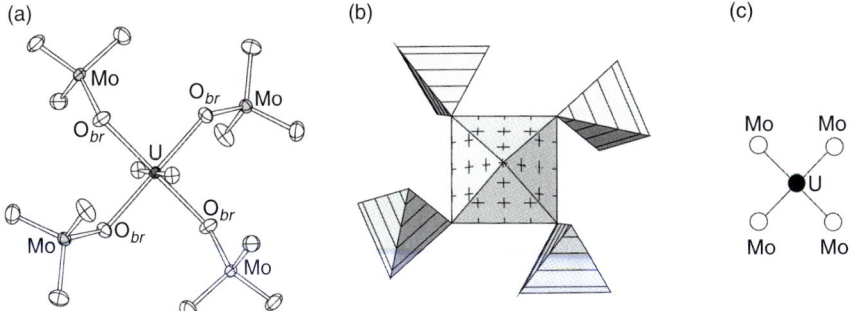

Fig. 2.1. $[(UO_2)(MoO_4)_4]^{6-}$ finite cluster from the structure of $Rb_6[(UO_2)(MoO_4)_4]$: (a) atomic structure (ellipsoids are drawn at 50% probability level); (b) polyhedral representation (U polyhedron is cross-hatched, MoO_4 tetrahedra are lined); (c) graphical representation (black node symbolizes U coordination polyhedron, and four white nodes correspond to four Mo tetrahedra).

three adjacent nodes each, respectively). The idealized version of this graph is shown in Fig. 2.2(d). This graph can be obtained from the more complete graph shown in Fig. 2.2(c) by deleting some of its white nodes and all edges associated with them. One may observe that the graph shown in Fig. 2.2(c) is very regular: it consists of 6-connected black and 3-connected white nodes (and therefore is (6;3)-connected). The elementary unit of this graph is a rhomb formed by two black and two white nodes. The **s** values of the nodes of the rhomb written in cyclic order is 3-6-3-6. Consequently, this graph is denoted {**3.6.3.6**}.

As we shall demonstrate in this chapter, the {**3.6.3.6**} graph is a *basic graph*, which means that it is parent to a number of other graphs corresponding to layers of coordination polyhedra observed in salts of inorganic oxoacids. There are several other basic graphs in the salts of inorganic oxoacids that will be discussed later.

As 2D black-and-white graphs are thought of as being infinite in the plane, they should contain cycles (or rings). The {**3.6.3.6**} graph itself contains only 4-membered rings. In contrast, the graph shown in Fig. 2.2(d) contains 4- and 6-membered rings in the ratio of 6:1. We define a *ring symbol* as a sequence $p_1^{r_1}p_2^{r_2}\ldots p_n^{r_n}$, where p_1, $p_2,\ldots p_n$ are numbers of nodes in a ring and $r_1, r_2,\ldots r_n$ are relative numbers of the corresponding rings in a graph. Thus, the ring symbol of the graph shown in Fig. 2.2(d) is $4^6 6^1$.

Traditionally, structural units in salts of inorganic oxoacids are described in the increasing order of their dimensionality (0D units or finite clusters are described first, then chains (1D units), sheets (2D units) and frameworks (3D units)) (see, e.g. Hawthorne *et al.* 2000). Here, we follow a slightly different philosophy. First, we explore 2D units and then show how 0D and 1D units can be obtained from corresponding 2D units using "cut-and-paste" operations.

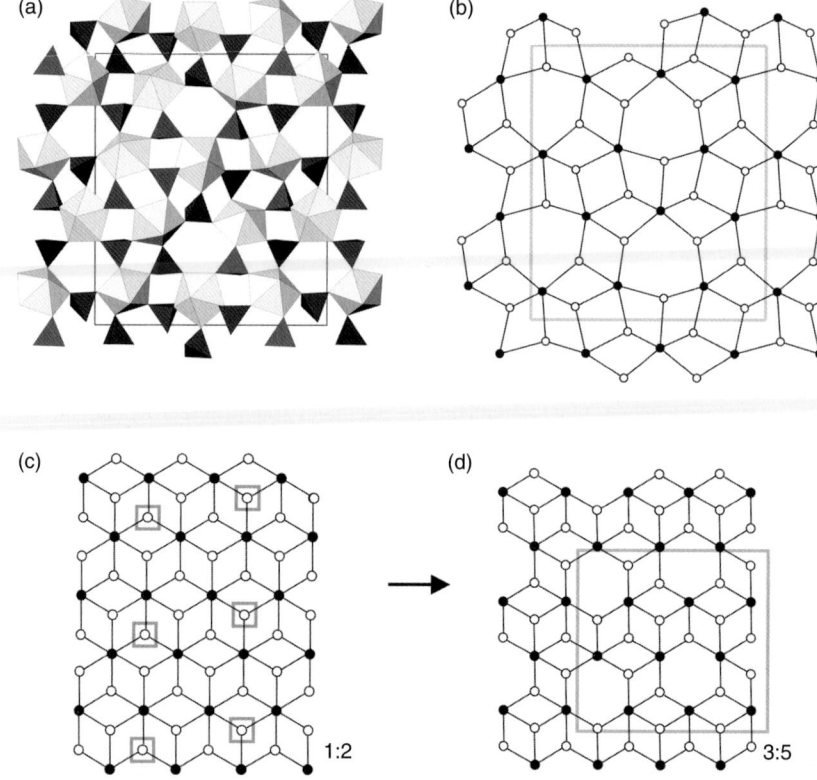

Fig. 2.2. $[(UO_2)_3(CrO_4)_5]^{4-}$ sheet observed in the structure of $Mg_2[(UO_2)_3(CrO_4)_5](H_2O)_{17}$; (b) its black-and-white graph; (c) the {**3.6.3.6**} graph; (d) an idealized version of the graph shown in (b). The graph in (d) can be obtained from the graph in (c) by eliminating marked white nodes and all links incident upon them.

2.2   2D topologies: graphs with *M–T* links only

2.2.1   *Basic graph {3.6.3.6} and its derivatives*

As was mentioned above, the {**3.6.3.6**} graph shown in Fig. 2.2(c) is a basic graph, i.e. it is parent to a number of 2D black-and-white graphs. Transformation of the {**3.6.3.6**} graph into its derivatives may involve: (i) deletion of a link and/or (ii) deletion of a vertex and all links incident upon it. Since the connectedness values (**s**) of black and white nodes in the {**3.6.3.6**} graph are 6 and 3, respectively, **s** values in the derivatives of the {**3.6.3.6**} graph may not exceed 6 and 3, respectively. The {**3.6.3.6**} derivatives can be easily classified according to the black:white ratio of its nodes. Obviously, this ratio corresponds to the *M:T* ratio of the respective heteropolyhedral units. In

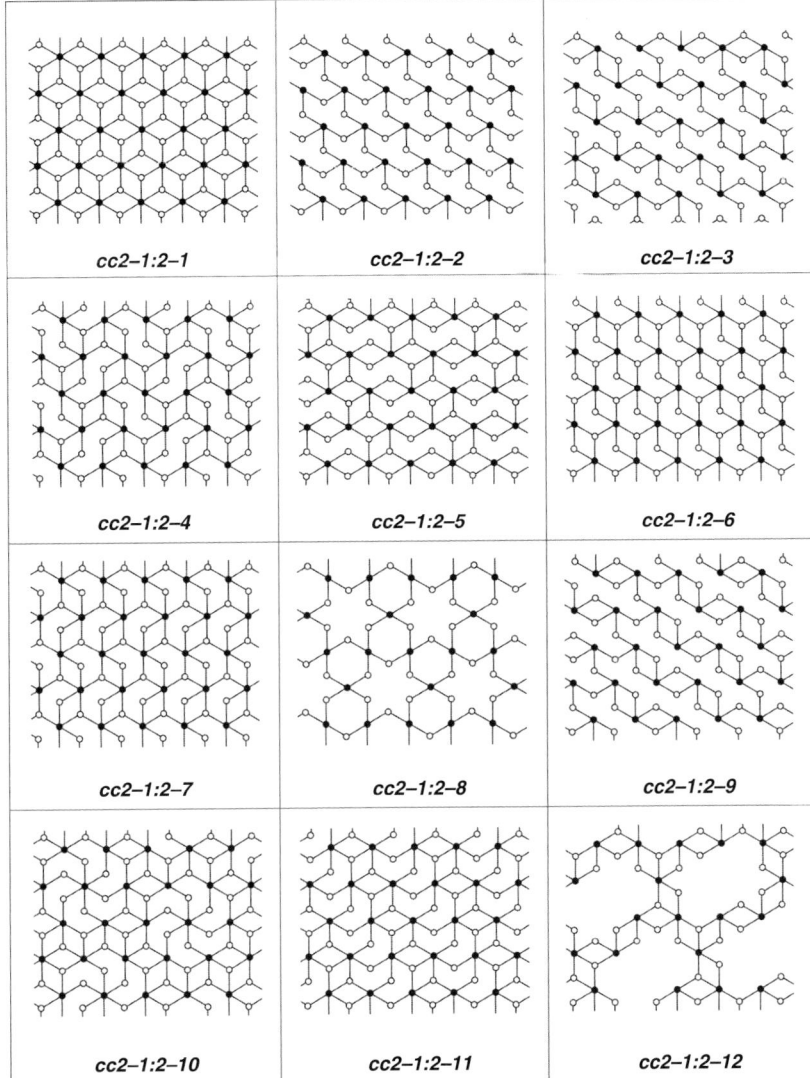

Fig. 2.3. Black-and-white 2D graphs derivative from the {**3.6.3.6**} graph. Graphs with *M:T* ratio of 1:2.

order to avoid misunderstandings, to each of the graphs described in this chapter, we assign a tentative index ***ccD–M:T–#***, where ***cc*** means "*c*ation-*c*entered", ***D*** indicates dimensionality (0 – finite complexes, 1 – chains, 2 – sheets, 3 – frameworks), ***M:T*** ratio, ***#*** – registration number of the unit. Thus, the {**3.6.3.6**} graph itself has the index ***cc2–1:2–1*** (Fig. 2.3).

2.2.1.1 $M{:}T = 1{:}2$

The {**3.6.3.6**} derivative graphs with the $M{:}T$ ratio of 1:2 are shown in Fig. 2.3. Table 2.1 provides a list of salts of inorganic oxoacids that contain sheets the topology of which can be characterized by the corresponding graphs.

The {**3.6.3.6**} graph itself (***cc2–1:2–1***) is quite common in inorganic compounds. It corresponds to octahedral–tetrahedral sheets occuring in a number of oxysalt structures (Table 2.1). Geometrical distortions of these units have been considered in more detail by Fleck and Kolitsch (2003).

The simplest {**3.6.3.6**} derivative is the ***cc2–1:2–2*** graph consisting of 4-connected black and 2-connected white vertices (Fig. 2.3). This graph is common for a large number of sheet topologies with the composition $[AnO_2(SO_4)_2(H_2O)]$. It is noteworthy that, in this group of layered structures, an interesting type of geometrical isomerism has been observed (Krivovichev *et al.* 2005a). Figures 2.4(a) and (b) show polyhedral diagrams of the $[UO_2(SeO_4)_2(H_2O)]^{2-}$ sheets observed in the structures of $[C_2H_{10}N_2]$ $[(UO_2)(SeO_4)_2(H_2O)](H_2O)$ and $[CH_6N_3]_2[(UO_2)(SeO_4)_2(H_2O)](H_2O)_{1.5}$, respectively. The black-and-white graphs of these sheets are shown in Figs. 2.4(c) and (d), respectively. The graphs are topologically the same and can be transformed one into the other by continuous topological transformation without breaking the edges. However, the detailed topological analysis reveals that the $[(UO_2)(SeO_4)_2(H_2O)]^{2-}$ sheets are different because of different positions of the H_2O molecules of the $[UO_6(H_2O)]^{6-}$ bipyramids. The difference can be visualized by adding the U→H_2O vectors to the black-and-white graphs. The idealized versions of the graphs with the U→H_2O vectors are shown in Figs. 2.4(e) and (f), respectively. It is clearly seen that the U→H_2O vectors in the structure of $[C_2H_{10}N_2][(UO_2)(SeO_4)_2(H_2O)](H_2O)$ are all pointing in the same direction, whereas, in the structure of $[CH_6N_3]_2[(UO_2)(SeO_4)_2(H_2O)](H_2O)_{1.5}$, two systems of the U→$H_2O$ vectors can be distinguished. The $[(UO_2)(SeO_4)_2(H_2O)]^{2-}$ sheets shown in Fig. 2.4 can be described as different geometrical isomers. However, in this case, geometrical isomerism is induced not by different orientation of tetrahedra but by the selective hydration of the uranyl coordination polyhedra.

The graphs ***cc2–1:2–2***, ***cc2–1:2–3*** and ***cc2–1:2–9*** (Fig. 2.3) have many features in common. In particular, they can be obtained from the {**3.6.3.6**} graph by deletion of links only, with arrangements of nodes being untouched. In addition, black and white nodes are 4- and 2-connected, respectively. It is noteworthy that these graphs correspond at the same time to chemically different sheets. For example, ***cc2–1:2–2*** and ***cc2–1:2–3*** graphs describe the structures of the $[UO_2(H_2O)(SO_4)_2]$ sheets observed in uranyl sulphates (Krivovichev *et al.* 2005a). These sheets are built up by corner sharing of $UO_2(H_2O)O_4$ pentagonal bipyramids and SO_4 tetrahedra. At the same time, the ***cc2–1:2–2*** and ***cc2–1:2–3*** graphs correspond to a topological structure of octahedral–tetrahedral units observed in structures of rhomboclase, $(H_5O_2)(Fe(SO_4)_2(H_2O)_2)$, and goldichite, $KFe(SO_4)_2(H_2O)_4$, respectively. In contrast, the ***cc2–1:2–2*** and ***cc2–1:2–9*** graphs are used to describe topological structure of the $[Al_2P_4O_{16}]$ sheet in $(NH_4)_3[Co(NH_3)_6]_3[Al_2P_4O_{16}]_2$, which consists of AlO_4 and PO_4 tetrahedra. This example demonstrates one important feature of a graphical

Table 2.1 Inorganic oxysalts containing 2D units based upon graphs with $M{:}T = 1{:}2$, derivatives from the {**3.6.3.6**} graph (see Fig. 2.3).

Graph	Ring symbol	Chemical formula	Reference
cc2–1:2–1	4^1	$Mn(HSO_4)_2(H_2O)$	Stiewe et al. 1998
		$Mg(HSO_4)_2(H_2O)$	Worzala et al. 1991
		$Cd(HSO_4)_2(H_2O)$	Kemnitz et al. 1996
		$KV(SO_4)_2$	Fehrmann et al. 1986
		$NaV(SO_4)_2$	Fehrmann et al. 1991
		$CsV(SO_4)_2$	Berg et al. 1993
		$H_3O(Al(SO_4)_2)$	Fischer et al. 1996b
		$Na[Al(MoO_4)_2]$	Kolitsch et al. 2003
		$NaFe(SeO_4)_2$	Giester 1993c
		$M(HSeO_4)_2(H_2O)$ M = Mn, Cd	Morosov et al. 1998
		$K_2Mn(SeO_3)_2$	Wildner 1992a
		$KMn^{3+}(SeO_4)_2$	Giester 1995
		$K_2Co(SeO_3)_2$	Wildner 1992b
		$ScH(SeO_4)_2(H_2O)_2$	Valkonen 1978
		$Ce(CrO_4)_2(H_2O)_2$	Lindgren 1977b
		$KA(MoO_4)_2$ (A = Al, Sc, Fe)	Klevtsova and Klevtsov 1970
		$KSc(WO_4)_2$	Klevtsova and Klevtsov 1970
		$Na_3Fe(PO_4)_2$	Belkhiria et al. 1998; Morozov et al. 2001
		α-$Ti(HPO_4)_2 \cdot H_2O$	Norlund Christensen et al. 1990; Bruque et al. 1995; Losilla et al. 1996
		α-$Zr(HPO_4)_2(H_2O)$	Clearfield and Smith 1969; Troup and Clearfield 1977
		$(NH_4)_2Zr(PO_4)_2H_2O$	Clearfield and Troup 1973
		$AZr(HPO_4)(PO_4)H_2O$ A = Na, K	Rudolf and Clearfield 1985a, b, 1989
		$Na_4[Zr(PO_4)_2]_2(H_2O)_6$	Poojary and Clearfield 1994
		$Na_3MgH(PO_4)_2$	Kawahara et al. 1995
		$Zr(HAsO_4)_2H_2O$	Clearfield and Duax 1969
		α-$M(HAsO_4)_2 \cdot (H_2O)$ (M = Ti, Sn, Pb)	Losilla et al. 1998
		yavapaiite $K[Fe(SO_4)_2]$	Graeber and Rosenzweig 1971; Anthony et al. 1972
		brianite $Na_2Ca[Mg(PO_4)_2]$	Alkemper and Fuess 1998
cc2–1:2–2	8^1	rhomboclase $(H_5O_2)^+(Fe(SO_4)_2(H_2O)_2)$	Mereiter 1974
		$HIn(SO_4)_2(H_2O)_4$	Tudo et al. 1979
		$(H_5O_2)[Mn(H_2O)_2(SO_4)_2]$	Chang et al. 1983
		$(H_5O_2)[Ti(H_2O)_2(SO_4)_2]$	Troyanov et al. 1996
		$(H_5O_2)[Al(H_2O)_2(SO_4)_2]$	Fischer et al. 1996c
		$AlH(SeO_3)_2 \cdot 2H_2O$	Morris et al. 1991
		$(Cu(HSeO_3)_2MCl_2(H_2O)_4)$, M(II) = Mn, Co, Ni, Cu, Zn	Lafront et al. 1995; Johnston and Harrison 2000
		$Cu(HSeO_3)_2$	Effenberger 1985
		$Cu(HSeO_3)_2(H_2O)$	Hiltunen et al. 1985
		$Co(HSeO_3)_2(H_2O)_2$	Gulya et al. 1994; Micka et al. 1994; Koskelinna et al. 1994

Table 2.1   *Continued*

Graph	Ring symbol	Chemical formula	Reference
		$(Co(NH_3)_6)NpO_2(SeO_4)_2(H_2O)_3$	Grigor'ev *et al.* 1992a
		$Fe(H_2PO_4)_2(H_2O)_2$	Guse *et al.* 1985
		$Sr_2Ga(HPO_4)(PO_4)F_2$	le Meins *et al.* 1998
		$Mn(H_2PO_4)_2(H_2O)_2$	Vasic *et al.* 1985
		$Na_5TiP_2O_9F$	Golubev *et al.* 1988
		$Co(H_2PO_4)_2(H_2O)_2$	Effenberger 1992
		$Mg(H_2PO_4)_2(H_2O)_2$	Hinsch 1985
		$Ba_2[NbOF(PO_4)_2]$	Wang *et al.* 2002d
		$CN_3H_6 \cdot Al(HPO_4)2 \cdot 2H_2O$	Bircsak and Harrison 1998a
		$(C_3N_2H_5)[M(HPO_4)_2(H_2O)_2]$ (C_5NH_6) $[M(HPO_4)_2(H_2O)_2]$ M = Al, Ga	Leech *et al.* 1998
		$[N_2C_3H_5][AlP_2O_8H_2 \cdot 2H_2O]$	Yu and Williams 1998
		$(H_3NCH_2CH_2NH_3)[(VO)(SeO_3)_2]$	Dai *et al.* 2003
		$[enH_2][CdCl_2(HSeO_3)_2]$	Pasha *et al.* 2003a
		$(C_3N_2H_5)[Fe(HPO_4)_2(H_2O)_2]$	Cowley and Chippindale 2000
		α-$Ce(SO_4)_2(H_2O)_4$	Lindgren 1977a
		β-$Ce(SO_4)_2(H_2O)_4$	Filipenko *et al.* 1998
		$Na_5Fe(PO_4)_2F_2$ (*cis–trans* isomer)	Rastsvetaeva *et al.* 1996
		olmsteadite $K(Fe^{2+})_2(H_2O)_2[NbO_2 (PO_4)_2]$	Moore *et al.* 1976
		$K(Mn^{2+})_2(H_2O)_2[NbO_2(PO_4)_2]$	Dunn *et al.* 1986
		$[C(NH_2)_3]_2[(UO_2)(SO_4)_2(H_2O)](H_2O)_2$	Baggio *et al.* 1977
		$Mg[(UO_2)(SO_4)_2(H_2O)](H_2O)_{10}$	Serezhkin *et al.* 1981a
		$K_2[(UO_2)(SO_4)_2(H_2O)](H_2O)$	Niisto *et al.* 1979
		$[(UO_2SO_4)(H_2SO_4)(H_2O)](H_2O)_4$	Alcock *et al.* 1982
		$[N_5C_8H_{28}]_2[(UO_2)_5(H_2O)_5(SO_4)_{10}](H_2O)$	Norquist *et al.* 2005a
		$[CN_4H_7]_2[UO_2(SO_4)_2(H_2O)]$	Medrish *et al.* 2005
		$Cs_2[(UO_2)(SeO_4)_2(H_2O)](H_2O)$	Mikhailov *et al.* 2001a
		$Rb_2[(UO_2)(SeO_4)_2(H_2O)](H_2O)$	Krivovichev and Kahlenberg 2005a
		$[C_2N_2H_{10}][UO_2(SeO_4)_2(H_2O)](H_2O)$	Krivovichev *et al.* 2007
		$[C_8H_{20}N]_2[(UO_2)(SeO_4)_2(H_2O)](H_2O)$	Krivovichev *et al.* 2007
		$[C_6H_{16}N_3]_2[(UO_2)(SeO_4)_2(H_2O)]$	Krivovichev *et al.* 2007
		$[C_4H_{14}N_2][(UO_2)(SeO_4)_2(H_2O)]$	Krivovichev *et al.* 2007
		$[C_4H_{12}N_2][(UO_2)(SeO_4)_2(H_2O)]$	Krivovichev *et al.* 2007
		$[C_8H_{26}N_4][(UO_2)(SeO_4)_2(H_2O)](H_2O)$	Krivovichev *et al.* 2007
		$(H_3O)[C_3H_5N_2][(UO_2)(SeO_4)_2(H_2O)]$	Krivovichev *et al.* 2007
		$[CH_6N_3]_2[(UO_2)(SeO_4)_2(H_2O)](H_2O)_{1.5}$	Krivovichev *et al.* 2005a
		$(H_3O)_2[(UO_2)(SeO_4)_2(H_2O)](H_2O)_2$	Krivovichev 2008a
		$UO_2(IO_3)_2(H_2O)$	Bean *et al.* 2001
		$(NH_4)_2[(UO_2)(SeO_4)_2(H_2O)](H_2O)_2$	Mikhailov *et al.* 1997a
cc2–1:2–3	4^112^1	$[C_6H_{22}N_4][C_2H_{10}N_2][Al_2P_4O_{16}]$	Wei *et al.* 2000
		$(NH_4)_2[(UO_2)(SO_4)_2(H_2O)](H_2O)$	Niisto *et al.* 1978
		$[N_2C_6H_{14}][(UO_2)(SO_4)_2(H_2O)]$	Norquist *et al.* 2002, 2003a
		$[C_5H_{14}N_2]_2[(UO_2)(SeO_4)_2(H_2O)]_2(H_2O)$	Krivovichev *et al.* 2007
		$[C_4H_{12}N]_2[(UO_2)(SeO_4)_2(H_2O)]$	Krivovichev *et al.* 2007
		$(H_3O)_2[(UO_2)(SeO_4)_2(H_2O)](H_2O)$	Krivovichev 2008a
		$NH_4[Fe(SO_4)_2(H_2O)](H_2O)_2$	Palmer *et al.* 1972

Table 2.1 *Continued*

Graph	Ring symbol	Chemical formula	Reference
		$NH_4[In(SO_4)_2(H_2O)_2](H_2O)_2$	Mukhtarova *et al.* 1978; Rastsvetaeva and Mukhtarova 1987
		$AIn(SO_4)_2(H_2O)_4$ (A = K, Rb)	Mukhtarova *et al.* 1979a,b; Rastsvetaeva and Mukhtarova 1987
		$(H_3O)In(SO_4)_2(H_2O)_4$	Caminiti *et al.* 1982a
		$NH_4In(SeO_4)_2(H_2O)_4$	Soldatov *et al.* 1979
		goldichite $KFe(SO_4)_2(H_2O)_4$	Graeber and Rosenzweig 1971
		$RbCe(SeO_4)_2 \cdot 5H_2O$	Ovanisyan *et al.* 1987a
		$[C_3H_{12}N_2][H_2ZnP_2O_8][H_4ZnP_2O_8]$	Harrison *et al.* 1998
cc2–1:2–4	4^38^1	$K_2(UO_2)(MoO_4)_2$	Sadikov *et al.* 1988
		$Rb_2(UO_2)(MoO_4)_2$	Krivovichev and Burns 2002a
		$Cs_2[(UO_2)(MoO_4)_2]$	Krivovichev and Burns 2005
		$Tl_2[(UO_2)(MoO_4)_2]$	Krivovichev *et al.* 2005c
		$Tl_2[(UO_2)(CrO_4)_2]$	Krivovichev *et al.* 2005c
		$K_2(UO_2)(MoO_4)_2(H_2O)$	Krivovichev *et al.* 2002a
		$Rb_2(UO_2)(MoO_4)_2(H_2O)$	Khrustalev *et al.* 2000
		$Cs_2(UO_2)(MoO_4)_2(H_2O)$	Rastsvetaeva *et al.* 1999; Krivovichev and Burns 2005
		$(NH_4)_2(UO_2)(MoO_4)_2(H_2O)$	Andreev *et al.* 2001
		$Na_2[(UO_2)(MoO_4)_2](H_2O)_4$	Krivovichev and Burns 2003b
		$(C_2H_{10}N_2)[(UO_2)(MoO_4)_2]$	Krivovichev and Burns 2003c
		$(C_5H_{14}N_2)(UO_2)(MoO_4)_2 \cdot H_2O$	Halasyamani *et al.* 1999
		$Na_2[(UO_2)(SeO_4)_2](H_2O)_4$	Mikhailov *et al.* 2001b
		$(NH_4)_2UO_2(SeO_3)_2 \cdot 0.5(H_2O)$	Koskenlinna *et al.* 1997
		$NH_4[(UO_2)(HSeO_3)(SeO_3)]$	Koskenlinna and Valkonen 1996
		$A[(UO_2)(HSeO_3)(SeO_3)]$ A = K, Rb, Cs, Tl	Almond and Albrecht-Schmitt 2002a
		$[C_6H_{14}N_2]_{0.5}[UO_2(HSeO_3)(SeO_3)]$ $(H_2O)_{0.5}(CH_3CO_2H)_{0.5}$	Almond and Albrecht-Schmitt 2003
		$(H_5O_2)_2[(UO_2)(SeO_4)_2]$	Krivovichev *et al.* 2007
		$[C_{12}N_2H_{30}][(UO_2)(SeO_4)_2](H_2O)_x$	Krivovichev *et al.* 2007
		$[C_{10}H_{26}N_2][(UO_2)(SeO_4)_2]$ $(H_2SeO_4)_{0.50}(H_2O)$	Krivovichev *et al.* 2007
		$[CH_6N_3]_3[(UO_2)_2(SeO_4)_3(HSeO_4)](H_2O)_2$	Krivovichev *et al.* 2005a
		$[C_5H_{14}N][(UO_2)(SeO_4)(SeO_2OH)]$	Krivovichev *et al.* 2005b
		$Al(H_2PO_4)(HPO_4)(H_2O)$	Kniep *et al.* 1978
cc2–1:2–5	4^38^1	$Na_2(UO_2)(MoO_4)_2$	Krivovichev *et al.* 2002a
		$K_3NpO_2(MoO_4)_2$	Grigor'ev *et al.* 1992b
		$Ag_2[(UO_2)(SeO_3)_2]$	Almond and Albrecht-Schmitt 2002a
		$(C_4H_{12}N_2)(UO_2)(MoO_4)_2$	Halasyamani *et al.* 1999
		$K_4(H_5O_2)(NpO_2)_3(MoO_4)_4(H_2O)_4$	Grigor'ev *et al.* 1991a
		$BaVO(PO_4)(H_2PO_4) \cdot (H_2O)$	Harrison *et al.* 1995a
		$BaVO(AsO_4)(H_2AsO_4) \cdot H_2O$	Cheng and Wang 1992
		$[dabcoH_2]_{0.5}[(VO)(HSeO_3)(SeO_3)](H_2O)$	Pasha *et al.* 2003b

Table 2.1 *Continued*

Graph	Ring symbol	Chemical formula	Reference
cc2–1:2–6	$4^1 6^1$	A(VO(SO$_4$)$_2$) A = K, NH$_4$	Richter and Mattes 1992
		K$_6$(VO)$_4$(SO$_4$)$_8$	Eriksen *et al.* 1996
		CsNbO(SO$_4$)$_2$	Kashaev and Sokolova 1973
		RbNbO(SO$_4$)$_2$, NH$_4$NbO(SO$_4$)$_2$	Kashaev *et al.* 1973; Kuznetsov *et al.* 1974
cc2–1:2–7	$4^1 6^1$	M(UO$_2$)[(UO$_2$)$_2$(MoO$_4$)$_4$](H$_2$O)$_8$ M = Mg, Zn	Tabachenko *et al.* 1983
		M$_2$(UO$_2$)$_6$(MoO$_4$)$_7$(H$_2$O)$_2$ M = Cs, NH$_4$, Rb	Krivovichev and Burns 2001a, 2002a
cc2–1:2–8	$6^2 12^1$	Na$_5$AlF$_2$(PO$_4$)$_2$	Arlt *et al.* 1987
		Na$_5$Cr(PO$_4$)$_2$F$_2$	Nagornyi *et al.* 1990
cc2–1:2–9	$4^3 20^1$	(NH$_4$)$_3$[Co(NH$_3$)$_6$]$_3$[Al$_2$P$_4$O$_{16}$]$_2$	Morgan *et al.* 1997
cc2–1:2–10	$4^3 6^2$	[C$_3$N$_2$H$_5$]$_3$In$_8$(HPO$_4$)$_{14}$(H$_2$O)$_{11}$(H$_3$O)	Chippindale *et al.* 1996
		[HN(CH$_2$CH$_2$)$_3$NH]$_3$[Fe$_8$(HPO$_4$)$_{12}$ (PO$_4$)$_2$(H$_2$O)$_6$]	Lii and Huang 1997a
cc2–1:2–11	$4^1 6^1$	[NpO$_2$(IO$_3$)$_2$](H$_2$O)	Bean *et al.* 2003
		[PuO$_2$(IO$_3$)$_2$](H$_2$O)	Runde *et al.* 2003
cc2–1:2–12	$4^5 16^1$	[N$_3$C$_6$H$_{18}$][(UO$_2$)$_2$(H$_2$O)(SO$_4$)$_3$(HSO$_4$)] (H$_2$O)$_{4.5}$	Norquist *et al.* 2003b
cc2–1:2–13	$4^7 12^1$	(H$_3$O)$_2$[C$_{12}$H$_{30}$N$_2$]$_3$[(UO$_2$)$_4$(SeO$_4$)$_8$] (H$_2$O)$_5$	Krivovichev *et al.* 2005d

representation of structural units: it describes the topology of connections of poly-hedra but does not contain any information about the chemistry and geometry of polyhedra. Obviously, the graphical approach ignores an important piece of chemical information. On the other hand, it allows investigation of the topology of chemically different structures and reveals the hidden relationships between them.

It is interesting to note that the *cc2–1:2–2*, *cc2–1:2–3* and *cc2–1:2–9* graphs can be easily transformed into the other graphs, as is shown in Figs. 2.5(a)–(c). This transformation does not involve any reconstruction of the graphs and is purely topo-logical. From a geometrical viepoint, it involves distortion of the local coordination of 4-connected black nodes, as shown in the middle of Fig. 2.5.

The *cc2–1:2–6* graph deserves special attention as well. It corresponds to the [(UO$_2$)(MoO$_4$)$_2$]$^{2-}$ sheet observed in the structures of M(UO$_2$)$_3$(MoO$_4$)$_4$(H$_2$O)$_8$ (M = Mg, Zn) and A$_2$[(UO$_2$)$_6$(MoO$_4$)$_7$(H$_2$O)$_2$] (A = NH$_4$, Rb, Cs). In the structures of M(UO$_2$)$_3$(MoO$_4$)$_4$(H$_2$O)$_8$, these sheets are linked by additional UO$_6$(H$_2$O) bipyramids, resulting in a framework. In A$_2$[(UO$_2$)$_6$(MoO$_4$)$_7$(H$_2$O)$_2$], two such sheets are linked together to form double sheets, which are further linked by additional UO$_6$(H$_2$O) bipyramids, resulting in a framework (Fig. 2.6).

The topology *cc2–1:2–13* (Fig. 2.7) corresponds to a complex uranyl selenate sheet observed in the structure of (H$_3$O)$_2$[C$_{12}$H$_{30}$N$_2$]$_3$[(UO$_2$)$_4$(SeO$_4$)$_8$] (H$_2$O)$_5$ (Krivovichev *et al.* 2005d). Along with the graph *cc2–1:2–13*, this is is the only graph that can be

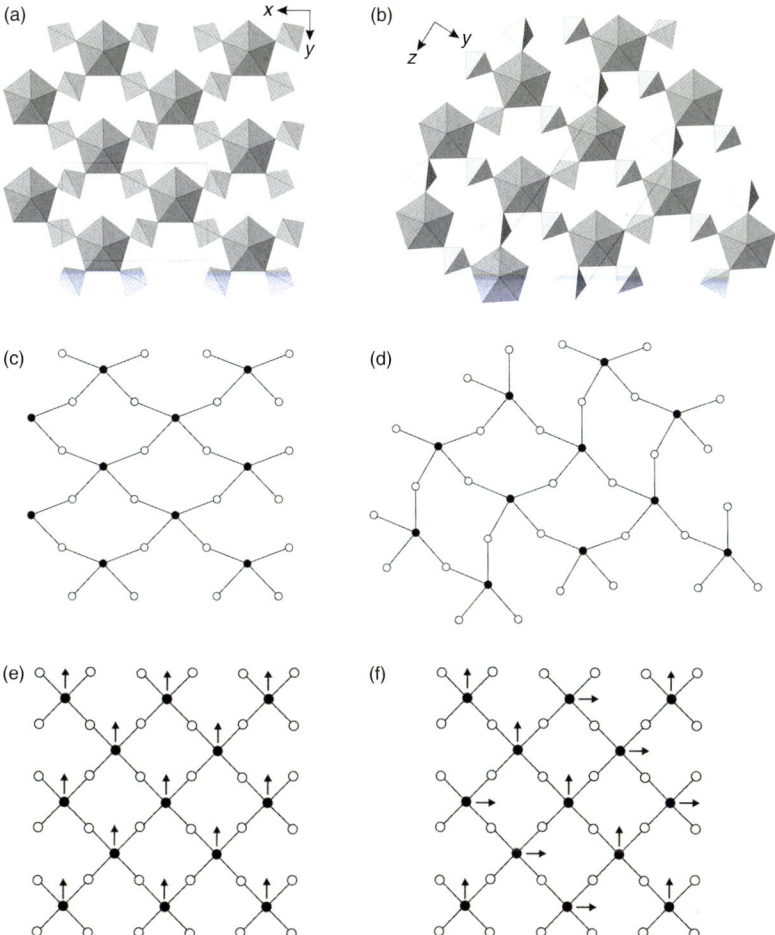

Fig. 2.4. The $[(UO_2)(SeO_4)_2(H_2O)]^{2-}$ sheets from the structures of $[C_2H_{10}N_2][(UO_2)(SeO_4)_2$ $(H_2O)](H_2O)$ (a) and $[CH_6N_3]_2[(UO_2)(SeO_4)_2(H_2O)](H_2O)_{1.5}$ (b). Black-and-white graph showing the topological connectivity of the U and Se polyhedra (symbolized by black and white vertices, respectively) ((c) and (d), respectively) and idealized graphs with the arrows symbolizing the U→H₂O vectors ((e) and (f), respectively).

obtained from the {**3.6.3.6**} graph by deleting black *and* white vertices together with edges that are incident upon them.

2.2.1.2 *Topological isomerism*

Let us consider in more detail the graphs ***cc2–1:2–4*** and ***cc2–1:2–5***. They describe, in particular, the topologies of the $[UO_2(MoO_4)_2]$ uranyl molybdate sheets that occur in

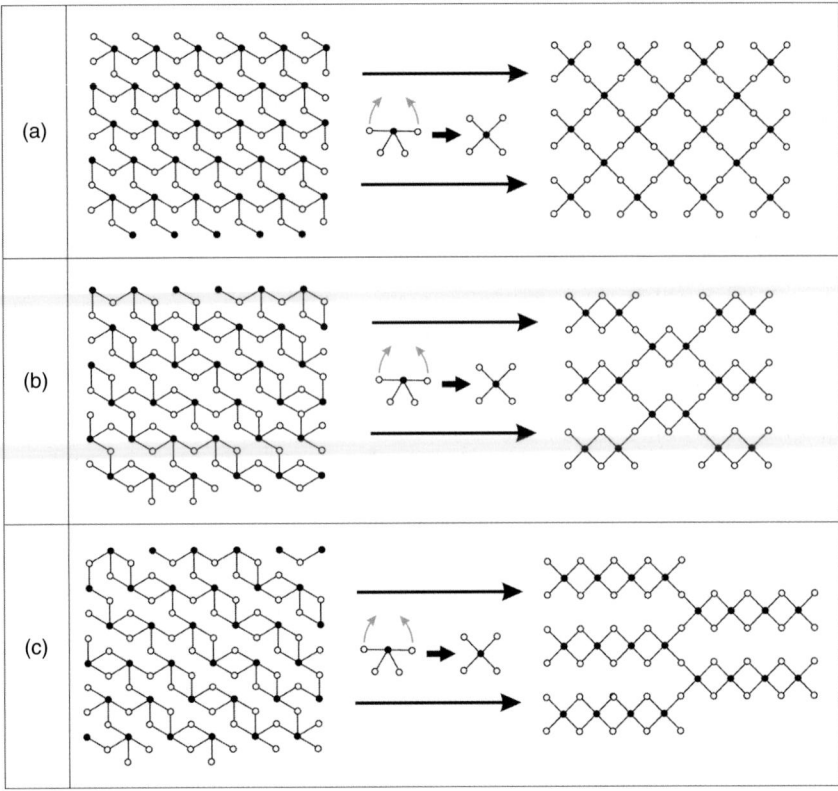

Fig. 2.5. Schemes of transformations of graphs that do not change their topological structure.

structures of alkali-metal uranyl molybdates with general formula $A_2[UO_2(MoO_4)_2]$ (A = Na, K, Rb, Cs). In these compounds, U^{6+} cations are coordinated by seven O atoms (pentagonal bipyramid), whereas Mo^{6+} cations are tetrahedrally coordinated by four O atoms. UO_7 and MoO_4 polyhedra share vertices to form continuous $[UO_2(MoO_4)_2]$ sheets. However, in the structure of $Na_2[UO_2(MoO_4)_2]$ (Krivovichev *et al.* 2002a), topology of the $[UO_2(MoO_4)_2]$ sheet corresponds to the *cc2–1:2–5* graph, whereas, in compounds $A_2[UO_2(MoO_4)_2]$ with A = K, Rb, Cs (Sadikov *et al.* 1988; Krivovichev and Burns 2002a, 2003b), the topology of chemically equivalent sheets corresponds to the *cc2–1:2–4* graph. Such chemically identical but topologically different structural units are usually called *topological or structural isomers* (see also: Krivovichev 2004).

It is noteworthy that the graphs *cc2–1:2–4* and *cc2–1:2–5* have the same connectivities of black and white vertices and the same ring symbols. However, the graphs are obviously topologically different. In order to qualify this difference, one

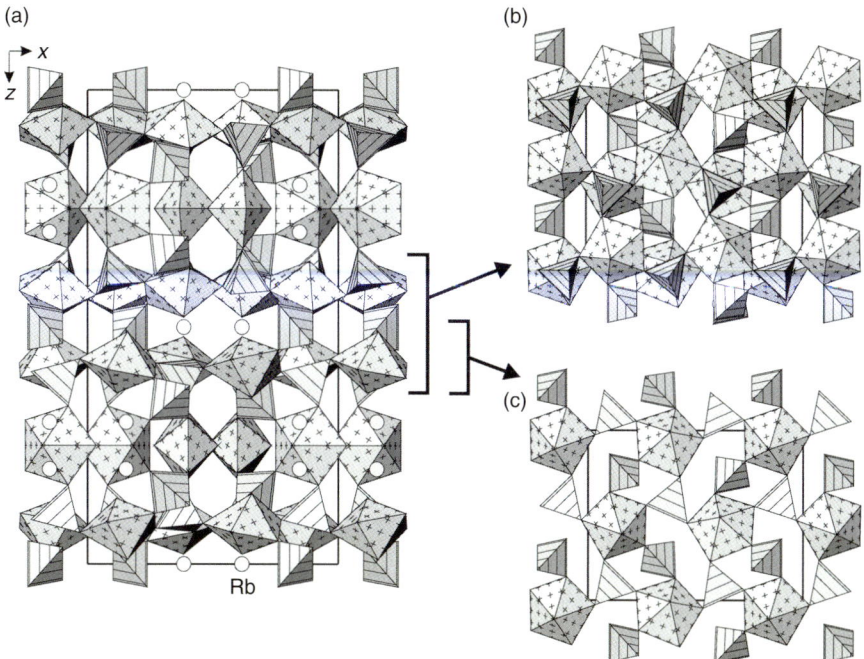

Fig. 2.6. The structure of $M_2[(UO_2)_6(MoO_4)_7(H_2O)_2]$ (M = NH$_4$, Rb, Cs) (a) as consisting of double uranyl molybdate sheets (b), and a projection of the single sheet (c).

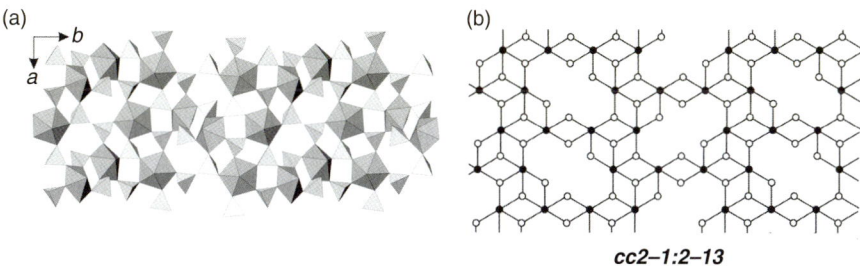

cc2–1:2–13

Fig. 2.7. Uranyl selenate sheet observed in $(H_3O)_2[C_{12}H_{30}N_2]_3[(UO_2)_4(SeO_4)_8](H_2O)_5$ (a) and its graph (b).

may explore the concept of a *coordination sequence* that is defined as a sequence of numbers N_k of vertices at a *topological distance* of k edges. Figure 2.8 shows how the graphs *cc2–1:2–4* and *cc2–1:2–5* can be distinguished by their coordination sequences of black vertices. In Fig. 2.8, vertices at different topological distances (i.e. separated from a black vertex by k edges) are separated from each other. For both

(a)   (b)

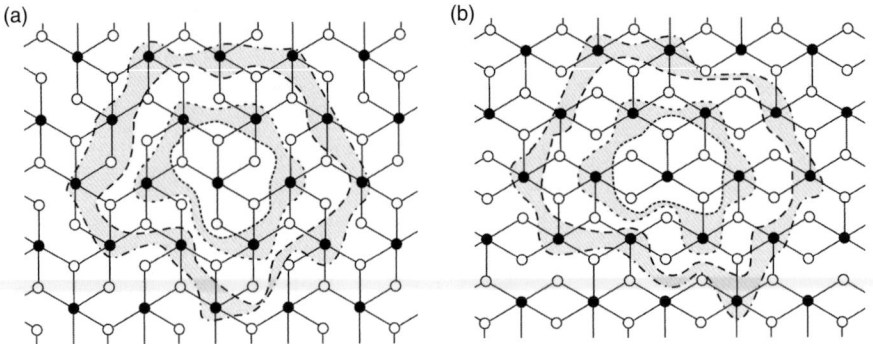

Fig. 2.8. Concept of coordination sequences. Two black-and-white graphs with a black:white ratio of 1:2. Four coordination spheres for a black vertex are selected.

graphs, $N_1 = 5$, $N_2 = 5$, $N_3 = 15$. However, $N_4 = 11$ for the graph **cc2–1:2–4** and $N_4 = 10$ for the graph **cc2–1:2–5**. The calculation of coordination sequences in order to distinguish different topologies might be useful if the latter are stored in a computer database.

2.2.1.3   M:T = 2:3

2D topologies of this type are shown in Fig. 2.9. They are quite common in the structures of aluminophosphates, as reviewed by Yu and Xu (2003). The list of compounds containing these units is presented in Table 2.2. Note that the topology **cc2–2:3–7** may be transformed into more regular topology by continuous rotation of the local coordination of its black nodes (Fig. 2.10).

Figures 2.11(a) and (c) show polyhedral diagrams of two $[(UO_2)_2(SeO_4)_3(H_2O)_2]^{2-}$ sheets observed in the structures of $(H_3O)_2[(UO_2)_2(SeO_4)_3(H_2O)_2](H_2O)_{3.5}$ and $Rb_2[(UO_2)_2(SeO_4)_3(H_2O)_2](H_2O)_4$, respectively. Both sheets correspond to the same **cc2–2:3–11** graph. This graph can be considered as consisting of chains cross-linked by 4-membered rhomb-shaped rings. The sheets shown in Figs. 2.11(a) and (c) differ by the sequence of "up" (= **u**) and "down" (= **d**) orientation of tetrahedra along the chains. The sequences are $(\mathbf{uudd})_\infty$ for the Rb compound (Fig. 2.11(b)) and $(\mathbf{ud})_\infty$ for the oxonium compound. Thus, the uranyl selenate sheets in these compounds should be considered as geometrical isomers.

2.2.1.4   M:T = 3:4

A catalog of the topologies with the $M:T = 3:4$ is presented in Fig. 2.12; Table 2.3 provides a list of relevant compounds.

Topology **cc2–3:4–1** is noteworthy in that it contains large 12-membered hexagonal rings. This topology is characteristic for the $[Al_3P_4O_{16}]^{3-}$ sheets of alternating

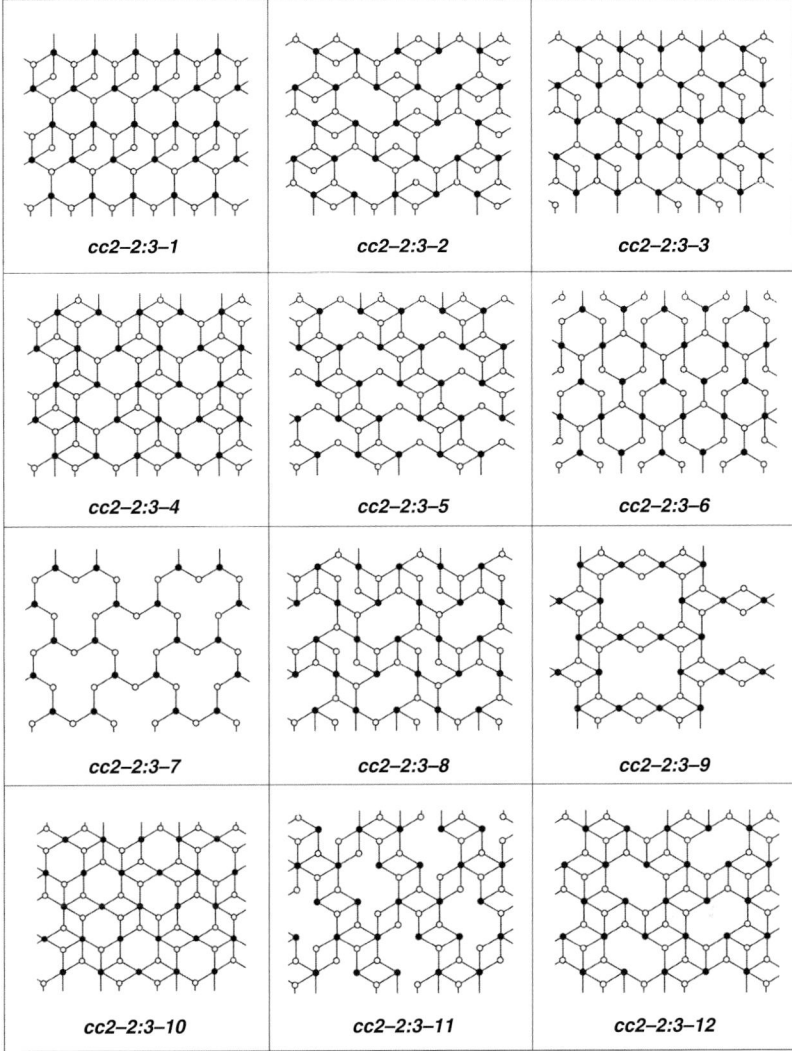

Fig. 2.9. Black-and-white 2D graphs derivative from the {3.6.3.6} graph. Graphs with *M:T* ratio of 2:3.

AlO_4 and PO_4 tetrahedra in aluminophosphates (corresponding to black and white nodes, respectively). The octahedral–tetrahedral version of this topology was observed by Choudhury *et al.* (1999) in the structure of $[NH_3(CH_2)_2NH_3]_{1.5}[Fe_3PO_4$ $(HPO_4)_3(C_2O_4)_{1.5}] \cdot xH_2O$ ($x = 1.5$–2). FeO_6 and PO_4 tetrahedra share corners to produce a sheet that can be described by the *cc2–3:4–1* graph. However, these sheets are further linked into a three-dimensional framework via C_2O_4 oxalate groups.

Table 2.2 Inorganic oxysalts containing 2D units based upon graphs with $M:T$ = 1:2, derivatives from the {**3.6.3.6**} graph (see Fig. 2.9).

Graph	Ring symbol	Chemical formula	Reference
cc2–2:3–1	$4^1 6^2$	$[C_5NH_{12}]_2[Al_2P_3O_{12}H]$	Oliver *et al.* 1996a
		$[C_4NH_{12}]_2[Al_2P_3O_{12}H]$	Oliver *et al.* 1996b
cc2–2:3–2	$4^3 6^2 8^1$	$[C_5NH_{12}]_2[Al_2P_3O_{12}H]$	Oliver *et al.* 1996a
cc2–2:3–3	$4^1 6^2$	$[BuNH_3]_2[Al_2P_3O_{12}H]$	Chippindale *et al.* 1992
		$[(CH_3)_3CNH_3]_2(H_2O)_{0.5}[Al_2P_3O_{11}(OH)]$	Marichal *et al.* 2006
cc2–2:3–4	$4^3 6^1$	$[N_2C_3H_{12}][(UO_2)_2(H_2O)(SO_4)_3]$	Thomas *et al.* 2003
		$[N_2C_4H_{14}][(UO_2)_2(H_2O)(SO_4)_3](H_2O)$	Doran *et al.* 2003d
		$[C_4H_{12}N]_2[(UO_2)_2(SeO_4)_3(H_2O)]$	Krivovichev *et al.* 2006a
		$[C_4H_{14}N_2][(UO_2)_2(SeO_4)_3(H_2O)](H_2O)_2$	Krivovichev *et al.* 2006a
		$[C_3H_{10}N]_2[(UO_2)_2(SeO_4)_3(H_2O)](H_2O)$	Krivovichev *et al.* 2006a
		$[C_5H_{16}N_2][(UO_2)_2(SeO_4)_3(H_2O)](H_2O)$	Krivovichev *et al.* 2006a
cc2–2:3–5	$4^2 8^1$	$(NH_4)_2(UO_2)_2(CrO_4)_3(H_2O)_6$	Mikhailov *et al.* 1997ba
		$K_2[(UO_2)_2(CrO_4)_3(H_2O)_2](H_2O)_4$	Krivovichev and Burns 2003a
		$[C_4H_{14}N_2]_2[H_6Zn_4P_6O_{24}]$	Natarajan *et al.* 1997
		$Sm_2(SO_4)_3(H_2O)_8$	Podberezskaya and Borisov 1976
		$Pr_2(SO_4)_3(H_2O)_8$	Gebert Sherry 1976; Ahmed Farag *et al.* 1981
		$Nd_2(SO_4)_3(H_2O)_8$	Bartl and Rodek 1983
		$Yb_2(SO_4)_3(H_2O)_8$	Hiltunen and Niinisto 1976a
		$M_2(SO_4)_3(H_2O)_8$ M = Ce, Lu	Junk *et al.* 1999
		$Er_2(SO_4)_3(H_2O)_8$	Wickleder 1999a
		$Gd_2(SO_4)_3(H_2O)_8$	Hummel *et al.* 1993
		$Yb_2(SeO_4)_3(H_2O)_8$	Hiltunen and Niinisto 1976b
		$Na_5(VO)_2(PO_4)_3(H_2O)$	Benhamada *et al.* 1992b
cc2–2:3–6	$6^1 8^1$	$Al_2(SO_4)_3(H_2O)_5$	Fischer *et al.* 1996a
		$In_2(SeO_4)_3(H_2O)_5$	Kadoshnikova *et al.* 1978
cc2–2:3–7	12^1	kornelite $[Fe^{3+}_2(H_2O)_6(SO_4)_3](H_2O)_{1.25}$	Robinson and Fang 1973
		newberyite $[Mg(PO_3OH)(H_2O)_3]$	Sutor 1967; Abbona *et al.* 1979; Bartl *et al.* 1983
		$[Mn(HPO_4)(H_2O)_3]$	Cudennec *et al.* 1989
cc2–2:3–8	$8^2 4^4$	$[C_9H_{20}N][H_2Al_2P_3O_{12}]$	Chippindale and Walton 1999
cc2–2:3–9	$12^1 4^5$	$[C_3H_{12}N_2][H_3Zn_2P_3O_{12}]$	Chavez *et al.* 1999; Liu *et al.* 2000; Choudhury *et al.* 2001
cc2–2:3–10	$6^1 4^3$	$[N_2C_3H_{12}][(NpO_2)_2(CrO_4)_3(H_2O)](H_2O)_3$	Budantseva *et al.* 2003
		$K_2[(NpO_2)_2(CrO_4)_3(H_2O)](H_2O)_3$	Grigor'ev *et al.* 2004
		$[C_4H_{10}NO]_2[(UO_2)_2(SeO_4)_3(H_2O)]$	Krivovichev *et al.* 2007
		$(H_3O)[C_6H_{16}N][(UO_2)_2(SeO_4)_3(H_2O)]$ (H_2O)	Krivovichev *et al.* 2006a
		$(H_5O_2)[C_6H_{16}N][(UO_2)_2(SeO_4)_3(H_2O)]$	Krivovichev *et al.* 2006a
		$[C_4H_{12}N](H_3O)[(UO_2)_2(SeO_4)_3(H_2O)]$	Krivovichev *et al.* 2006a
		$[C_6H_{18}N_2][(UO_2)_2(SeO_4)_3(H_2O)]$	Krivovichev *et al.* 2006a
		$[C_7H_{20}N_2][(UO_2)_2(SeO_4)_3(H_2O)](H_2O)$	Krivovichev *et al.* 2006a
		$[C_5H_{16}N_2][(UO_2)_2(SeO_4)_3(H_2O)]$	Krivovichev *et al.* 2006a
		$[C_8H_{22}N_2][(UO_2)_2(SeO_4)_3(H_2O)]$	Krivovichev *et al.* 2006a
cc2–2:3–11	$12^1 4^5$	$(H_3O)_2[(UO_2)_2(SeO_4)_3(H_2O)_2](H_2O)_{3.5}$	Krivovichev and Kahlenberg 2005d
		$Rb_2[(UO_2)_2(SeO_4)_3(H_2O)_2](H_2O)_4$	Krivovichev and Kahlenberg 2005a
cc2–2:3–12	$8^2 4^7$	$[C_8H_{26}N_4]_{0.5}[(UO_2)_2(SO_4)_3(H_2O)]$	Doran *et al.* 2004

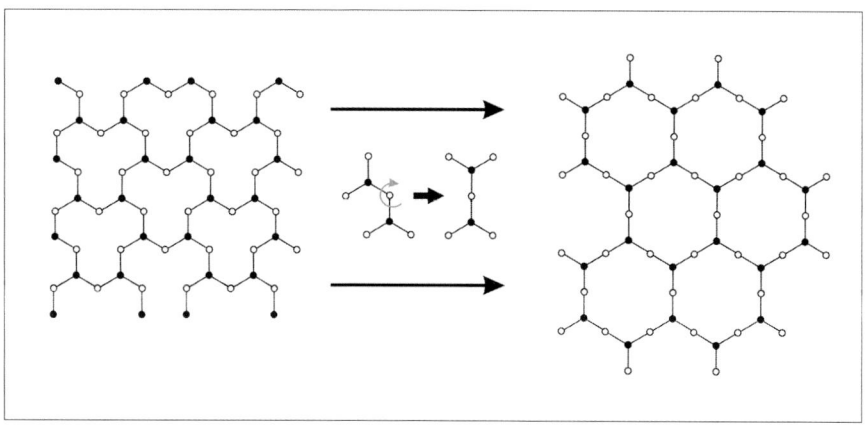

Fig. 2.10. Transformation of the *cc2–2:3–7* graph into more regular topology by continuous rotation of local coordination of its black nodes.

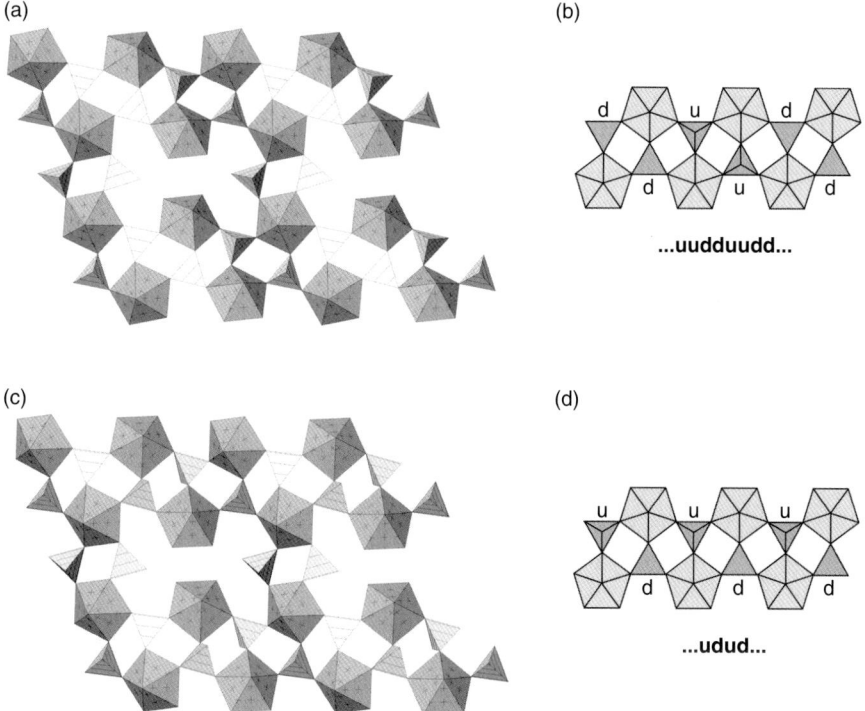

Fig. 2.11. Description of geometrical isomerism of the $[(UO_2)_2(SeO_4)_3(H_2O)_2]^{2-}$ sheets with the *cc2–2:3–11* topology. See text for details.

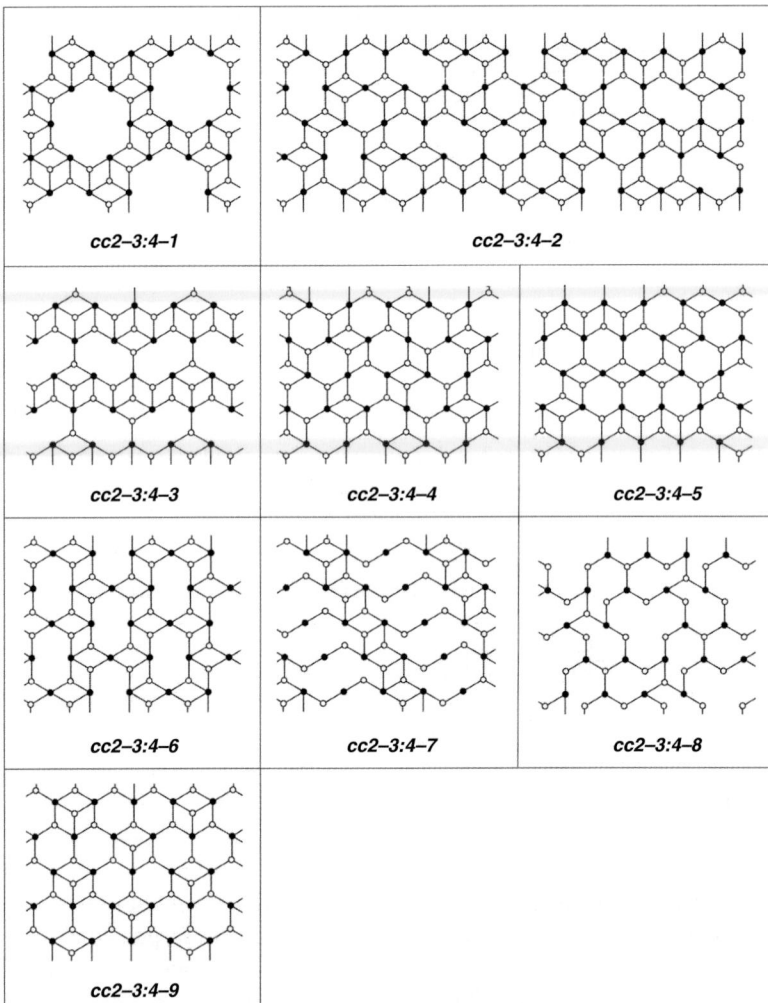

Fig. 2.12. Black-and-white 2D graphs derivative from the {**3.6.3.6**} graph. Graphs with $M{:}T$ ratio of 4:3.

2.2.1.5 $M{:}T = 13{:}18$, $3{:}8$, $5{:}8$, $3{:}5$, $1{:}1$ and $1{:}3$

2D black-and-white graphs with these $M{:}T$ ratios are shown in Figs. 2.13, 2.14 and 2.15; a list of compounds is given in Table 2.4.

The *cc2–13:18–1* graph is noteworthy by its somewhat unusual $M{:}T$ ratio. It corresponds to a complex sheet of AlO_4 and PO_4 tetrahedra that has been observed in the structure of $[C_9H_{24}N_2]_7[HAl_{13}P_{18}O_{72}](H_2O)_8$ by Feng *et al.* (2000).

Table 2.3 Inorganic oxysalts containing 2D units based upon graphs with $M{:}T = 3{:}4$, derivatives from the {**3.6.3.6**} graph (see Fig. 2.12).

Graph	Ring symbol	Chemical formula	Reference
cc2–3:4–1	$4^9 12^1$	$[NH_3(CH_2)_4NH_3]_3[Al_3P_4O_{16}]_2$	Thomas *et al.* 1992
		$[NH_3(CH_2)_2NH_3]_{1.5}[Fe_3PO_4(HPO_4)_3(C_2O_4)_{1.5}] \cdot xH_2O\ x = 1.5\text{–}2$	Choudhury *et al.* 1999
		$[BuNH_3]_3[Al_3P_4O_{16}]$	Chippindale *et al.* 1997
		$(C_3H_{10}NO)_3[Al_3P_4O_{16}]$	Yuan *et al.* 2000
cc2–3:4–2	$4^{14}6^4 8^1$	$[C_6H_{21}N_4][Al_3P_4O_{16}]$	Yao *et al.* 1999
cc2–3:4–3	$4^4 8^1$	$[C_2H_{10}N_2][C_2H_6O_2][C_2H_7O_2][Al_3P_4O_{16}]$	Jones *et al.* 1991
		$[CH_3CH_2NH_3]_3[Al_3P_4O_{16}]$	Gao *et al.* 1997
		$[C_5H_{12}N]_2[C_4H_{10}N][Al_3P_4O_{16}]$	Oliver *et al.* 1996c
		$[CH_3CH_2CH_2NH_3]_3[Al_3P_4O_{16}]$	Togashi *et al.* 1998
		$[C_5H_{12}N][C_5H_{16}N_2][Al_3P_4O_{16}]$	Jones *et al.* 1994
		$[C_4N_2H_{12}]_{4.5}[Al_3P_4O_{16}]_3(H_2O)_5$	Tuel *et al.* 2001
		$[(CH_3)_2NHCH_2CH_2NH(CH_3)_2][H_3O][Al_3P_4O_{16}]$	Yan *et al.* 2002
		$[Ni(C_4N_3H_{13})(C_4N_3H_{14})H_2O][Al_3P_4O_{16}]$	Wang *et al.* 2006
		$[H_3N(CH_2)_3NH_3][Zn_3(HPO_4)_4](H_2O)$	Phillips *et al.* 2002
		$[C_6H_{16}N_2][Zn_3(HPO_4)_4](H_2O)$	Wang *et al.* 2003a
cc2–3:4–4	$4^3 6^2$	$[NH_3CHMeCH_2NH_3]_3[Al_3P_4O_{16}]$	Williams *et al.* 1996
cc2–3:4–5	$4^3 6^2$	$[C_5N_2H_9]_2[NH_4][Al_3P_4O_{16}]$	Yu *et al.* 1999a
cc2–3:4–6	$4^4 8^1$	$[CH_6N_3][Al_3P_4O_{16}]$	Vidal *et al.* 1999a, b
		$[NH_4]_2[H_3N(CH_2)_2NH_3]_2[Zr_3(OH)_6(PO_4)_4]$	Wang *et al.* 2000a
		$Ba_3[Nb_3O_3F(PO_4)_4](HPO_4)(H_2O)_7$	Wang *et al.* 2002
cc2–3:4–7	$4^2 12^1$	hannayite $Mg_3(NH_4)_2(HPO_4)_4(H_2O)_8$	Catti and Franchini-Angela 1976
cc2–3:4–8	$8^3 12^1$	$Ba_{2.5}(VO_2)_3(SeO_3)_4(H_2O)$	Sivakumar *et al.* 2006
cc2–3:4–9	$6^1 4^3$	$(C_4H_{14}N_2)_{1.5}[Al_3P_4O_{16}]$	Tuel *et al.* 2005

It is interesting to note that deletion of a black vertex and all links associated with it results in formation of a star-shaped region that is easily recognized in the graphs *cc2–1:2–8*, *cc2–3:8–1* and *cc2–1:3–1*. In contrast, elimination of a white vertex results in formation of an empty hexagon clearly seen in many graphs given herein.

The *cc2–1:1–1* graph (Fig. 2.14) is used to describe the topological structure of the $[Sn^{2+}PO_4]$ sheet observed in $[C_6N_2H_{14}][Sn^{2+}PO_4]_2(H_2O)$ (Ayyapan *et al.* 1998). The $[Sn^{2+}PO_4]$ sheet consists of $Sn^{2+}O_3$ triangular pyramids (Sn^{2+} has a stereoactive lone pair of electrons) and PO_4 tetrahedra.

The graphs with the *M:T* ratio equal to 1:3 and their corresponding sheets are shown in Fig. 2.15. It is easy to see that these graphs differ from each other by just two edges, whereas all vertices remain in place. As a result, all black vertices in the graph *cc2–1:3–1* are 6-connected, whereas all black vertices in the graph *cc2–1:3–2* are 5-connected. This topological transition is due to the coordination requirements of the respective *M* cations: the Sc^{3+} cation has an octahedral coordination, whereas the U^{6+} cation has a pentagonal bipyramidal coordination with five equatorial bonds in the same plane (Fig. 2.15(c)).

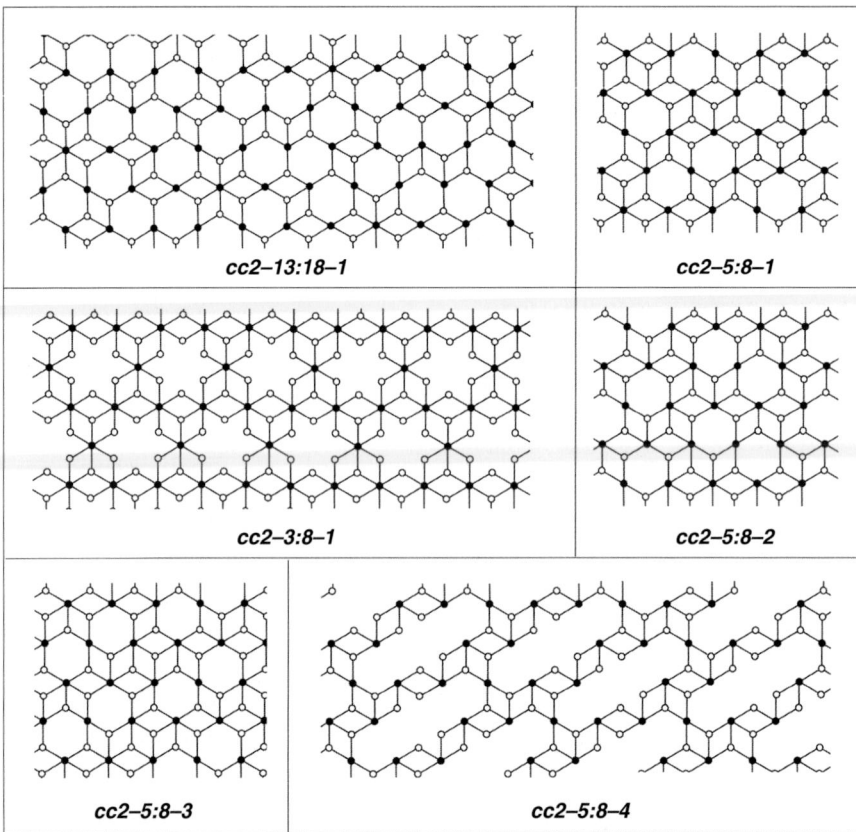

Fig. 2.13. Black-and-white 2D graphs derivative from the {**3.6.3.6**} graph. Graphs with *M:T* ratio of 13:18, 3:8, and 5:8.

2.2.1.6 *Geometrical isomerism and orientation matrix*

Figures 2.16 (a) and (b) show the $[(UO_2)_3(CrO_4)_5]^{4-}$ sheets observed in the structures of $Mg_2[(UO_2)_3(CrO_4)_5](H_2O)_{17}$ and $Ca_2[(UO_2)_3(CrO_4)_5](H_2O)_{19}$, respectively. These sheets are built up by the corner sharing between UO_7 pentagonal bipyramids and CrO_4 tetrahedra. The topology of these sheets corresponds to the same graph *cc2–3:5–2*. However, a detailed analysis of the polyhedral diagrams shown in Fig. 2.16 shows that these two sheets are geometrical isomers and cannot be transformed, one into the other, without the breaking of chemical bonds. Figure 2.17(b) shows a black-and-white graph corresponding to the $[(UO_2)_3(CrO_4)_5]^{4-}$ sheet in $Ca_2[(UO_2)_3(CrO_4)_5]$ $(H_2O)_{19}$, with the letters **u** (up) and **d** (down) written near the white nodes. With the graph oriented as in Fig. 2.17(b), the sequences of tetrahedral orientations along the horizontal line may be written in rows. Thus, the first row of the **u** and **d** symbols near the white nodes may be written as…**ududud**…. The second row contains vacant

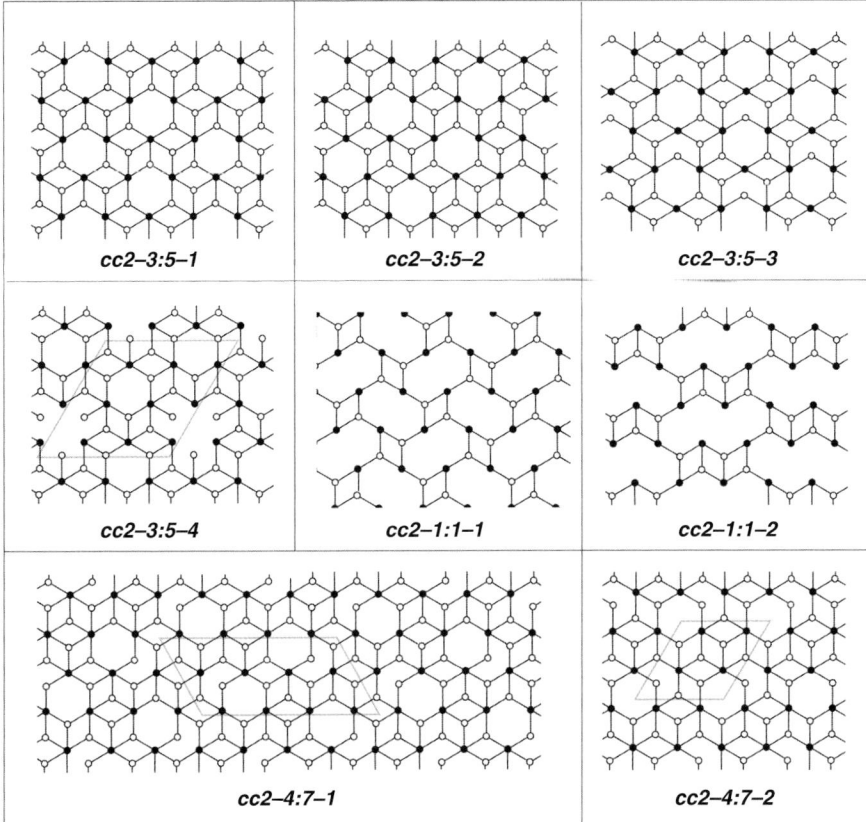

Fig. 2.14. Black-and-white 2D graphs derivative from the {**3.6.3.6**} graph. Graphs with *M:T* ratio of 3:5, 1:1 and 4:7.

sites for white nodes. Adopting the symbol □ for a vacancy, the sequence of the **u**, **d**, and □ symbols for the second row is written as ... **du**□**ud**□**du**□**ud** .... The third and fourth rows repeat the first and second rows, respectively. Thus, one may write **u**, **d**, and □ symbols for a given row in a tabular form, as shown in Fig. 2.17(e) for the graph given in Fig. 2.17(b) (it is important to maintain the vertical order of the rows: the symbol of the next row should be exactly below the corresponding symbol of the preceding row). As the graph shown in Fig. 2.17(b) is thought to be infinite in the horizontal and vertical directions, the corresponding table of the **u**, **d**, and □ symbols is also infinite within the plane of the figure. In this symbolic table, one can choose an orthogonal part (elementary unit) that may be used to reproduce the entire table by independent translations along the horizontal and vertical directions. The corresponding elementary unit for the table of symbols shown in Fig. 2.17(b) is indicated by the bold line. This elementary unit consists of six columns and two rows

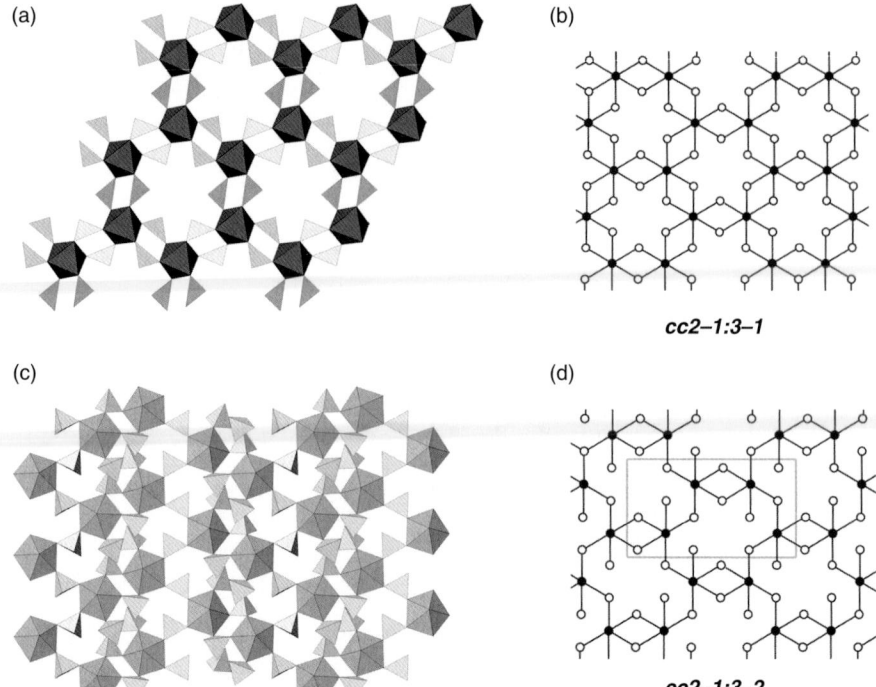

Fig. 2.15. 2D structural units with the *M*:*T* ratio of 1:3 and their graphs.

and may be used to construct the entire table by vertical and horizontal translations. We call this unit an orientation matrix of tetrahedra. Obviously, this matrix can also be written in a row as (ud☐du☐)(ududud) implying that the first six symbols in brackets correspond to the first row and the second six symbols in brackets corres-pond to the second row of the matrix. Figure 2.17(a) shows black-and-white graphs with **u** and **d** symbols written near the white nodes for the $[(UO_2)_3(CrO_4)_5]^{4-}$ sheet in $Mg_2[(UO_2)_3(CrO_4)_5](H_2O)_{17}$. The corresponding 2D table of the **u**, **d**, and ☐ sym-bols is shown in Fig. 2.17(d). The orientation matrix of tetrahedra for the sheet is different from that observed for the Ca compound and may be written as **(ddduuu)** **(ud☐ud☐)(uuuddd)(du☐du☐)**. In contrast to the 2 × 6 matrix of the sheet in the Ca chromate, this matrix has dimensions 4 × 6, which indicates that the structure of the sheet in the Mg compound is more complicated. Figure 2.17(c) shows the graph that corresponds to the structure of the $[(UO_2)_3(CrO_4)_5]^{4-}$ sheet in $K_4[(UO_2)_3(CrO_4)_5]$ $(H_2O)_8$. Its structure is related to but distinct from that of the uranyl chromate sheet in the Ca chromate. The corresponding orientation matrix of tetrahedra is **(dduuud)** **(ud☐ud☐)(uuuddd)(du☐du☐)**. It differs from the matrix of the sheet in the Ca compound by the first row only.

Table 2.4 Inorganic oxysalts containing 2D units based upon graphs with $M{:}T$ = 1:1, 13:18, 5:8, 3:5 and 1:3, derivatives from the {**3.6.3.6**} graph (see Figs. 2.13 and 2.14).

Graph	Ring symbol	Chemical formula	Reference
cc2–13:18–1	$6^8 4^{15}$	$[C_9H_{24}N_2]_7[HAl_{13}P_{18}O_{72}](H_2O)_8$	Feng *et al.* 2000
cc2–5:8–1	$6^2 4^9$	$[C_6H_{14}N_2]_3[(UO_2)_5(MoO_4)_8](H_2O)_4$	Krivovichev and Burns 2003c
cc2–5:8–2	$6^2 4^9$	$[N_3C_6H_{18}]_2[(UO_2)_5(H_2O)(SO_4)_8]$	Norquist *et al.* 2003b
cc2–5:8–3		$[C_3H_{12}N_2][(UO_2)_5(SeO_4)_8(H_2O)]$ $(H_2SeO_3)_{0.74}(H_2O)_4$	Krivovichev *et al.* 2007
cc2–5:8–4		$(H_3O)_6[(UO_2)_5(SeO_4)_8(H_2O)_5](H_2O)_5$	Krivovichev and Kahlenberg 2005e
cc2–3:8–1	$12^1 4^6$	$(NH_4)[Fe_3(H_2PO_4)_6(HPO_4)_2](H_2O)$	Mgaidi *et al.* 1999
		$(H_3O)Fe_3(HPO_4)_2(H_2PO_4)_6(H_2O)_4$	Bosman *et al.* 1986
		$KFe_3(HPO_4)_2(H_2PO_4)_6 \cdot 4(H_2O)$	Anisimova *et al.* 1997
		$(H_3O)(Al_3(H_2PO_4)_6(HPO_4)_2) \cdot (H_2O)_4$	Brodalla and Kniep 1980
cc2–3:5–1	$6^1 4^6$	$(NH_3(CH_2)_3NH_3)(H_3O)_2[(UO_2)_3(MoO_4)_5]$	Halasyamani *et al.* 1999
		$Na_6[(Np^{5+}O_2)_2(Np^{6+}O_2)(MoO_4)_5](H_2O)_{13}$	Grigoriev *et al.* 2003
		$\alpha\text{-}Mg_2[(UO_2)_3(SeO_4)_5](H_2O)_{16}$	Krivovichev and Kahlenberg 2004
		$M_2[(UO_2)_3(SeO_4)_5](H_2O)_{16}$, M = Co, Zn	Krivovichev and Kahlenberg 2005f
cc2–3:5–2	$6^1 4^6$	$K_4[(UO_2)_3(CrO_4)_5](H_2O)_8$	Krivovichev and Burns 2003a
		$Mg_2[(UO_2)_3(CrO_4)_5](H_2O)_{17}$	Krivovichev and Burns 2003e
		$Ca_2[(UO_2)_3(CrO_4)_5](H_2O)_{19}$	Krivovichev and Burns 2003e
		$\beta\text{-}Mg_2[(UO_2)_3(SeO_4)_5](H_2O)_{16}$	Krivovichev and Kahlenberg 2004
		$Zn_2[(UO_2)_3(SeO_4)_5](H_2O)_{17}$	Krivovichev and Kahlenberg 2005g
		$Cu_2[(UO_2)_3((S,Cr)O_4)_5](H_2O)_{17}$	Krivovichev and Burns 2004
cc2–3:5–3	$6^2 4^4$	$Rb_4[(UO_2)_3(SeO_4)_5(H_2O)]$	Krivovichev and Kahlenberg 2005a
		$[C_5H_{14}N]_4[(UO_2)_3(SeO_4)_4(HSeO_3)(H_2O)]$ $(H_2SeO_3)(HSeO_4)$	Krivovichev *et al.* 2006b
		$(H_3O)[C_5H_{14}N]_2[(UO_2)_3(SeO_4)_4(HSeO_4)$ $(H_2O)]$	Krivovichev *et al.* 2006c
		$(H_3O)[C_5H_{14}N]_2[(UO_2)_3(SeO_4)_4(HSeO_4)$ $(H_2O)](H_2O)$	Krivovichev *et al.* 2006c
cc2–3:5–4	$12^1 6^2 4^{11}$	$[C_9H_{24}N_2]_2[(UO_2)_3(SeO_4)_5(H_2O)_2](H_2O)_{12}$	Krivovichev *et al.* 2007
cc2–4:7–1	$6^2 4^7$	$[C_3H_{10}N]_8[(UO_2)_4(SeO_4)_7](HSeO_4)(NO_3)$ $(H_2O)_4$	Krivovichev *et al.* 2007
cc2–4:7–2	$6^2 4^7$	$K_6[(UO_2)_4(CrO_4)_7](H_2O)_6$	Sykora *et al.* 2004
cc2–1:1–1	$8^1 4^1$	$[C_6N_2H_{14}][Sn^{2+}PO_4]_2(H_2O)$	Ayyapan *et al.* 1998
		$Ba[MoO_3SeO_3]$	Harrison *et al.* 1996a
		$[C_{10}H_{10}N_2]_2[Zn_2(HPO_3)_2]$	Fan *et al.* 2005
cc2–1:1–2	$12^1 4^3$	$[C_{14}H_{14}N_4][Zn_2(HPO_3)_2](H_2O)_{0.4}$	Fan *et al.* 2005
cc2–1:3–1	$12^1 3^3$	$Sc(IO_3)_3$	Hector *et al.* 2002
cc2–1:3–2	$12^1 4^1$	$[CH_3NH_3][(UO_2)(H_2AsO_4)_3]$	Alekseev *et al.* 2008

Following Moore (1970a, 1975), we call chemically and topologically identical but geometrically different structural units *geometrical isomers*. There are several types of geometrical isomerism. In the case of uranyl chromates, we suggest to call the kind of isomerism observed in their structures *an orientational geometrical isomerism*. The presence of orientational geometrical isomers in uranyl chromates is induced by

(a)   (b)

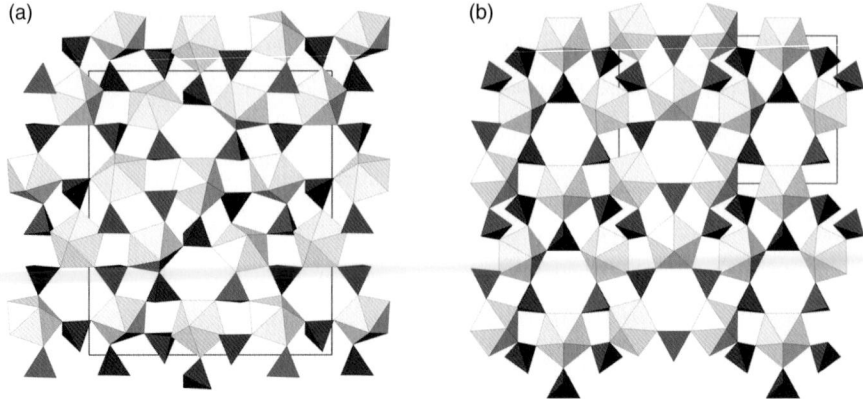

Fig. 2.16. $[(UO_2)_3(CrO_4)_5]$ sheets observed in the structures of $Mg_2[(UO_2)_3(CrO_4)_5](H_2O)_{17}$ and $Ca_2[(UO_2)_3(CrO_4)_5](H_2O)_{19}$ ((a) and (b), respectively).

the presence of 3-connected CrO_4 tetrahedra with non-shared O corners. To distinguish between these isomers, one may use the orientation matrix of tetrahedra first introduced by Krivovichev and Burns (2003a).

An interesting isomeric variation of the *cc2–3:5–2* topology has been observed in β-$Mg_2[(UO_2)_3(SeO_4)_5](H_2O)_{16}$. The polyhedral diagram of the $[(UO_2)_3(SeO_4)_5]^{4-}$ sheet from this structure is shown in Fig. 2.18(b). Analysis of its black-and-white graph (Fig. 2.18(d)) indicates that this graph does correspond to the *cc2–3:5–2* topology. However, the system of tetrahedral orientations is fundamentally different from those observed in uranyl chromates discussed above. The point is that one of the tetrahedra within the sheet has a disordered "up-and-down" orientation for which we adopt the symbol **m**. Figures 2.19(c) and (f) provide a description of the tetrahedral orientations in graphical and tabular forms, respectively. The orientation matrix is rather complicated as it has 12×2 dimensions and can be written as (**dd□dd□uu□uu□/ uuduumddduddm**). This high degree of complexity is most probably the result of the strongly modulated geometrical shape of the uranyl selenate sheet (Fig. 2.20(b)). It is noteworthy that β-$Mg_2[(UO_2)_3(SeO_4)_5](H_2O)_{16}$ is unstable under atmospheric conditions.

In contrast to the β-form, α-$Mg_2[(UO_2)_3(SeO_4)_5](H_2O)_{16}$ is stable in air. Its structure contains flat $[(UO_2)_3(SeO_4)_5]^{4-}$ sheets with the *cc2–3:5–2* topology (Figs. 2.19(a) and (c)). The same topology is also characteristic for the actinyl molybdate sheets in $[C_3N_2H_{12}](H_3O)_2[(UO_2)_3(MoO_4)_5]$ and $Na_6[(Np^{5+}O_2)_2(Np^{6+}O_2)(MoO_4)_5](H_2O)_{13}$. However, the sheets in selenate and molybdate compounds correspond to different geometrical isomers (Fig. 2.19). The uranyl selenate sheet has the (**uumdd□/ dd□uum**) the orientation matrix (in compact form), whereas, in molybdates, orientation matrix is (**ddudd□/uu□uud**).

Figures 2.21(a) and (b) depict two $[(UO_2)_3(SeO_4)_5(H_2O)]^{4-}$ sheets observed in the structures of $Rb_4[(UO_2)_3(SeO_4)_5(H_2O)]$ and $(H_3O)[C_5H_{14}N]_2[(UO_2)_3(SeO_4)_4(HSeO_4)$

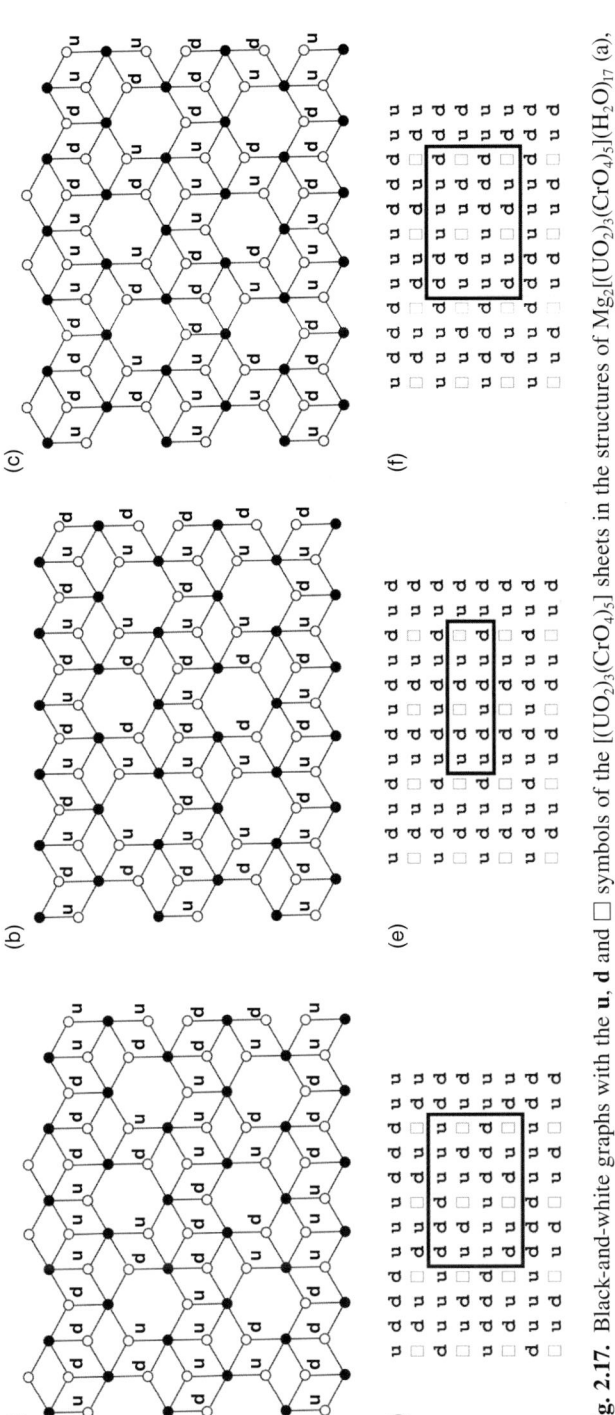

Fig. 2.17. Black-and-white graphs with the **u**, **d** and □ symbols of the [(UO₂)₃(CrO₄)₅] sheets in the structures of Mg₂[(UO₂)₃(CrO₄)₅](H₂O)₁₇ (a), Ca₂[(UO₂)₃(CrO₄)₅](H₂O)₁₉ (b) and K₄[(UO₂)₃(CrO₄)₅](H₂O)₈ (c); their corresponding **u**, **d** and □ symbolic tables ((d), (e) and (f), respectively). The orientation matrices of tetrahedra are indicated in (d), (e) and (f) by bold lines.

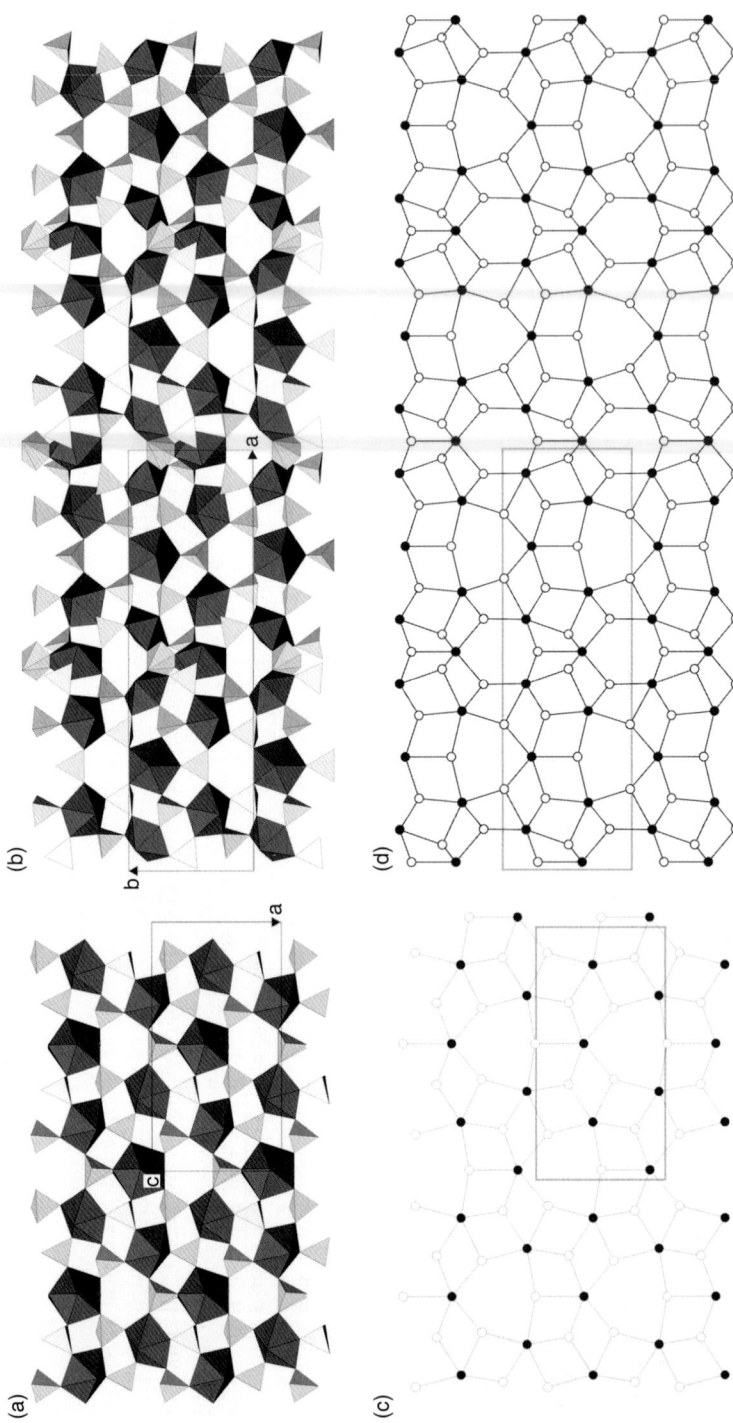

Fig. 2.18. [(UO$_2$)$_3$(SeO$_4$)$_5$]$^{4-}$ sheets in the crystal structures of α- (a) and β-Mg$_2$[(UO$_2$)$_3$(SeO$_4$)$_5$](H$_2$O)$_{16}$ (b) and their nodal representations ((c) and (d), respectively). Legend: [UO$_7$]$^{8-}$ bipyramids = black circles; [SeO$_4$]$^{2-}$ tetrahedra = white circles.

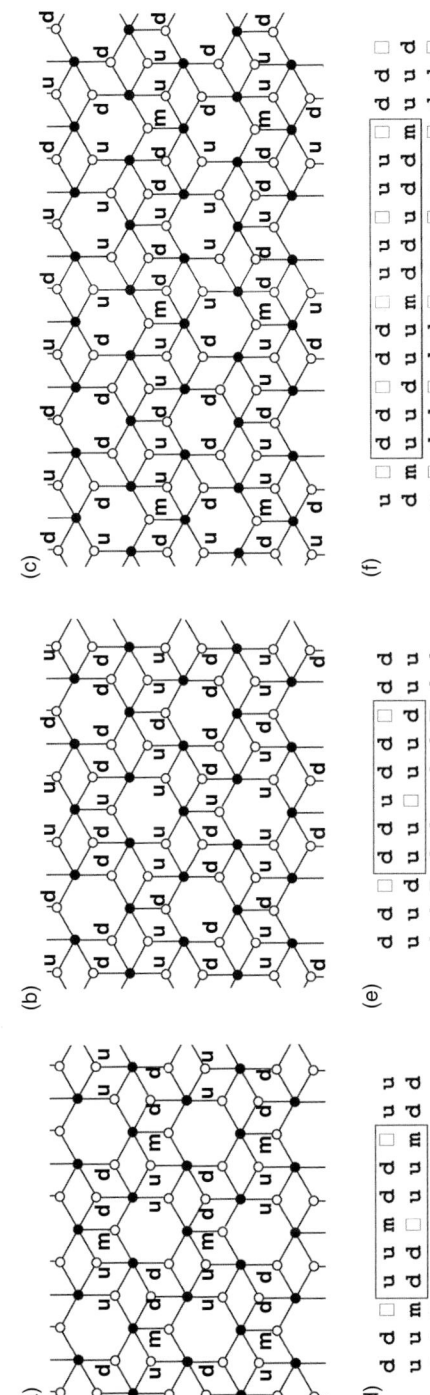

Fig. 2.19. Idealized versions of the graphs shown in Figs. 2.18(a) and (b) ((a) and (c), respectively) can be produced from the {**3.6.3.6**} graph by elimination of one sixth of its white vertices and all edges incident upon them. Idealized black-and-white graph of the $[(UO_2)_3(SeO_4)_5]^{4-}$ sheets observed in α- (d) and β-$Mg_2[(UO_2)_3(SeO_4)_5](H_2O)_{16}$ (f), and $[(UO_2)_3(MoO_4)_5]^{4-}$ sheet observed in $(NH_3(CH_2)_3NH_3)(H_3O_2[(UO_2)_3(MoO_4)_5]$ (e) together with the tetrahedral orientation symbols written next to the white vertices; tabular forms of orientation symbols for these sheets ((g), (h) and (i), respectively). See text for details.

(H_2O)](H_2O), respectively. The sheets are topologically characterized by the ***cc2–3:5–3*** graph shown in Fig. 2.14. Both sheets consist of two types of uranyl cations. One is coordinated by five O atoms of the SeO_4 tetrahedra, thus forming UO_7 pentagonal bipyramids, whereas another is coordinated by four O atoms and one H_2O molecule, which results in a $UO_6(H_2O)$ bipyramid. As can be seen from Figs. 2.21(a) and (b), the sheets are obviously different by the orientation of the $UO_6(H_2O)$ bipyramids. This difference can be visualized by adding the U→H_2O vectors to the black-and-white graphs of the sheets. The idealized versions of the graphs with the U→H_2O vectors are shown in Figs. 2.21(c) and (d), respectively. In $Rb_4[(UO_2)_3(SeO_4)_5(H_2O)]$, all U→$H_2O$ vectors whithin the sheet have the same direction, whereas, in (H_3O) $[C_5H_{14}N]_2[(UO_2)_3(SeO_4)_4(HSeO_4)(H_2O)](H_2O)$, U→$H_2O$ vectors with different orientations alternate. As was in the case with the ***cc2–1:2–2***-type sheets (see above), the sheets shown in Figs. 2.21(a) and (b) can be considered as different geometrical isomers induced by selective hydration of the U polyhedra.

Further analysis of the ***cc2–3:5–3*** graph indicates that its white vertices are either 2- or 3-connected. Figure 2.22 shows the uranyl selenate–selenite sheet from the structure of $[C_5H_{14}N]_4[(UO_2)_3(SeO_4)_4(HSeO_3)(H_2O)](H_2SeO_3)(HSeO_4)$. The topology of this sheet can also be described by the ***cc2–3:5–3*** graph, however, with 2-connected white vertices corresponding to $Se^{4+}O_3^{2-}$ selenite groups. The selenite group has the configuration of triangular pyramid with its apical vertex occupied by the Se^{4+} cation possessing a stereochemically active lone-electron pair. The 3-connected white vertices of the ***cc2–3:5–3*** graph correspond to tetrahedral SeO_4^{2-} oxoanions. Once again, the presence of tridentate tetrahedra gives rise to the occurence of orientational geometrical isomerism. Figure 2.23(a) shows the ***cc2–3:5–3*** graph with u and d symbols written near the white vertices. This diagram corresponds to the system of orientation of tetrahedra in the uranyl selenate–selenite sheet shown in Fig. 2.22. In principle, if 2-connected white vertices correspond to tetrahedral oxoanions, their orientation relative to the plane of the sheet does not really matter since bidentate tetrahedra possess enough freedom to change their orientation without affecting the topological structure of the sheet. However, in the case when 2-connected vertices correspond to $Se^{4+}O_3^{2-}$ groups, one has to distinguish between the non-shared O apical corner and another corner that is occupied by the lone-electron pair. For this reason, the diagram in Fig. 2.23(a) provides orientation symbols for the 2-connected vertices as well. The system of tetrahedral orientations in tabular form is given in Fig. 2.23(d); it corresponds to the matrix (**ududud**)(**ud**□**du**□). Figures 2.23(b) and (c) describe the topological structures of the uranyl selenate sheets in the structures of (H_3O) $[C_5H_{14}N]_2[(UO_2)_3(SeO_4)_4(HSeO_4)(H_2O)]$ and (H_3O)$[C_5H_{14}N]_2[(UO_2)_3(SeO_4)_4(HSeO_4)$ (H_2O)](H_2O), respectively. It is of interest that these compounds are uranyl selenates templated by protonated *N*-methylbutylamine molecules and their compositions differ by one H_2O molecule only. However, their uranyl sheets are topologically different and correspond to two different geometrical isomers. Since in this case, 2-connected white vertices symbolize bidentate SeO_4 tetrahedra, their orientation relative to the plane of the sheet is not important, which is symbolized by the **m** symbol. Tables of the orientations of tetrahedra are given in Figs. 2.23(e) and (f). The uranyl selenate

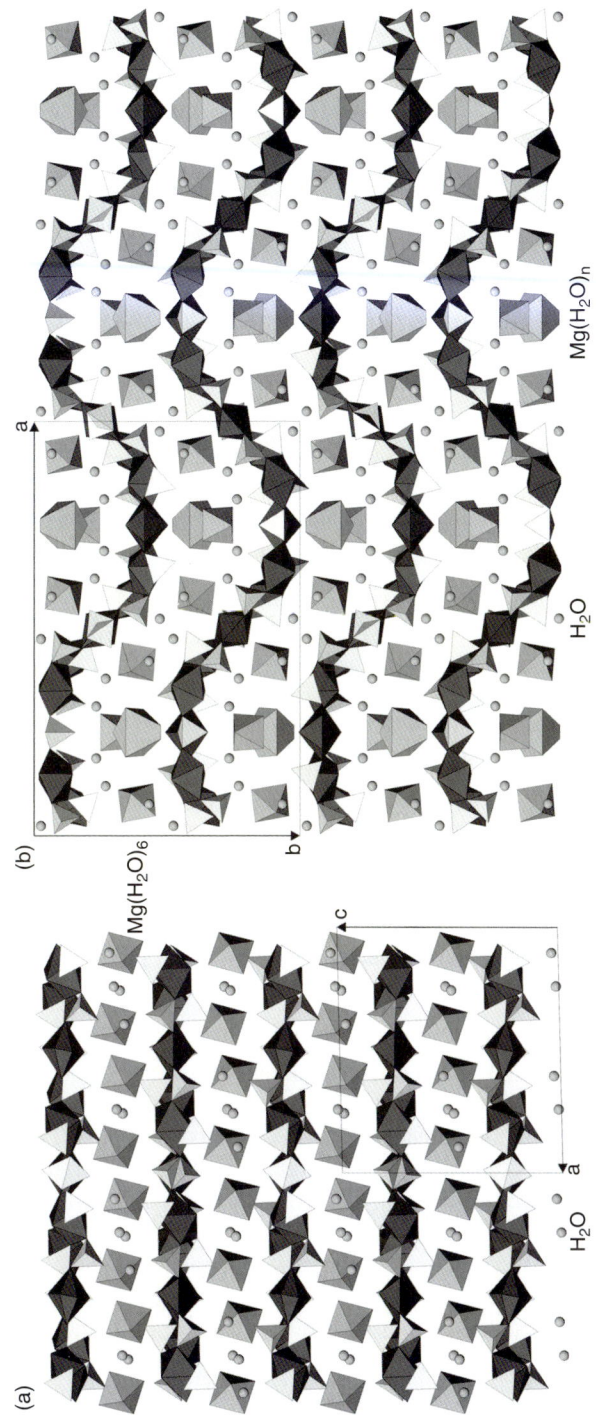

Fig. 2.20. Projections of the structures of α- (a) and β-$Mg_2[(UO_2)_3(SeO_4)_5](H_2O)_{16}$ (b) along the b- or c-axis, respectively.

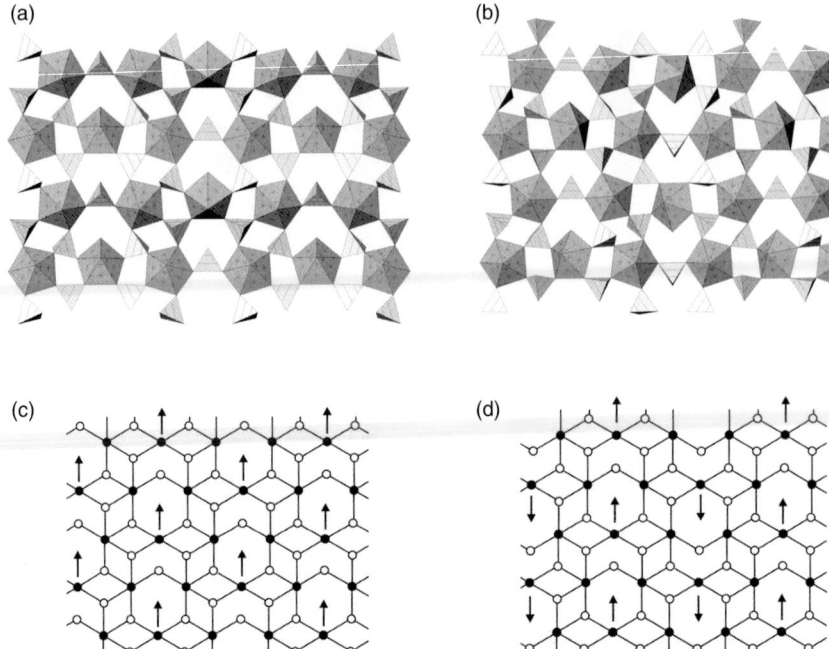

Fig. 2.21. $[(UO_2)_3(SeO_4)_5(H_2O)]^{4-}$ sheets observed in the structures of $Rb_4[(UO_2)_3(SeO_4)_5$ $(H_2O)]$ (a) and $(H_3O)[C_5H_{14}N]_2[(UO_2)_3(SeO_4)_4(HSeO_4)(H_2O)](H_2O)$ (b), respectively. The sheets are different by the orientations of their $U{\to}H_2O$ vector systems ((c) and (d)).

sheet observed in the low hydrate has a 3×2 orientation matrix **(dum)(du☐)**. The matrix of the sheet in the higher hydrate has 6×4 dimensions and can be written as **(dumdum)(du☐ud☐)(udmudm)(du☐ud☐)**.

2.2.1.7   Cis–trans geometrical isomerism and connectivity diagrams

Figures 2.24(a), (b) and (c) show three octahedral–tetrahedral sheets from the structures of $Fe(H_2PO_4)_2(H_2O)_2$ (Guse *et al.* 1985), olmsteadite $K(Fe^{2+})_2(H_2O)_2[NbO_2$ $(PO_4)_2]$ (Moore *et al.* 1976), and $Na_5[Fe(PO_4)_2F_2]$ (Rastsvetaeva *et al.* 1996). The topology of the three sheets corresponds to the graph **11/2b**. However, the arrangement of shared and non-shared vertices of octahedra are different in all three structures. First, it should be noted that the black nodes in the **11/2b** graph are 4-connected, i.e. four corners of the MO_6 octahedra are shared with tetrahedra (which are symbolized by white nodes). There are just two possibilities of arrangements of four shared vertices in an octahedron and these two are shown in Figs. 2.24(d) and (e). To visualize these arrangements, we use a *connectivity diagram* that represents a projection of a coordination polyhedron onto a two-dimensional plane. Connectivity diagrams

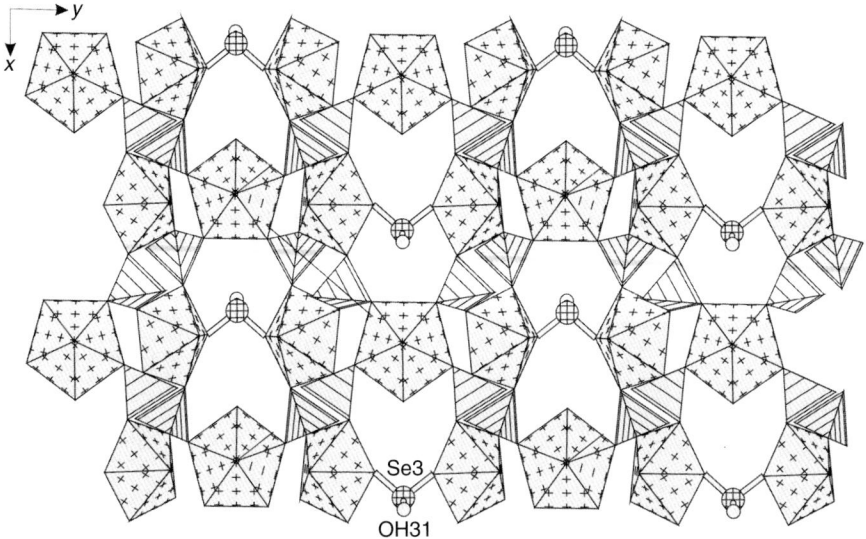

Fig. 2.22. Uranyl selenate–selenite sheet in the structure of $[C_5H_{14}N]_4[(UO_2)_3(SeO_4)_4(HSeO_3)$ $(H_2O)](H_2SeO_3)(HSeO_4)$.

may be based upon Schlegel diagrams that have been used in crystal chemistry for a long time (Moore 1970b; Hoppe and Koehler 1988; Krivovichev 1997; Krivovichev *et al.* 1998). In the case of octahedral structures, the projections of octahedra shown in Figs. 2.24(d), (e), and (f) are more convenient (Krivovichev *et al.* 2002b, 2003a). A vertex of an octahedron shared with an adjacent polyhedron is marked by a black circle, whereas an edge common to two polyhedra is marked by a bold line. The connectivity diagram in Fig. 2.24(d) shows the arrangement of shared vertices of the MO_6 octahedra that can be called a *trans*-arrangement, whereas the arrangement shown in Fig. 2.24(e) can be called a *cis*-arrangement. It is interesting that the sheet depicted in Fig. 2.24(c) contains octahedra of two types. Octahedra of the **A** type have a *trans*-arrangement of shared vertices, whereas octahedra of the **B** type have a *cis*-arrangement of shared vertices. Since three octahedral–tetrahedral sheets corres- pond to the same **I1/2b** graph, they should be called geometrical isomers. This kind of geometrical isomerism is usually called a *cis–trans geometrical isomerism*. It is quite common in heteropolyhedral units in salts of inorganic oxoacids.

2.2.2 *Autunite topology and its derivatives*

A black-and-white graph corresponding to the autunite topology is shown in Fig. 2.25(a). This graph is basic because it can be considered as parent to a number of other topologies shown in Fig. 2.25. Its topological description is {**4.4.4.4**} that means that

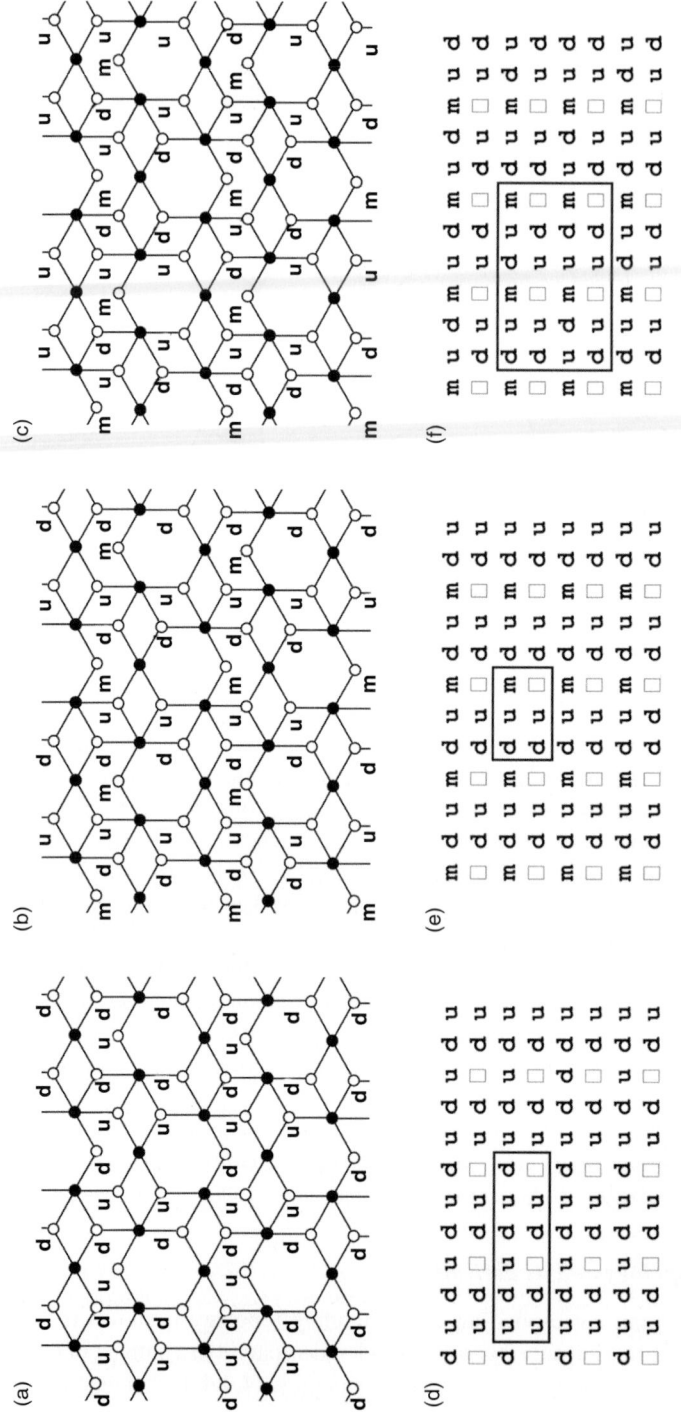

Fig. 2.23. The *cc2–3:5–3* graph with **u**, **d**, **m** and □ symbols symbols written near the white vertices for the sheets in the structures of [C₅H₁₄N]₄[(UO₂)₃(SeO₄)₄(HSeO₃)(HSeO₄)] (a), (H₃O)[C₅H₁₄N]₂[(UO₂)₃(SeO₄)₄(HSeO₄)(H₂O)] (b) and (H₃O)[C₅H₁₄N]₂[(UO₂)₃(SeO₄)₄ (HSeO₄)(H₂O)][(H₂O)] (c), and their corresponding **u**, **d** and □ symbolic tables ((d), (e) and (f), respectively). The orientation matrices of tetrahedra are indicated in (d), (e) and (f) by bold lines.

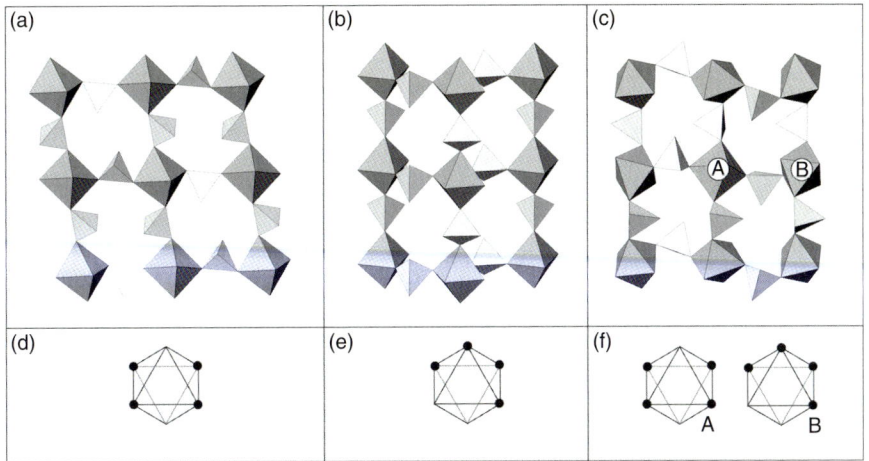

Fig. 2.24. Polyhedral diagrams of octahedral–tetrahedral sheets from the structures of $Fe(H_2PO_4)_2(H_2O)_2$ (a), olmsteadite $K(Fe^{2+})_2(H_2O)_2[NbO_2(PO_4)_2]$ (b), and $Na_5[Fe(PO_4)_2F_2]$ (c). Connectivity diagrams of the octahedra of these sheets ((d), (e) and (f), respectively) show the arrangements of vertices shared with adjacent tetrahedra. See text for details.

the elementary unit of the graph has four 4-connected nodes. A list of compounds containing heteropolyhedral sheets of this topology is given in Table 2.5. An example of a heteropolyhedral unit corresponding to this topology is shown in Fig. 2.26(a). This topology is characteristic for $[UO_2(TO_4)]$ sheets (T = P, As) observed in a large number of uranyl minerals of the autunite and meta-autunite groups. It is noteworthy that, until recently, only a few structures of the minerals comprising these groups were known. The main problem is their rapid dehydratation under atmospheric conditions. In recent ground-breaking studies, Locock and Burns (2003a, b) applied a capillary technique that allowed them to refine many new structures of these groups (see Locock (2007) for a complete review).

The autunite topology has a number of variations. Figure 2.26(b) shows a side view of the octahedral–tetrahedral sheet depicted in Fig. 2.26(a). Note that the apical corners of the octahedra of the sheet are non-shared. In the structure of $Na_3[(VO)(PO_4)(HPO_4)]$ (Schindler *et al.* 1999), An additional tetrahedron is attached to one apical corner of each octahedron of the sheet, resulting in a *branched* topology. The side view of the $[(VO)(PO_4)(HPO_4)]$ sheet in this structure is shown in Fig. 2.26(c) (topology *cc2–1:2–14*: Fig. 2.27(a)). The additional tetrahedra in $Na_3[(VO)(PO_4)(HPO_4)]$ are 1-connected. In contrast, in β-$Ti(PO_4)(H_2PO_4)$ (Krogh Andersen *et al.* 1998) and γ-$Zr(AsO_4)(H_2AsO_4)\cdot2(H_2O)$ (Rodriguez *et al.* 1999), additional tetrahedra are 2-connected: they link two octahedra within the sheet and cause its substantional distortion (Fig. 2.26(d) (topology *cc2–1:2–14*: Fig. 2.27(b))).

The structure of $K_2(VO)_2(HPO_4)_3(H_2O)_{1.125}$ (Lii and Tsai 1991) consists of complex sheets that are composed of slabs of autunite-type octahedral sheet (Fig. 2.26(e);

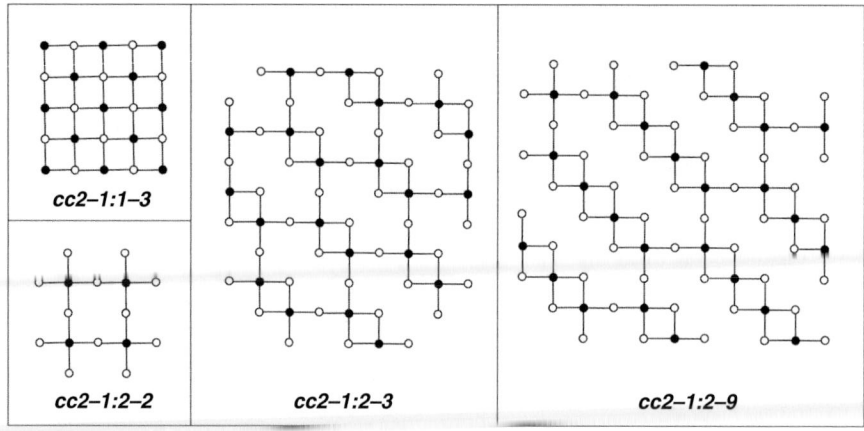

Fig. 2.25. Black-and-white graphs corresponding to the autunite-type topology (a) and its derivatives ((b), (c) and (d)).

Table 2.5 Inorganic oxysalts containing 2D structural units based upon autunite topology (graph *cc2–1:1–3*) and its derivatives

Graph	Mineral name	Chemical formula	Reference
cc2–1:1–3	saléeite	$Mg[(UO_2)(PO_4)]_2(H_2O)_{10}$	Miller and Taylor 1986
	threadgoldite	$Al[(UO_2)(PO_4)]_2(OH)(H_2O)_8$	Piret *et al*. 1979; Khosrawan-Sazedj 1982a
	metauranocircite	$Ba[(UO_2)(PO_4)]_2(H_2O)_6$	Khosrawan-Sazedj 1982b
	abernathyite	$K[(UO_2)(AsO_4)](H_2O)_3$	Ross and Evans 1964
	sincosite	$Ca[VOPO_4]_2 \cdot 4(H_2O)$	Franke *et al*. 1997
	torbernite	$Cu[(UO_2)(PO_4)]_2(H_2O)_{12}$	Locock and Burns 2003a
	metatorbernite	$Cu[(UO_2)(PO_4)]_2(H_2O)_8$	Locock and Burns 2003a
	zeunerite	$Cu[(UO_2)(AsO_4)]_2(H_2O)_{12}$	Locock and Burns 2003a
	metazeunerite	$Cu[(UO_2)(AsO_4)]_2(H_2O)_8$	Locock and Burns 2003a
	autunite	$Ca[(UO_2)(PO_4)]_2(H_2O)_{11}$	Locock and Burns 2003b
		$A_{0.5}(VO)(PO_4)(H_2O)_2$ (A = Na, K)	Wang *et al*. 1991
		$Ni_{0.5}VOPO_4 \cdot 2H_2O$	Lii *et al*. 1993b
		$M(VOPO_4)_2 \cdot 4(H_2O)$ M = Ca, Ba, Cd, Pb, Sr, Co	Kang *et al*. 1992a; Roca *et al*. 1997; le Fur *et al*. 1999; le Fur and Pivan 1999
		$Pb_2VO(PO_4)_2$	Shpanchenko *et al*. 2006b
cc2–1:2–14		$Na_3[(VO)(PO_4)(HPO_4)]$	Schindler *et al*. 1999b
cc2–1:2–15		$\beta\text{-}[Ti(PO_4)(H_2PO_4)]$	Krogh Andersen *et al*. 1998
		$[C_6N_2H_{16}]_{0.5}[Ti_2(H_2PO_4)(HPO_4)(PO_4)_2]$	Chen *et al*. 2004
		$\gamma\text{-}[Zr(AsO_4)(H_2AsO_4)](H_2O)_2$	Rodriguez *et al*. 1999
cc2–1:2–16		$Rb[Nd(SeO_4)_2(H_2O)_3]$	Gasanov *et al*. 1985
cc2–2:3–13		$K_2[(VO)_2(HPO_4)_3](H_2O)_{1.125}$	Lii and Tsai 1991a
cc2–3:4–10		$K_2[(VO)_2V(PO_4)_2(HPO_4)(H_2PO_4)(H_2O)_2]$	Haushalter *et al*. 1993a

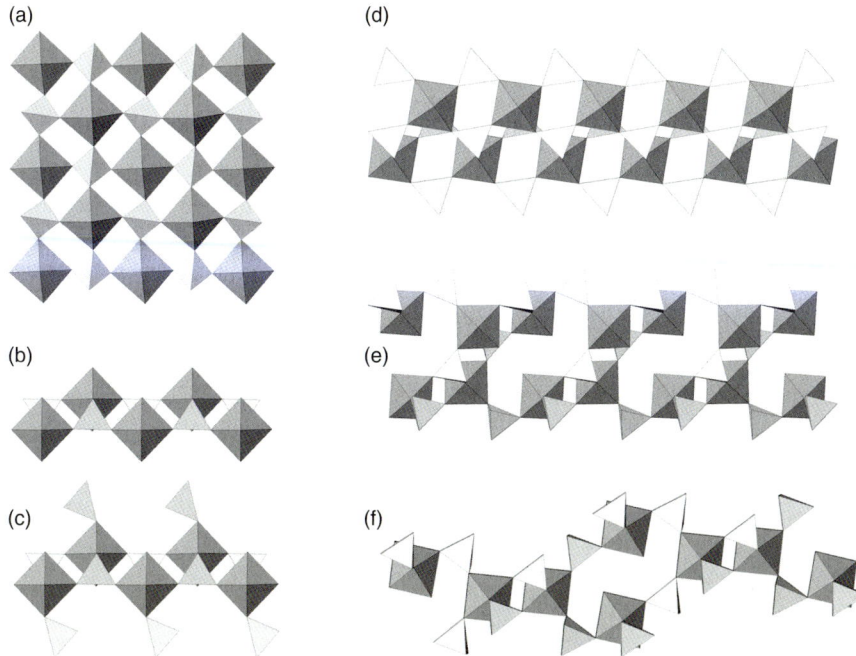

Fig. 2.26. An example of an autunite-type octahedral–tetrahedral sheet (a), its side view (b); side views of the octahedral–tetrahedral sheets in the structures of $Na_3[(VO)(PO_4)(HPO_4)]$ (c), β-$Ti(PO_4)(H_2PO_4)$ (d), γ-$Zr(AsO_4)(H_2AsO_4) \cdot 2(H_2O)$ (e), $K_2((VO)_2V(PO_4)_2(HPO_4)(H_2PO_4)(H_2O)_2)$ (f).

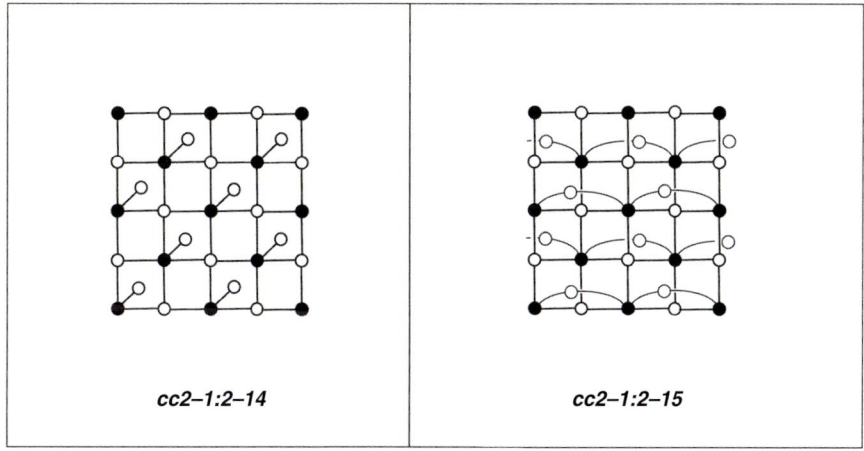

cc2–1:2–14   cc2–1:2–15

Fig. 2.27. Autunite-derivative topologies.

topology $cc2-2:3-13$: Fig. 2.28(a)). An even more complex autunite derivative sheet is observed in $K_2((VO)_2V(PO_4)_2(HPO_4)(H_2PO_4)(H_2O)_2)$ (Haushalter et al. 1993). It consists of 2- and 1-membered chains of the autunite-type octahedral–tetrahedral sheet combined into a $[(VO)_2V(PO_4)_2(HPO_4)(H_2PO_4)(H_2O)_2]^{2-}$ complex porous layer encapsulating K^+ cations (Fig. 2.26(f)). The topology of the layer can be described using the idealized graph shown in Fig. 2.28(b) (topology $cc2-3:4-10$).

Another example of a distorted derivative of the autunite topology is shown in Fig. 2.29. It corresponds to the heteropolyhedral layer in the structure of $Rb[Nd(SeO_4)_2(H_2O)_3]$ (Fig. 2.29(a)). In this compound, each Nd^{3+} cation is coordinated by five O atoms of the selenate groups and three H_2O molecules. The layer can be described as based upon chains excised from the autunite sheet (Figs. 2.29(b) and (c)) and further interlinked into a complex porous graph (Fig. 2.29(d)).

It should be noted that the autunite-derived graphs shown in Fig. 2.25 are topologically equivalent to the {**3.6.3.6**} graph derivatives shown in Fig. 2.3 according to the transformation scheme given in Figs. 2.4(a), (b) and (c).

2.2.3   Other basic graphs and their derivatives

In addition to the basic graphs {**3.6.3.6**} and {**4.4.4.4**}, there are two other basic graphs that consist of black and white nodes with single links and with no black–black or white–white connections.

Figure 2.30 shows a black-and-white graph of relatively complex topological structure. It consists of two elementary rhombs and can be described by the symbol {**5.3.5.3**} {**5.3.5.4**}. Heteropolyhedral units corresponding to this graph occur exclusively in

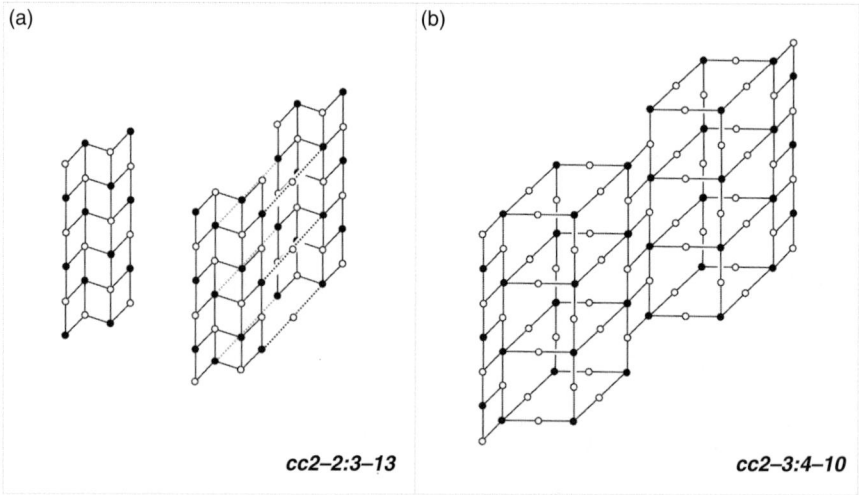

(a)   (b)

$cc2-2:3-13$ $cc2-3:4-10$

Fig. 2.28. Idealized graphs of porous autunite-related topologies.

uranyl compounds (see Table 2.6). Figure 2.30(b) shows the $[(UO_2)_2(MoO_4)_3]$ sheet from the structure of β-$Cs_2[(UO_2)_2(MoO_4)_3]$ (Krivovichev *et al.* 2002c). It consists of UO_7 pentagonal bipyramids and MoO_4 tetrahedra sharing corners with each other. The structure of α-$Cs_2[(UO_2)_2(MoO_4)_3]$ consists of a framework of the same subunits that is strongly related to the β-modification (see Chapter 3 and Krivovichev *et al.* 2002c for more details).

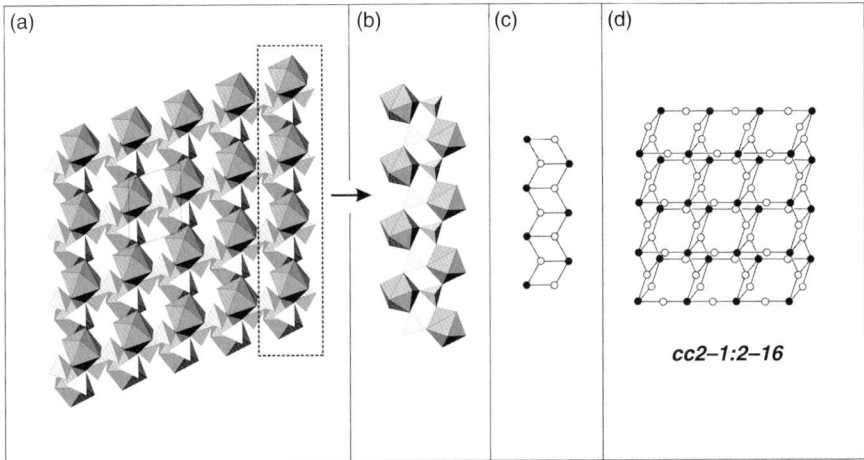

Fig. 2.29. Sheet composed from $(NdO_3(H_2O)_5)$ polyhedra and (SO_4) tetrahedra in the structure of $Rb[Nd(SeO_4)_2(H_2O)_3]$ (a), its basic chain in polyhedral (b) and graphical (c) aspects, and construction of graph of the sheet (d).

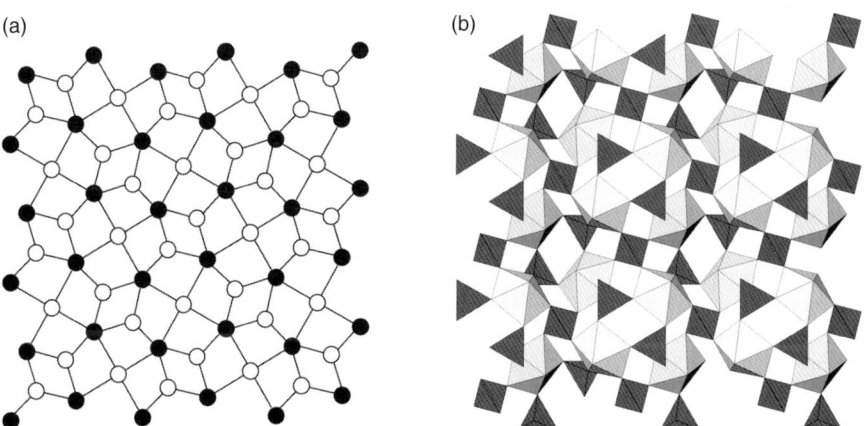

Fig. 2.30. The {5.3.5.3}{5.3.5.4} graph (a) and corresponding heteropolyhedral unit from the structure of β-$Cs_2[(UO_2)_2(MoO_4)_3]$.

Table 2.6 Inorganic oxysalts based upon 2D units derived from graphs of different types

Graph	Chemical formula	Reference
cc2–2:3–14	β-Cs$_2$[(UO$_2$)$_2$(MoO$_4$)$_3$]	Krivovichev *et al.* 2002c
	Cs$_2$[(UO$_2$)$_2$(SO$_4$)$_3$]	Ross and Evans 1960
	[N$_4$C$_6$H$_{22}$][(UO$_2$)$_2$(SO$_4$)$_3$]$_2$(H$_2$O)	Norquist *et al.* 2003a
cc2–2:3–15	[C$_6$NH$_8$][Al$_2$P$_3$O$_{10}$(OH)$_2$]	Yu *et al.* 1998
cc2–2:3–16	Sr$_2$Fe$_2$(HPO$_4$)(PO$_4$)$_2$F$_2$	le Meins *et al.* 1998
cc2–1:4–1	Na$_6$[Mg(SO$_4$)$_4$] vanthoffite	Fischer and Hellner 1964
cc2–1:3–3	Na$_8$(VO)$_2$(SO$_4$)$_6$	Nielsen *et al.* 1999
cc2–1:3–4	[Co(en)$_3$][Ga$_3$(H$_2$PO$_4$)$_6$(HPO$_4$)$_3$]	Wang *et al.* 2003b
cc2–5:8–5	[Co(H$_2$O)$_6$]$_3$[(UO$_2$)$_5$(SO$_4$)$_8$(H$_2$O)](H$_2$O)$_5$	Krivovichev *et al.* 2007

Figure 2.31(a) shows a black-and-white graph that can be given a symbol {**6.2.6.2**} {**4.6.4.6**}. To our knowledge, this graph itself does not correspond to any known topology in salts of inorganic oxoacids. However, it has derivatives that have their realizations in heteropolyhedral units. Figure 2.31(b) shows a black-and-white graph that corresponds to the octahedral–tetrahedral sheet observed in the structure of vanthoffite, Na$_6$[Mg(SO$_4$)$_4$] (Fischer and Hellner 1964). The graphs depicted in Figs. 2.31(c) and (d) describe the topologies of heteropolyhedral sheets observed in the structures of Na$_8$(VO)$_2$(SO$_4$)$_6$ (Nielsen *et al.* 1999) and Na$_3$[(VO)(PO$_4$)(HPO$_4$)] (Schindler *et al.* 1999), respectively. It is noteworthy that the [(VO)(PO$_4$)(HPO$_4$)] sheet from Na$_3$[(VO) (PO$_4$)(HPO$_4$)] can be described as a derivative of both autunite topology (see above) and the topology of the {**6.2.6.2**}{**4.6.4.6**} graph.

2.2.4 *Modular approach to complex 2D topologies*

Within the last decade, a modular approach to the description of complex structures of minerals and inorganic compounds became very popular in mineralogy and structural chemistry (Merlino 1997; Eddaoudi *et al.* 2001; Ferraris *et al.* 2004). Using this approach, the structure is considered as being constructed from modules (or secondary building blocks) of different dimensionality. Usually, the same modules occur in different structures in different combinations. Here, we describe several 2D topologies with modular structure. Tentatively, they can be subdivided into: (i) topologies derived from the same basic graph; (ii) topologies derived from two or more basic graphs.

2.2.4.1 *Topologies derived from the same basic graph*

Figure 2.32 demonstrates transition from the basic graph {**3.6.3.6**} to the graph *cc2– 2:3–15*, which has a modular structure. The transition can be achieved through an intermediate graph *cc2–1:2–6* that can be obtained from {**3.6.3.6**} by deletion of some of its edges. The *cc2–1:2–6* graph is cut off into 1D modules, white vertices

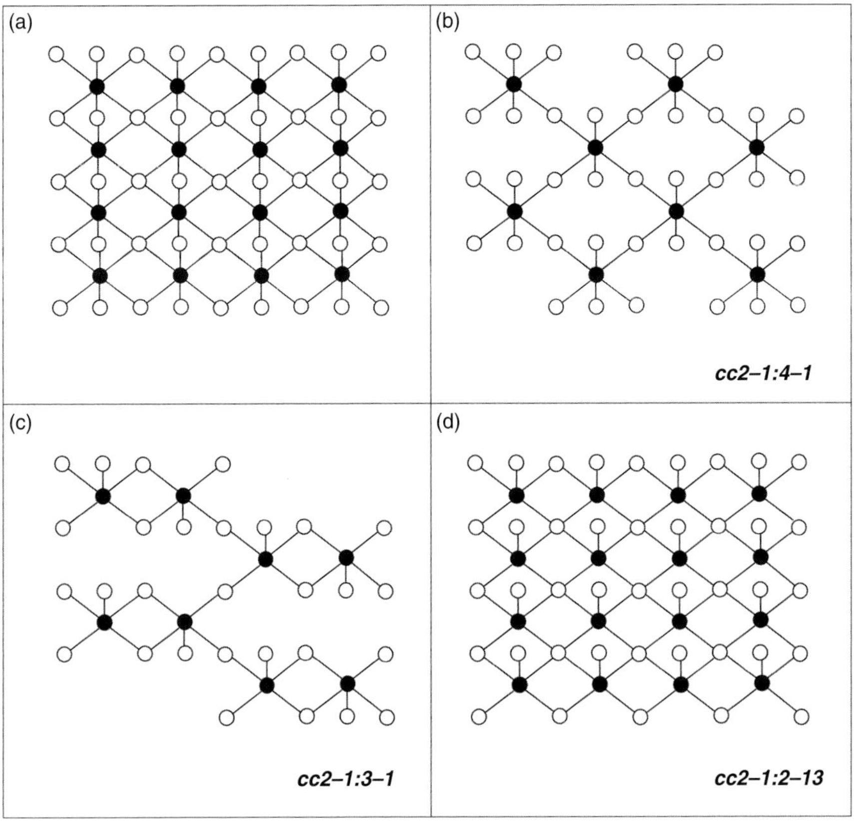

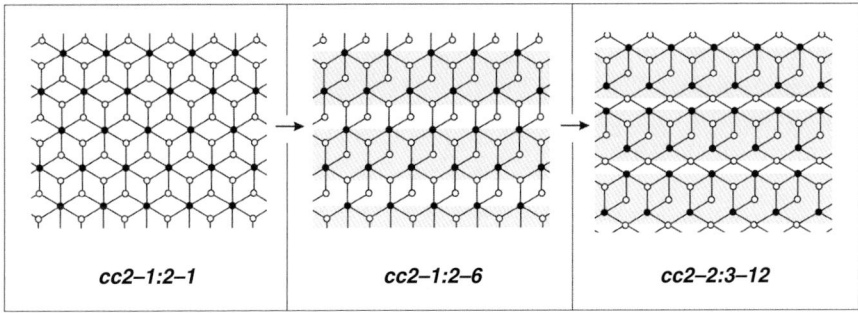

(a)

(b)

$cc2$–1:4–1

(c)

$cc2$–1:3–1

(d)

$cc2$–1:2–13

Fig. 2.31. The {**6.2.6.2**}{**4.6.4.6**} basic graph (a) and its derivatives ((b), (c), (d)).

$cc2$–1:2–1 $cc2$–1:2–6 $cc2$–2:3–12

Fig. 2.32. Transition from the basic graph {**3.6.3.6**} to the modularly structured graph *cc2–2:3–12* through the graph *cc2–1:2–6*.

outside the modules are deleted and the modules are merged into the 2D graph $cc2$–$2{:}3$–15. It is important that, as a result of the transition, the $cc2$–$2{:}3$–15 graph possess 4-connected white vertices that are absent in the initial graph. In other words, the $cc2$–$2{:}3$–15 graph is *not* a derivative of {**3.6.3.6**}. This graph is an underlying topology for the aluminophosphate tetrahedral sheet in the structure of $[Al_2P_3O_{10}(OH)_2]$ $[C_6NH_8]$ (Yu *et al.* 1998).

Figure 2.33 illustrates a more complex case of transformation of graphs. The $cc2$–$2{:}3$–16 shown here can be obtained from the autunite ideal graph (Fig. 2.33(a)) by a series of operations. First, the autunite graph is cutted into two types of tapes: one shown in Fig. 2.33(b) and another shown in Fig. 2.33(c). The latter is transformed into a strongly corrugated chain (Figs. 2.33(d) and (e)). The chains are further merged to produce a complex topology (Fig. 2.33(f)) that is known as a topology of an octahedral–tetrahedral sheet in the structure of $Sr_2Fe_2(HPO_4)(PO_4)_2F_2$ (le Meins *et al.* 1998).

Another example of 2D topology derived from the same basic graph is shown in Fig. 2.34. We have found this topology in the $[Ga_3(H_2PO_4)_6(HPO_4)_3]$ octahedral–tetrahedral sheet that occurs in the structure of $[Co(en)_3][Ga_3(H_2PO_4)_6(HPO_4)_3]$ recently reported by Wang *et al.* (2003b) (the authors described the sheets as composed from propellane-like chiral motifs). The topology of the sheet can be derived from the graph {**6.2.6.2**}{**4.6.4.6**} (Fig. 2.34(a)) by the following steps. First, the {**6.2.6.2**}{**4.6.4.6**} graph is cut into chains as is shown in Fig. 2.34(a). The resulting chain is depicted in Fig. 2.34(b). These chains are merged together into 2D sheet as shown in Fig. 2.34(c). Note that, within the sheet, chains occur in two different versions related by a mirror plane: each chain has on its sides its mirror images. Thus,

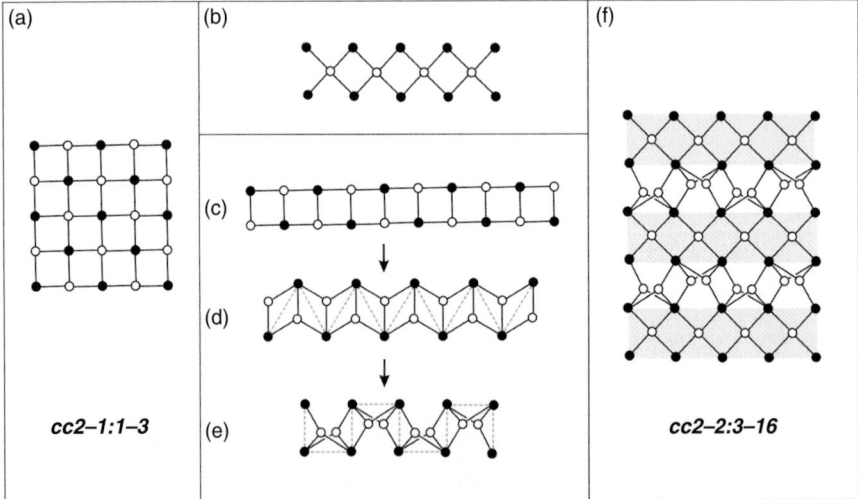

Fig. 2.33. The $cc2$–$2{:}3$–13 graph as a derivative of autunite topology. See text for details.

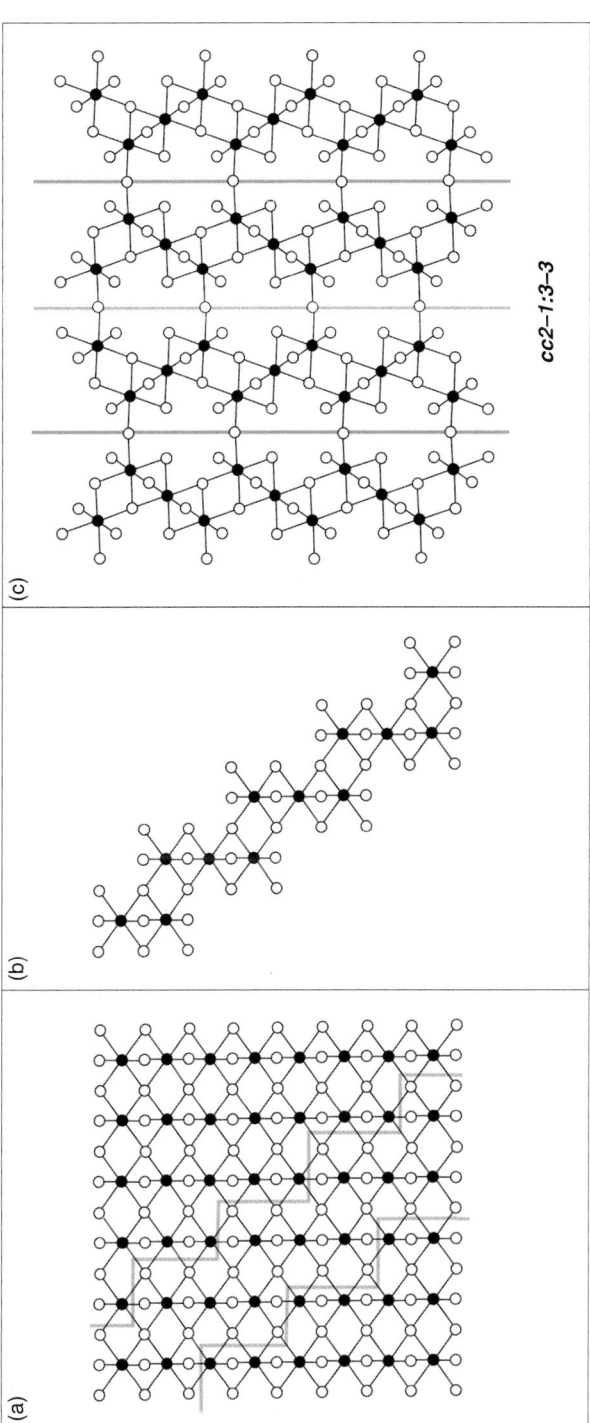

cc2–1:3–3

Fig. 2.34. Construction of topology of the $[Ga_3(H_2PO_4)_6(HPO_4)_3]$ octahedral–tetrahedral sheet from the structure of $[Co(en)_3][Ga_3(H_2PO_4)_6(HPO_4)_3]$. The $\{6.2.6.2\}\{4.6.4.6\}$ basic graph has to be terminated as is shown by bold lines (a) to obtain a chain shown in (b). Gluing of chains results in the topology needed (c).

the line separated two adjacent chains acts as an imaginary mirror plane and in fact can be considered as a *twin boundary*. One can say that the topology shown in Fig. 2.34(c) can be produced from the {**6.2.6.2**}{**4.6.4.6**} graph by a repetitive twinning. It should be noted that twinning has been frequently considered as a structure-building mechanism in crystals (see, e.g. Takeuchi 1997).

2.2.4.2   *Topologies derived from more than one basic graph*

An example of 2D topology that can be derived from two different basic graphs is given in Fig. 2.35. It was found in the structure of $[Co(H_2O)_6]_3[(UO_2)_5(SO_4)_8(H_2O)]$ $(H_2O)_5$ (Krivovichev and Burns 2007). This structure is based upon the $[(UO_2)_5(SO_4)_8$ $(H_2O)]$ sheets consisting of corner-sharing UO_7 pentagonal bipyramids and SO_4 tetrahedra (Fig. 2.35(a)). The black-and-white graph corresponding to the topological structure of the uranyl sulphate sheet is shown in Fig. 2.35(b). The way of constructing of this graph from the parent {**3.6.3.6**} and {**5.3.5.3**}{**5.3.5.4**} graphs is given in Fig. 2.36. Terminating the {**4.5.3.5**} graph by the gray lines in Fig. 2.36(a) gives a complex chain, which is designated **A** and shown in Fig. 2.36(c). This chain is an important constituent of the $[(UO_2)_5(SO_4)_8(H_2O)]$ sheet in $[Co(H_2O)_6]_3[(UO_2)_5(SO_4)_8(H_2O)](H_2O)_5$. On terminating the {**3.6.3.6**} graph, as shown by the gray lines, chain **B** is obtained, as shown in Fig. 2.36(d). The graph corresponding to the $[(UO_2)_5(SO_4)_8(H_2O)]$ sheet is obtained by linkage of the chains shown in Figs. 2.36(c) and 4.36(d) through their black nodes (Fig. 2.36(e)). The sequence of chains in the graph is ... **ABABAB** ... In general, the $[(UO_2)_5(SO_4)_8(H_2O)]$ sheet in the structure of $[Co(H_2O)_6]_3[(UO_2)_5(SO_4)_8(H_2O)]$ $(H_2O)_5$ has a *modular* structure that consists of two 1-dimensional modules derived from two different basic graphs ({**3.6.3.6**} and {**5.3.5.3**}{**5.3.5.4**}).

2.3   2D topologies: graphs with *M=T* links and without *M–M* links

The presence of the *M=T* links in a graph indicates sharing of an edge between two coordination polyhedra. Because of Pauling's third rule, this leads to the repulsive forces between the central cations of the polyhedra and therefore edge sharing is avoided in the case of small and high-valent cations. However, the edge sharing can be afforded in the case of large and relatively low-charged cations such as rare earths and actinides, for example.

Many topologies in this group can be obtained from the graphs listed previously by the procedure that can be conveniently described as edge-doubling. Figure 2.37 shows the 2D unit observed in the structure of $RbEu(SO_4)_2$. Its graph (*cc2–1:2–17*) is an edge-doubled derivative of the {**3.6.3.6**} graph. From six edges that are incident upon a black vertex in the basic graph, exactly two are doubled.

Figure 2.38 shows four graphs that can be derived from the *cc2–1:2–2* and *cc2– 1:2–3* graphs by the edge-doubling. The respective list of compounds is given in Table 2.7. It is important that, despite the fact that the white vertices in these graphs are 2-connected (bidentate), edge sharing generates a possibility for an orientational isomerism. This is clearly illustrated by Figs. 2.39 and 2.40.

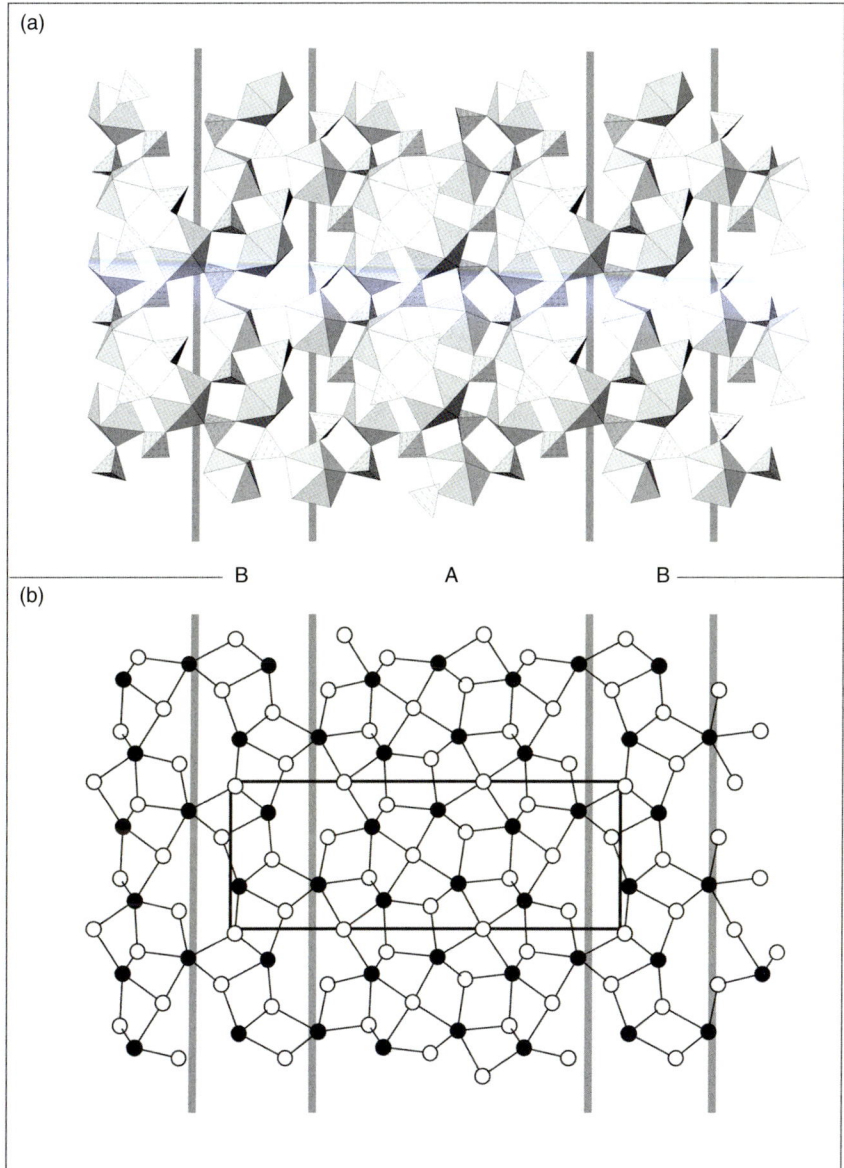

Fig. 2.35. The [(UO$_2$)$_5$(SO$_4$)$_8$(H$_2$O)] sheet in the structure of [Co(H$_2$O)$_6$]$_3$[(UO$_2$)$_5$(SO$_4$)$_8$(H$_2$O)] (H$_2$O)$_5$ (a), and its graphical representation (b). The **A** and **B** chains are delineated by gray lines.

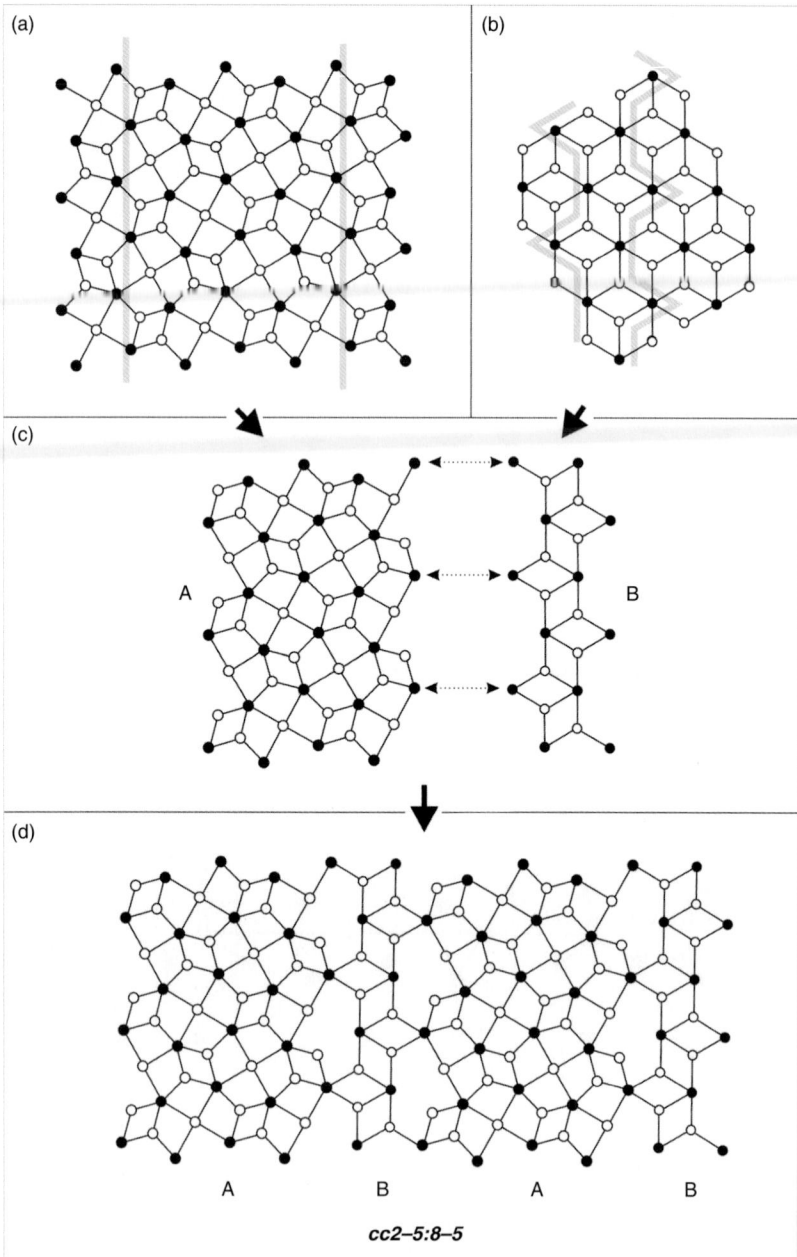

Fig. 2.36. (a) The {5.3.5.3}{5.3.5.4} graph; (b) the {3.6.3.6} graph; (c) the **A** chain from the graph shown in (a) as delineated by gray lines; (d) the **B** chain from the graph shown in (b) as delineated by gray lines; (e) the composite graph obtained by merging the **A** and **B** chains shown in (c) and (d), respectively. Note that the graph shown in (e) is an idealized version of that shown in Fig. 2.35.

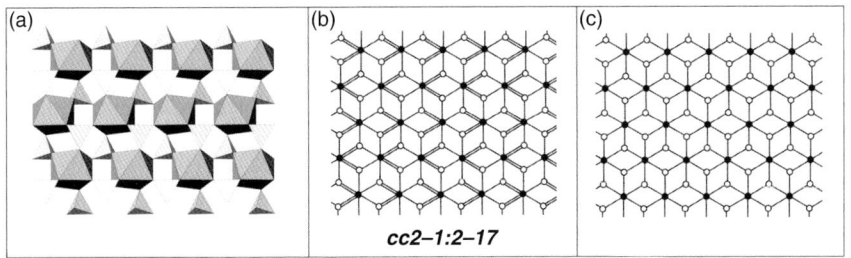

Fig. 2.37. 2D unit observed in the structure of RbEu(SO₄)₂ (a) and its graph (*cc2–1:2–17*) (b) as an edge-doubled derivative of the {**3.6.3.6**} graph (c).

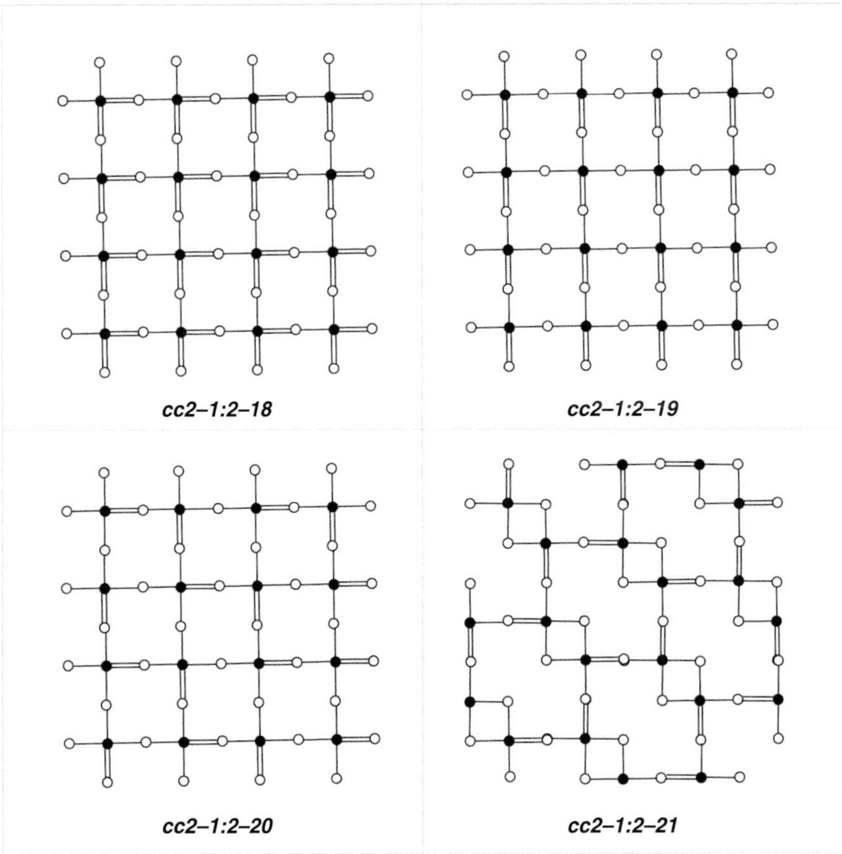

cc2–1:2–18

cc2–1:2–19

cc2–1:2–20

cc2–1:2–21

Fig. 2.38. Autunite-derivative graphs with double links between white and black vertices.

Table 2.7 Inorganic oxysalts based upon 2D units with graphs containing $M=T$ double edges

Graph	Chemical formula	Reference
cc2–1:2–17	$RbEu(SO_4)_2$	Sarukhanyan *et al.* 1983
cc2–1:2–18	$Cs(Pr(H_2O)_3(SO_4)_2)(H_2O)$	Bukovec and Golic 1975
	$Cs(La(H_2O)_3(SO_4)_2)(H_2O)$	Saf'yanov *et al.* 1975
	$Rb(Sm(H_2O)_3(SO_4)_2)(H_2O)$	Jasty *et al.* 1991
	$Rb(Pr(H_2O)_3(SO_4)_2)(H_2O)$	Iskhakova *et al.* 1981
	$Cs(Nd(H_2O)_3(SeO_4)_2)(H_2O)$	Ovanisyan *et al.* 1987
	$NH_4(M(H_2O)_3(SO_4)_2)(H_2O)$	
	M = La, Tb	Junk *et al.* 1999
cc2–1:2–19	$Cs(Lu(SO_4)_2(H_2O)_3)(H_2O)$	Bukovec *et al.* 1979
cc2–1:2–20	$[N_4C_{10}H_{28}][(UO_2)_2(SO_4)_4]$	Doran *et al.* 2003a
	$[N_4C_6H_{22}][UO_2(SO_4)_2]_2$	Norquist *et al.* 2005a
	$[N_2C_4H_{14}][UO_2(SO_4)_2]$	Doran *et al.* 2003d
cc2–1:2–21	$[N_2C_5H_{16}][UO_2(SO_4)_2]$	Norquist *et al.* 2003c
	$Cs_2[NpO_2(SO_4)_2]$	Fedoseev *et al.* 1999
	$Ba[(UO_2)(SeO_3)_2]$	Almond *et al.* 2002
cc2–2:3–17	$[C_7H_{20}N_2][(UO_2)_2(SO_4)_3(H_2O)]$	Norquist *et al.* 2005b
cc2–1:3–4	$Cs_3[Gd(SO_4)_3]$	Chibiskova *et al.* 1984
	$Cs_3Nd(CrO_4)_3$	Gasanov *et al.* 1990
cc2–1:3–5	$(H_5O_2)(H_3O)_2Nd(SO_4)_3$	Wickleder 1999b
cc2–1:4–1	$(H_3O)_2Nd(HSO_4)_3SO_4$	Wickleder 1999b

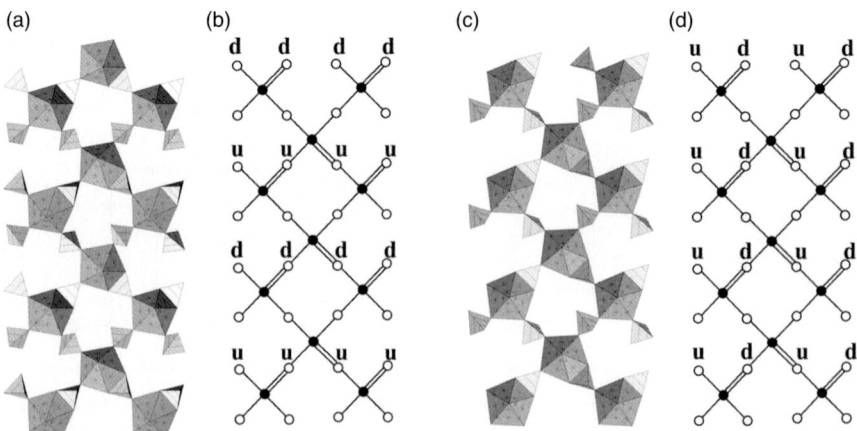

(a)   (b)   (c)   (d)

Fig. 2.39. Description of geometrical isomerism of the $[(UO_2)(SO_4)_2]^{2-}$ sheets with edge-sharing between U polyhedra and S tetrahedra. See text for details.

Figure 2.39 shows two examples of the 2D units with UO_7 bipyramids sharing edges with SO_4 tetrahedra (found in $[N_4C_6H_{22}][UO_2(SO_4)_2]_2$ (Fig. 2.39(a)) and $[N_2C_4H_{14}][UO_2(SO_4)_2]$ (Fig. 2.39(c)). Each UO_7 bipyramid shares one edge and three corners with the sulphate tetrahedra. In turn, the sulphate tetrahedra belong to the

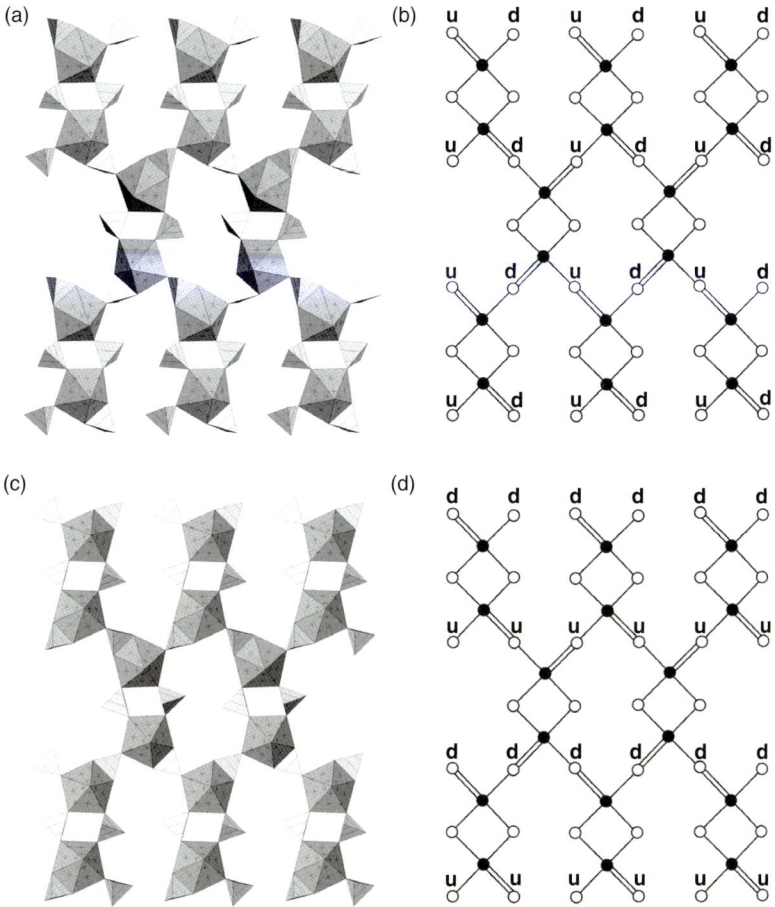

Fig. 2.40. Description of geometrical isomerism of the $[(AnO_2)(SO_4)_2]^{2-}$ sheets with edge-sharing between An polyhedra and S tetrahedra. See text for details.

two types. Tetrahedra of the first type are bidentate and share two of their corners with UO_7 bipyramids. Tetrahedra of the second type are again bidentate but they share one edge and one corner each with the uranyl polyhedra. Thus, one corner remains non-shared and can be oriented either up or down relative to the plane of the sheet. The possibility of different orientations induces the appearance of the orientational geometrical isomers shown in Fig. 2.39. By analogy with the geometrical isomers described above, the isomers under discussion can be described using **u** and **d** symbols of tetrahedra orientation. The isomer shown in Figs. 2.39(a) and (b) can be described as a **(u)(d)** isomer (its orientation matrix has the 1×2 dimensions), whereas the isomer shown in Figs. 2.39(c) and (d) is a **(ud)** isomer (matrix dimensions are 2×1). The same type of isomerism is observed for the actinyl sulphate

(a)

(b)

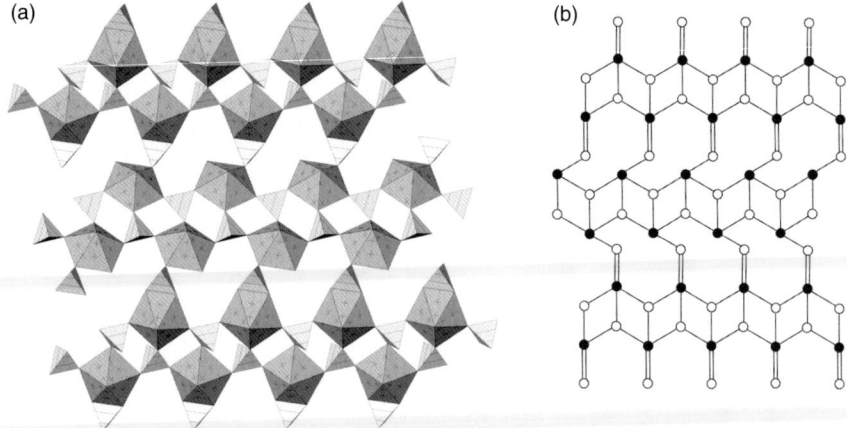

Fig. 2.41. Uranyl sulphate sheet in $[C_7H_{20}N_2][(UO_2)_2(SO_4)_3(H_2O)]$ (a) and its 2D graph (b).

sheets shown in Fig. 2.40 (found in $Cs_2[NpO_2(SO_4)_2]$ (Fig. 2.40(a)) and $[N_2C_5H_{16}]$ $[UO_2(SO_4)_2]$ (Fig. 2.40(c)).

Another example of edge-doubling transition is the graph *cc2–2:3–17* (Fig. 2.41), which is an edge-doubled variation of the *cc2–2:3–5* graph. It is realized in a complex topology of the uranyl sulphate sheet in the structure of $[C_7H_{20}N_2]$ $[(UO_2)_2(SO_4)_3(H_2O)]$.

In some structures, the edge-doubling procedure is associated with the addition of branches, i.e. additional 1-connected white vertices. Figure 2.42 shows two graphs, *cc2–1:3–4* and *cc2–1:3–5*, that can be obtained from the graphs *cc2–1:2–4* and *cc2–1:2–5* via a two-stage process. First, some edges in the initial graph are doubled. Second, new white vertices are added. It is noteworthy that both topologies shown in Fig. 2.42 are known for rare-earth sulphates and chromates only.

Finally, there are structures with the *M=T* links that have no obvious relations to other topologies. Figure 2.43(a) shows a 2D sheet composed from the NdO_9 and SO_4 polyhedra. The graph of this sheet (Fig. 2.43(b)) includes single and double edges and this is the only graph of this kind we are aware of that contains 9-connected black vertices.

2.4   2D topologies: graphs with *M–M, M=M,* or *M≡M* links

2.4.1   *Graphs with M–M links*

Figure 2.44(a) shows a 2D structural unit present in the structure of $K_2Mn(SO_4)_2(H_2O)_{1.5}$ (Borene and Solery 1972). Each Mn^{2+} in this structure is octahedrally coordinated by five O atoms of sulphate groups and one H_2O molecule. This molecule is bridging between two adjacent Mn^{2+} centers so that the $MnO_5(H_2O)$ octahedra form

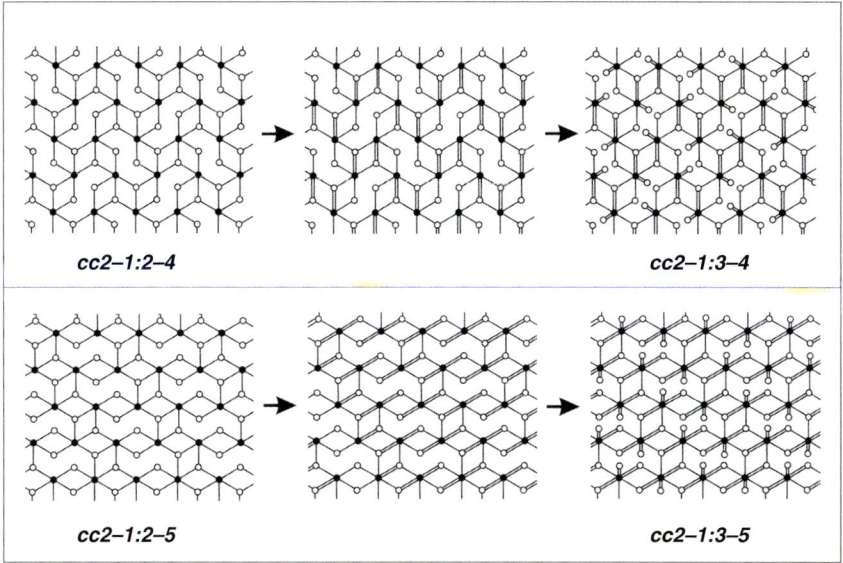

cc2–1:2–4

cc2–1:3–4

cc2–1:2–5

cc2–1:3–5

Fig. 2.42. *cc2–1:3–4* and *cc2–1:3–5* graphs can be obtained from the graphs *cc2–1:2–4* and *cc2–1:2–5* via a two-stage process. First, some edges in the initial graph are doubled. Second, new white vertices are added.

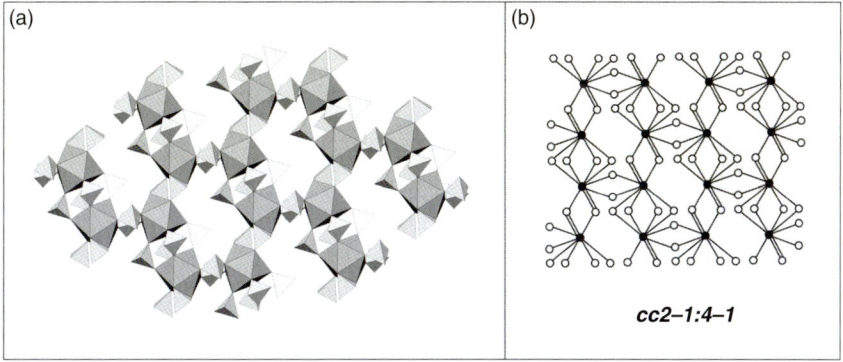

(a)

(b)

cc2–1:4–1

Fig. 2.43. 2D structural unit in $(H_3O)_2Nd(HSO_4)_3SO_4$ (a) and its graph (b).

$Mn_2O_{10}(H_2O)$ dimers. The dimers are linked via SO_4 tetrahedra into a sheet. The idealized graph of the sheet is shown in Fig. 2.44b. Since Mn^{2+} coordination centers are linked via bridging ligand, the graph possesses pairs of black vertices linked by an edge. The graph in Fig. 2.44(b) can be obtained from the highly symmetrical basic graph depicted in Fig. 2.45(a). This graph can be denoted as {**3.12.12**}, which reflects

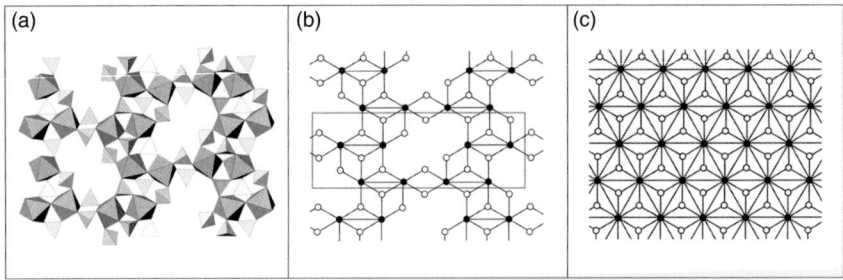

Fig. 2.44. Octahedral–tetrahedral 2D unit in $K_2Mn(SO_4)_2(H_2O)_{1.5}$ (a) and its topology (b) as a derivative of the {**3.12.12**} graph.

the fact that it consists of triangles formed by two 12-connected black vertices and one 3-connected white vertex. The {**3.12.12**} graph is remarkable in the fact that all black vertices are connected into an infinite 2D subgraph. It is worthy of note that the {**3.12.12**} graph itself has not been found in any inorganic structure so far. Some derivatives of this graphs are shown in Figs. 2.45 and 4.46. As is clearly seen from the diagrams, subgraphs of black vertices may have different dimensionalities, i.e. may be 0-, 1-, and 2-dimensional. It is therefore natural to use this feature as an additional criterion for classication and systematic description.

2.4.1.1   *Graphs with finite subgraphs of black vertices*

Inorganic compounds containing 2D units with these topologies are listed in Table 2.8. The graphs are shown in Fig. 2.45. In most of the graphs, black vertices form linear dimers (graphs *cc2–1:2–22, cc2–1:1–4, cc2–1:1–5*, and *cc2–2:3–18*) or trimers (graph *cc2–3:4–11*). Figure 2.47(a) shows the $[Al_2F(H_2O)_4(PO_4)_2]^-$ sheet observed in the structure of minyulite, $K[Al_2F(H_2O)_4(PO_4)_2]$. Two $AlF(H_2O)_2O_3$ octahedra share one F^- anion to form octahedral dimers stabilized by two PO_4 tetrahedra. The dimers are linked via additional PO_4 tetrahedron into a 2D sheet.

Figure 2.47(b) demonstrates the unique topology of a uranyl fluoride sulphate sheet discovered in the structure of $[N_2C_6H_{18}]_2[UO_2F(SO_4)]_4(H_2O)$ (graph *cc2–1:1–6*). Here, U^{6+}-centered pentagonal bipyramids share F^- anions to form elegant tetramers that are further linked by SO_4 tetrahedra into a 2D unit.

2.4.1.2   *Graphs with 1D subgraphs of black vertices*

Graphs with black vertices forming 1D subgraphs are shown in Fig. 2.46. Sheets with these topologies are especially rich in geometrical isomeric variations. The graphs *cc2–1:1–8* and *cc2–1:1–9* deserve special attention.

In 1975, P. B. Moore described geometrical isomerism of octahedral–tetrahedral sheets in the laueite, pseudolaueite, stewartite and metavauxite structure types. Fig. 2.48(a) shows the $[Fe_2(PO_4)_2(OH)_2(H_2O)_2]^{2-}$ octahedral–tetrahedral sheet

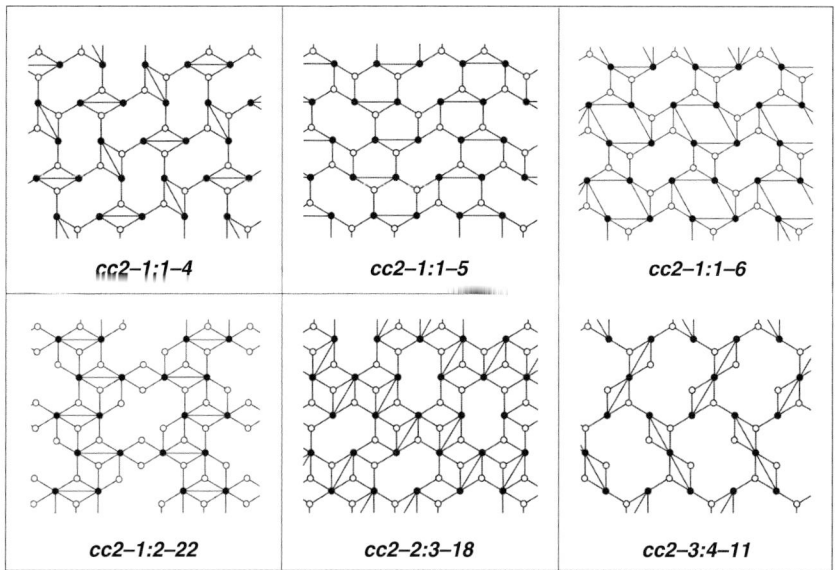

Fig. 2.45. Black-and-white 2D graphs derivative from the {**3.12.12**} graph.

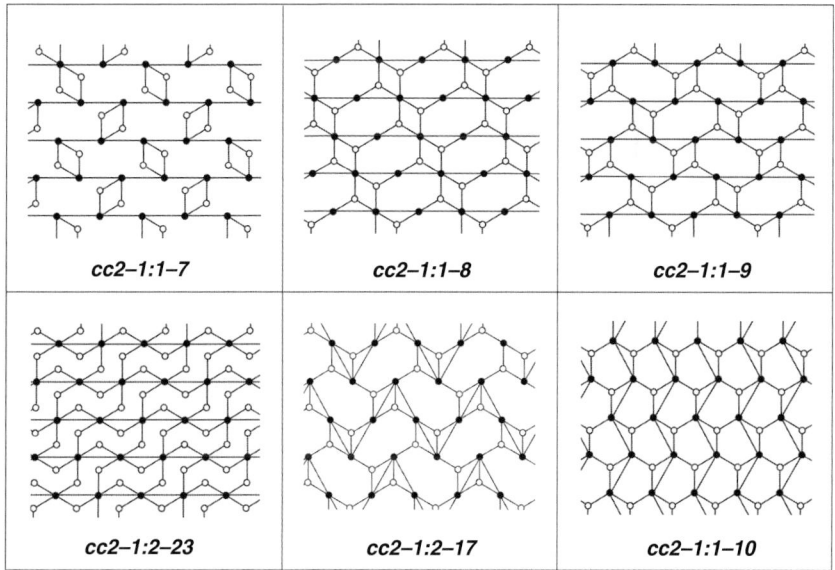

Fig. 2.46. Black-and-white 2D graphs derivative from the {**3.12.12**} graph.

Table 2.8   Inorganic oxysalts containing sheets with topologies derivative of the {**12.12.3**} graph (except graphs *cc2–1:1–9*, *cc2–1:1–10*, *cc2–3:2–1*, and *cc2–3:1–1*)

Graph	Chemical formula	Reference
Compounds with 0D subgraphs of black vertices		
cc2–1:2–22	$K_2M(SO_4)_2(H_2O)_{1.5}$ M = Mn, Cd	Borene and Solery 1972
cc2–1:1–4	minyulite $K[Al_2F(H_2O)_4(PO_4)_2]$	Kampf 1977; Dumas *et al.* 2001
	$K[Al_2(OH)(H_2O)_4(PO_4)_2]$	Dick *et al.* 1997
cc2–1:1–5	$[CH_3NH_2CH_3][Al_3P_3O_{12}(OH)]$	Kongshaug *et al.* 1999
cc2–1:1–6	$[N_2C_6H_{18}]_2[UO_2F(SO_4)]_4(H_2O)$	Doran *et al.* 2005b
cc2–2:3–18	$[N_4C_6H_{21}][Al_2F_2(HPO_4)_3](H_2PO_4)$	Simon *et al.* 1999a
cc2–3:4–11	$[H_3N(CH_2)_6NH_3]_2[Ga_3F_2(OH)_4(H_2PO_4)$	Livage *et al.* 2001
	$(HPO_4)_3](H_2O)_{3.5}$	
Compounds with 1D subgraphs of black vertices		
cc2–1:1–7	$[N_2C_5H_{14}][UO_2F(H_2O)(SO_4)]_2$	Doran *et al.* 2005b
cc2–1:2–23	$M(VO)(TO_4)_2$ M = Sr, Ba; T = P, As	Wadewitz, Müller-Buschbaum
		1996a, b; Müller-Buschbaum,
		Wadewitz 1996
cc2–2:3–19	$K_2Co_3(OH)_2(SO_4)_3(H_2O)_2$	Effenberger, Langhof 1984
cc2–1:1–10	$Rb[UO_2(SO_4)F]$	Mikhailov *et al.* 2002
	$K[(UO_2)F(HPO_4)](H_2O)$	Ok *et al.* 2006
	$[N_2C_4H_{14}][UO_2F(SO_4)]_2$	Doran *et al.* 2005b
Compounds with 2D subgraphs of black vertices		
	$[N_2C_3H_{12}][UO_2F(SO_4)]_2(H_2O)$	Doran *et al.* 2005b
cc2–5:6–1	slavikite $Na[Fe_5(H_2O)_6(OH)_6(SO_4)_6]$	Süsse 1973, 1975
	$[Mg(H_2O)_6]_2(SO_4)(H_2O)_{15}$	

observed in the structure of laueite, $MnFe_2(PO_4)_2(OH)_2(H_2O)_8$ (Moore 1965). Within this sheet, $Fe\varphi_6$ octahedra (φ = O, OH, H_2O) share trans-vertices to form chains that are further interlinked by PO_4 tetrahedra. Each tetrahedron is 3-connected (i.e. is linked to three octahedra), whereas $Fe\varphi_6$ octahedra are either 6-connected (linked to two octahedra and four tetrahedra) or 4-connected (linked to two octahedra and two tetrahedra). The black-and-white graph corresponding to the sheet is the *cc2–1:1–8* graph.

Figure 2.49(a) shows the $[Fe_2(OH)_2(H_2O)_2(PO_4)_2]^{2-}$ octahedral–tetrahedral sheet observed in the structure of pseudolaueite, a polymorph of laueite (Baur 1969a). This sheet is chemically identical to the sheet in laueite. However, its topological structure is different. Its black-and-white graph is the *cc2–1:1–9* graph. This graph is different from *cc2–1:1–8* in that all black vertices are 5-connected, whereas, in the graph *cc2–1:1–8*, they are either 4- or 6-connected. Therefore, the sheets in laueite and pseudolaueite should be considered as *topological* isomers.

However, laueite has another polymorph, stewartite (Moore and Araki 1974a). It is also based upon the $[Fe_2(OH)_2(H_2O)_2(PO_4)_2]^{2-}$ shown in Fig. 2.48(b). Its black-and-white graph is isomorphous to the *cc2–1:1–8* graph. However, detailed examination of the orientations of tetrahedra within the sheets reveal that the sheets

(a)   (b)

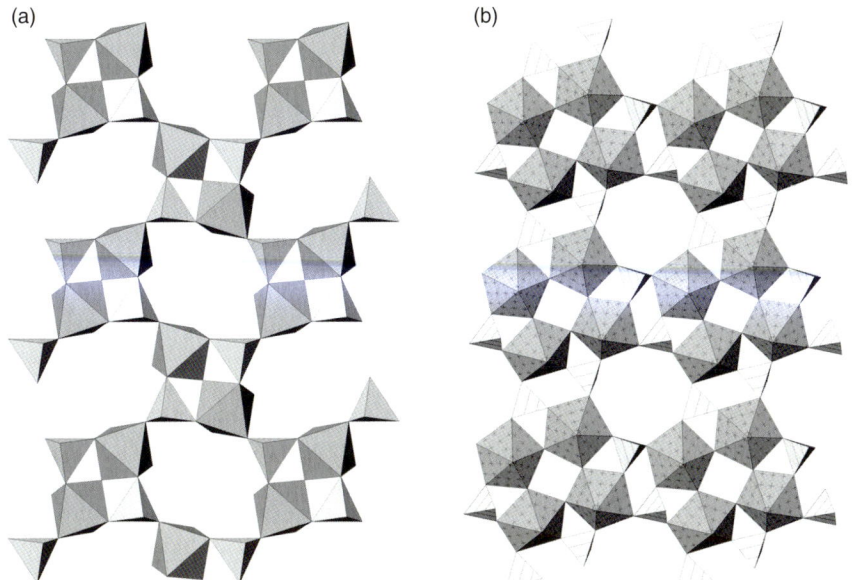

Fig. 2.47. The $[Al_2F(H_2O)_4(PO_4)_2]^-$ sheet in minyulite, $K[Al_2F(H_2O)_4(PO_4)_2]$ (a) and uranyl fluoride sulphate sheet discovered in $[N_2C_6H_{18}]_2[UO_2F(SO_4)]_4(H_2O)$ (b).

are geometrically different and therefore should be regarded as *geometrical* isomers. The geometrical isomerism is caused by the presence of 3-connected PO_4 tetrahedra with the fourth corner oriented either up or down relative to the plane of the sheet. To distinguish between the two sheets shown in Figs. 2.48(a) and (b), we use the concept of an orientation matrix that has been described in detail early in this chapter. Following this approach, the orientation matrix of the laueite sheet can be written as **(ud☐☐)(☐ud☐)(☐☐ud)(d☐☐u)**, whereas one of the stewartite sheet can be written as **(u☐☐ud☐☐d)(ud☐☐du☐☐)** **(☐du☐☐ud☐)(☐☐ud☐☐du)(d☐☐du☐☐u)** **(du☐☐ud☐☐)(☐ud☐☐du☐)(☐☐du☐☐ud)** (Fig. 2.48). In the case of stewartite, the matrix is much more complex than that for laueite and has dimensions of 8 × 8.

The *cc2–1:1–9* graph is typical for the 2D sheets in the structures of pseudolaueite and metavauxite, which contain two different geometrical isomers of the same topology. This is clearly demonstrated by Fig. 2.49. According to the method of description used, the orientation matrices of tetrahedra for the sheets shown in Figs. 2.49(a) and (b) are **(ud☐☐)(☐☐du)** and **(ud☐☐)(☐☐ud)**, respectively. The list of compounds with the *cc2–1:1–8* and *cc2–1:1–9* topologies is given in Table 2.9.

Geometrical isomerism has also been observed for the uranyl sulphate sheets based upon the *cc2–1:1–10* graph (Fig. 2.46). Note that white vertices in this graph correspond to tridentate sulphate tetrahedra that may be turned either up or down relative to the plane of the sheet (Fig. 2.50(a)). The presence of tridentate tetrahedra indicates

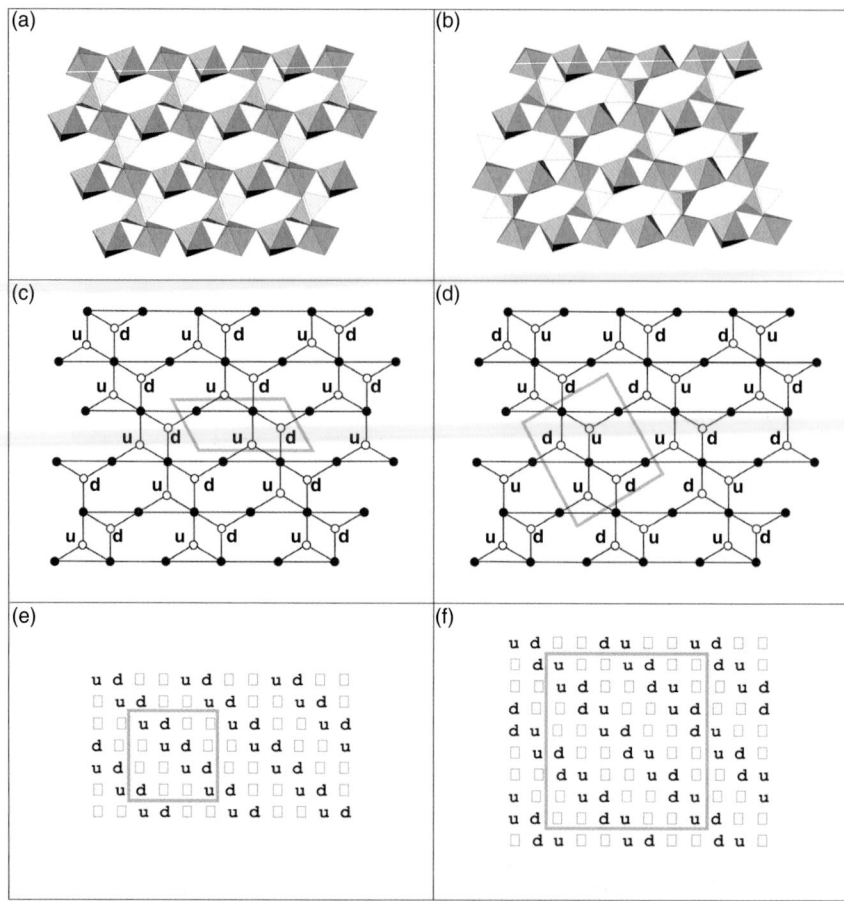

Fig. 2.48. Octahedral–tetrahedral sheets in the structures of laueite (a) and stewartite (b), their 2D black-and-white graphs with orientations of non-shared vertices of tetrahedra written near the white circles ((c) and (d) for laueite and stewartite sheets, respectively) (legend: black circles = octahedra; white circles = tetrahedra). Tables of tetrahedra orientations for the sheets shown in (a) and (b) are given in (e) and (f), respectively.

that orientational geometrical isomerism is possible. This is what indeed takes place in the structures of F-bearing uranyl sulphates. Figures 2.50(b) and (c) show two black-and-white graphs with **d** and **u** symbols denoting orientation of tetrahedra. In the structure of $Rb[UO_2(SO_4)F]$, the sequence of orientations can be described as (**uuduud**) [note that it cannot be reduced to (**uud**) because the graph periodicity along the chain of black vertices is 2] (Fig. 2.50(b)). In the structures of $[N_2C_4H_{14}]$ $[UO_2F(SO_4)]_2$ and $[N_2C_3H_{12}][UO_2F(SO_4)]_2(H_2O)$, the up- and down-orientations alternate and the sequence can be written as (**ud**) (Fig. 2.50(c)).

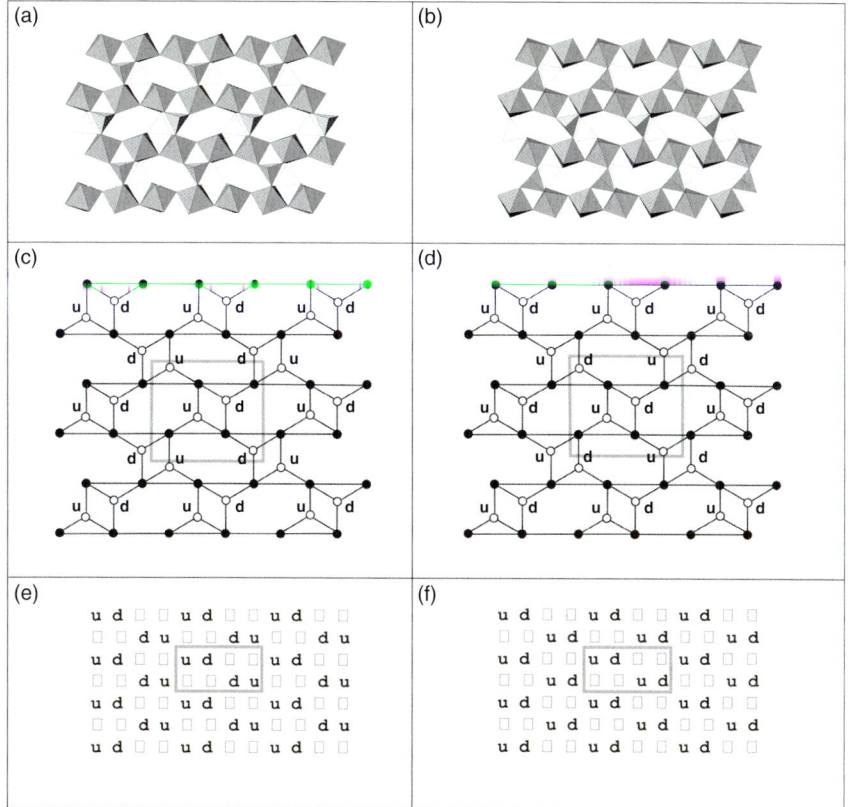

Fig. 2.49. Octahedral–tetrahedral sheets in the structures of pseudolaueite (a) and metavauxite (b), their 2D black-and-white graphs with orientations of non-shared vertices of tetrahedra written near the white circles ((c) and (d), respectively) (legend as in Fig. 2.1). Tables of tetrahedra orientations for the sheets shown in (a) and (b) are given in (e) and (f), respectively.

2.4.1.3 *Graphs with 2D subgraphs of black vertices*

In graphs of this type, black vertices are linked into a 2D subgraph that dominates the topological structure. The role of white vertices is to incrustate the backbone of black vertices. Figure 2.51(a) shows a complex sheet observed in the structure of slawikite, $Na[Fe_5(H_2O)_6(OH)_6(SO_4)_6]$ $[Mg(H_2O)_6]_2(SO_4)(H_2O)_{15}$ (Süsse 1973, 1975). The sheet consists of the $Fe\varphi_6$ octahedra and SO_4 tetrahedra and has the composition $[Fe_5(H_2O)_6(OH)_6(SO_4)_6]$. The graph of this sheet (*cc2–5:6–1*: Fig. 2.51(b)) contains large 12-membered rings formed by black vertices linked via single edges. The white vertices are 2-connected.

The graph *cc2–3:2–1* (Fig. 2.52) is an underlying topology for a large number of minerals of the alunite supergroup (Table 2.10). The black vertices form a Kagomé

Table 2.9 Inorganic oxysalts containing 2D structural units based upon the *cc2–1:1–8* and *cc2–1:1–9* graphs and their isomeric variations

Mineral name/Code	Chemical formula	Reference
Topological isomer based upon the *cc2–1:1–8* graph		
Geometrical isomer (ud☐☐)(☐ud☐)(☐☐ud)(d☐☐u)		
Laueite	$MnFe_2(PO_4)_2(OH)_2(H_2O)_8$	Moore 1965
Gordonite	$MgAl_2(PO_4)_2(OH)_2(H_2O)_6(H_2O)_2$	Leavens and Rheingold 1988
Mangangordonite	$MnAl_2(PO_4)_2(OH)_2(H_2O)_6(H_2O)_2$	Leavens and Rheingold 1988
Paravauxite	$FeAl_2(PO_4)_2(OH)_2(H_2O)_6(H_2O)_2$	Baur 1969b
Sigloite	$FeAl_2(PO_4)_2(OH)_3(H_2O)_5(H_2O)_2$	Hawthorne 1988
Ushkovite	$MgFe_2(PO_4)_2(OH)_2(H_2O)_6(H_2O)_2$	Galliski *et al.* 2002
Curetonite	$Ba(Al,Ti)(PO_4)(OH,O)F$	Cooper and Hawthorne 1994a
–	$SrFePO_4F_2$	le Meins *et al.* 1997
ULM-10	$(enH)[Fe^{3+}Fe^{2+}F_2(HPO_4)_2(H_2O)_2]$	Cavellec *et al.* 1994
–	$(enH_2)[Fe^{3+}Fe^{2+}F_2(HAsO_4)(AsO_4)(H_2O)_2]$	Ekambaram and Sevov 2000
–	$(enH_2)[CoIn(PO_4)_2H(H_2O)_2F_2]$	Yu *et al.* 1999b
–	$(enH_2)[Fe_2F_2(SO_4)_2(H_2O)_2]$	Paul *et al.* 2003
–	$(enH_2)[NbFeOF(PO_4)_2(H_2O)_2]$	Wang *et al.* 2000c
–	$(enH_2)[NbCoOF(PO_4)_2(H_2O)_2]$	Wang *et al.* 2000c
–	$(enH_2)[Ti(Fe,Cr)(F,O)(H_{0.3}PO_4)_2(H_2O)_2]$	Wang *et al.* 2000c
Geometrical isomer (u☐☐ud☐☐d)(ud☐☐du☐☐) (☐du☐☐ud☐)(☐☐ud☐☐du)(d☐☐du☐☐u)		
(du☐☐ud☐☐)(☐ud☐☐du☐)(☐☐du☐☐ud)		
Stewartite	$MnFe_2(OH)_2(H_2O)_6(PO_4)_2(H_2O)_2$	Moore and Araki 1974a
Kastningite	$(Fe_{0.5}Mn_{0.5})(Al_2(OH)_2(H_2O)_2(PO_4)_2)(H_2O)_6$	Adiwidjadja *et al.* 1999
Topological isomer based upon the *cc2–1:1–9* graph		
Geometrical isomer (ud☐☐)(☐☐du)		
Pseudolaueite	$MnFe_2(PO_4)_2(OH)_2(H_2O)_8$	Baur 1969a
–	$\alpha\text{-}(VO)(HPO_4)(H_2O)_2$	Worzala *et al.* 1998
UiO-15-as	$(enH_2)[Al_2(OH)_2(PO_4)_2(H_2O)](H_2O)$	Kongshaug *et al.* 1999
UiO-15-125	$(enH_2)[Al_2(OH)_2(PO_4)_2(H_2O)]$	Kongshaug *et al.* 1999
APDAP$_{12}$-150	$(C_3N_2H_{12})[Al_2(PO_4)_2(OH)_2](H_2O)$	Tuel *et al.* 2000
Geometrical isomer (ud☐☐)(☐☐ud)		
Metavauxite	$Fe(H_2O)_6Al_2(PO_4)_2(OH)_2(H_2O)_2$	Baur and Rao 1967
Strunzite	$MnFe_2(PO_4)_2(OH)_2(H_2O)_6$	Fanfani *et al.* 1978

en = ethylenediamine, $NH_2(C_2H_5)NH_2$

net consisting of regular triangles and hexagons. The white vertices are in the centers of the triangles and are 3-connected. The deficient version of this graph with exactly half of the white vertices removed is the *cc2–3:1–1* graph (Fig. 2.52(c)) that is realized in the structures of $M_2[(MoO_3)_3(SeO_3)]$, where M = Cs, NH$_4$ (Harrison *et al.* 1994).

2.4.1.4 *Graphs with modular structure*

Figures 2.53(a) and (b) shows 2D sheets observed in the structures of mitryae-vaite, $Al_5(PO_4)_2[(P,S)O_3(O,OH))]_2F_2(OH)_2(H_2O)_8 \cdot 6.48H_2O$, and kingite, $Al_3(PO_4)_2$

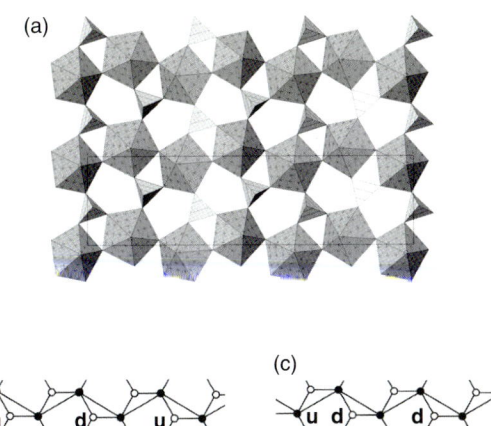

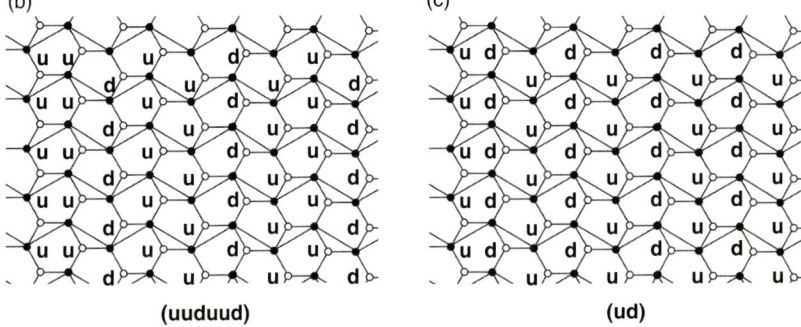

(uuduud)   **(ud)**

Fig. 2.50. Uranyl fluoride sulphate sheet in the structure of $Rb[UO_2(SO_4)F]$ (a) and geometrical isomers of the sheets with the *cc2–1:1–10* topology.

$(F,OH)_2 \cdot 8(H_2O,OH)$, two rare aluminum fluoride phosphates, respectively. The sheets are composed from Al-centered octahedra linked via fluoride ions into linear pentamers (mitryaevaite) and trimers (kingite) further interlinked by PO_4 tetrahedra. The corresponding graphs for these sheets are shown in Figures 2.53(c) and (d), respectively. Note that these graphs are "real" images of the experimentally observed topologies, i.e. not idealized from the geometrical viewpoint.

Both mitryaevaite and kingite topologies can be obtained from the ideal graphs that are derivatives of the {**12.12.3**} graph.

The mitryaevaite graph *cc2–5:4–1* is derivative of the laueite *c2–1:1–8* graph and can be produced from it by a series of topological operations (Fig. 2.54). First, the *c2–1:1–8* graph is cut into tapes such that the tapes consist of linear pentamers of five black vertices (Figs. 2.54(a) and (b)). Then, the tapes are linked together by additional edges between black and white vertices so that the white vertices in the peripheral parts of the tapes become 4-connected (Fig. 2.54(c)). The resulting graph is isomorphous to the graph shown in Fig. 2.53(c). Note that the tapes excised from the *c2–1:1–8* graph have the system of tetrahedra orientations of the laueite-type geometrical isomer.

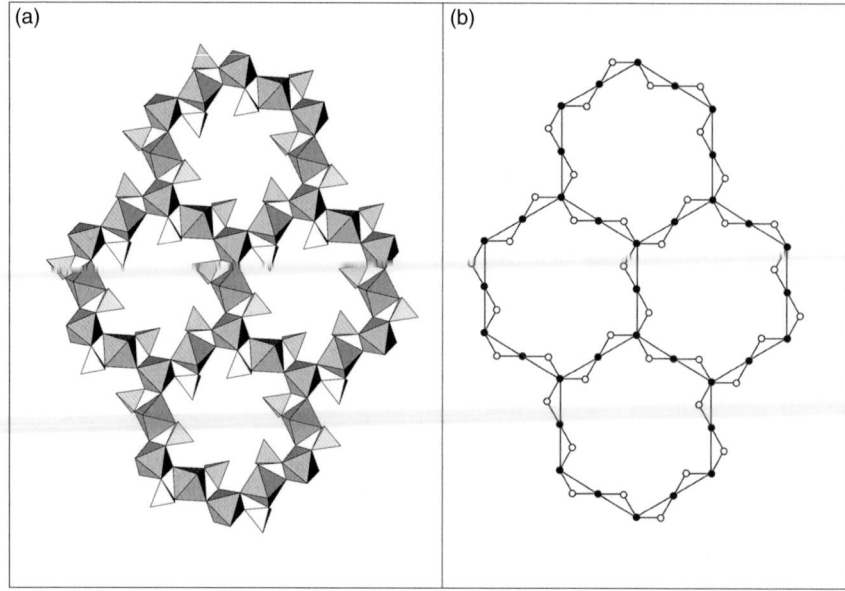

Fig. 2.51. 2D unit in slavikite (a) and its graph (b).

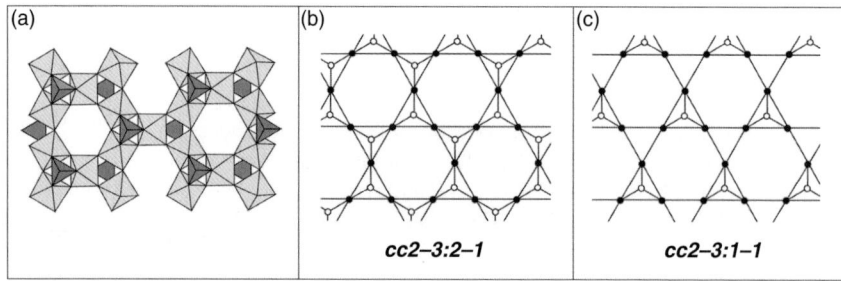

Fig. 2.52. 2D unit in the structures of the alunite- and crandallite-group minerals (a), its graph (b) and related topology (c).

In contrast to the mitryaevaite graph, the kingite graph *cc2–3:2–2* is a derivative of the *c2–1:1–9* graph and, more precisely, its pseudolaueite isomer (Figs. 2.49(a) and (c)). The "production process" in this case also involves cutting the initial graph into 1D tapes and linking them by additional edges between black and white vertices (Fig. 2.55). The tape consists of linear trimers of black vertices. The linkage of the tapes initiates the formation of 4-connected white vertices, as in the case of mitryaevaite.

Table 2.10 Inorganic oxysalts containing 2D structural units based upon the *cc2–3:2–1* and *cc2–3:2–1* graphs

Graph	Mineral name	Chemical formula	Reference
cc2–3:2–1	argentojarosite	$Ag[Fe_3(SO_4)_2(OH)_6]$	Groat *et al.* 2003
	alunite	$K[Al_3(OH)_6(SO_4)_2]$	Menchetti and Sabelli (1976a)
	ammonioalunite	$(NH_4)[Al_3(OH)_6(SO_4)_2]$	Lengauer *et al.* 1994
	natroalunite	$Na[Al_3(OH)_6(SO_4)_2]$	Okada *et al.* 1982
	jarosite	$K[Fe_3(OH)_6(SO_4)_2]$	Menchetti and Sabelli 1976a
	ammoniojarosite	$(NH_4)[Fe_3(OH)_6(SO_4)_2]$	Lengauer *et al.* 1994
	natrojarosite	$Na[Fe_3(OH)_6(SO_4)_2]$	Lengauer *et al.* 1994
	argentojarosite	$Ag[Fe_3(OH)_6(SO_4)_2]$	Lengauer *et al.* 1994
	hydronium jarosite	$(H_3O)[Fe_3(OH)_6(SO_4)_2]$	Lengauer *et al.* 1994
	beaverite	$Pb[(Fe,Cu,Al)_3(OH)_6(SO_4)_2]$	Breidenstein *et al.* 1992
	dorallcharite	$Tl_{0.8}K_{0.2}[Fe_3(OH)_6(SO_4)_2]$	Balic Zunic *et al.* 1994
	osarizawaite	$Pb[(Al,Cu,Fe)_3(OH)_6(SO_4)_2]$	Giusepetti and Tadini 1980
	schlossmacherite	$(H_3O,Ca)[Al_3(OH)_6(AsO_4)(SO_4)]$	Menchetti and Sabelli 1976a
	woodhouseite	$Ca[Al_3(OH)_6((P_{0.5}S_{0.5})O_4)]$	Kato 1977
	svanbergite	$Sr[Al_3(OH)_6((P_{0.5}S_{0.5})O_4)]$	Kato and Miura 1977
	hinsdalite	$Pb[Al_3(OH,H_2O)_6((P_{0.69}S_{0.31})O_4)]$	Kolitsch *et al.* 1999a
	beaudantite	$Pb[Fe_3(OH)_6((As,S)O_4)]$	Szymanski 1988
	hidalgoite	$Pb[Al_3(OH)_6(AsO_4)(SO_4)]$	Menchetti and Sabelli 1976a
	kemmlitzite	$(Sr,Ce)[Al_3(OH)_6(AsO_4)(SO_4)]$	Menchetti and Sabelli 1976a
	weilerite	$Ba[Al_3(OH)_6(AsO_4)(SO_4)]$	Menchetti and Sabelli 1976a
	plumbojarosite	$Pb_{0.5}[Fe_3(OH)_6(SO_4)_2]$	Szymanski 1985
	minamiite	$(Na,Ca)_{1-x}[Al_3(OH)_6(SO_4)_2]$	Ossaka *et al.* 1982
	huangite	$Ca_{0.5}[Al_3(OH)_6(SO_4)_2]$	Li *et al.* 1992
	walthierite	$Ba_{0.5}[Al_3(OH)_6(SO_4)_2]$	Li *et al.* 1992
	corkite	$Pb[Fe_3(OH)_6(PO_4)(SO_4)]$	Giusepetti and Tadini 1987
	gallobeudantite	$Pb[Ga_3(OH)_6(AsO_4)(SO_4)]$	Jambor *et al.* 1996
		$Na[V_3(SO_4)_2(OH)_6]$	Dobley *et al.* 2000
	benauite	$HSr[Fe_3(PO_4)_2(OH)_6]$	Walenta *et al.* 1996
	crandallite	$Ca[Al_3(PO_3(O_{0.5}(OH)_{0.5}))(OH)_6]_2$	Blount 1974
	eylettersite	$(Th,Pb)_{1-x}[Al_3(PO_4,SiO_4)_2(OH)_6]$	van Wambeke 1972
	florensite-(Ce)	$(Ce,La,Nd)[Al_3(PO_4)_2(OH)_6]$	Kato 1990
	florensite-(La)	$La[Al_3(PO_4)_2(OH)_6]$	Lefebre, Gasparrini, 1980
	florensite-(Nd)	$Nd[Al_3(PO_4)_2(OH)_6]$	Milton, Bastron, 1971
	gorceixite	$Ba[Al_3(PO_4)(PO_3(OH))(OH)_6]$	Radoslovich, Slade, 1980; Radoslovich 1982
	plumbogummite	$Pb[Al_3((P,As)O_4)_2((OH)_5(H_2O)]$	Kolitsch *et al.* 1999a
	waylandite	$(Bi,Ca)[Al_3(PO_4,SiO_4)_2(OH)_6]$	Clark *et al.* 1986
	zairite	$Bi[(Fe^{3+},Al)(PO_4)_2(OH)_6]$	van Wambeke 1975
	dussertite	$Ba[Fe_3(AsO_4)_2(OH)_6]$	Kolitsch *et al.* 1999b
		$[NH(CH_2)_6NH][Fe^{3+}{}_2Fe^{2+}{}_2F_6(SO_4)_2](H_3O)$	Paul *et al.* 2002b
		$[H_3N(CH_2)_6NH_3][Fe^{II}{}_{1.5}F_3(SO_4)](H_2O)_{0.5}$	Rao *et al.* 2004
		$(NH_4)(VO_2)_3(SeO_3)_2$	Vaughey *et al.* 1994
		$K(VO_2)_3(SeO_3)_2$	Harrison *et al.* 1995c
		$Cs(VO_2)_3(TeO_3)_2$	Harrison and Buttery, 2000
cc2–3:1–1		$M_2[(MoO_3)_3(SeO_3)] M = Cs, NH_4$	Harrison *et al.* 1994a

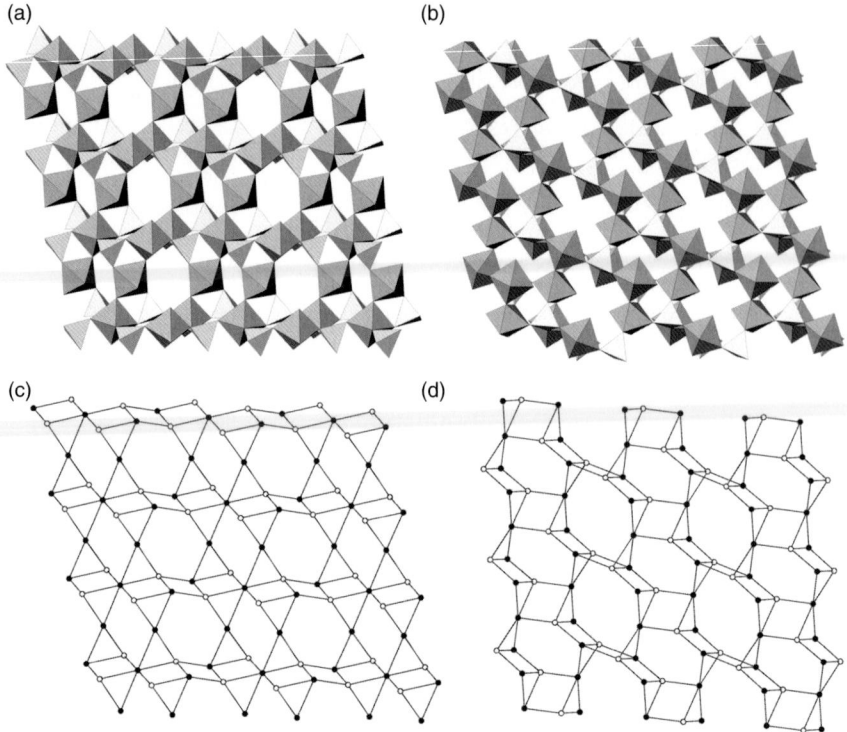

Fig. 2.53. 2D sheets observed in the structures of mitryaevaite, $Al_5(PO_4)_2[(P,S)O_3(O,OH))]_2F_2$ $(OH)_2(H_2O)_8 \cdot 6.48H_2O$ (a) and kingite, $Al_3(PO_4)_2(F,OH)_2 \cdot 8(H_2O,OH)$ (b), and their topologies ((c), (d), respectively).

The topology of a 2D unit in another Al phosphate, $[N_2C_3H_{12}]Al_2(PO_4)(OH_x,F_{5-x})$ $(x \sim 2)$ (Simon *et al.* 1999b), is demonstrated in Fig. 2.56. The sheet (Fig. 2.56(a)) is a derivative of the alunite-type graph *cc2–3:2–1* (Fig. 2.56(b)). By analogy with mitryaevaite and kingite, the transition from the parent graph involves its partition into 1D tapes and linkage of the tapes by additional edges (Fig. 2.56(c)).

The *cc2–1:1–11* graph is a direct derivative of the {**3.12.12**} graph (Fig. 2.57). To obtain the former, the latter has to be partitioned into 1D tapes (Figs. 2.57(b) and (c) and linked by additional edges (Fig. 2.57(d)).

2.4.1.5   *Other graphs*

In addition to the {**3.12.12**} derivatives, there are several other topologies that contain the *M–M* links. Figure 2.58(a) shows the octahedral–tetrahedral sheet in the structure of montgomeryite, $Ca_4Mg(H_2O)_{12}[Al_4(OH)_4(PO_4)_6]$ (Moore, Araki, 1974). The sheet consists of chains of corner-linked $Al\varphi_6$ octahedra and PO_4 tetrahedra. It is important that the structure of chains is different from those observed in the laueite-group

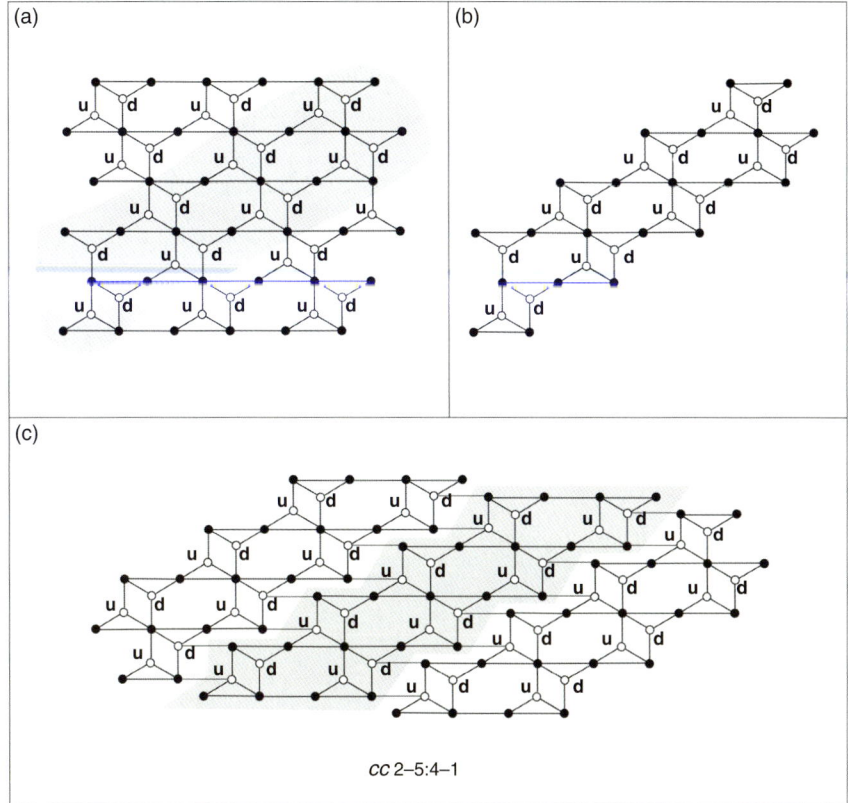

Fig. 2.54. Topology of the 2D unit in mitryaevaite (c) as a derivative of the laueite graph ((a), (b)).

minerals (see Table 2.9). In the latter, octahedra share *trans*-vertices only, but, in montgomeryite, *trans*- and *cis*-linkages alternate (Fig. 2.58(c)). The corresponding graph is shown in Fig. 2.58(b). In some respect, it may be considered as being derived from the autunite graph by splitting half of its white vertices into two.

Three closely related graphs are shown in Fig. 2.59. All three are based upon chains of black vertices (corresponding to chains of corner-sharing octahedra in real structures) with different mode of linkages. These graphs are idealized versions of topologies observed in some V and Ti phosphates (Table 2.11).

2.4.2 *Graphs with M=M links*

Most common among graphs of this type are those based upon pairs of double-linked black vertices. Some of the observed topologies are shown in Figs. 2.60 and 2.61. The list of compounds is given in Table 2.12.

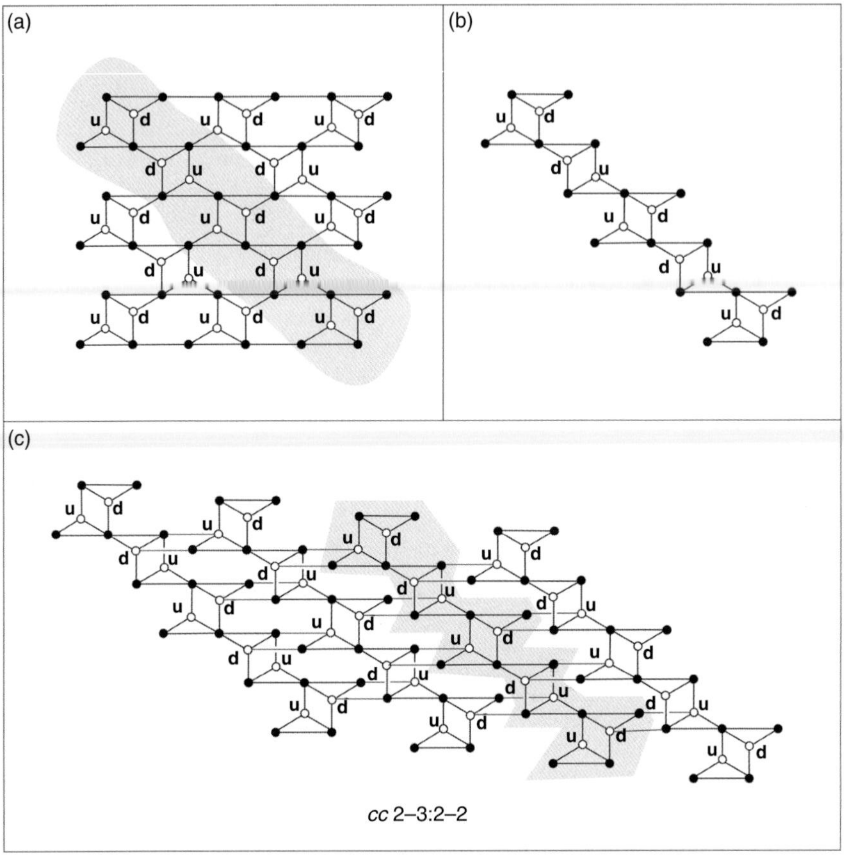

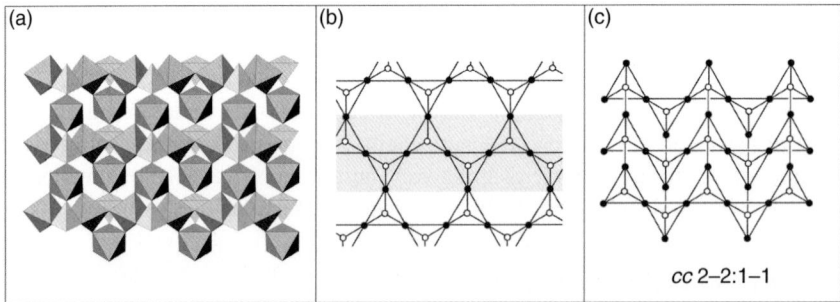

Fig. 2.55. Topology of the 2D unit in kingite (c) as a derivative of the pseudolaueite graph ((a), (b)).

Fig. 2.56. 2D unit in $[N_2C_3H_{12}]Al_2(PO_4)(OH_x,F_{5-x})$ $(x \sim 2)$ (a) has the topology (b) that can be constructed from the 1D fragments of the alunite graph (c).

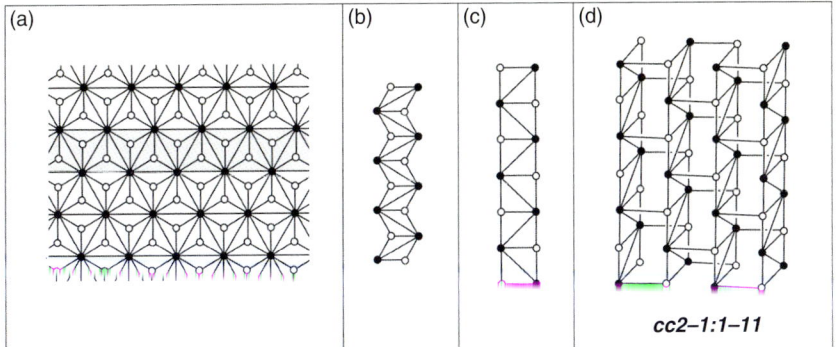

Fig. 2.57. *cc2–1:1–8* graph can be obtained by joining chains excised from the {**3.12.12**} graph.

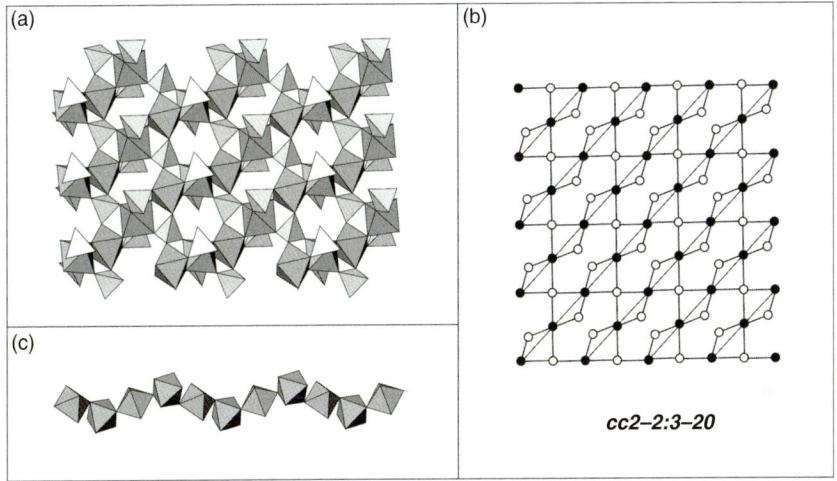

Fig. 2.58. 2D unit from the structure of montgomeryite (a), its octahedral chain (b) and graph (c).

As in the case of graphs with *M=T* links, some graphs with *M=M* links may be obtained from those with the *M–M* links by the doubling-edge procedure. For instance, graph *cc2–1:1–13* shown in Fig. 2.60(b) can obviously be obtained from the minyulite-type graph *cc2–1:1–4* doubling edges linking its black vertices.

In the structure of tsumcorite-group minerals (Table 2.12), *M*-centered octahedra share *trans*-edges to form infinite chains interlinked by 3-connected tetrahedra (Fig. 2.62(a)). The resulting graph contains chains of double-linked black vertices (Fig. 2.62(b)).

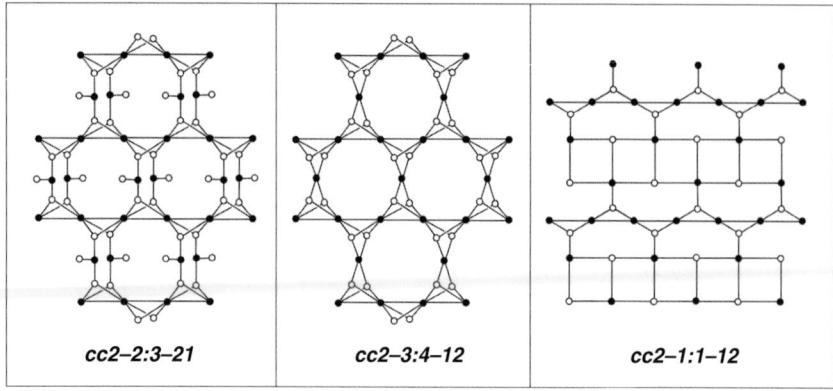

| cc2–2:3–21 | cc2–3:4–12 | cc2–1:1–12 |

Fig. 2.59. Graphs with linear chains of connected black vertices.

Table 2.11   Inorganic oxysalts based upon graphs containing $M–M$ links and having a modular structure or a structure that is not a derivative of the {**3.12.12**} graph

Graph	Chemical formula / name	Reference
cc2–5:4–1	$Al_5(PO_4)_2[(P,S)O_3(O,OH))]_2F_2(OH)_2(H_2O)_8 \cdot 6.48H_2O$ mitryaevaite	Cahill *et al.* 2001
cc2–3:2–2	$Al_3(PO_4)_2(F,OH)_2 \cdot 8(H_2O,OH)$ kingite	Wallwork *et al.* 2003, 2004
cc2–2:1–1	$[N_2C_3H_{12}]Al_2(PO_4)(OH_x,F_{5-x})$ ($x \sim 2$)	Simon *et al.* 1999b
cc2–1:1–11	$[NH_3(CH_2)_3NH_3]_{0.5}[M(OH)AsO_4]$ (M = Ga, Fe)	Liao *et al.* 2000
cc2–2:3–20	$Ca_4Mg(H_2O)_{12}[Al_4(OH)_4(PO_4)_6]$ montgomeryite	Moore and Araki 1974b; Fanfani *et al.* 1976
cc2–2:3–21	$(NH_3CH_2CH_2NH_3)_{1.5}[(VO)_2(HPO_4)_2(PO_4)]$	Zima and Lii 2003
cc2–3:4–12	$Ti_3O_2X_2(HPO_4)_x(PO_4)_y \cdot (N_2C_nH_{2n+6})_z \cdot (H_2O)_2$ $n = 2, 3; x = 0, 2; y = 4, 2; z = 3, 2; X = F, OH$	Serre *et al.* 2002
cc2–1:1–12	$(C_4H_{12}N_2)[V_4O_6(HPO_4)_2(PO_4)_2]$	Zima and Lii 2003

With the appearance of edge-sharing in heteropolyhedral complexes, the number of topologically possible structures increases significantly. It is not our purpose here to provide a comprehensive list of the observed topologies. Rather, our goal is to demonstrate how this vast array of structures can be systematically described in a coherent and unified approach.

2.4.3   *Graphs with M≡M links*

As can be expected, structures with face-sharing polyhedra with reasonably low coordination numbers of central cations are very rare. Usually, they occur if face-sharing polyhedral dimers are stabilized by metal–metal bonding between the central atoms. Figure 2.63(a) shows the octahedral–tetrahedral sheet in the structure of

Table 2.13 Minerals and inorganic oxysalts based upon 1D chains with $M{:}T = 1{:}1$

Graph	Mineral name	Chemical formula	Reference
ccl–1:1–1	chalcanthite	$[Cu(H_2O)_4(SO_4)](H_2O)$ *trans*-isomer	Bacon and Titterton 1975
	pentahydrite	$[Mg(H_2O)_4(SO_4)](H_2O)$ *trans*-isomer	Baur and Rolin 1972
	siderotil	$[Fe(H_2O)_4(SO_4)](H_2O)$ *trans*-isomer	Peterson *et al.* 2003
	jokokuite	$[Mn(H_2O)_4(SO_4)](H_2O)$ *trans*-isomer	Caminiti *et al.* 1982b
	brassite	$MgAsO_4(H_2O)_4$ *trans*-isomer	Protas and Gindt, 1976
		$K[VO_2(SO_4)(H_2O)_2]H_2O$ *trans*-isomer	Richter and Mattes 1991
		$CoSeO_4(H_2O)_5$ *trans*-isomer	Mestres *et al.* 1985
		$Al_2(SO_4)_3(H_2O)_{10.5}$ *trans*-isomer	Fischer *et al.* 1996d
		$MgMoO_4(H_2O)_5$ *trans*-isomer	Bars *et al.* 1977
		$NH_4In(SeO_4)_2(H_2O)_4$ *cis*-isomer	Soldatov *et al.* 1978
ccl–1:1–2		$[(UO_2)(SO_4)(H_2O)_3] \cdot 0.5(18\text{-crown-}6)$	Rogers *et al.* 1991
		$[(UO_2)(SO_4)(H_2O)_2]_2(H_2O)$	Van den Putten and Loopstra 1974
		$\alpha\text{-}[(UO_2)(SO_4)(H_2O)_2]_2(H_2O)_3$	Zalkin *et al.* 1978
		$\beta\text{-}[(UO_2)(SO_4)(H_2O)_2]_2(H_2O)_3$	Brandenburg and Loopstra 1973
		$[NC_4H_{12}][(UO_2)(SO_4)(H_2O)_2]Cl$	Serezhkina and Trunov 1989
		$[CH_2ClCONH_2][(UO_2)(SO_4)(H_2O)_2]$	Mikhailov *et al.* 1995
		$[(UO_2)(SO_4)(H_2O)_2] \cdot 0.5(12\text{-crown-}4)$ (H_2O)	Rogers *et al.* 1991
		$[(UO_2)(SO_4)(H_2O)_2] \cdot 0.5$ $(benzo\text{-}15\text{-crown-}5)(H_2O)$	Rogers *et al.* 1991
		$(UO_2)(CrO_4)(H_2O)_2$	Krivovichev and Burns 2003g
		$[(UO_2)(CrO_4)(H_2O)_2](H_2O)$	Krivovichev and Burns 2003g
		$[(UO_2)(CrO_4)(H_2O)_2]_4(H_2O)_9$	Krivovichev and Burns 2003g
		$[(UO_2)(CrO_4)(H_2O)_2]_2(H_2O)_7$	Serezhkin and Trunov 1981
		$[(UO_2)(SeO_4)(H_2O)_2](H_2O)_2$	Serezhkin *et al.* 1981b
		$VO(HPO_4)(H_2O)_4$	Leonowicz *et al.* 1985
		$((VO(OH)(H_2O))(SeO_3))_4(H_2O)_2$	Dai *et al.* 2003
		$(C_2H_{10}N_2)[Ga_{0.98}Cr_{0.02}(HPO_3)F_3]$	Fernandez-Armas *et al.* 2004
		$(C_2H_{10}N_2)[M(HPO_3)F_3]$ $[M^{III}{=}V$ and Cr$]$	Fernandez *et al.* 2003
		$[VO(H_2O)_2(HPO_3)](H_2O)_3$	Zakharova *et al.* 1994
ccl–1:1–3		$InOHSO_4(H_2O)_2$	Johansson 1961
		$Mn(OH)SO_4(H_2O)_2$	Mereiter 1979
		$M(VO_2)(HPO_4)(M = NH_4, K, Rb)$	Amoros *et al.* 1988; Amoros and leBail 1992
		$[C_4N_2H_{12}][FeF_3(SO_4)]$	Paul *et al.* 2003
	butlerite	$[Fe(OH)(H_2O)_2(SO_4)]$	Fanfani *et al.* 1971
	parabutlerite	$[Fe(OH)(H_2O)_2(SO_4)]$	Borene 1970
	uklonskovite	$Na[MgF(H_2O)_2(SO_4)]$	Sabelli 1985
	fibroferrite	$[Fe(OH)(H_2O)_2(SO_4)]$	Scordari 1981a
ccl–1:1–4		$K[VO_2(SO_4)(H_2O)]$	Richter and Mattes 1991
		$Cs_2[(UO_2)O(MoO_4)]$	Alekseev *et al.* 2007
ccl–1:1–5	botryogen	$[MgFe(OH)(H_2O)_6(SO_4)_2](H_2O)$	Süsse 1968
	zincobotryogen	$[(Zn, Mg, Mn)Fe(OH)(H_2O)_6(SO_4)_2]$ (H_2O)	Hexiong and Pingqiu 1988

Table 2.13 *Continued*

Graph	Mineral name	Chemical formula	Reference
ccl–1:1–6	destinezite	$[Fe_2(OH)(H_2O)_5(PO_4)(SO_4)](H_2O)$	Peacor *et al.* 1999
ccl–1:1–7		$[H_3N(CH_2)_2NH_3]_2[CdCl_2(SO_4)](SO_4)$	Paul *et al.* 2002a
		(H_2O)	
		$[HN(CH_2)_6NH][CdBr_2(SO_4)]$	Paul *et al.* 2002a
		$[HN(CH_2)_6NH][CdCl_2(SO_4)]$	Paul *et al.* 2002a
	linarite	$Pb[Cu(OH)_2(SO_4)]$	Effenberger 1987
	schmiederite	$Pb_2Cu_2(OH)_4(SeO_3)(SeO_4)$	Effenberger 1987
ccl–1:1–8	caledonite	$Pb_5[Cu_2(OH)_6(SO_4)_2](SO_4)(CO_3)$	Giacovazzo *et al.* 1973
ccl–1:1–9	chlorothionite	$K_2[CuCl_2(SO_4)]$	Giacovazzo *et al.* 1976a
ccl–1:1–10	amarantite	$[Fe_2O(H_2O)_4(SO_4)_2](H_2O)_3$	Süsse 1967
	hohmannite	$[Fe_2O(H_2O)_4(SO_4)_2](H_2O)_4$	Scordari 1978
ccl–1:1–11		$[NH_3(CH_2)_2NH(CH_2)_2NH_3]$	Chakrabarti *et al.* 2004
		$[Fe_2F_4(HAsO_4)_2]$	

causes the occurrence of orientational geometrical isomers. Indeed, this is the case for uranyl compounds containing $[(UO_2)(TO_4)(H_2O)_2]$ chains (T = S, Cr and Se). Figure 2.67 shows three different geometrical isomers of these chains. The isomers differ from each other by the sequence of up-and-down orientations of non-shared corners of TO_4 tetrahedra. As in the case of 2D topologies, the concept of an orientation matrix can be applied to their description. Obviously, for chains of this kind, the orientation matrix consists of one row of **u** and **d** symbols. All tetrahedra of the chain shown in Fig. 2.67(a) have the same orientation; the orientation matrix of this chain is therefore…**uuuu**…or (**u**)$_\infty$. Such chains occur in the structures of $[(UO_2)(CrO_4)(H_2O)_2]_4(H_2O)_9$, $[(UO_2)(SO_4)(H_2O)_2]_2(H_2O)$, and α-$[(UO_2)(SO_4)(H_2O)_2](H_2O)_3$. The (**ud**)$_\infty$ isomer (Fig. 2.67(b)) is most common and is observed in the structures of $[(UO_2)(CrO_4)(H_2O)_2]$, $[(UO_2)(CrO_4)(H_2O)_2](H_2O)$, $[(UO_2)(CrO_4)(H_2O)_2]_4(H_2O)_9$, $[(UO_2)(CrO_4)(H_2O)_2]_2(H_2O)_7$, and $[(UO_2)(SeO_4)(H_2O)_2](H_2O)_2$. The (**uudd**)$_\infty$ isomer is shown in Fig. 2.67(c). The only known representative of this isomer is the structure of β-$[(UO_2)(SO_4)(H_2O)_2](H_2O)_3$. It is noteworthy that the structure of $[(UO_2)(CrO_4)(H_2O)_2]_4(H_2O)_9$ contains uranyl chromate chains representing two isomers: (**u**)$_\infty$ and (**uudd**)$_\infty$. The chemical reasons for the origin of different orientational geometrical isomers in uranyl oxysalt hydrates is not absolutely clear. It is likely that the occurence of a certain type of isomer is impacted by the number of water molecules present in the structure, i.e. by features of the hydrogen-bond system. Figure 2.68 shows the structures of four uranyl chromate hydrates with the general composition $[UO_2(CrO_4)(H_2O)_2](H_2O)_n$ projected along the chain extension [n = 0 (a), 1 (c), 2.25 (b), and 3.5 (d)]. In all four structures, the (**ud**)$_\infty$ isomers are present, but, in the structure with n = 2.25, the (**u**)$_\infty$ are present as well. It seems that the occurence of isomers is related to the number of water molecules present in the compound, i.e. by features of the hydrogen-bonding system that is responsible for the three-dimensional integrity of the structure.

Table 2.14 Minerals and inorganic oxysalts based upon 1D chains with $M:T = 1:2$

Type	Mineral name	Chemical formula	Reference
ccl–1:2–1-trans*	kröhnkite	$Na_2[Cu(H_2O)_2(SO_4)_2]$	Hawthorne and Ferguson, 1975
	brandtite	$Ca_2[Mn(AsO_4)_2(H_2O)_2]$	Dahlman, 1952
	collinsite	$Ca_2[Mg(PO_4)_2(H_2O)_2]$	Brotherton et al. 1974; Yakubovich et al. 2003a
	cassidyite	$Ca_2[Ni(PO_4)_2(H_2O)_2]$	White et al. 1967
	talmessite	$Ca_2(Mg,Co)(AsO_4)_2 \cdot (H_2O)_2$	Catti et al. 1977
	roselite	$Ca_2(Co,Mg)(AsO_4)_2(H_2O)_2$	Hawthorne and Ferguson, 1977
	fairfieldite	$Ca_2[Mn(PO_4)_2(H_2O)_2]$	Fanfani et al. 1970a
	hillite	$Ca_2[Zn(PO_4)_2(H_2O)_2]$	Yakubovich et al. 2003b
	–	$A[Al(CrO_4)_2(H_2O)_2]$ A = Na, K	Cudennec and Riou, 1977
		$K_2Fe(SO_4)_2 \cdot 2(H_2O)$	Ishigami et al. 1999
		$KMg(H(SO_4)_2)(H_2O)_2$	Maciček et al. 1994
		$K_2Cd(SeO_4)_2(H_2O)_2$	Peytavin et al. 1973, 1974
		$M(HSeO_3)_2 \cdot 4(H_2O)$ (M = Mg, Co, Ni, Zn)	Engelen et al. 1995; Micka et al. 1996
		$K(VO)H(SeO_3)_2$	Kim et al. 1996
		$Sr_3[(UO_2)(TeO_3)_2](TeO_3)$	Almond and Albrecht-Schmitt 2002b
		$NaIn(CrO_4)_2 \cdot 2H_2O$	Kolitsch 2006
		$\beta\text{-}Tl_2[(UO_2)(TeO_3)_2]$	Almond and Albrecht-Schmitt 2002b
	deloryite	$Cu_4[(UO_2)(MoO_4)_2](OH)_6$	Tali et al. 1993; Pushcharovskii et al. 1996
	walpurgite	$Bi_4O_4[(UO_2)(AsO_4)_2](H_2O)_2$	Mereiter 1982
	orthowalpurgite	$Bi_4O_4[(UO_2)(AsO_4)_2](H_2O)_2$	Krause et al. 1995
		$Cu_2[(UO_2)(PO_4)_2]$	Guesdon et al. 2002
		$Li_2[(UO_2)(MoO_4)_2]$	Krivovichev and Burns 2003f
		$(CN_3H_6)(VO)(H_2O)(HPO_4)(H_2PO_4)(H_2O)$	Bircsak et al. 1999
		$NH_4Pr(SeO_4)_2(H_2O)_5$	Iskhakova 1995
		$Sm_2(SeO_4)_3(H_2O)_8$	Ovanisyan et al. 1988
		$[N_2C_4H_{12}][UO_2(H_2O)(SO_4)_2]$	Norquist et al. 2002
		$[N_2C_3H_{12}][UO_2(H_2O)(SO_4)_2]$	Thomas et al. 2003
		$[N_2C_5H_{12}][UO_2(H_2O)(SO_4)_2]$	Stuart et al. 2003
		$[N_2C_2H_{10}][UO_2(H_2O)(SO_4)_2]$	Norquist et al. 2003c
		$[C_2N_2H_{10}][(UO_2)(SO_4)_2(H_2O)]$	Mikhailov et al. 2000
		$[N_2C_3H_{12}][UO_2(H_2O)(SO_4)_2]$	Norquist et al. 2003a
		$Mn[(UO_2)(SO_4)_2(H_2O)](H_2O)_5$	Tabachenko et al. 1979
		$M[(UO_2)(SeO_4)_2(H_2O)](H_2O)_4$ M = Mg, Zn	Krivovichev and Kahlenberg 2005b
		$[C_5H_{14}N_2][(UO_2)(SeO_4)_2(H_2O)]$	Krivovichev et al. 2007
		$[C_5H_{16}N_2]_2[(UO_2)(SeO_4)_2(H_2O)](NO_3)_2$	Krivovichev et al. 2007
		$[C_4H_{12}N_2][(UO_2)(SeO_4)_2(H_2O)]$	Krivovichev et al. 2007
		$[C_4H_{14}N_2][(UO_2)(SeO_4)_2(H_2O)]$	Krivovichev et al. 2007
		$[C_5H_{14}N_2][(UO_2)(SeO_4)_2(H_2O)]$	Krivovichev et al. 2007
		$[C_{10}H_{26}N_2][(UO_2)(SeO_4)_2(H_2O)](H_2SeO_4)_{0.85}(H_2O)_2$	Krivovichev et al. 2007

Table 2.14 *Continued*

Type	Mineral name	Chemical formula	Reference
		$[C_3H_{12}N_2][(UO_2)(SeO_4)_2(H_2O)]$	Krivovichev *et al.* 2007
		$[C_6N_2H_{18}]_{0.5}[Fe(SO_4)_2(H_2O)_2]$	Fu *et al.* 2006
		$[(UO_2)(H_2PO_4)_2(H_2O)](H_2O)_2$	Mercier *et al.* 1985
		$[(UO_2)(H_2AsO_4)_2(H_2O)]$	Gesing and Rüscher 2000
		$(UO_2)(HSeO_3)_2(H_2O)$	Mistryukov and Mikhailov 1983
		$Ba_2VO(PO_4)_2 \cdot (H_2O)$	Harrison *et al.* 1994b
		$Sr_2(VO)(AsO_4)_2$	Wadewitz and Mueller-Buschbaum 1996a
		$Ba_2(VO)(PO_4)_2$	Wadewitz and Mueller-Buschbaum 1996a
		$Na_4VO(PO_4)_2$	Shpanchenko *et al.* 2006a
		$Ba_2TeO(PO_4)_2$	Ok and Halasyamani 2006
		$K_2Fe[H(HPO_4)_2]F_2$	Mi *et al.* 2005
*ccl–1:2–1-cis**		$K_2MoO_2(SO_4)_2$	Noerbygaard *et al.* 1998
		$Mg(H_2PO_4)_2 \cdot 4(H_2O)$	Miyake *et al.* 1998
		$[enH_2][Cd(H_2O)_2(SeO_4)_2]$	Pasha *et al.* 2003c
		$CsTl^{3+}(SO_4)_2(H_2O)_2$	Manoli *et al.* 1972
		$CsIn(SeO_4)_2(H_2O)_2$	Saf'yanov *et al.* 1975
ccl–1:2–2	krausite	$K[Fe(H_2O)_2(SO_4)_2]$	Effenberger *et al.* 1986
		$Sc(NH_3OH)(SO_4)_2 \cdot 1.5(H_2O)$	Mirceva and Golic 1995
		$Ba_8(VO)_6(PO_4)_2(HPO_4)_{11} \cdot 3(H_2O)$	Harrison *et al.* 1995b
		$Ba_3(NbO)_2(PO_4)_4$	Wang *et al.* 2000b
		$(H_2Bipy)[Sc(H_2O)(SO_4)_2]_2 \cdot 2H_2O$	Petrosyants *et al.* 2005
ccl–1:2–3		$Ba_2(VO_2)(PO_4)(HPO_4) \cdot H_2O$	Bircsak and Harrison, 1998b
ccl–1:2–4		$[N_2C_5H_{14}][UO_2(H_2O)(SO_4)_2]$	Norquist *et al.* 2002
ccl–1:2–5		$Al_3(HSO_4)(SO_4)_4 \cdot 9H_2O$	Fischer *et al.* 1996e
ccl–1:2–6		$Rb[UO_2(CrO_4)(IO_3)(H_2O)]$	Sykora *et al.* 2002b
ccl–1:2–7		$(NpO_2)_2(TcO_4)_4(H_2O)_3$	Fedoseev *et al.* 2003
		$(UO_2)_2(ReO_4)_4(H_2O)_3$	Karimova and Burns 2007
ccl–1:2–8		$[C_6N_4H_{22}]_{0.5}[Zn(HPO_4)_2]$	Choudhury *et al.* 2002
		$[C_4H_{14}N_2][(UO_2)(SeO_4)_2(H_2O)]$	Krivovichev and Kahlenberg 2005c
		$[C_6H_{16}N_2][(UO_2)(SeO_4)_2(H_2O)]$	Krivovichev *et al.* 2007
		$[N_4C_6H_{22}][UO_2(H_2O)(SO_4)_2]_2(H_2O)_2$	Norquist *et al.* 2003a
		$[C_2N_2H_{10}][AlH(PO_4)_2]$	Wang *et al.* 2003c
ccl–1:2–9		$[C_6N_2H_{14}][Fe^{III}_2F_2(HPO_4)_2(H_2PO_4)_2] \cdot 2H_2O$	Mahesh *et al.* 2002
	sideronatrite	$Na_2[Fe(OH)(SO_4)_2](H_2O)_3$	Scordari 1981c
	metasideronatrite	$Na_4[Fe_2(OH)_2(SO_4)_4](H_2O)_3$	Scordari *et al.* 1982
		$[C_4N_2H_{12}][Fe_2F_2(HPO_4)_2(H_2PO_4)_2]$	Choudhury and Rao 2002
		$[H_3N(CH_2)_3NH_3][GaF(HPO_4)_2](H_2O)_2$	Walton *et al.* 2000a, b
		$[C_6N_2H_{14}][GaF(HPO_4)_2]$	Bonhomme *et al.* 2001
		$[C_6H_{10}(NH_3)_2][Ga(OH)(HPO_4)_2](H_2O)$	Lin and Lii 1998
		$Na_3[Fe(PO_4)_2](Na_{1.548}H_{0.452}O)$	Bridson *et al.* 1998
		$Na_4[Al(PO_4)_2(OH)]$	Attfield *et al.* 1995
		$Na_3[M(OH)(HPO_4)(PO_4)]$ $M = Al, Ga$	Lii and Wang 1997
		$[CN_3H_6]_2[FeF(SO_4)_2]$	Paul *et al.* 2003
		$[C_4N_3H_{16}][FeF(SO_4)_2]$	Paul *et al.* 2003

Table 2.14 *Continued*

Type	Mineral name	Chemical formula	Reference
		$[NH_3CH_2CH_2CH(NH_3)CH_2CH_3]$ $[FeF(HPO_4)_2]$	Mandal *et al.* 2003
		$[NH_2(CH_2)_6NH_2][Al(OH)(H(HPO_3))_2]$	Li and Xiang 2002
		$[Al(OH)(H(HPO_3))_2](H_2O)_2$	Li and Xiang 2002
		$[C_2N_2H_{10}][Fe(SO_4)_2(OH)] \cdot H_2O$	Fu *et al.* 2006
		$[C_6N_4H_{22}]_{0.5}[Fe(SO_4)_2(OH)] \cdot 2H_2O$	Fu *et al.* 2006
		$Hg_4(VO)(PO_4)_2$	le Fur and Pivan 2001
		$Na_4[TiO(AsO_4)_2]$	Yaakoubi and Jouini 1998
		$[NH_3(CH_2)_2NH_3]_5[H_3O]_2[TiO(PO_4)_2]_3$ *cis*-isomer	Guo *et al.* 2001
ccl–1:2–10		$A_6[(UO_2)_2O(MoO_4)_4]$ A = Na, K, Rb	Krivovichev and Burns 2001b, 2002a
ccl–1:2–11		$[(H_3NC_2H_4NH_3)_3]$ $[Sc_3(OH)_2(PO_4)_2(HPO_4)_3(H_2PO_4)]$	Miller *et al.* 2005
ccl–1:2–12		$[N_2C_{10}H_{10}][UO_2(SO_4)_2](H_2O)$	Norquist *et al.* 2003a
		$[N_2C_6H_{18}][UO_2(SO_4)_2](H_2O)$	Norquist *et al.* 2003c
		$[C_6H_{16}N_2][UO_2(SO_4)_2](H_2O)$	Doran *et al.* 2003b
		$[Co(NH_3)_6][NpO_2(SO_4)_2](H_2O)_2$	Grigor'ev *et al.* 1991b
		$Na_3[(NpO_2)(SO_4)_2](H_2O)_{2.5}$	Forbes and Burns 2005
		$CaZn_2[(NpO_2)_2(SO_4)_4](H_2O)_{10}$	Forbes and Burns 2005
		$[C_4H_{12}N][UO_2(NO_3)(SO_4)]$	Doran *et al.* 2003c
		$[C_4H_{13}N][(UO_2)(SeO_4)(NO_3)]$	Krivovichev and Kahlenberg 2005c
ccl–1:2–12	wherryite	$Pb_7[Cu(OH)(SO_4)(SiO_4)]_2(SO_4)_2$	Cooper and Hawthorne 1994b
	tsumebite	$Pb_2[Cu(OH)(PO_4)(SO_4)]$	Fanfani and Zanazzi 1967

* only selected compounds are listed; for more detailed discussion on compounds with kröhnkite-type chains see Fleck *et al.* (2002) and Fleck and Kolitsch (2003).

2.5.3 *Lone-electron-pair-induced geometrical isomerism*

In the structures of minerals and compounds with kröhnkite-type chains (*ccl–1:2–1* graph in Fig. 2.65), another interesting type of geometrical isomerism had been observed. This isomerism is induced by the presence of cations with stereoactive lone electron pairs (LEP). If a tetrahedron in a "classic" kröhnkite-like chain is replaced by a TO_3 triangular pyramid (T = Se^{4+}, Te^{4+}), the latter can be considered as a $TO_3\Psi$ pseudo-tetrahedron, where Ψ is a LEP. Within the chain, non-shared vertices of the $TO_3\Psi$ pseudo-tetrahedron are chemically non-equivalent: one is occupied by O, whereas another is occupied by Ψ. This inequivalency of non-shared vertices results in the appearance of geometrical isomers that differ from each other in the orientations of the LEPs of the $TO_3\Psi$ pseudo-tetrahedra. As in the case of orientational geometrical isomers with 3-connected tetrahedra, orientation of LEPs in kröhnkite-like chains can be described using (**u**+**d**) formalism. Thus, the $[M(HSeO_3)_2(H_2O)_2]$ octahedral–pseudo-tetrahedral chain shown in Fig. 2.69a [it has been found in structures

Table 2.15 Minerals and inorganic oxysalts based upon 1D chains with $M{:}T = 1{:}3$, $1{:}4$, $2{:}3$, $3{:}4$, and $2{:}1$

Graph	Mineral name	Chemical formula	Reference
ccl–1:3–1	ferrinatrite	$Na_3[Fe^{3+}(SO_4)_3](H_2O)_3$	Mereiter 1976, Scordari 1977
		$(NH_4)_3In(SO_4)_3$	Jolibois *et al.* 1980, 1981
		$Na_3In(SO_4)_3(H_2O)_3$	Mukhtarova *et al.* 1979c
		$Na_3V(SO_4)_3$	Boghosian *et al.* 1994
		$Cs_3[Yb(SO_4)_3]$	Samartsev *et al.* 1980
		$(NH_4)_3Sc(SeO_4)_3$	Valkonen and Niinisto, 1978
		$Sr_2M(PO_4)_2(H_2PO_4)$ M = V, Fe	Lii *et al.* 1993a
	kaatialaite	$Fe(H_2AsO_4)_3(H_2O)_5$	Boudjada and Guitel 1981
		$Cs[Al(H_2AsO_4)_2(HAsO_4)]$	Schwendtner and Kolitsch 2007
		$CsM(H_{1.5}AsO_4)_2(H_2AsO_4)$ M = Ga, Cr	Schwendtner and Kolitsch 2005
		$(C_2H_{10}N_2)[Fe(HAsO_4)_2(H_2AsO_4)](H_2O)$	Bazán *et al.* 2000
ccl–1:3–2		$[C_2N_2H_{10}]_{1.5}[Fe(SO_4)_3] \cdot 2H_2O$	Fu *et al.* 2006
		$Na_4[(UO_2)(CrO_4)_3]$	Krivovichev and Burns 2003h
		$K_5[(UO_2)(CrO_4)_3](NO_3)(H_2O)_3$	Krivovichev and Burns 2003a
		$Na_3Tl_5[(UO_2)(MoO_4)_3]_2(H_2O)_3$	Krivovichev and Burns 2003b
		$Na_{13-x}Tl_{3+x}[(UO_2)(MoO_4)_3]_4(H_2O)_{6+x}$ $(x = 0.1)$	Krivovichev and Burns 2003b
	demesmaekerite	$Pb_2Cu_5[(UO_2)(SeO_3)_3](OH)_6(H_2O)_2$	Ginderow and Cesbron 1983
ccl–1:3–3		$K_2[UO_2(CrO_4)(IO_3)_2]$	Sykora *et al.* 2002a
		$Rb_2[UO_2(CrO_4)(IO_3)_2]$	Sykora *et al.* 2002a
		$Cs_2[UO_2(CrO_4)(IO_3)_2]$	Sykora *et al.* 2002a, b
		$K_2[UO_2(MoO_4)(IO_3)_2]$	Sykora *et al.* 2002a
ccl–1:3–4		$[H_2N_2C_6H_{12}][Zn(H_2PO_4)_2(HPO_4)]$	Patarin *et al.* 2004
ccl–1:4–1		$Mn(HSO_4)_2(H_2SO_4)_2$	Stiewe *et al.* 1998
		$Mg(HSO_4)_2(H_2SO_4)_2$	Troyanov *et al.* 1990
		$Cd(HSO_4)_2(H_2SO_4)_2$	Simonov *et al.* 1988
		$Zn(HSO_4)_2(H_2SO_4)_2$	Kemnitz *et al.* 1996
		$(NH_4)_4Cd(HSe^{IV}O_3)_2(Se^{VI}O_4)_2$	Kolitsch 2004
		$Na_9Zr(PO_4)_4Cl$	Genkina *et al.* 1987
ccl–2:3–1		$M[(VO)_2(IO_3)_3O_2]$ (M = NH_4, Rb, Cs)	Sykora *et al.* 2002c
		$[C_4N_2H_{12}][Fe_2(SO_4)_3(OH)_2(H_2O)_2] \cdot H_2O$	Fu *et al.* 2006
ccl–2:3–2	copiapite	$[Fe_2(OH)(H_2O)_4(SO_4)_3]_2[Fe(H_2O)_6](H_2O)_6$	Fanfani *et al.* 1973
	magnesio copiapite	$[Fe_2(OH)(H_2O)_4(SO_4)_3]_2[Mg(H_2O)_6](H_2O)_6$	Süsse 1970
	cuprocopiapite	$[Fe_2(OH)(H_2O)_4(SO_4)_3]_2[Cu(H_2O)_6](H_2O)_6$	Gaines *et al.* 1997
	ferricopiapite	$[Fe_4O(OH)(H_2O)_8(SO_4)_6][Cu(H_2O)_6](H_2O)_6$	Bayliss and Atencio 1985
	calciocopiapite	$[Fe_2(OH)(H_2O)_4(SO_4)_3]_2[Ca(H_2O)_6](H_2O)_6$	Gaines *et al.* 1997
	zincocopiapite	$[Fe_2(OH)(H_2O)_4(SO_4)_3]_2[Zn(H_2O)_6](H_2O)_6$	Gaines *et al.* 1997
	aluminocopiapite	$[(Fe,Al)_2(OH)(H_2O)_4(SO_4)_3]_2$ $[(Mg,Fe)(H_2O)_6](H_2O)_6$	Jolly and Foster 1967
		$[C_6N_2H_{18}]_{0.5}[Fe_2(SO_4)_3(H_2O)_4(OH)] \cdot H_2O$	Fu *et al.* 2006
ccl–3:4–1		$[H_2NC_2H_4NH_2]_{2.5}[Sn^{4+}_3O_2(H_2O)(HPO_4)_4]$ $(H_2O)_2$	Serre and Férey 2003
ccl–2:1–1	bøggildite	$Sr_2Na_2[Al_2PO_4F_9]$	Hawthorne 1982

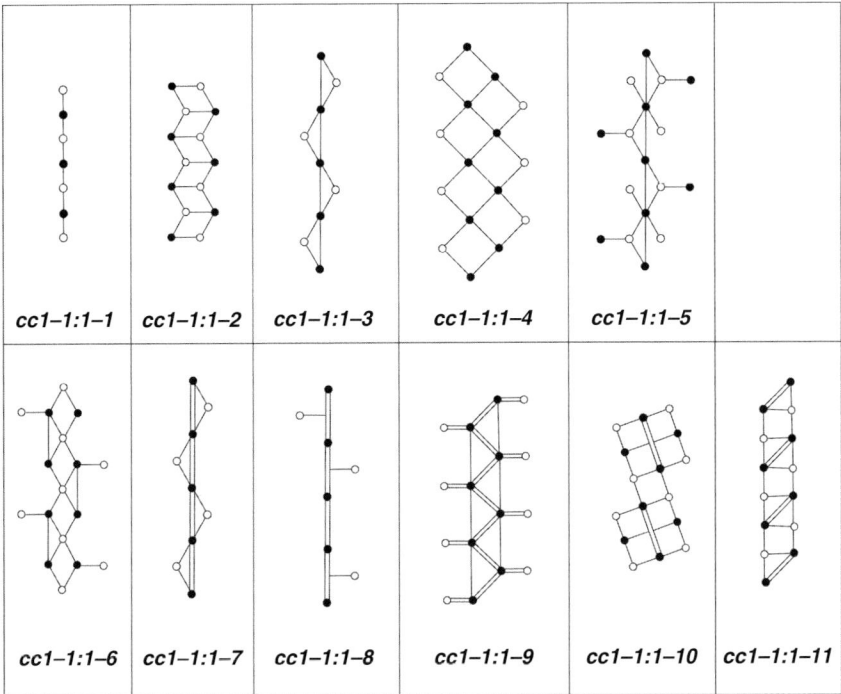

| cc1–1:1–1 | cc1–1:1–2 | cc1–1:1–3 | cc1–1:1–4 | cc1–1:1–5 |

| cc1–1:1–6 | cc1–1:1–7 | cc1–1:1–8 | cc1–1:1–9 | cc1–1:1–10 | cc1–1:1–11 |

Fig. 2.64. Graphs of 1D chains in the structures of inorganic oxysalts. $M{:}T = 1{:}1$.

of compounds $M(HSeO_3)_2 \cdot 4(H_2O)$ (M = Mg, Co, Ni, Zn) (Engelen *et al.* 1995; Micka *et al.* 1996)], has the ... **ududud** ... orientation of LEPs. The chain has two sides (i.e. two rows of $TO_3\Psi$ pseudo-tetrahedra) and therefore the orientation matrix has two rows. In the case of the chain shown in Fig. 2.69(a), it can be written as **(ud)(du)**. There are two other geometrical isomers of this type shown in Figs. 2.69(b) and (c). Their orientation matrices are **(u)(d)** and **(u)(u)**, respectively. Obviously, the type of geometrical isomerism described here is induced by the presence of stereoactive LEPs on Se^{4+} and Te^{4+} cations.

2.5.4 *Cis–trans isomerism*

Cis–trans geometrical isomerism is common for octahedral–tetrahedral chains. Figures 2.70(b) and (c) show two octahedral–tetrahedral chains corresponding to the *c1–1:1–1* graph depicted in Fig. 2.70(a). The chains are different in the arrangements of shared corners of their octahedra. This can be clearly seen in the connectivity diagrams of the octahedra of these chains shown in Figs. 2.70(b) and (c) (*trans-* and *cis*-arrangements, respectively). A similar effect has also been observed

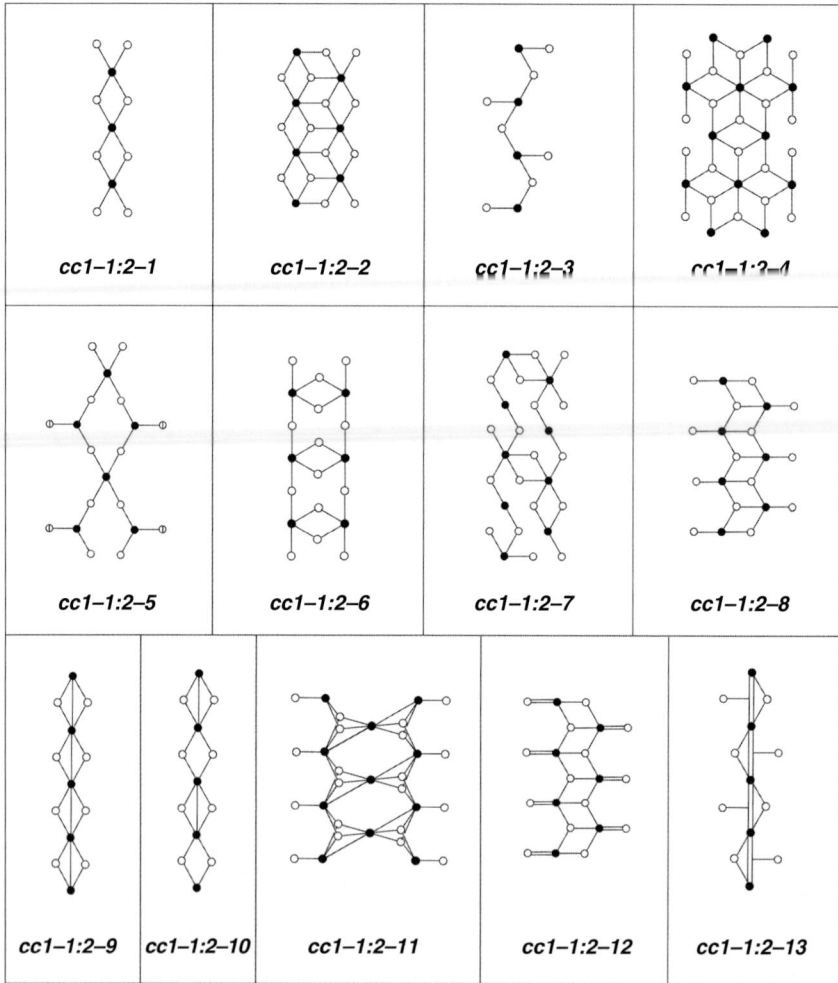

Fig. 2.65. Graphs of 1D chains in the structures of inorganic oxysalts. $M:T = 1:2$.

for the kröhnkite-like chains (**c1–1:2–1** graph in Fig. 2.70(d)). In this case, each octahedron shares four corners with four adjacent tetrahedra (Figs. 2.70(e) and (f)). Connectivity diagrams represent *trans-* and *cis*-arrangements of the *non-shared* corners and are similar to the diagrams shown in Fig. 2.24(f). *Cis–trans* isomerism can also be observed for chains with the *M–M* links (Fig. 2.70(g)). Figures 2.70(h) and (i) show two chains consisting of a linear backbone of corner-sharing FeO_6 octahedra incrustated by SO_4 sulphate tetrahedra. It is of interest that the two chains have the same chemical formula, $[Fe(SO_4)_2(OH)]^{2-}$, and occur in the structures of $[C_2N_2H_{10}]$

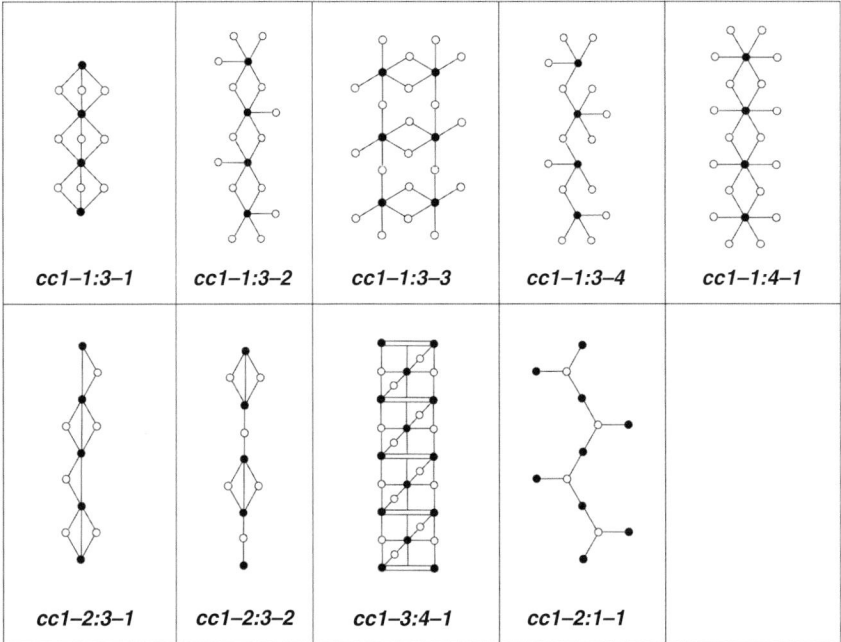

cc1–1:3–1	cc1–1:3–2	cc1–1:3–3	cc1–1:3–4	cc1–1:4–1
cc1–2:3–1	cc1–2:3–2	cc1–3:4–1	cc1–2:1–1	

Fig. 2.66. Graphs of 1D chains in the structures of inorganic oxysalts. $M{:}T = $ 1:3, 1:4, 2:3, 3:4, 2:1.

$[Fe(SO_4)_2(OH)] \cdot H_2O$ and $[C_6N_4H_{22}]_{0.5}[Fe(SO_4)_2(OH)] \cdot 2H_2O$, respectively (Fu *et al.* 2006). The chain observed in the latter compound (Fig. 2.70(i)) has a *cis*-linkage of Fe-centered octahedra, whereas the former one (Fig. 2.70(h)) shows a *trans*-linkage of octahedra. As the result of the *cis*-arrangement of the octahedral linkage, the chain in $[C_6N_4H_{22}]_{0.5}[Fe(SO_4)_2(OH)] \cdot 2H_2O$ has a spiral conformation that is sometimes seen in *cis*-isomers. The complexity of chain topology and conformation correlates with the complexity of the template: the structure of $[C_2N_2H_{10}][Fe(SO_4)_2(OH)] \cdot H_2O$ contains protonated ethylenediamine molecules, whereas $[C_6N_4H_{22}]_{0.5}[Fe(SO_4)_2(OH)] \cdot 2 H_2O$ is templated by more complex triethylenetetramine molecules. Thus, the isomerism originates from the difference of the structure of template and is controlled by molecular packing and the pecularities of the hydrogen-bonding system.

2.6 0D topologies: finite clusters

Topological diagrams of the 0D structural units in inorganic oxysalts with some examples of finite clusters are given in Figs. 2.71, 2.72 and 2.73. A list of respective compounds is provided in Table 2.16. The number of different finite cluster topologies

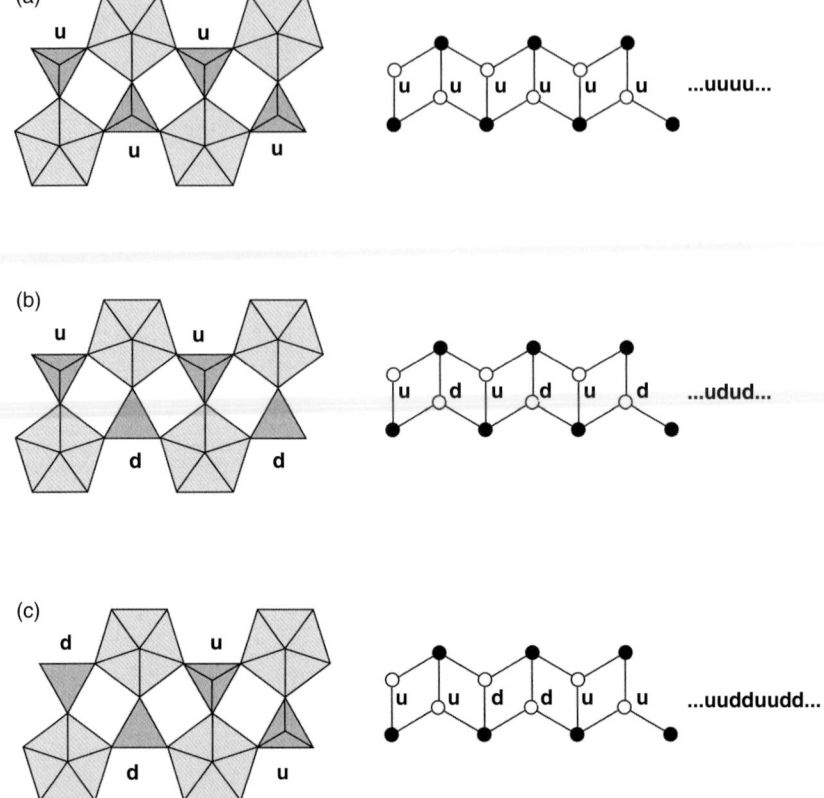

Fig. 2.67. Description of geometrical isomerism of [AnO$_2$(TO$_4$)(H$_2$O)$_2$] chains.

is markedly lower than the number of known 2D and 1D topologies. The most plausible explanation for this fact is that finite clusters are more reactive in comparison to sheets and chains. Most of the topologies of the finite clusters can be derived from parent higher-dimensional graphs. In fact, 0D units play a role of pre-nucleation building blocks in crystal growth processes. It was proposed that single and branched 4-membered ring (4MR) (topologies ***cc0–1:1–2, cc0–1:2–3, cc0–1:3–1, cc0–1:4–1***) exist in hydrothermal solutions and polymerize to form structural units of higher dimensionality. The crucial point in favor of this hypothesis was the identification of intermediate products containing isolated 4MRs. This trend in crystallization had been proposed for a number of systems, including metal phosphates (Neeraj *et al.* 2000; Rao *et al.* 2001a, b; Dan *et al.* 2003), uranyl selenates (Krivovichev *et al.* 2007a), etc. Polymerization of 4-MRs results in formation of chains that in turn may transform into chains with different topology, sheets and frameworks (see, e.g. Oliver *et al.* 1998; Walton *et al.* 2000, 2001; Ayi *et al.* 2001; Millange *et al.* 2002a, b;

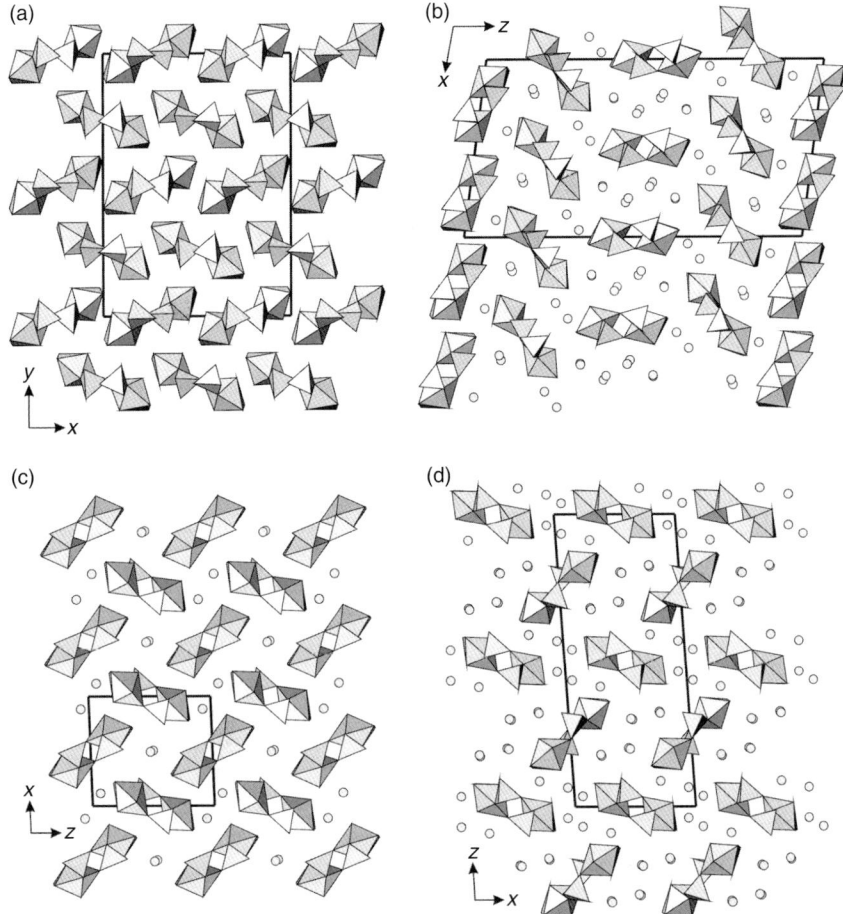

Fig. 2.68. Structures of four uranyl chromate hydrates with the general composition $[UO_2(CrO_4)(H_2O)_2](H_2O)_n$ projected along the chain extension [$n = 0$ (a), 1 (c), 2.25 (b), and 3.5 (d)].

Norquist and O'Hare, 2004; Loiseau *et al.* 2004; Wang *et al.* 2003). Yang *et al.* (2007) isolated a remarkable oxyfluorotitanophosphate cluster $[Ti_{10}P_4O_{16}F_{44}]^{16-}$ (Fig. 2.73(a)) that occurs in the structure of hydrothermally synthesized $K_{16}[Ti_{10}P_4O_{16}F_{44}]$. The cluster consists of three 4MRs and has a branched structure. Quite unusually, the branches are Ti-centered octahedra coordinated by F^- anions that probably prevent further polymerization of the clusters. The topology of the cluster (Fig. 2.73(b)) can be derived from the *c2–1:1–3* topology by the "cut-and-paste" procedure (Fig. 2.73(c)). The $[Ti_{10}P_4O_{16}F_{44}]^{16-}$ cluster can also be considered as a result of polymerization of two branched 4MRs.

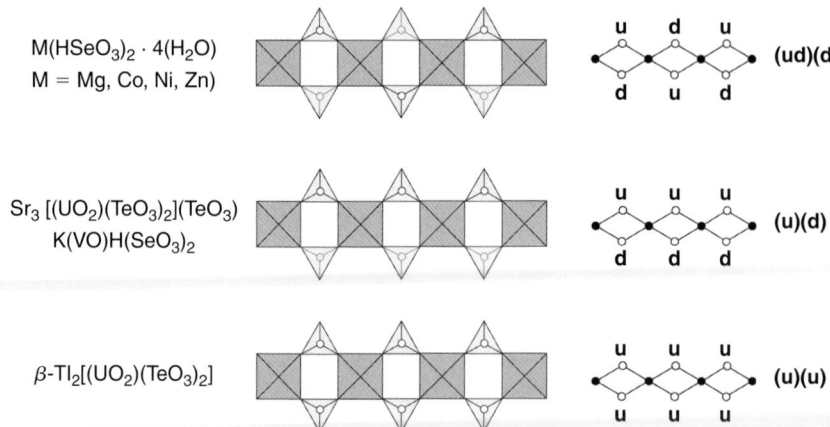

Fig. 2.69. Schemes illustrating geometrical isomerism of kröhnkite-like chains induced by the presence of cations with stereoactive lone pairs of electrons (from left to right: list of compounds containing the chains; view from the top on the chain, graphical representation with **u** and **d** indices corresponding to the orientations of the tetrahedra up and down relative to the plane of the chain, short description of orientation matrices).

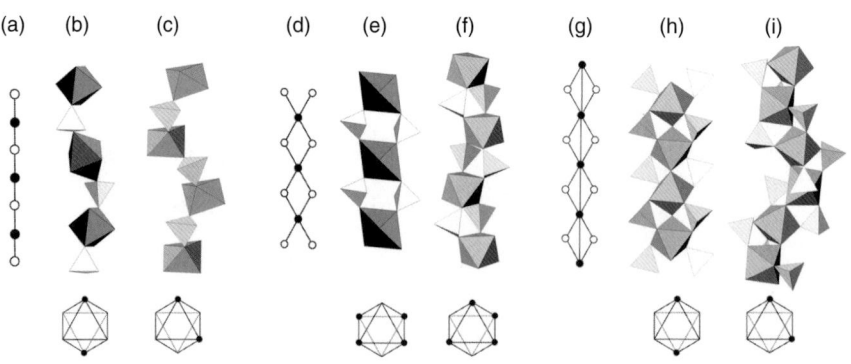

Fig. 2.70. Schemes illustrating *cis–trans* geometrical isomerism observed for octahedral–tetrahedral chains: (a) and (d): black-and-white graphs describing topology of the chains; (b), (c), (e), and (f): polyhedral diagrams of the chains and connectivity diagrams of their octahedra.

It is important to note that 0D finite clusters existing in crystallization solutions may change their topology when a solid of high dimensionality is formed. Taulelle *et al.* (1999) described isomeric evolution of aluminophosphate 4MRs (with the **cc0–1:1–2** topology as a backbone) during the formation of microporous solid $AlPO_4$-CJ2 into clusters with the **cc0–1:1–3** topology via polymerization of the Al centers across the rings.

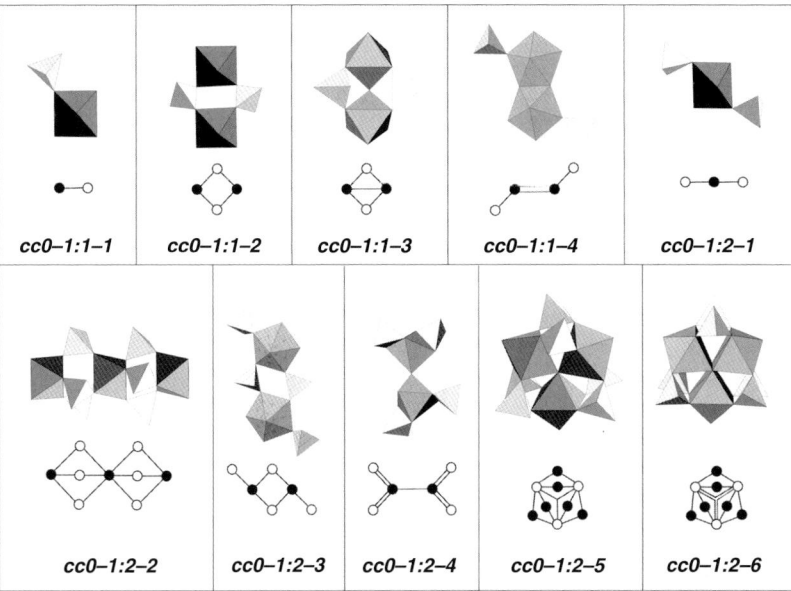

Fig. 2.71. Finite heteropolyhedral clusters in inorganic oxysalts and their black-and-white graphs.

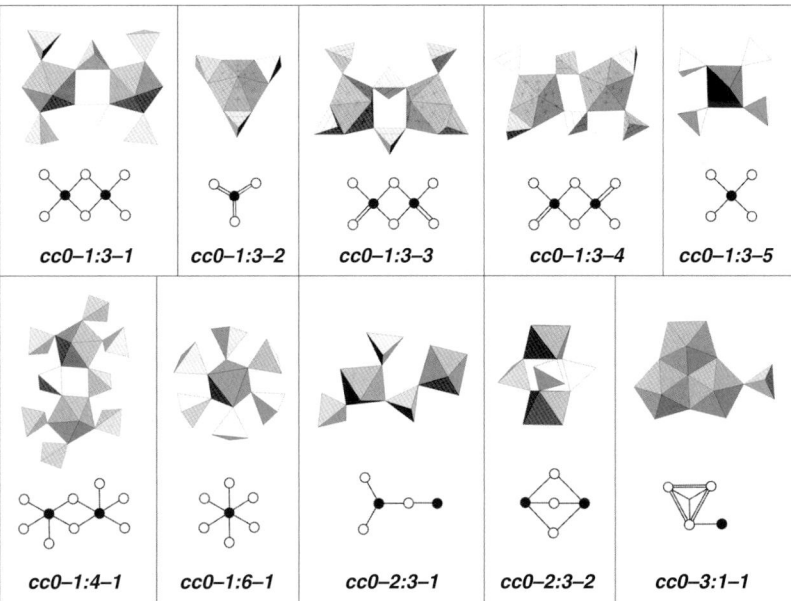

Fig. 2.72. Finite heteropolyhedral clusters in inorganic oxysalts and their black-and-white graphs.

(a)   (b)   (c)

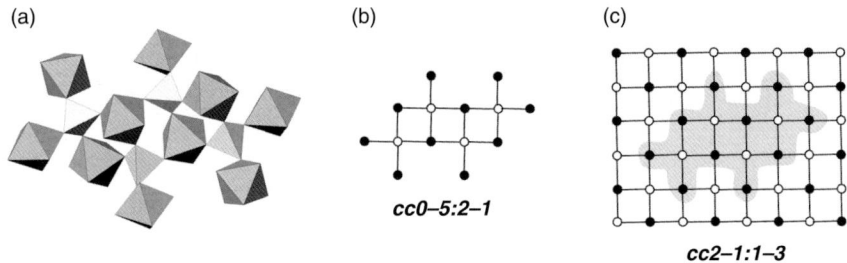

cc0–5:2–1

cc2–1:1–3

Fig. 2.73. The oxyfluorotitanophosphate cluster $[Ti_{10}P_4O_{16}F_{44}]^{16-}$ (a) from the structure of $K_{16}[Ti_{10}P_4O_{16}F_{44}]$, its topology (b) and scheme of its derivation from the autunite *c2–1:1–3* graph.

It is interesting that the graphs depicted in Figs. 2.71 and 2.72 do not contain 3-connected nodes. For this reason, orientational geometric isomerism is not characteristic for these units. However, *cis–trans* isomerism is obviously possible and is observed for octahedral-tetrahedral $[M(H_2O)_4(TO_4)]$ finite clusters described by the *cc0–1:2–1* graph (M = Fe, Mg,; T = S, P). *Trans*-isomers are observed, e.g. in the structures of blödite and anapaite, whereas *cis*-isomers are present in the structures of roemerite and quenstedtite. It is noteworthy that the structure of quenstedtite contains two types of octahedral–tetrahedral clusters described by the graphs *cc0–1:1–1* and *cc0–1:2–1*.

An interesting type of *topological* isomerism of finite clusters is observed in the structures of uranyl sulphates (Fig. 2.72, topologies *cc0–1:3–3* and *cc0–1:3–4*). The $[(UO_2)_2(SO_4)_6]$ clusters consist of two UO_7 pentagonal bipyramids each sharing corners and edges with six SO_4 tetrahedra. The clusters are obviously topological isomers as their graphs cannot be transformed one into another without breaking of their edges. The most reasonable description of the difference of the graphs is that of their *combinatorial symmetry*: the *cc0–1:3–3* graph has a mirror symmetry, whereas the *cc0–1:3–4* graph does not.

2.7   Nanoscale low-dimensional units in inorganic oxysalts: some examples

Since the discovery of carbon nanotubes in 1991 (Iijima 1992), a great deal of attention was attracted to inorganic nanotubes as promising modules for nanotechnology applications. Oxidic nanotubes are of special interest because of their unique atomic structure and interesting physical properties. Krivovichev *et al.* (2005f, g, 2007f, g) and Alekseev *et al.* (2008c) have recently reported the synthesis and structures of uranium(VI) oxysalts containing nanometer-sized tubules formed by corner sharing of $U^{6+}O_7$ pentagonal bipyramids and SeO_4 tetrahedra. Such nanometer-scale tubules formed by two types of coordination polyhedra are new in the realm of inorganic oxosalts; probably, the most closely related but yet distinct are elliptic porous

Table 2.16   Minerals and inorganic oxysalts based upon 0D structural units

Type	Mineral name	Chemical formula	Reference
cc0–1:1–1	minasragrite	$[VO(SO_4)(H_2O)_4](H_2O)$	Tachez *et al.* 1979
	anortho-minasragrite	$[VO(SO_4)(H_2O)_4](H_2O)$	Cooper *et al.* 2003
	xitieshanite	$[Fe(H_2O)_4Cl(SO_4)](H_2O)_2$	Zhou *et al.* 1988
	apjohnite	$[Mn(SO_4)(H_2O)_5][Al(H_2O)_6]_2(SO_4)(H_2O)_5$	Menchetti and Sabelli 1976b
	quenstedtite	$[Fe(H_2O)_4(SO_4)_2][Fe(H_2O)_5(SO_4)](H_2O)_2$	Thomas *et al.* 1974
cc0–1:1–2	rozenite	$[Fe(SO_4)(H_2O)_4]$	Baur 1960
	starkeyite	$[Mg(SO_4)(H_2O)_4]$	Baur 1962
	aplowite	$[(Co,Mn,Ni)(SO_4)(H_2O)_4]$	Kellersohn 1992
	bobjonesite	$VOSO_4(H_2O)_3$	Theobald and Galy 1973; Tachez and Theobald 1980, Tachez *et al.* 1982; Schindler *et al.* 2003
cc0–1:1–3	morinite	$Ca_2Na[Al_2F_4(PO_4)_2(OH)(H_2O)_2]$	Hawthorne 1979
cc0–1:1–4	–	$(UO_2)_4(ReO_4)_2O(OH)_4(H_2O)_{12}$	Karimova and Burns 2007
cc0–1:2–1 *trans*	blödite	$Na_2[Mg(H_2O)_4(SO_4)_2]$	Rumanova and Malitskaya 1959
	anapaite	$[Fe^{2+}(PO_4)_2(H_2O)_4]$	Rumanova and Znamenskaya 1960, Catti *et al.* 1979
	schertelite	$[Mg(PO_3OH)_2(H_2O)_4]$	Khan and Baur 1972
	leonite	$K_2[Mg(H_2O)_4(SO_4)_2]$	Jarosch 1985
	mereiterite	$K_2[Fe(H_2O)_4(SO_4)_2]$	Giester and Rieck 1995
	chenite	$Pb_4(OH)_2[Cu(OH)_4(SO_4)_2]$	Hess *et al.* 1988
		$(NH_4)_3(VO_2(SO_4)_2(OH_2)_2) \cdot 1.5(H_2O)$	Hashimoto *et al.* 2000
		$Na_2Mg(SO_4)_2 \cdot 4(H_2O)$	Vizcayno and Garcia-Gonzalez 1999
		$NaIn(SeO_4)_2(H_2O)_6$	Mukhtarova *et al.* 1977
cc0–1:2–1 *cis*	roemerite	$[Fe(SO_4)_2(H_2O)_4][Fe(H_2O)_6]$	Fanfani *et al.* 1970b
	quenstedtite	$[Fe(H_2O)_4(SO_4)_2][Fe(H_2O)_5(SO_4)](H_2O)_2$	Thomas *et al.* 1974
cc0–1:2–2	coquimbite	$[Fe_3(SO_4)_6(H_2O)_6][Fe(H_2O)_6](H_2O)_6$	Fang and Robinson 1970
	paracoquimbite	$[Fe_3(SO_4)_6(H_2O)_6][Fe(H_2O)_6](H_2O)_6$	Robinson and Fang 1971
		$(Fe(H_2O)_6)(Fe_3(H_2O)_6(SeO_4)_6)(H_2O)_6$	Giester and Miletich 1995
cc0–1:2–3	–	$[C_3H_{12}N_2][(UO_2)(SeO_4)_2(H_2O)_2](H_2O)$	Krivovichev and Kahlenberg 2005c
	–	$(C_3H_{12}N_2)_2[UO_2(H_2O)_2(SO_4)_2]_2(H_2O)_2$	Doran *et al.* 2005a
cc0–1:2–4	–	$Cs_4(VO)_2O(SO_4)_4$	Nielsen *et al.* 1993
cc0–1:2–5	metavoltine	$K_2Na_6[Fe_3O(SO_4)_6(H_2O)_3][Fe(H_2O)_6] (H_2O)_6$	Giacovazzo *et al.* 1976b
	Maus salts	$A_5[Fe_3O(SO_4)_6(H_2O)_3](H_2O)_n$ $A = Na, K, Rb, Tl, H_2O, H_3O, NH_4$	Mereiter, 1980, 1990; Mereiter, Voellenkle, 1978, 1980; Scordari, 1980, 1981; Scordari and Milella, 1983; Scordari and Stasi, 1990a
cc0–1:2–6	–	$K_4H_5O_2[Nb_3O_2(SO_4)_6(H_2O)_3](H_2O)_5$	Bino, 1980, 1982
	–	$(NH_4)_3(H_3O)_2(Nb_3O_2(SO_4)_6(H_2O)_3)(H_2O)_3$	Cotton *et al.* 1986, 1988
cc0–1:3–1	–	$Na[UO_2(ReO_4)(H_2O)_2]$	Karimova and Burns 2007
	–	$[(C_5NH_5)(C_4N_2H_{10})][Zn(H_2PO_4)_2(HPO_4)]$	Natarajan *et al.* 2003
	–	$[C_6N_2H_{18}][Zn(HPO_4)(H_2PO_4)_2]$	Neeraj *et al.* 2000
	–	$[N(CH_3)_4][Zn(H_2PO_4)_3]$	Harrison and Hannooman 1997

Table 2.16 *Continued*

Type	Mineral name	Chemical formula	Reference
cc0–1:3–2	–	$[N_2C_5H_{14}]_2[UO_2(SO_4)_3]$	Doran *et al.* 2003d
cc0–1:3–3	–	$K_4[UO_2(SO_4)_3]$	Mikhailov *et al.* 1977
cc0–1:3–4	–	$Na_{10}[UO_2(SO_4)_4](SO_4)_2(H_2O)_3$	Burns and Hayden 2002
	–	$KNa_5[UO_2(SO_4)_4](H_2O)$	Hayden and Burns 2002a
	–	$Na_6[UO_2(SO_4)_4](H_2O)_2$	Hayden and Burns 2002b
	–	$[N_4C_6H_{22}]_2[(UO_2)_2(SO_4)_6](H_2O)$	Norquist *et al.* 2003a
	–	$[N_2C_4H_{14}]_2[UO_2(SO_4)_3](H_2O)_2$	Norquist *et al.* 2003c
	–	$[Co(NH_3)_6](H_8O_3)[NpO_2(SO_4)_3]$	Grigor'ev *et al.* 1991b
cc0–1:4–1	polyhalite	$K_2Ca_2[Mg(SO_4)_4(H_2O)_2]$	Schlatti *et al.* 1970
	–	$Na_6(Zn(SO_4)_4(H_2O)_2)$	Heeg *et al.* 1986
	–	$Na_2Co(H_2PO_4)_4 \cdot 4(H_2O)$	Guesmi *et al.* 2000
	–	$K_8[(UO_2)(CrO_4)_4](NO_3)_2$	Krivovichev and Burns 2003a
	–	$Rb_6[(UO_2)(MoO_4)_4]$	Krivovichev and Burns 2002a
	–	$Cs_6[(UO_2)(MoO_4)_4]$	Krivovichev and Burns 2002b
cc0–1:4–2	–	$Na_6[(UO_2)(MoO_4)_4]$	Krivovichev and Burns 2001b
	–	$Na_3Tl_3[(UO_2)(MoO_4)_4]$	Krivovichev and Burns 2003b
cc0–1:6–1	ungemachite	$K_3Na_8[Fe(SO_4)_6](NO_3)_2(H_2O)_6$	Groat and Hawthorne 1986
	humberstonite	$K_3Na_6(Na, Mg)_2[Mg(SO_4)_6](NO_3)_2(H_2O)_6$	Burns and Hawthorne 1994
	–	$K_7M(SO_4)_6$ M = Nb, Ta	Borup *et al.* 1990
cc0–2:3–1	–	$Al_2(SO_4)_3(H_2O)_8$	Fischer *et al.* 1996f
cc0–2:3–2	–	$Al_2(SeO_3)_3 \cdot 6H_2O$	Morris *et al.* 1991, 1992
	–	$Al_2(H_2PO_4)_3(H_2O)_6(PO_4)$	Kniep and Wilms 1979
cc0–3:1–1	–	$(UO_2)_4(ReO_4)_2O(OH)_4(H_2O)_{12}$	Karimova and Burns 2007
cc0–5:2–1	–	$K_8[Ti_5(PO_4)_2F_{22}]$	Yang *et al.* 2007

nanorods in the structure of yuksporite, a natural material from the Kola peninsula, Russia (see below). The uranyl selenate nanotubules in uranyl selenates have circular cross-sections with outer diameters of either 17 and ~25 Å (=1.7 and 2.5 nm). The crystallographic free diameters of the tubules are 4.7 and 12.7 Å, respectively.

The black-and-white graph corresponding to the topological structure of the $[(UO_2)_3(SeO_4)_5]^{4-}$ tubule in $K_5[(UO_2)_3(SeO_4)_5](NO_3)(H_2O)_{3.5}$ (Fig. 2.74; Krivovichev *et al.* 2005f) and $(H_3O)_2K[(H_3O)@(18$-crown-6)]$[(UO_2)_3(SeO_4)_5](H_2O)_4$ (Alekseev *et al.* 2008c) is shown in Fig. 2.75(a). Its idealized unfolded version is given in Fig. 2.75(b) and it is obviously the *cc2–3:5–2* graph described above (see Table 2.4 and Fig. 2.16). To obtain the planar graph corresponding to the$[(UO_2)_3(SeO_4)_5]^{4-}$ tubule, one has to cut the graph into tapes along the lines indicated in Fig. 2.75(b), to fold the tape and to glue its corresponding sides (Fig. 2.75(c)). The same procedure can be used to investigate the local topology of the tubules observed in the structure of $(C_4H_{12}N)_{14}[(UO_2)_{10}(SeO_4)_{17}(H_2O)]$ (Krivovichev *et al.* 2005g). The black-and-white graph corresponding to the topological structure of the $[(UO_2)_{10}(SeO_4)_{17}(H_2O)]^{14-}$ tubule in the latter compound is shown in Fig. 2.76(a). Its idealized unfolded version is given in Fig. 2.76(b).

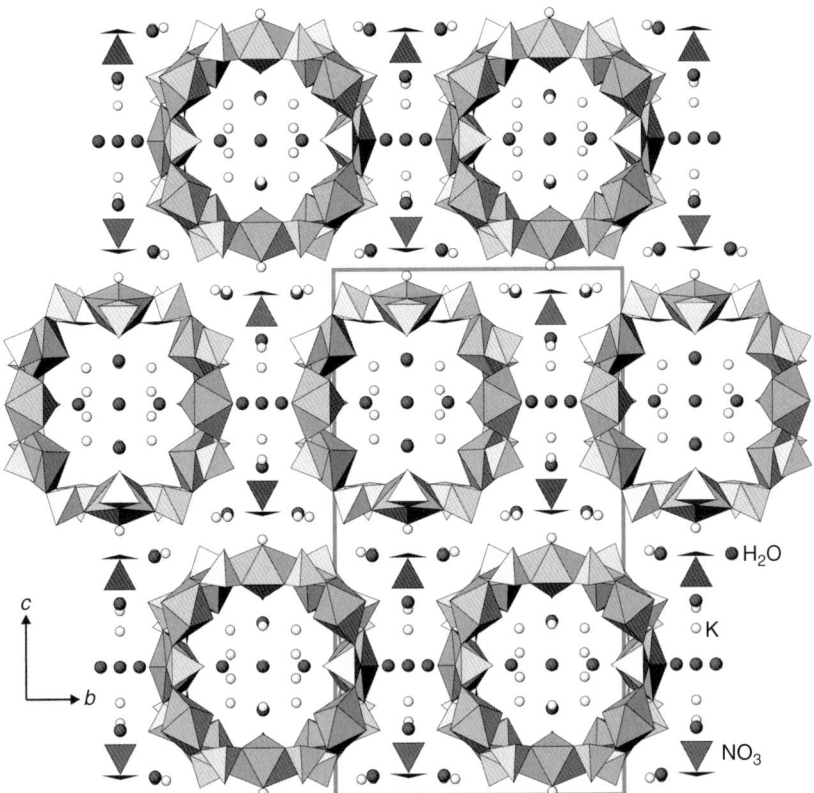

Fig. 2.74. Crystal structure of $K_5[(UO_2)_3(SeO_4)_5](NO_3)(H_2O)_{3.5}$ projected along the extension of uranyl selenate nanotubules.

Most of the known inorganic nanotubes have a prototype lamellar material from which they can be (at least theoretically) obtained by exfoliation and folding of single-layer sheets into a tube. The same holds for the uranyl selenate tubules shown in Fig. 2.75(a). The *cc2–3:5–2* graph shown is an underlying topology for the $[(UO_2)_3(SeO_4)_5]^{4-}$ sheets found in the structures of a number of uranyl selenates (see Table 2.4). However, there are some complications that arise due to the richness of isomeric variations in this class of 2D units. We recall that isomerism in this class of structural units arises due to the possibility of different orientations of non-shared corners of tetrahedra relative to the plane of the sheet. In planar (2D) units, the ratio **u:d** of the "up" (**u**) and "down" (**d**) orientations is 1:1. In contrast, for the selenate tetrahedra forming nanotubules, the obtained **u:d** ratio is 4:1 (here **u** orientations are those pointing *outside* the tubule). Thus, only one out of five tetrahedra is oriented inside the tubule. The possible reason is the tendency of the tubule to contain as much additional species (e.g. cations) as possible that results in lower negative charge of the "filled" tube.

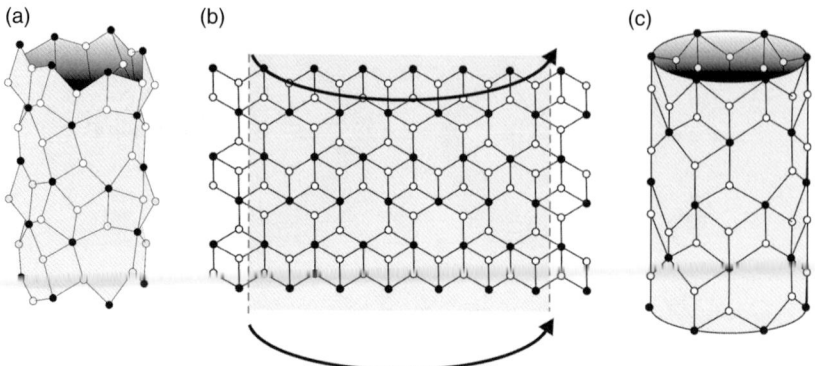

Fig. 2.75. Tubular graph describing topology of the uranyl selenate nanotubule in $K_5[(UO_2)_3$ $(SeO_4)_5](NO_3)(H_2O)_{3.5}$ (a), its unfolded version (b), and ideal version obtained from the 2D graph by the folding-and-gluing procedure (c).

The mechanism of formation of uranyl selenate tubules in aqueous media is probably controlled for $(C_4H_{12}N)_{14}[(UO_2)_{10}(SeO_4)_{17}(H_2O)]$ by the presence of protonated butylamine $(C_4H_{12}N)^+$ cations. In this regard, it can be similar to the process that leads to the formation of highly undulated uranyl selenate sheets in the structure of $(H_3O)_2[C_{12}H_{30}N_2]_3[(UO_2)_4(SeO_4)_8](H_2O)_5$ (Krivovichev *et al.* 2005d) (Fig. 2.77(a)). Here, the inorganic substructure consists of the UO_7 bipyramids and SeO_4^{2-} tetrahedra that share corners to form $[(UO_2)(SeO_4)_2]^{2-}$ sheets depicted in Fig. 2.7(a) (graph ***cc2–1:2–13***). The sheets are parallel to (001) and are strongly undulated along the *c*-axis. The undulation vector is parallel to [010] and equals to $b = 24.804$ Å. The undulation amplitude is about 25 Å. The undulations in the adjacent sheets have an antiphase character so the large eliptical channels are created along the *a*-axis. The organic substructure consists of micelles of protonated 1,12-dodecanediamine molecules oriented parallel to the *a*-axis. The micelles occupy channels created by the packing of the $[(UO_2)(SeO_4)_2]^{2-}$ sheets. The scheme of assembly of the 1,12-dodecanediamine chain molecules within the micelle is shown in Fig. 2.77(b). The molecules are arranged into sublayers approximately parallel to (−102). The planes of these sublayers are not perpendicular to the micelle axis but form an angle of about 60°. In each layer, there are three $[C_{12}H_{30}N_2]^{2+}$ molecules that are parallel to each other. The molecules in the adjacent layers are at a 30° angle relative to each other, and this results in the elliptical form of the perpendicular section of the micelle. The lateral dimensions of the micelle are about 20×24 Å, i.e. are at the level of nanometers. The interactions between organic and inorganic substructures involve N⋯O hydrogen bonds to oxygen atoms of uranyl ions and terminal oxygen atoms of selenate tetrahedra.

The protonated amine molecules with long-chain structure are known to form cylindrical micelles in aqueous solutions that involves self-assembly governed by competing hydrophobic/hydrophillic interactions. The flexible inorganic complexes present in the reaction mixture could then form around cylindrical micelles to produce

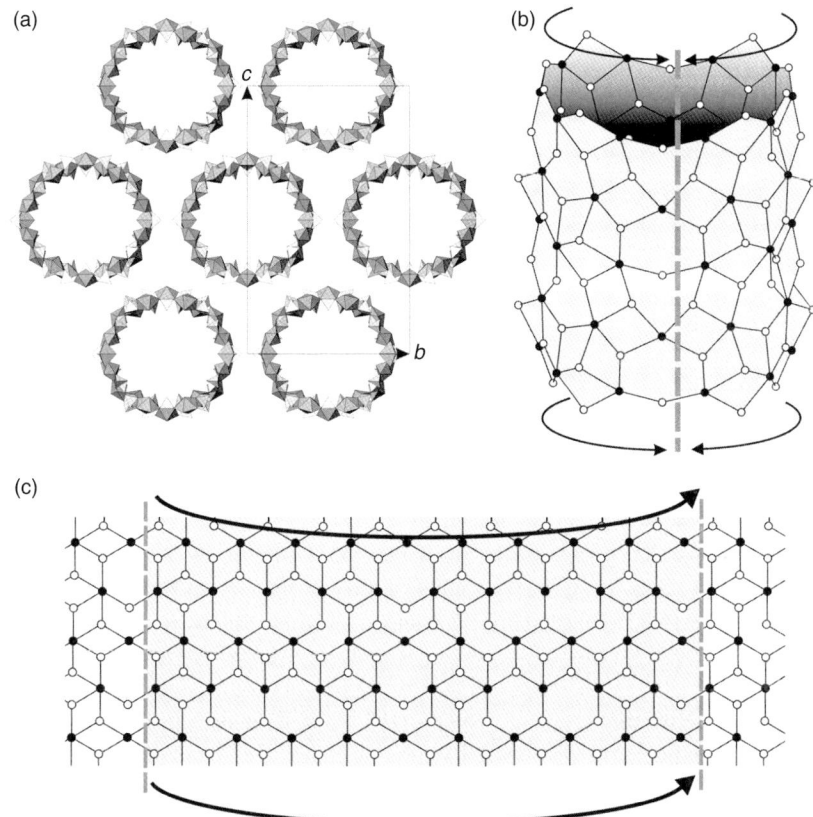

Fig. 2.76. Orientation of uranyl selenate nanotubules in the structure of $(C_4H_{12}N)_{14}[(UO_2)_{10}$ $(SeO_4)_{17}(H_2O)]$ (a), tubular graph describing topology of the uranyl selenate nanotubule (b) and its unfolded version (c). Note that the graph shown in (c) is locally topologically isomorphous to the graph shown in (b).

an inorganic structure that reflects the cylindrical form of the micelles. In the case of $(H_3O)_2[C_{12}H_{30}N_2]_3[(UO_2)_4(SeO_4)_8](H_2O)_5$, the inorganic structure forms a strongly undulated sheet, whereas, in the case of $(C_4H_{12}N)_{14}[(UO_2)_{10}(SeO_4)_{17}(H_2O)]$, formation of highly porous uranyl selenate nanotubules has been observed (see above).

Another interesting example of nanoscale 1D structural units is yuksporite, a rare mineral from the Kola peninsula, Russia (Krivovichev *et al.* 2004e). The structure of yuksporite (Fig. 2.78) is based upon complex rods consisting of corner-sharing TiO_6 octahedra and SiO_4 tetrahedra (Fig. 2.79). The rods are parallel to the *a*-axis and have an elliptical cross-section measuring ca. 16×19 Å $= 1.6 \times 1.9$ nm. The structure of the rods is remarkable and is unique among both synthetic and natural inorganic oxysalts. Silicate tetrahedra form condensed silicate anions of two different types.

(a)   (b)

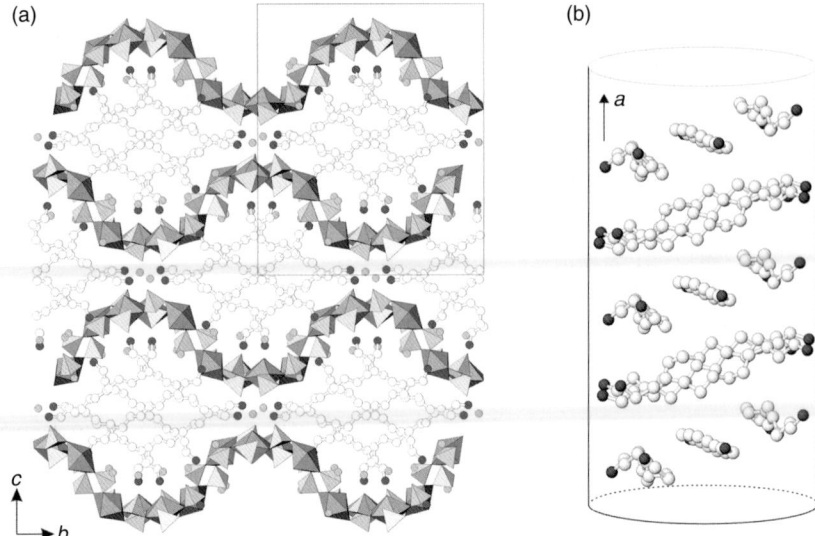

Fig. 2.77. Structure of $(H_3O)_2[C_{12}H_{30}N_2]_3[(UO_2)_4(SeO_4)_8](H_2O)_5$ (a) and the scheme of self-assembly of the 1,12-dodecanediamine chain molecules within the micelle (b).

The xonotlite-like double chains $^1_\infty[Si_6O_{17}]$ (Fig. 2.79(c)) are parallel to the a-axis and oriented perpendicular to the c-axis. Two double chains are linked into a rod via TiO_6 and double Si_2O_7 tetrahedra (Fig. 2.79(e)). Another Si_2O_7 group is located within the rod and is oriented parallel to the b-axis. This group provides linkage of two opposite walls of the rod along the b-axis (Fig. 2.79(b)). A total composition of the rods is $\{(Ti,Nb)_4(O,OH)_4[Si_6O_{17}]_2[Si_2O_7]_3\}$. The nanorods are porous. The internal pores are defined by eight-membered rings (8MR) oriented perpendicular to (100). Their free diameter, estimated as a distance between oxygen atoms across the ring minus 2.7 Å (two oxygen radii taken as 1.35 Å) is 3.2 Å. These 8MRs consist of two octahedra and six tetrahedra each (Fig. 2.79(b)). Within the rod, the 8MRs are arranged in two parallel internal channels separated by the Si_2O_7 groups. From the outside, rods are bounded by 8MRs formed solely by silicate tetrahedra (Fig. 2.79(c)). The plane of these rings is parallel to (001); their apertures have an elliptical section with free diameters of 2.3×4.4 Å. Other outside rings are 6MRs shown in Fig. 2.79(e). These rings consist of two octahedra and four tetrahedra and have a free diameter of ca. 2.3 Å. The interior of the titanosilicate nanorods in yuksporite is occupied by alkali-metal cations (Na1, Na2, K1-K5) and H_2O molecules. In the structure, the $\{(Ti,Nb)_4(O,OH)_4[Si_6O_{17}]_2[Si_2O_7]_3\}$ rods are separated by walls built up from Ca coordination polyhedra. The walls are parallel to (010) and provide linkage of the rods into a three-dimensional structure. The structural formula of yuksporite is rather complex and can be written as $(Sr,Ba)_2K_2(Ca,Na)_{14}(\square,Mn,Fe)\{(Ti,Nb)_4(O,OH)_4[Si_6O_{17}]_2[Si_2O_7]_3\}$ $(H_2O,OH)_n$, where $n \sim 3$ ($\square$ = vacancy).

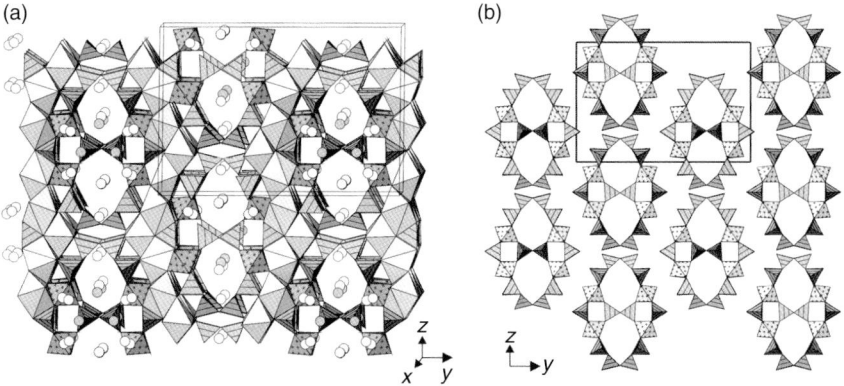

Fig. 2.78. Crystal structure of yuksporite (a) and arrangement of titanosilicate nanorods along the *a*-axis (b). Legend: TiO_6 octahedra = cross-hatched; SiO_4 tetrahedra = lined; $Ca\phi_n$ polyhedra = shaded; K, Sr, and Na cations are shown as white circles; H_2O groups are shown as gray circles.

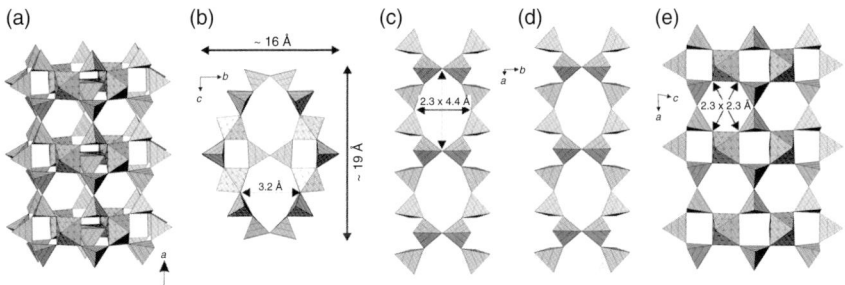

Fig. 2.79. Structure of titanosilicate nanorods in yuksporite: view from the side (a) and from the top (b); two xonotlite-like $[Si_6O_{17}]$ chains ((c), (d)) form walls of nanorods perpendicular to the *c*-axis; TiO_6 octahedra, Si_2O_7 groups and halves of the xonotlite chains form walls of nanorods perpendicular to the *b*-axis. Legend: TiO_6 octahedra = cross-hatched; SiO_4 tetrahedra = lined.

We note that many of the clusters described above in Section 2.6 can also be considered as being at the nanoscale. It is probably their self-assembly that leads to the formation of complicated 3D topologies that will be described in the next chapter. With the introduction of more sophisticated experimental techniques, there is a strong hope that more nanoscale units in inorganic oxysalts will be determined that will shed new light into the complexity of self-organization processes in this class of chemical compounds.

3

Topology of framework structures
in inorganic oxysalts

Framework solids are of interest both from the viewpoint of their industrial applications as porous materials and from the viewpoint of topological chemistry. Before the 1980s, the most studied framework oxysalts were zeolites and related tetrahedral materials. The shift of interest from zeolites and related structures to oxysalts with non-tetrahedral cations (with coordination numbers of five or more) opened a wide range of new possibilities for synthesis and exploration. The chemical range of possible compositions was significantly extended and, from the structural point of view, this chemical extension resulted in endless topological opportunities of framework construction.

The aim of this chapter is to provide a review of the possible approaches that can be used for topological and geometrical analysis of framework structures in inorganic oxysalts. Zeolites and related tetrahedral materials are not considered here, since their detailed topological descriptions can be found in many reviews and reference books (Smith 1988, 2000; Baerlocher *et al.* 2001, etc.).

3.1 Regular and quasiregular nets

There are several important high-symmetrical 3D nets that represent the underlying topologies of a number of inorganic and metal-organic frameworks. Delgado-Friedrichs *et al.* (2003) considered 3D nets with the following properties: (i) all vertices are of the same type; (ii) a convex hull of the coordination figure of a vertex in a net is a regular polygon or a regular polyhedron (i.e. one of five Platonic solids); (iii) site symmetry of a vertex in the net is at least the rotation symmetry of that regular polygon or polyhedron. Nets that fulfill these three conditions are called *regular*. There are exactly five types of regular nets that correspond to some regular polygons and polyhedra: equilateral triangle, square, tetrahedron, octahedron and cube (Fig. 3.1). These nets are designated as **pcu, bcu, srs, nbo** and **dia**. The **pcu** and **bcu** nets correspond to the **p**rimitive **cu**bic lattice and **b**ody-centered **cu**bic lattice, respectively. The **srs, nbo,** and **dia** nets correspond to structural topologies of SrS, NbO, and diamond, respectively. The **fcu** net shown in Fig. 3.1 is *quasiregular* and corresponds to the **f**ace-centered **cu**bic lattice.

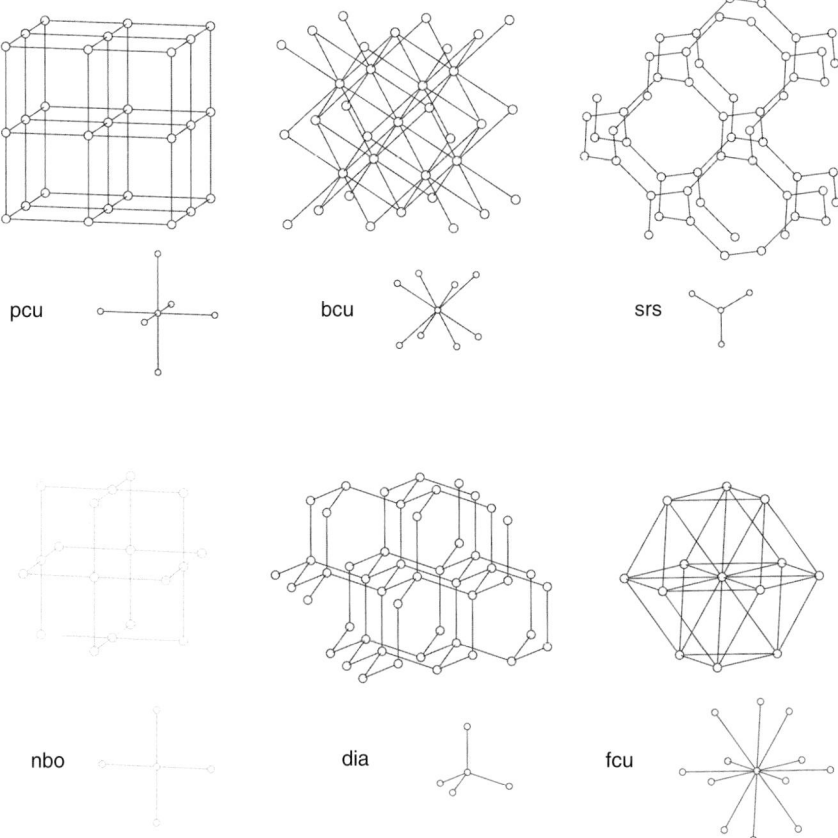

Fig. 3.1. Regular and quasiregular nets.

The regular and quasiregular nets themselves are relatively dense. However, the distances between the vertices of these nets can be considerably increased by topological operations identified by O'Keeffe *et al.* (2000) as *decoration, augmentation* and *expansion.*

Decoration is replacement of a vertex by a group of vertices.

Augmentation is a special case of decoration and involves replacement of vertices of *n*-connected net by a group of *n* vertices.

Expansion is increasing of the length of an edge by a group of edges. This process is very common for metal-organic frameworks (O'Keeffe *et al.* 2000).

Schindler *et al.* (1999) considered many examples of correspondence (in mathematical terminology, homeomorphisms) between simple 3D nets and complex inorganic frameworks. They suggest to call the complex structures that can be obtained by transformations of simple 3D nets *metastructures*. The transformations considered by Schindler *et al.* (1999) are in fact those recognized by O'Keeffe *et al.* (2000) as decoration, augmentation and expansion.

Regular and quasiregular nets are those of the simplest 3D nets. These nets dominate the topologies of metal-organic solids with extremely large pores as discussed by O'Keeffe *et al.* (2000), Eddaoudi *et al.* (2001). However, they are not so common in structures of inorganic oxysalts owing to the high complexity of their polyhedral arrangements.

As an example of the expansion process, let us consider the regular **pcu** net (Fig. 3.1). Insertion of a white vertex at the midpoint between two black vertices at the nodes of this net results in an expanded **pcu** net shown in Fig. 3.2(a). In fact, this net is an underlying topology for octahedral–tetrahedral frameworks observed in $Cs_3Fe_3H_{15}(PO_4)_9$ (Anisimova *et al.* 1996) and $NaFe(HPO_4)(H_2PO_4)_2 \cdot H_2O$ (Anisimova *et al.* 1998) (Fig. 3.2(c)), as well as for octahedral–triangular frameworks in $M(HSeO_3)_3$ (M = Sc (Valkonen and Leskelae 1978), Fe (Muilu and Valkonen 1987) and $Ca_3Fe_2(SeO_3)_6$ (Giester 1996). Note that, for these frameworks, the $M:T$ ratio is 1:3. Insertion of an edge consisting of two white vertices (Fig. 3.2(b)) results in a more porous network – this graph is the basis of the zirconosilicate framework in $Sr_7Zr(Si_2O_7)$ with Zr:Si = 1:6 (Plaisier *et al.* 1994) (Fig. 3.2(d)).

3.2   Heteropolyhedral frameworks: classification principles

Within a nodal approach to complex heteropolyhedral frameworks, each 3D framework is symbolized by a 3D graph with n different vertices, where n is the number of topologically and geometrically different coordination polyhedra. In the simple case of $n = 2$, we have vertices of two types only: black and white. The simplest example is an octahedral–tetrahedral framework that corresponds to a 3D net with black and white vertices symbolizing octahedra and tetrahedra, respectively.

In each black-and-white 3D net associated with an octahedral–tetrahedral framework we can subdivide a connected subnet of black vertices (that corresponds to an octahedral substructure within the framework) and a connected subnet of white vertices (= tetrahedral substructure). For instance, if there are no edges connecting black vertices, the subnet of black vertices is a single vertex and the octahedral substructure is a single octahedron. If octahedra are linked into a chain, the corresponding black vertices form a chain.

We recall that the dimensionality (D) of a graph is defined as the number of dimensions in which it has an infinite extension. For a 3D net corresponding to an octahedral–tetrahedral framework, we define D_{nt} as a dimensionality of a non-tetrahedral substructure (black subgraph) and D_{tetr} as a dimensionality of a tetrahedral substructure (white subgraph). Thus, we can classify 3D nets according to the pairs of values $(D_{nt}; D_{tetr})$. The classification can be represented as a 4 × 4 table where rows correspond to particular values of D_{nt} and columns to particular values of D_{tetr}. Table 3.1 presents an example of the classification of minerals and inorganic compounds

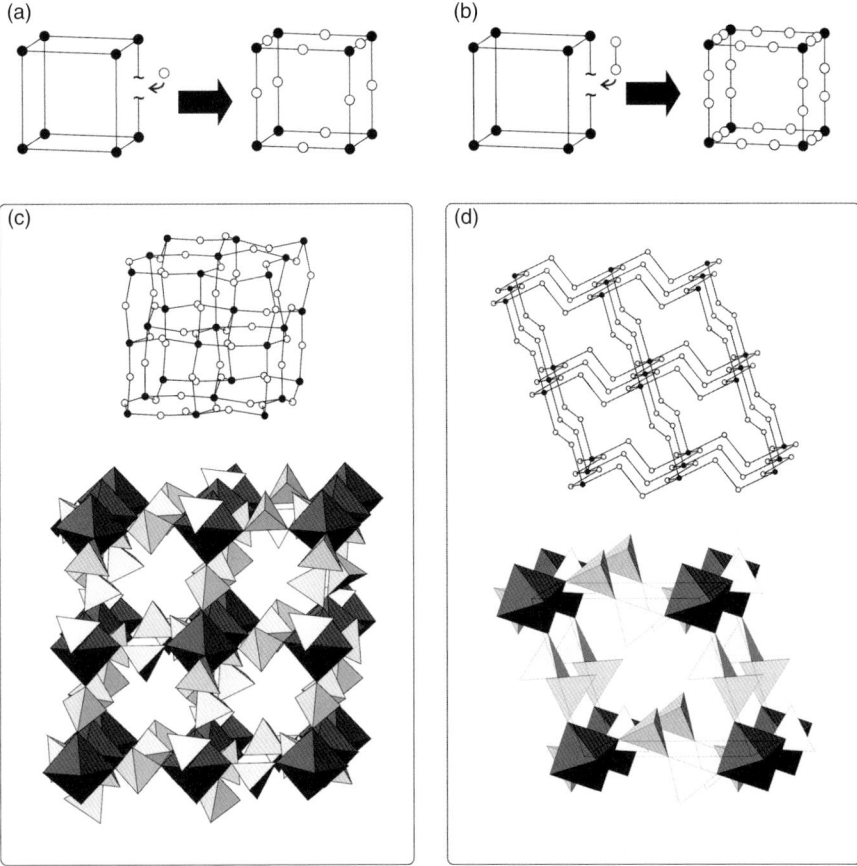

Fig. 3.2. Expansion of the **pcu** regular net by insertion single (a) and double (b) white vertices into the edges. The obtained nets are underlying topologies for heteropolyhedral frameworks in $Cs_3Fe_3H_{15}(PO_4)_9$ (c) and $Sr_7Zr(Si_2O_7)$ (d).

with mixed frameworks according to the (D_{nt}; D_{tetr}) pairs. Table 3.1 provides only very general and formal classification of mixed frameworks. For detailed topological systematics, one has to find an appropriate description for each 3D net that provides its simple construction and allows its comparison with other nets. It is convenient to subdivide frameworks given in Table 3.1 into five major groups: (i) frameworks based upon fundamental building blocks (FBBs); (ii) frameworks assembled from polyhedral units; (iii) frameworks based upon 1D units (fundamental chains or tubular units), (iv) frameworks based upon 2D units, and (v) frameworks consisting of broken tetrahedral subframeworks.

Table 3.1. Classification of materials based upon frameworks of tetrahedra and non-tetrahedrally coordinated high-valent metal cations. Some heteropolyhedral framework silicates are given as examples.

$D_{tetr}D_{nt}$	0	1	2	3
0	wadeite, benitoite, kostylevite, petarasite, keldyshite	hilairite, umbite, terskite, vlasovite, elpidite, gaidonnayite, zektzerite	lemoynite, armstrongite, $Cs_2(ZrSi_6O_{15})$, $Na_{1.8}Ca_{1.1}{}^{VI}Si({}^{IV}Si_5O_{14})$, penkvilksite	–
1	belkovite, labuntsovite-group minerals	batisite-shcherbakovite, zorite, ETS-4	USH-8 = $[(CH_3)_4N]$ $[(C_5H_5NH)_{0.8}((CH_3)_3NH)_{0.2}]$ $(UO_2)_2[Si_9O_{19}]F_4$	ETS-10
2	komarovite	–	–	–
3	–	–	–	–

3.3   Frameworks based upon fundamental building blocks (FBBs)

3.3.1   *Some definitions*

A preferred approach to describe a crystal structure is in terms of coordination polyhedra of cations. A coordination polyhedron can be considered as a *basic building unit* (BBU) following the IUPAC nomenclature (McCusker *et al.* 2001) or a *structural subunit* if one follows the terminology recommended by the International Union of Crystallography (Lima-de-Faria *et al.* 1990). In some cases, crystal structure can be conveniently described as based upon clusters of coordination polyhedra usually called *fundamental building blocks* (FBBs) (Hawthorne 1994). An alternative term is *a composite building unit* (CBU) that is used as a second level of structural hierarchy (the first is a BBU, which is a single polyhedron) (McCusker *et al.* 2001). Description of inorganic structures in terms of FBBs is especially justified if the same FBB occurs in many related structures. This indicates the possible role of FBBs as clusters pre-existing in a crystallization media (Férey 2001). FBBs (or CBUs) are effectively used for the description of Mo phosphates (Haushalter and Mundi 1992) and a large family of open-framework Fe, Al and Ga phosphates with Fe, Al and Ga in non-tetrahedral coordinations (materials known as ULM-*n* and MIL-*n*; see Férey 1995, 1998, 2001; Riou-Cavellec *et al.* 1999 for reviews).

An interesting feature of frameworks based upon FBBs is that FBBs are frequently arranged according to positions of nodes of simple regular nets discussed above. Below, we consider some examples of frameworks based upon FBBs of various shapes and compositions.

3.3.2   *Leucophosphite-type frameworks*

Figures 3.3(a) and (b) show clusters consisting of four octahedra and six tetrahedra. The octahedra share edges and corners to form octahedral tetramers surrounded

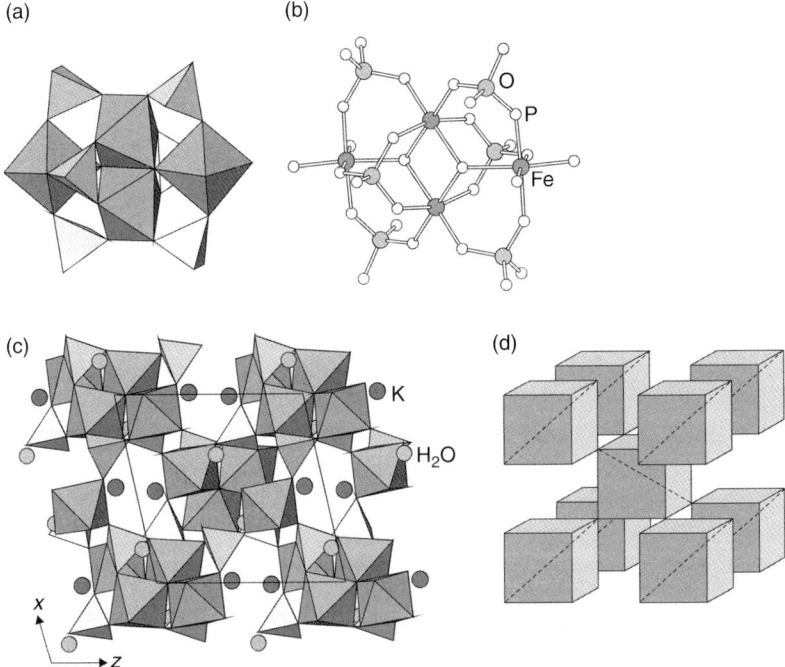

Fig. 3.3. "Butterfly-shaped" cluster (FBB) consisting of four octahedra and six tetrahedra shown in polyhedral (a) and ball-and-stick (b) aspects. The structure of leucophosphite, $K_2[Fe_4(OH)_2(H_2O)_2(PO_4)_4](H_2O)_2$ (c) and the scheme demonstrating arrangement of FBBs in leucophosphite (d).

by tetrahedra sharing their corners with octahedra. This "butterfly-shaped" cluster occurs as a FBB in a number of framework phosphates (Férey 1995, 1998, 2001). Figure 3.3(c) shows the structure of leucophosphite, $K_2[Fe_4(OH)_2(H_2O)_2(PO_4)_4]$ $(H_2O)_2$ (Moore 1972). In this structure, "butterfly-shaped" FBBs have composition $[Fe_4(OH)_2(H_2O)_2(PO_4)_6]^{8-}$ and are linked into the framework by sharing corners of PO_4 tetrahedra with FeO_6 octahedra of adjacent clusters. The arrangement of FBBs in leucophosphite is schematically shown in Fig. 3.3(d). As is clearly seen, its topology corresponds to the **bcu** regular net (Fig. 3.1). It is noteworthy that the leucophosphite structure type based upon the highly symmetrical regular net is common among minerals and inorganic compounds. It has been observed for oxysalts chemically analogous to leucophosphite: tinsleyite $K_2[Al_4(OH)_2(H_2O)_2(PO_4)_4](H_2O)_2$ (Dick 1999), spheniscidite $(NH_4)[Fe_2(OH)(H_2O)(PO_4)_2](H_2O)$ (Yakubovich and Dadachov 1992), $NH_4[M_2(OH)(H_2O)(PO_4)_2](H_2O)$ (M = Al, (Pluth and Smith 1984), V (Soghomonian et al. 1998), Ga (Loiseau and Férey 1994)), $[Ga(PO_4)(H_2O)](H_2O)$ (Mooney-Slater 1966), etc. It is also common for alkali-metal Mo phosphates of general formula

$A_n[Mo_2O_2(PO_4)_2](H_2O)_m$ ($n = 1$–2; $m = 0$–1) (King *et al.* 1991a; Guesdon *et al.* 1993; Leclaire *et al.* 1994).

3.3.3 *Frameworks with oxocentered tetrahedral cores*

Figure 3.4 shows the structure of $(C_2H_{10}N_2)_2[Fe_4O(PO_4)_4](H_2O)$ (de Bord *et al.* 1997; Song *et al.* 2003) that is based upon a pentahedral–tetrahedral cluster consisting of four FeO_5 trigonal bipyramids and four PO_4 tetrahedra (Figs. 3.4(a) and (b)). It is noteworthy that four FeO_5 polyhedra share the same O atom that can be considered as being at the center of a tetrahedron formed by four Fe atoms (Fig. 3.4(c)). The $[Fe_4O(PO_4)_4]$ FBBs are linked by sharing O corners with adjacent clusters (Fig. 3.4(d)). The framework topology corresponds to the **bcu** regular net.

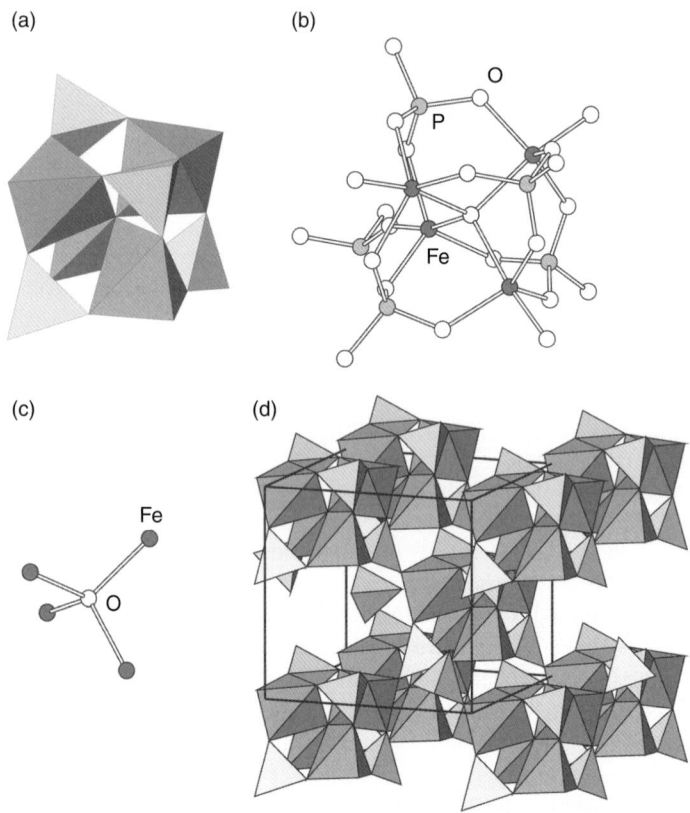

Fig. 3.4. Pentahedral–tetrahedral cluster consisting of four FeO_5 trigonal bipyramids and four PO_4 tetrahedra shown in polyhedral (a) and ball-and-stick (b) aspects, and the Fe_4O oxo-centered tetrahedron (c) as a core of the cluster. Arrangement of the $[Fe_4O(PO_4)_4]$ clusters in the structure of $(C_2H_{10}N_2)_2[Fe_4O(PO_4)_4](H_2O)$ (d).

Harrison *et al.* (1996, 2000) reported a series of microporous materials with general formula $M_3[Zn_4O(XO_4)_3](H_2O)_n$ (M = Na, K, Rb, Cs; X = P, As; n = 3–6). The structures of these materials are based upon the clusters shown in Figs. 3.5(a) and (b). The core of the $[Zn_4O(XO_4)_6]$ cluster is an oxocentered OZn_4 tetrahedron similar to the OFe_4 tetrahedron in $(C_2H_{10}N_2)_2[Fe_4O(PO_4)_4](H_2O)$ (de Bord *et al.* 1997) (Fig. 3.5(c)). However, in contrast to the latter structure, Zn atoms in $M_3[Zn_4O(XO_4)_3]$ $(H_2O)_n$ are tetrahedrally coordinated. The $[Zn_4O(XO_4)_6]$ clusters are linked through common XO_4 groups to form a 3D framework with topology of the **pcu** regular net (Figs. 3.5(d) and (e)). Note that the function of the XO_4 groups in the organization of the $[Zn_4O(XO_4)_3]$ is to link adjacent OZn_4 tetrahedra into a 3D network. Replacement of the XO_4 groups by 1,4-benzenedicarboxylate (BDC) groups, $C_8H_4O_4$ leads to the

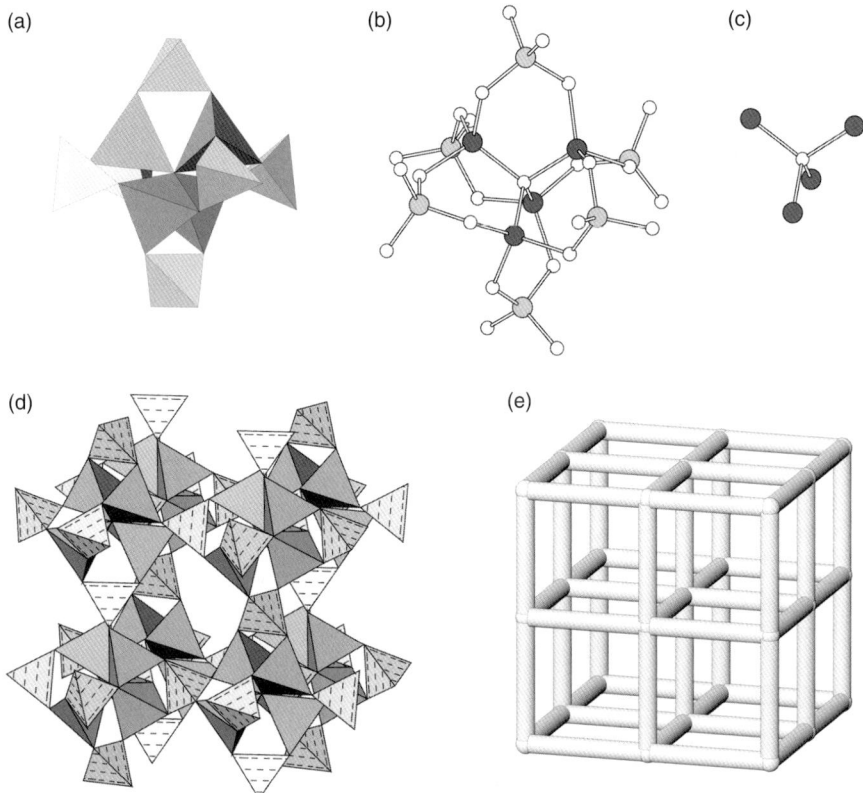

(a)　　　　　　　　　(b)　　　　　　　　　(c)

(d)　　　　　　　　　(e)

Fig. 3.5. The $[Zn_4O(XO_4)_6]$ cluster from the structures of $M_3[Zn_4O(XO_4)_3](H_2O)_n$ (M = Na, K, Rb, Cs; X = P, As; n = 3–6) shown in polyhedral (a) and ball-and-stick (b) aspects, and the Zn_4O oxocentered tetrahedron (c) as a core of the cluster. The $[Zn_4O(XO_4)_6]$ clusters are linked through common XO_4 groups to form a 3D framework (d) with topology of the **pcu** regular net (e).

formation of an *expanded* framework material $Zn_4O(BDC)_3$ with unusually large pores (Li *et al.* 1999).

3.3.4 *Pharmacosiderite-related frameworks*

Minerals of the pharmacosiderite group have the general formula $A_x[M_4(OH)_4(AsO_4)_3]$ $(H_2O)_y$, where A = K, Na, Ba; x = 0.5–1; M = Al, Fe^{3+}, y = 6–7 (Strunz and Nickel 2001). The structure of pharmacosiderite (A = K, M = Fe^{3+}) was reported by Zemann (1948) and refined by Duugei *et ul.* (1967). It is based upon an octahedral–tetrahedral framework that consists of octahedral tetramers $M_4(O,OH)_{16}$ linked via isolated AsO_4 tetrahedra. The FBB of this structure is shown in Figs. 3.6(a) and (c). It is interesting to note that the core of this FBB is a cubic $M_4(OH)_4^{8+}$ polycation consisting of four M^{3+} cations and four OH^- groups. The FBBs are linked into the 3D framework of the **pcu** regular net topology (Fig. 3.6(d)). Zemann (1959) pointed out that pharmacosiderite is isostructural to a family of open-framework germanates with general

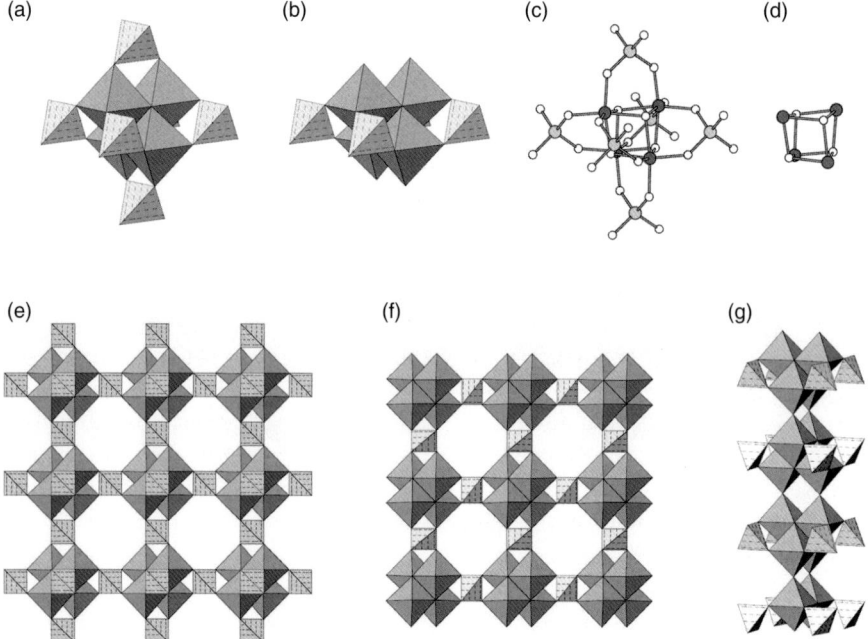

(a)   (b)   (c)   (d)

(e)   (f)   (g)

Fig. 3.6. Octahedral–tetrahedral clusters in pharmacosiderite-related frameworks ((a), (b), (c)) and $M_4(OH)_4^{8+}$ polycation consisting of four M^{3+} cations and four OH^- groups at the core of these clusters (d). The structures of the minerals of the pharmacosiderite group are based upon the 3D framework of the **pcu** regular net topology (e). $[M_4(O,OH)_4(XO_4)_4]$ FBBs (b) in the structures of $A_2[Ti_2O_3(SiO_4)](H_2O)_n$ (A = Na, H) are linked into a 3D framework (e) consisting of chains shown in (f).

formula $A_xH_y[Ge_7O_{16}](H_2O)_z$ (A = NH_4, Na, K, Rb, Cs; $x+y$ = 4; z = 0–4) (Nowotny and Wittman 1954; Sturua *et al.* 1978; Bialek and Gramlich 1992; Roberts *et al.* 1995; Roberts and Fitch 1996). In germanates, the framework contains Ge atoms in both tetrahedral and octahedral coordinations; thus, its formula can be written as $[^{VI}Ge_4O_4(^{IV}GeO_4)_3]$. The pharmacosiderite-type framework was also observed in the structures of $A_3[Mo_4O_4(PO_4)_3]$ (A = NH_4, Cs) (Haushalter 1987; King *et al.* 1991b). In 1990, Chapman and Roe prepared a number of titanosilicate analogs of pharmaco-siderite, including Cs, Rb and exchanged protonated phases. Harrison *et al.* (1995a) reported structure of $Cs_3H[Ti_4O_4(SiO_4)_3](H_2O)_4$. Behrens *et al.* (1996), Behrens and Clearfield (1997) and Dadachov and Harrison (1997) provided data on the prepar-ation, structures and properties of $A_3H[Ti_4O_4(SiO_4)_3](H_2O)_n$ (A = H, Na, K, Cs). Structures and ion-exchanged properties of $HA_3[M_4O_4(XO_4)_3](H_2O)_4$ (A = K, Rb, Cs; M = Ti, Ge; X = Si, Ge) were reported by Behrens *et al.* (1998). These compounds were considered as potential materials for the selective removal of Cs and Sr from wastewater solutions.

In the pharmacosiderite framework with the M:X = 4:3, each $M_4(O,OH)_4$ cluster is surrounded by six XO_4 tetrahedra, thus forming $[M_4(O,OH)_4(XO_4)_6]$ FBB. Removal of two opposing tetrahedra from this FBB results in another FBB with composition $[M_4(O,OH)_4(XO_4)_4]$ (Fig. 3.6(b)). This FBB serves as a basis for open frameworks in the structures of phosphovanadylite, $(Ba,K,Ca,Na)_x[(V,Al)_4P_2(O,OH)_{16}](H_2O)_{12}$ (Medrano *et al.* 1998), and $[(CH_3)_4N]_{1.3}(H_3O)_{0.7}[Mo_4O_8(PO_4)_2](H_2O)_2$ (Haushalter *et al.* 1989). However, removal of two XO_4-links results in a change of framework topology. Instead of the **pcu** regular net topology of pharmacosiderite M:X = 4:3 framework, the M:X = 2:1 framework topology in phosphovanadylite is that of the **nbo** regular net. Note that this topology is more open in terms of its framework dens-ity calculated as a number of metal atoms per 1000 Å^3.

The $[M_4(O,OH)_4(XO_4)_4]$ FBBs (Fig. 3.6(b)) may also link through non-shared cor-ners of their M(O,OH)$_6$ octahedra (Figs. 3.6(e) and (f)). This situation is realized in the structure of sitinakite $(Na_2(H_2O)_2)K[Ti_4(OH)O_5(SiO_4)_2](H_2O)_2$ (Sokolova *et al.* 1989) and its synthetic analogs $A_2[Ti_2O_3(SiO_4)](H_2O)_n$ (A = Na, H) (Poojary *et al.* 1994; Clearfield *et al.* 2000). In these structures, octahedral tetramers are linked to form chains running along the *c*-axis. Linkage of the chains along the *a*- and *b*-axis is of the same type as in pharmacosiderite 4:3 frameworks. More details about the structures and properties of open-framework pharmacosiderite titanosilicates may be found in (Clearfield 2001).

3.3.5 *Nasicon, langbeinite and related frameworks*

"Nasicon" is the name of the solid with chemical formula $Na_4Zr_2(SiO_4)_3$. The Nasicon structure type is one of the most important in oxysalts, since many compounds crys-tallizing in this structure type possess important physical properties such as fast ionic conductivity, luminescence, ion exchange, etc. These compounds may be described by the general formula $A_nM_2(TO_4)_3$, where T = Si, Ge, P, As, S, Se, Mo, etc. There are also a number of chemically and structurally related compounds, from which langbeinite $K_2Mg_2(SO_4)_3$ (Mereiter 1979) and $Li_3Fe_2(PO_4)_3$ (Maksimov *et al.* 1986; Bykov *et al.* 1990, etc.) are probably the most well known due to their technological importance.

The structure of Nasicon is based upon the octahedral–tetrahedral framework shown in Fig. 3.7(a). Each octahedron shares vertices with six adjacent tetrahedra and each tetrahedron shares vertices with four adjacent octahedra. The topology of the framework may be rationalized by graphical approach as based upon FBB shown in Fig. 3.7(b) (its graph is shown in Fig. 3.7(c)). This FBB is in fact a finite cluster observed as a separate moiety in the structure coquimbite and, in the following, we will refer to it as a coquimbite cluster. The linkage modes of coquimbite clusters are represented in Figs. 3.7(d)–(g). If a cluster is symbolized by a vertex (note that this is the second hierarchical level of the symbolic description of structural complexity), the double linkage of the clusters (Fig. 3.7(d)) can be symbolyzed by the black line (Fig. 3.7(e)), whereas single linkage (Fig. 3.7(f)) can be symbolized by the white line (Fig. 3.7(g)). Using this approach, a graph consisting of coquimbite clusters can be visualized as a second-level graph. For example, the 2D array of double-linked clusters shown in Fig. 3.8(a) corresponds to a simple square net with black links.

According to the model developed, the Nasicon framework corresponds to a **pcu** regular net with black links (Fig. 3.8(b)), whereas the langbeinite framework is a **fcu**

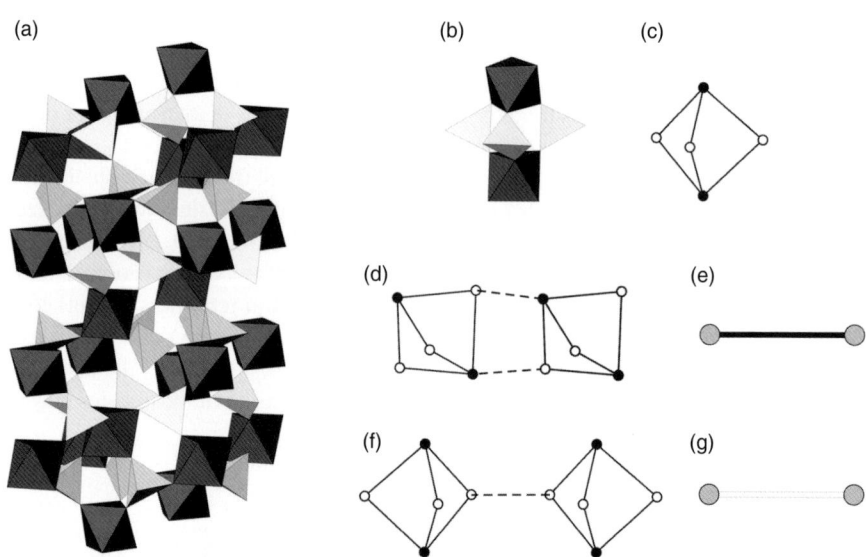

Fig. 3.7. Octahedral–tetrahedral framework in the structure of Nasicon (a) can be considered as being based upon FBB representing the coquimbite cluster (b; its graph is given in (c)). This FBB can be symbolized on the next level of structural complexity as a gray node, and linkage of the nodes is either double ((d); can be identified by a black link (e)) or single ((f); can be identified by a white link).

quasiregular net with white links (Fig. 3.8(c)). The $Li_3Fe_2(PO_4)_3$-type framework is a combination of black and white links. First, coquimbite clusters are linked by double linkages into 2D arrays shown in Fig. 3.8(a). Then, the obtained 2D square nets are interlinked by white links so that the whole graph looks as depicted in Fig. 3.8(d). It is remarkable that the resulting second-level graph is again the **fcu** quasiregular with links of different colors. Note that the $Li_3Fe_2(PO_4)_3$-framework type has also been observed for some simple oxysalts such as $Fe_2(SeO_4)_3$ (Giester and Wildner 1991) and $In_2(SO_4)_3$ (Krause and Gruehn 1995).

3.4   Frameworks based upon polyhedral units

Description of frameworks in terms of polyhedral units is widely used in crystal chemistry of zeolites and related tetrahedral structures (Smith 1988, 2000; Baur and

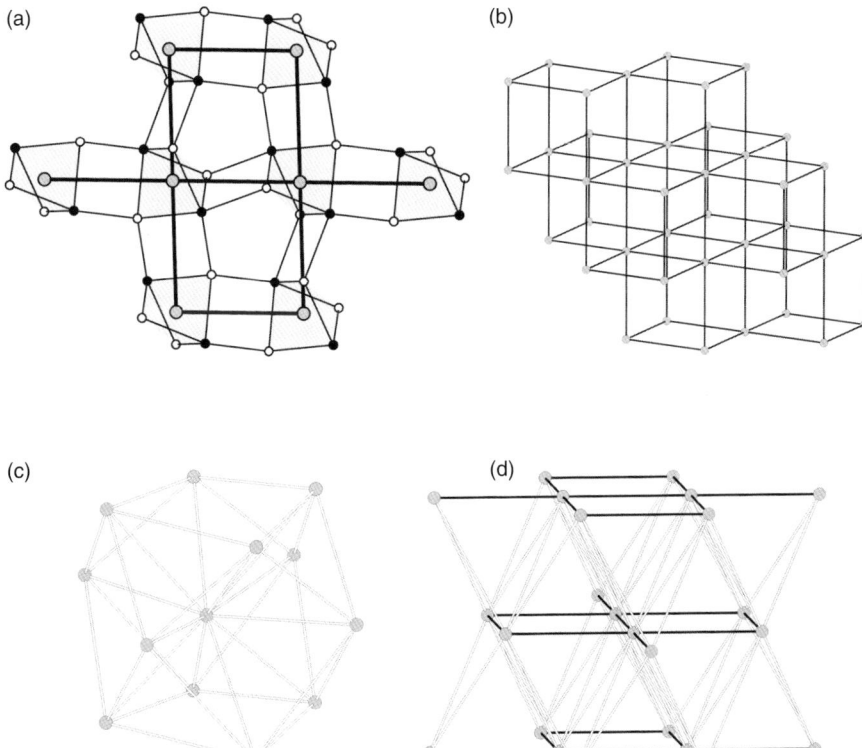

Fig. 3.8. Example of second-level symbolic description of connectivity of doubly linked coquimbite-type FBB (a; FBBs are highlighted); symbolic description of Nasicon, langbein-ite, and $Li_3Fe_2(PO_4)_3$-type frameworks ((b), (c), (d), respectively).

Fischer 2000, 2002; Liebau 2003). However, it is rarely employed in descriptions of more complex frameworks, e.g. those based upon linked octahedra and tetrahedra. Below, after introducing some important topological concepts, we provide some examples of the description of titanosilicate octahedral–tetrahedral frameworks in terms of assemblies of polyhedral units.

3.4.1 *Polyhedra*

Some 3D graphs can be described as being assembled from polyhedral units of different shape and topology. In this regard, the following topological concepts and equations are relevant.

Face symbol is a number of polygonal faces comprising the polyhedron specified by the number of their vertices; for instance, the face symbol for a cube is 4^6 [it has 6 (superscript) square faces]; the face symbol for a hexagonal prism is $6^2 4^6$.

Vertex symbol is a number of edges meeting at vertices; the vertex symbol for a cube is 3^8 as it has eight (superscript) 3-valent vertices; the vertex symbol for a hexagonal prism is 3^{12}.

Schläfli symbol of a polyhedron provides list of numbers n of N-gonal faces meeting at the same vertex. The Schläfli symbol of a tetrahedron is 3^3 (three triangular faces meeting at a vertex), whereas the Schläfli symbol of a cube is 4^3 (three square faces meeting at a vertex).

Euler equation. For a polyhedron with v vertices, e edges and f faces, the following equation holds: $v + f - e = 2$. This equation was derived (in a more general context) by the Swiss mathematician Leonhard Euler.

Extended face symbol can be introduced for description of polyhedra with two types of vertices (e.g. black and white). It provides a sequence of black (b) and white (w) vertices of a polygonal face taken in a cyclic order. For example, the polygonal face shown in Fig. 3.9(a) has the sequence bbwwbbww or $\mathbf{b^2w^2b^2w^2}$. The extended face symbol for the cage shown in Fig. 3.9(b) is $\mathbf{(b^2w^2b^2w^2)^2(bw^2bw^2)^4(w^4)^2(b^2w)^4}$.

3.4.2 *Tilings*

Tiling of 3D Euclidean space is a countable number of 3D bodies (*tiles*) that cover space without gaps and overlaps. Here we shall consider only tilings with polyhedral tiles. Tiling is called *face-to-face* if adjacent polyhedral tiles either share vertices, whole edges, whole faces or have no points in common. If a tiling has a finite number n of topologically and geometrically different tiles, it is called *n-hedral*. If $n = 1$, tiling is *isohedral*; if $n = 2$, tiling is *dihedral*; if $n = 3$, tiling is *trihedral*, etc.

A special class of isohedral face-to-face tiling of 3D space is that in which all tiles are in the same orientation. Such tiles are called *parallelohedra*. For 3D Euclidean space, there are exactly five types of parallelohedra derived by Fedorov (1885): cube, hexagonal prism, truncated octahedron, rhombic dodecahedron and elongated rhombic dodecahedron. Tiles of isohedral face-to-face tiling of 3D space with not necessarily parallel orientations of tiles are called *stereohedra*. The complete list of stereohedra for a 3D Euclidean space is still unknown.

(a)   (b)

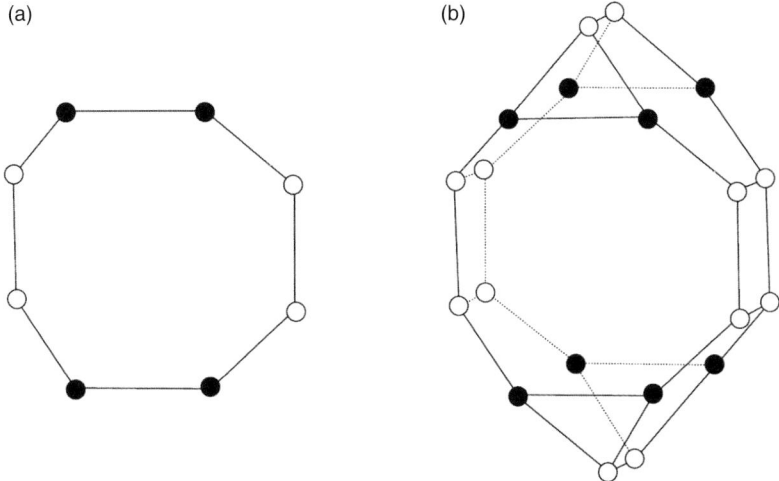

Fig. 3.9. Polygonal face with an extended symbol **bbwwbbww** or $b^2w^2b^2w^2$ that gives a sequence of black and white vertices (a) and polyhedral unit with the extended face symbol $(b^2w^2b^2w^2)^2(bw^2bw^2)^4(w^4)^2(b^2w)^4$.

3.4.3 *Example 1: minerals of the labuntsovite group*

Labuntsovite-group minerals have recently attracted considerable attention from mineralogists owing to the high variability of their structure and chemical composition (Chukanov *et al.* 2002, 2003; Armbruster *et al.* 2004, etc.). The basis of their structures is an octahedral–tetrahedral framework shown in Fig. 3.10(a). The $M\phi_6$ octahedra (M = Ti, Nb; ϕ = O, OH) share corners to produce single chains that are interlinked by Si_4O_{12} four-membered silicate rings. The resulting framework has channels oriented perpendicular to the octahedral chains and occupied by low-valent cations and H_2O molecules. Note that the framework shown in Fig. 3.10(a) has $D_{nt} = 1$ (octahedra are linked to form a chain) and $D_{tetr} = 0$ (silicate tetrahedra form 4-membered rings (4MRs)). A nodal representation of the labuntsovite-type framework is shown in Fig. 3.10(b). It can be considered to consist of two types of polyhedral units shown in Figs. 3.10(c) and (d). The face and vertex symbols of the unit shown in Fig. 3.10(c) are $8^26^44^23^4$ and 3^{20}, respectively. The face and vertex symbols of the unit shown in Fig. 3.10(d) are 6^23^4 and 3^8, respectively. Extended face symbols (see above) of the polyhedral units shown in Figs. 3.10(c) and (d) are $(b^2w^2b^2w^2)^2(b^2wb^2w)^4(w^4)^2(b^2w)^4$ and $(bw^2bw^2)^2(b^2w)^4$, respectively. The $8^26^44^23^4$ units share 8-membered faces to form tunnels.

Owing to the fact that the labuntsovite-type 3D net can be assembled from polyhedral units of two types, it is possible to construct a *dihedral* tiling of 3D space that corresponds to this net. This tiling is demonstrated in Fig. 3.11. It consists of two polyhedra that are not topologically equivalent to the polyhedral units shown in Fig. 3.10 and can be obtained from the latter by addition of edges linking two black vertices

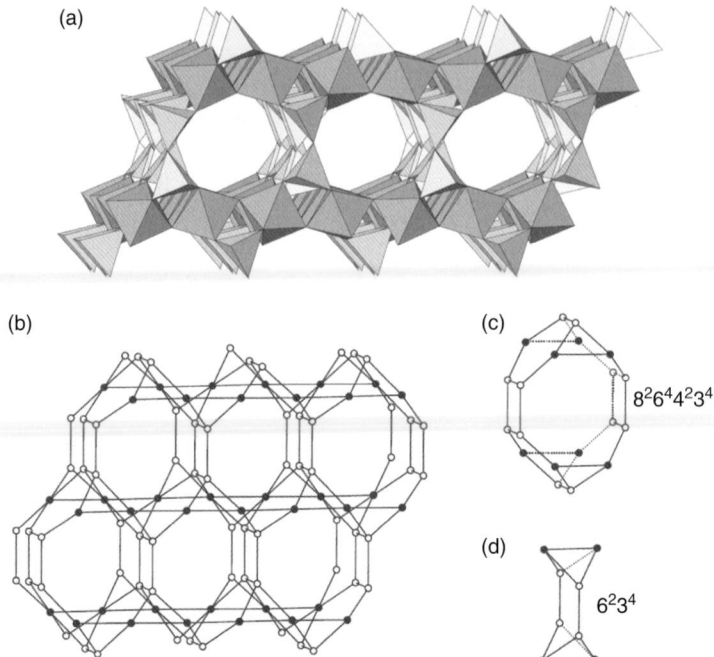

Fig. 3.10. Octahedral–tetrahedral framework in the structures of labuntsovite-group minerals (a) and its nodal representation. The labuntsovite 3D net consists of two types of polyhedral units with extended face symbols $(\mathbf{b^2w^2b^2w^2})^2(\mathbf{b^2wb^2w})^4(\mathbf{w^4})^2(\mathbf{b^2w})^4$ (c) and $(\mathbf{bw^2bw^2})^2(\mathbf{b^2w})^4$ (d).

(oriented vertically in Fig. 3.10). The polyhedra of the tiling are arranged in the following way. First, polyhedra of the type shown in Fig. 3.11(b) are arranged into layers by sharing faces on their sides (Fig. 3.11(c)). These layers have polyhedral hollows that are perfectly suited for the polyhedra of the type shown in Fig. 3.11(a) (Fig. 3.11(d)). Whole tiling is obtained by assembling layers of larger tiles one under another and by filling gaps with the smaller tiles (Fig. 3.11(e)).

3.4.4 *Example 2: shcherbakovite–batisite series*

Minerals of the shcherbakovite–batisite series have the general formula NaA_2 $(Ti,Nb)_2O_2[Si_4O_{12}]$, where A = K, Ba for shcherbakovite and Ba, K for batisite. The two minerals are isotypic with structures based upon chains of corner-linked MO_6 octahedra (M = Ti, Nb) and chains of SiO_4 tetrahedra (Fig. 3.12(a)) (Nikitin and Belov 1962; Schmahl and Tillmans 1987; Rastsvetaeva *et al.* 1997; Uvarova *et al.* 2003; Krivovichev *et al.* 2004d). The nodal representation of the shcherbakovite–batisite

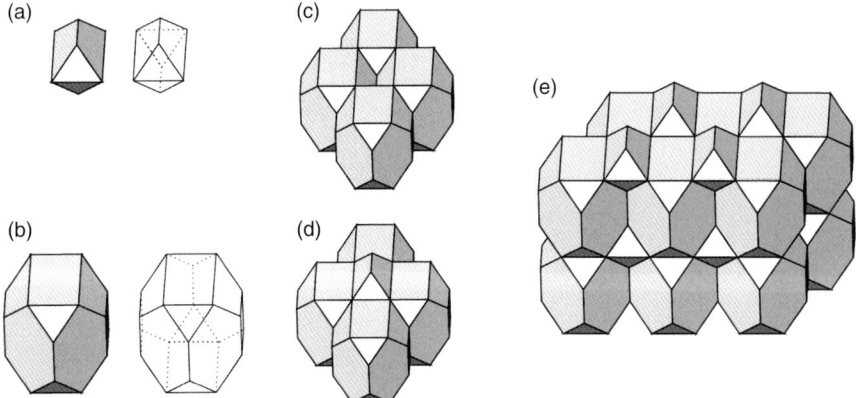

Fig. 3.11. Dihedral tiling of 3D space that corresponds to the labuntsovite net. See text for details.

framework is shown in Fig. 3.12(b). This 3D black-and-white net can be constructed by assembling three different polyhedral units characterized by the following face symbols: $6^4 4^2$ (Fig. 3.12(c)), 4^6 (Fig. 3.12(d)) and $6^2 4^2$ (Fig. 3.12(e)). Note that the $6^4 4^2$ unit contains 6-membered faces of two types: **bw²bw²** and **b²w⁴**. The 4^6 and $6^2 4^2$ units are closely related and can be obtained one from another by adding (deleting) two edges linking white vertices.

Using simple topological and geometrical operations, the $6^4 4^2$ unit can be transformed into a hexagonal prism, whereas the 4^6 and $6^2 4^2$ units can be transformed into square prisms. Thus the 3D net shown in Fig. 3.12(b) can be considered as a modified version of the tiling of 3D space into hexagonal and square prisms shown in Fig. 3.13.

3.4.5 *Combinatorial topology of polyhedral units*

A detailed consideration of polyhedral units that occur in different structures reveals some interesting topological relationships. About half of the units can be produced from some simple convex polyhedra by a series of topological operations including deleting edges, inserting new vertices and edges and merging vertices. As an example, we consider a hexagonal prism (Fig. 3.14). Its face and vertex symbols are $6^2 4^6$ and 3^{12}, respectively. Let us number vertical edges of the prism (edges that link two hexagonal bases) in a cyclic order (Fig. 3.14(a)). If two vertical edges of the prism are deleted, the result is a *two-edges-deleted* hexagonal prism. If two edges are not of the same square face, there are two ways to delete them: (i) to delete edges 1 and 4 (Fig. 3.14(b)) and (ii) to delete edges 1 and 3 (Fig. 3.14(c)). The results are polyhedral units shown in Figs. 3.14(d) and (e), respectively. Note that they have the same face

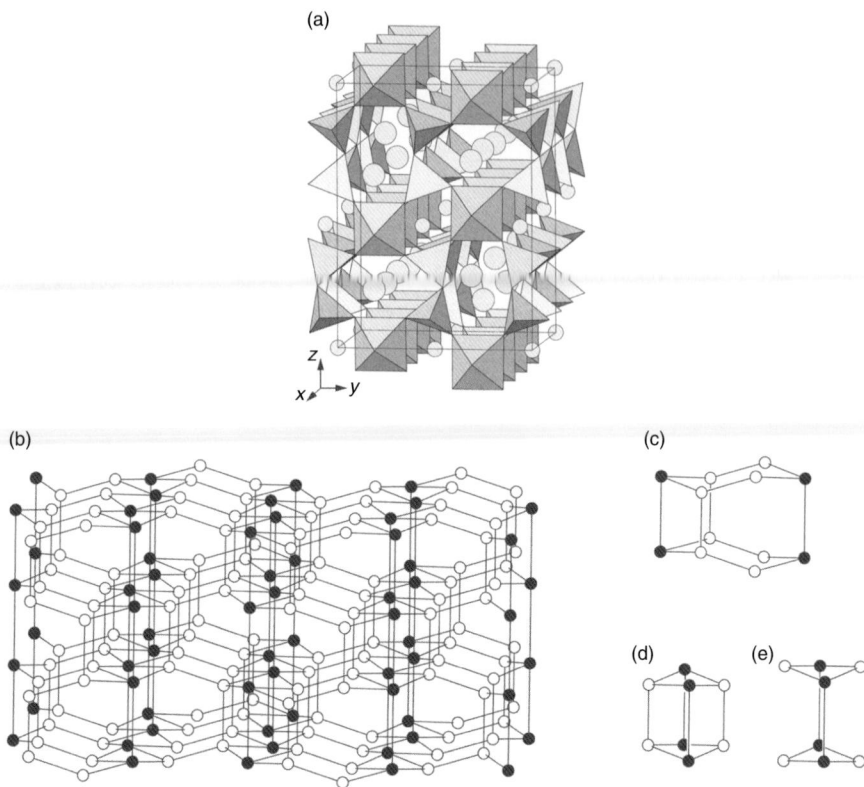

Fig. 3.12. Octahedral–tetrahedral framework in minerals of the shcherbakovite–batisite series (a) and its nodal representation (b). The 3D black-and-white net can be constructed by assembling three different polyhedral units characterized by the following face symbols: $6^4 4^2$ (c), 4^6 (d) and $6^2 4^2$ (e).

and vertex symbols: $6^4 4^2$ and $2^4 3^8$, respectively. As the vertices of 3D black-and-white nets can be either black or white, there are many ways to color vertices of these two units and all can be easily enumerated. However, we consider only those cages that are realized in the real structures. The units shown in Figs. 3.14(f) and (g) are hexagonal prisms with 1,4-deleted edges. However, their black-and-white coloring is different: vertices that are black in one cage are white in another and *vice versa*. The units shown in Figs. 3.14(h) and (i) are 1,3-edge-deleted hexagonal prisms with different coloring of their vertices. From the four units considered, the first (Fig. 3.14(f)) occurs in the structure of zektzerite, $NaLiZrSi_6O_{15}$ (Ghose and Wan 1978), the second and third (Figs. 3.14(g) and (h)) in $K_4Nb_8O_{14}(PO_4)_4(SiO_4)$ (Leclaire *et al.* 1992), and

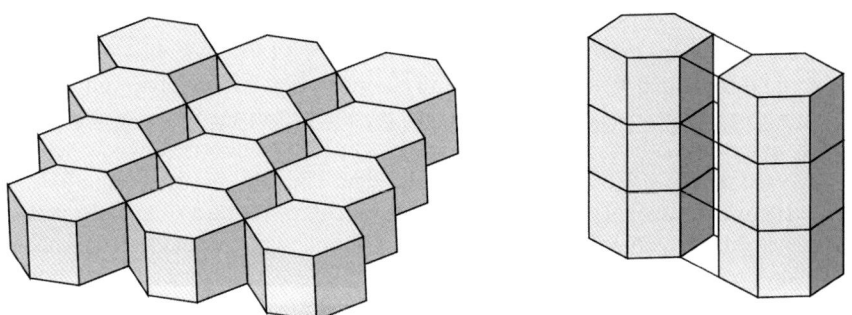

Fig. 3.13. Dihedral tiling of 3D space into hexagonal and square prisms that corresponds to the shcherbakovite–batisite 3D net.

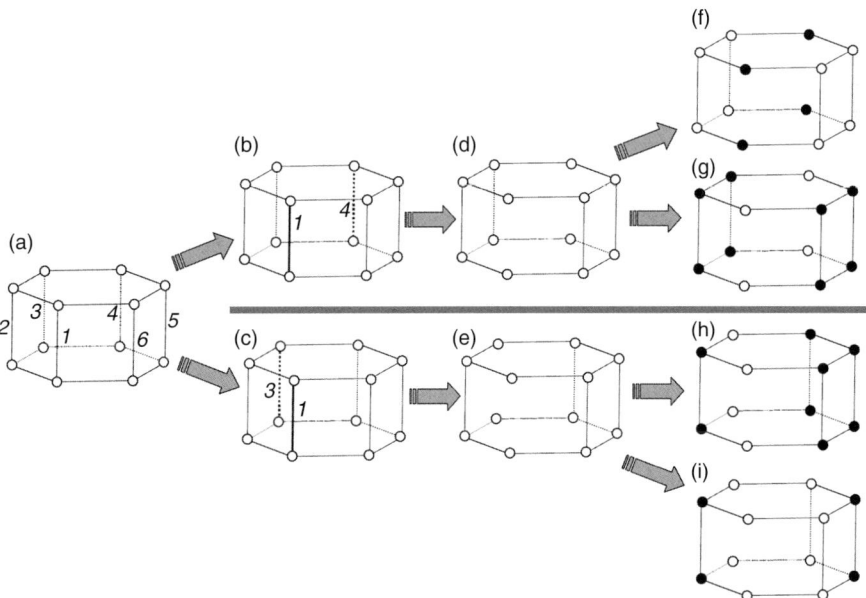

Fig. 3.14. Topological evolution of hexagonal prism into polyhedral units that occur in 3D nets corresponding to octahedral–tetrahedral frameworks. Note that the paths shown include only deletion of edges and color changes for the vertices. See text for details.

the fourth (Fig. 3.14(i)) is the $6^4 4^2$ unit from the structures of minerals of the batisite–shcherbakovite series (see above).

Another type of topological operations used to relate different polyhedral units is inserting a vertex into an edge (*stellation* according to Smith (2000)) or inserting an edge into an edge (adding *a handle*). Inserting an edge into the vertical edge of the hexagonal prism results in the unit shown in Fig. 3.15(b), which is observed in zektzerite (Ghose and Wan 1978). The unit shown in Fig. 3.15(c) (observed again in zektzerite) can be produced from a hexagonal prism by inserting three vertices, deleting two edges and *contracting an edge*. The latter procedure corresponds to deleting an edge and merging vertices incident to this edge. The unit shown in Fig. 3.15(d) is derived from a hexagonal prism by deleting four edges, inserting a vertex into an edge and addition of two diagonal edges. This can be found in the octahedral–tetrahedral framework of the high-pressure phase $Na_{1.8}Ca_{1.1}{}^{VI}Si({}^{IV}Si_5O_{14})$ (Gasparik *et al.* 1995).

It is noteworthy that similar topological relationships between different polyhedral units has been observed for tetrahedral frameworks by Smith (1988, 2000).

3.4.6 *Topological complexity of polyhedral units: petarasite net*

It is important to note that not all polyhedral units can be considered as a result of topological transformations of simple convex polyhedra. As an example of topological complexity, we consider a polyhedral unit with 22 faces observed in the structure of petarasite, $Na_5Zr_2Si_6O_{18}(Cl,OH)\cdot 2H_2O$ (Ghose *et al.* 1980) (Fig. 3.16). This unit is noteworthy because (i) it has a maximal number of faces observed in zircono-, niobo- and titanosilicates, (ii) it can be realized as a convex polyhedron, (iii) it is quite large (6.6 × 9.6 × 14.4 Å^3). Note that this polyhedral unit has 4-valent vertices that is impossible for polyhedral units in zeolites and other tetrahedral frameworks (Smith 2000).

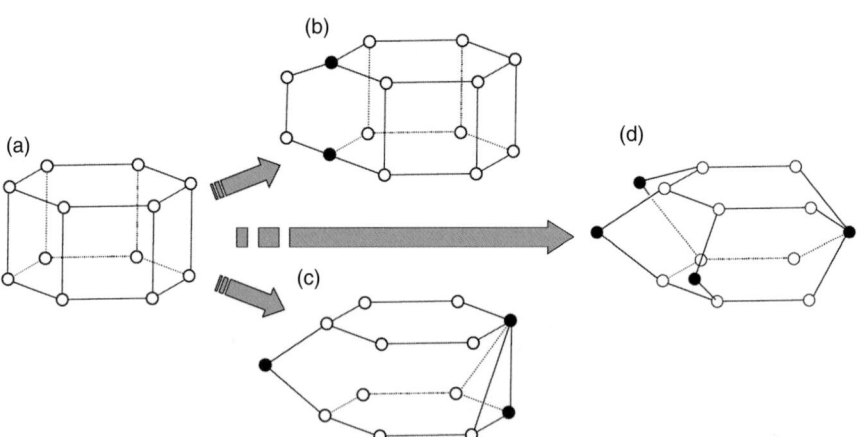

Fig. 3.15. Topological evolution of hexagonal prism into polyhedral units that occur in 3D nets corresponding to octahedral–tetrahedral frameworks. The paths shown include stellation, adding a handle, contracting an edge, addition of edges. See text for details.

3.5   Frameworks based upon infinite chains

3.5.1   *Fundamental chains as bases of complex frameworks: an example*

Liebau (1985) suggested that the description of silicate sheets and frameworks can be based upon simple silicate chains (*fundamental* chains) and formulated special rules to subdivide these chains in 2D and 3D silicate anions. A similar approach can be applied to heteropolyhedral frameworks as well. For example, the titanosilicate framework observed in the labuntsovite-group minerals can be described as based upon chains of corner-sharing octahedra interlinked by silicate 4MRs. The octahedral–tetrahedral framework in the shcherbakovite–batisite series consists of linked octahedral and tetrahedral chains. The choice of one structure description or another is again dictated by the goal, simplicity and convenience.

The concept of a fundamental chain is especially convenient in the case when structure contains no obvious 0D or 2D units. As an example, we consider a family of chiral open-framework uranyl molybdates based upon corner-linked UO_7 pentagonal bipyramids and MoO_4 tetrahedra. The general chemical formula of these compounds can be written as $A_r[(UO_2)_n(MoO_4)_m(H_2O)_p](H_2O)_q$, where A is either an inorganic or organic (e.g. protonated amine) cation. There are three topological types of chiral open-framework uranyl molybdates that are characterized by different U:Mo $= n{:}m$ ratios of 6:7 (Krivovichev *et al.* 2005i,j), 5:7 (Krivovichev *et al.* 2003c) and 4:5 (Krivovichev *et al.* 2005h). The structure of the only known 5:7 compound, $(NH_4)_4[(UO_2)_5(MoO_4)_7](H_2O)$, consists of a three-dimensional framework of composition $[(UO_2)_5(MoO_4)_7]^{4-}$ (Fig. 3.17(a)). The framework contains a three-dimensional system of channels. The largest channel is parallel to [001] and has the dimensions 7.5 × 7.5 Å, which result in a crystallographic free diameter (effective pore width) of 4.8 × 4.8 Å (based on an oxygen radius of 1.35 Å). Smaller channels run parallel to [100], [110], [010], [$\bar{1}$10], [1$\bar{1}$0] and [$\bar{1}\bar{1}$0] and have dimensions 5.2 × 6.3 Å (giving an effective pore width of 2.5 × 3.6 Å). Four symmetrically unique NH_4^+ cations and H_2O molecules are located in the framework channels. The $[(UO_2)_5(MoO_4)_7]^{4-}$ framework is unusually complex. Its nodal representation

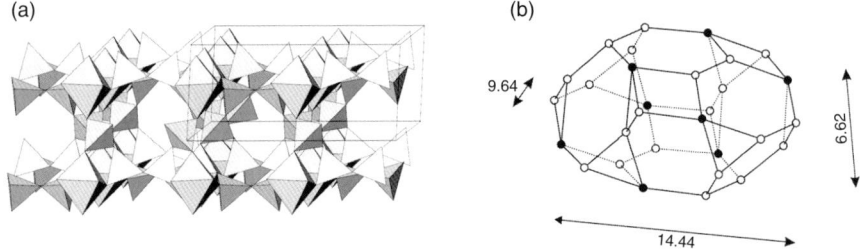

(a)   (b)

9.64   6.62

14.44

Fig. 3.16. Octahedral–tetrahedral framework in the structure of petarasite, $Na_5Zr_2Si_6O_{18}$ (Cl,OH)·$2H_2O$ (a) and its topologically complex polyhedral unit with 22 faces (b).

is shown in Fig. 3.17(b). Each node corresponds to a UO_7 bipyramid (black) or a MoO_4 tetrahedron (white). All black vertices are 5-connected and all white vertices are either 3- or 4-connected. Using the nodal representation, the uranyl molybdate framework in the structure of $(NH_4)_4[(UO_2)_5(MoO_4)_7](H_2O)$ can be described in terms of fundamental chains. The nodal representation of the fundamental chain corresponding to the uranyl molybdate framework is shown in Fig. 3.17(c). The chain is a sequence of 3- and 4-connected MoO_4 tetrahedra (white vertices) linked through one, two or three UO_7 pentagonal bipyramids (black vertices). The graph shown in Fig. 3.17(c) can be further reduced to the simplified isomorphic graph shown in Fig. 3.17(d). This reduction preserves all topological linkages between the nodes. Note that the graph shown in Fig. 3.17(d) is periodic and its identity unit includes seven white vertices, whereas, in the real structure, the identity period of the fundamental chain includes 21 white vertices. Thus, the topological structure of the fundamental chain is simpler than its geometrical realization.

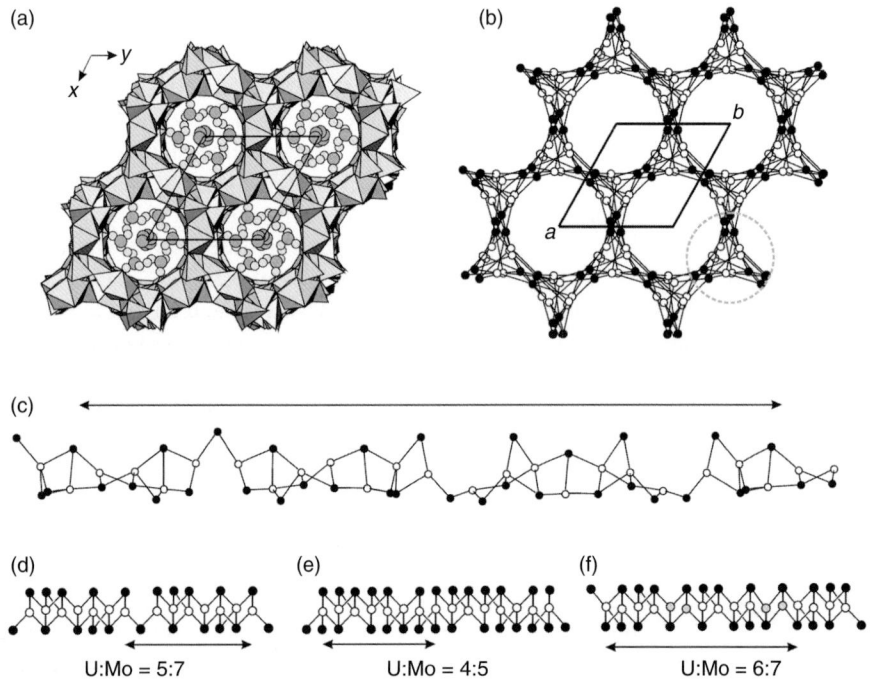

(a)

(b)

(c)

(d)

(e)

(f)

U:Mo = 5:7 U:Mo = 4:5 U:Mo = 6:7

Fig. 3.17. The structure of $(NH_4)_4[(UO_2)_5(MoO_4)_7](H_2O)$ projected along the c-axis (a), nodal representation of its $[(UO_2)_5(MoO_4)_7]$ framework (b), nodal representation of its fundamental chain (c), and graphs isomorphous to nodal representations of fundamental chains of chiral uranyl molybdate frameworks with the U:Mo ratio of 5:7, 4:5 and 6:7 ((d), (e) and (f), respectively).

Distinction of a fundamental chain permits the comparison of the 5:7 uranyl molybdate framework in $(NH_4)_4[(UO_2)_5(MoO_4)_7](H_2O)$ with the 6:7 and 4:5 chiral uranyl molybdate frameworks. The two latter frameworks can also be described as being based upon fundamental chains of UO_7 and MoO_4 polyhedra (Krivovichev et al. 2005h,i,j). The reduced black-and-white graphs of these chains are shown in Figs. 3.17(e) and (f) for 6:7 and 4:5 frameworks, respectively. Detailed examination of these graphs demonstrated that they cannot be transformed one into another without significant topological reconstruction. Thus, the fundamental chains that form the bases for the chiral uranyl molybdate frameworks with U:Mo = 5:7, 4:5 and 6:7 are topologically different, though closely related.

3.5.2 *Frameworks with non-parallel orientations of fundamental chains*

We note that, in the uranyl molybdate frameworks discussed above, all fundamental chains are parallel to each other. Of course, non-parallel orientations of fundamental chains are also possible. Figure 3.18 shows the complex octahedral–tetrahedral framework observed by Lii and Huang (1997) in the structure of $[H_3N(CH_2)_3NH_3]_2$ $[Fe_4(OH)_3(HPO_4)_2(PO_4)_3]\cdot xH_2O$ (Fig. 3.18(a)). This structure is based upon a "butterfly-type" octahedral–tetrahedral cluster, shown in Figure 3.3(a) and (b). In the structure, these clusters are linked to form chains oriented parallel to [100] and [010] (Fig. 3.18(b)). The chains are cross-linked to form a framework, as schematically shown in Fig. 3.18(c).

Another example of a structure with non-parallel orientation of fundamental chains is that of $[(UO_2)_3(PO_4)O(OH)(H_2O)_2](H_2O)$ recently reported by Burns et al. (2004). In this structure, fundamental chains are ribbons of edge-sharing UO_8 hexagonal bipyramids, UO_7 pentagonal bipyramids and PO_4 tetrahedra (Fig. 3.19(a)). The ribbons are cross-linked to form an open framework with cavities filled by H_2O molecules (Fig. 3.19(b)). It is interesting that topologically identical ribbons form continuous sheets in the structures of the phosphuranylite-group (see Chapter 4 for details) (Burns et al. 1996).

3.5.3 *Frameworks with no M–M and T–T linkages*

3.5.3.1 *Frameworks consisting of kroehnkite chains*

In their reviews on kroehnkite-type compounds, Fleck et al. (2002) and Fleck and Kolitsch (2003) noticed that there are a number of octahedral–tetrahedral frameworks that can be described on the basis of linked kroehnkite-type chains (see Chapter 2). Some examples of inorganic oxysalts with such a structure are given in Table 3.2. Heteropolyhedral frameworks based upon condensed kroehnkite-type chains have the *M:T* ratio of 1:2, equal to the *M:T* ratio observed for the chains themselves. Condensation of the chains may occur along different pathways that is reflected in different topologies of the resulting frameworks.

Figure 3.20 shows two frameworks that are based upon single kroehnkite chains. In the structure of $Na_3[NbO(AsO_4)_2]$ (Fig. 3.20(c)), chains are oriented parallel to

(a)

(b)   (c)

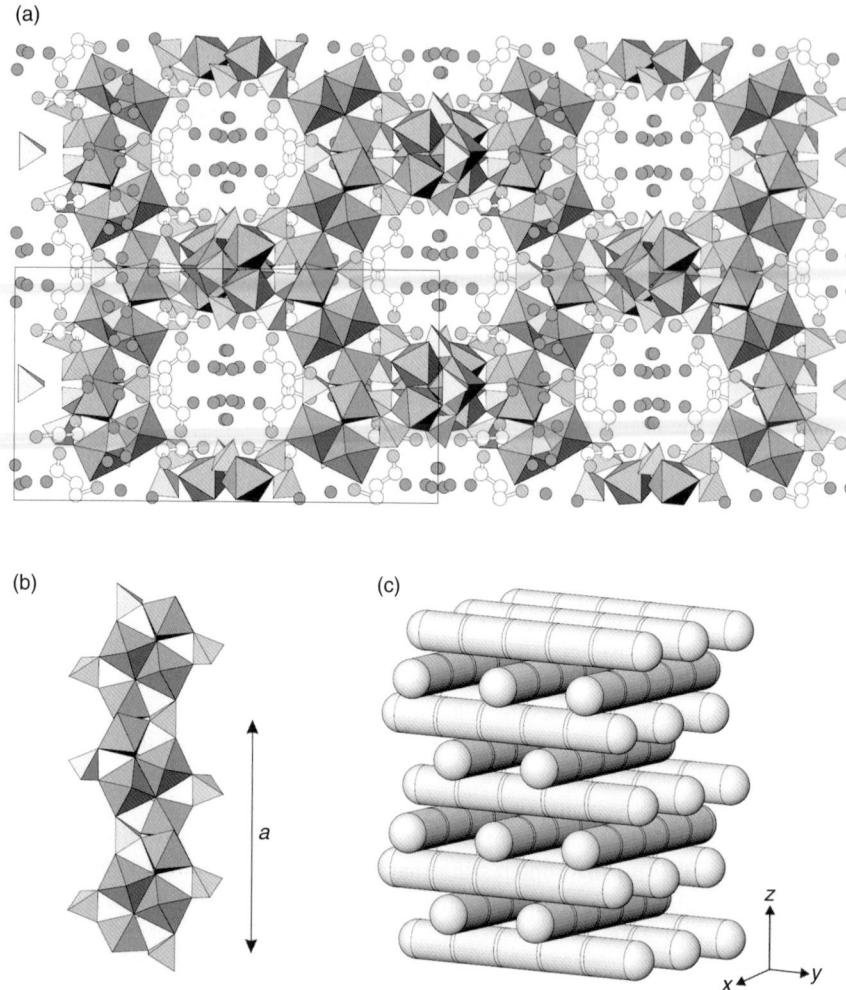

Fig. 3.18. Structure of $(C_3H_{12}N_2)[Fe_4(OH)(H_2O)_2(PO_4)_5](H_2O)_4$ projected along the a-axis (a), its fundamental chain consisting of "butterfly-type" octahedral–tetrahedral clusters (b), and schematic representation of linkage of the chains in the framework (c).

each other. One of the octahedral corners of Nb-centered octahedra is not bonded to As^{5+} and is occupied by O^{2-} anions forming a strong $Nb^{5+}=O$ bond. In another framework (Fig. 3.20(f)), all octahedral ligands are bonded to the T atoms; however, each tetrahedron has one terminal corner. This corner can be either protonated (as in $Zr(HPO_4)_2$) or occupied by a lone pair of electrons. The latter is the case for triangular pyramidal selenite anions, $SeO_3\Psi$, that explains the frequency of this topology in selenites. In the structure of $Ba[(VO)_2(SeO_3)_2(HSeO_3)_2]$ (Fig. 3.21), half of the selenite

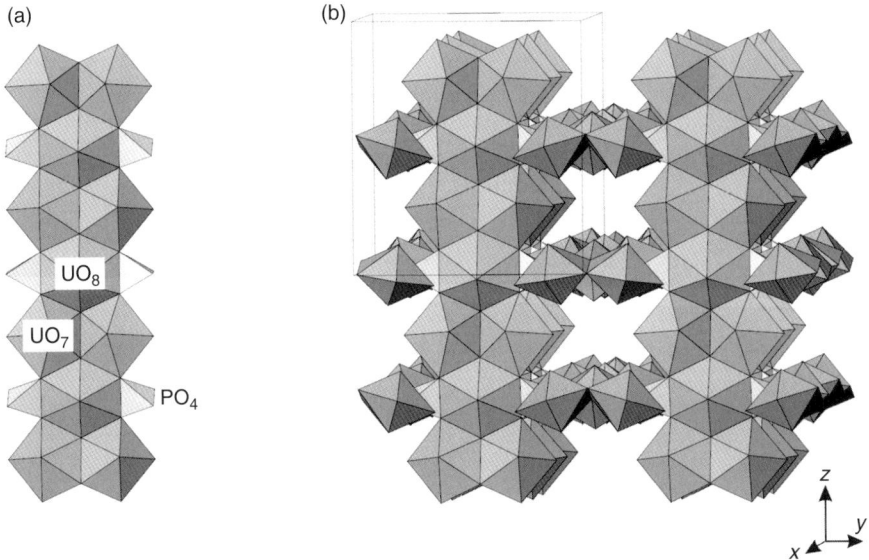

Fig. 3.19. Ribbons of edge-sharing UO_8 hexagonal bipyramids, UO_7 pentagonal bipyramids and PO_4 tetrahedra (a) as fundamental chains in the structure of $[(UO_2)_3(PO_4)O(OH)(H_2O)_2]$ (H_2O).

Table 3.2. Heteropolyhedral $MX(TO_4)_2$ frameworks based upon kroehnkite-type chains (examples)

M:T ratio	Framework formula	Compound	Reference
1:2	$MX(TO_4)_2$	$Na_3[NbO(AsO_4)_2]$	Hizaoui *et al.* 1999
	$M(TO_4)_2$	$Zr(HPO_4)_2$	Norby 1997; Krogh Andersen and Norby 2000
	$M(SeO_3)_2$	$[Mg(HSeO_3)_2]$	Boldt *et al.* 1997
		$Ba[Fe_2(SeO_3)_4]$	Giester 2000
		$Pb[Fe_2(SeO_3)_4]$	Johnston and Harrison 2004
		$Li[Fe(SeO_3)_2]$	Giester 1994
		$Sr[Co_2(SeO_2OH)_2(SeO_3)_2]$	Giester and Wildner 1996a
	$MX(SeO_3)_2$	$Ba[(VO)_2(SeO_3)_2(HSeO_3)_2]$	Harrison *et al.* 1995b
	$M(TO_4)_2$	$(H_3O)[Fe(HPO_4)_2]$	Vencato *et al.* 1989
		$Na_3[In(PO_4)_2]$	Lii 1996a
		$Pb[Fe(AsO_4)(AsO_3(OH))]$	Effenberger *et al.* 1996
	$M(SeO_3)_2$	$K[Fe(SeO_3)_2]$	Giester 1993
		$NaFe(SeO_3)_2$	Giester and Wildner 1996b
		$BaCo(SeO_3)_2$	Giester and Wildner 1996b
5:4	$M'_3M''_2X_8(TO_4)_4$	$Co_3(NbO)_2(PO_4)_4(H_2O)_{10}$	Wang *et al.* 2000
1:2	$M(TO_4)_2$	$Ba[V_2(HPO_4)_4](H_2O)$	Wang *et al.* 1993

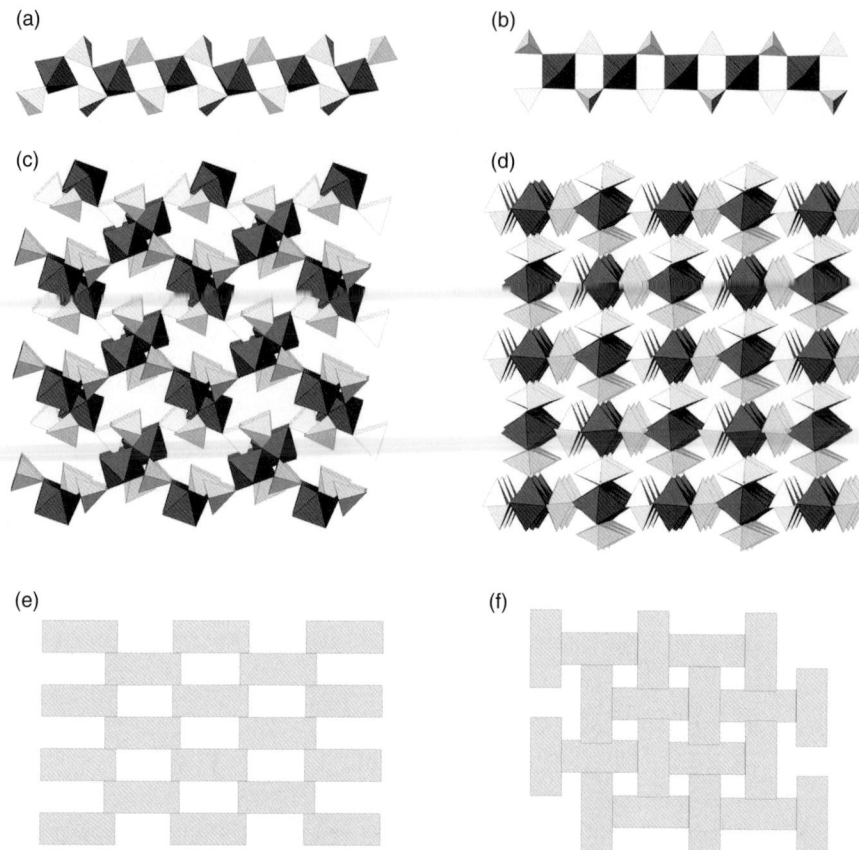

Fig. 3.20. Kroehnkite-type chains in the structures of $Na_3[NbO(AsO_4)_2]$ (a) and $Zr(HPO_4)_2$ (b), octahedral–tetrahedral frameworks in these compounds ((c) and (d), respectively), and schemes of the chain orientations ((e) and (f), respectively; orthogonal domains symbolize cross-sections of the chains).

anions are protonated and the respective triangular pyramids are bidentate, whereas those that are not protonated are tridentate.

The structure of $(H_3O)[Fe(HPO_4)_2]$ (Fig. 3.22) contains octahedral–tetrahedral framework resulting from condensation of double kroehnkite chains, i.e. chains that occur in krausite (see Chapter 2). This framework is built solely from the kroehnkite chains; another situation is observed in $Co_3(NbO)_2(PO_4)_4(H_2O)_{10}$ (Fig. 3.23). Here, NbO_6 octahedra and PO_4 tetrahedra form double kroehnkite chains that are further linked into a 3D framework by $CoO_4(H_2O)_2$ and $CoO_2(H_2O)_4$ octahedra.

In kroehnkite-type chains, only four octahedral corners are bridging between M and T atoms, whereas two other corners are either terminal or involved in polymerization with adjacent chain. Usually, the terminal octahedral corners have a

(a)   (b)

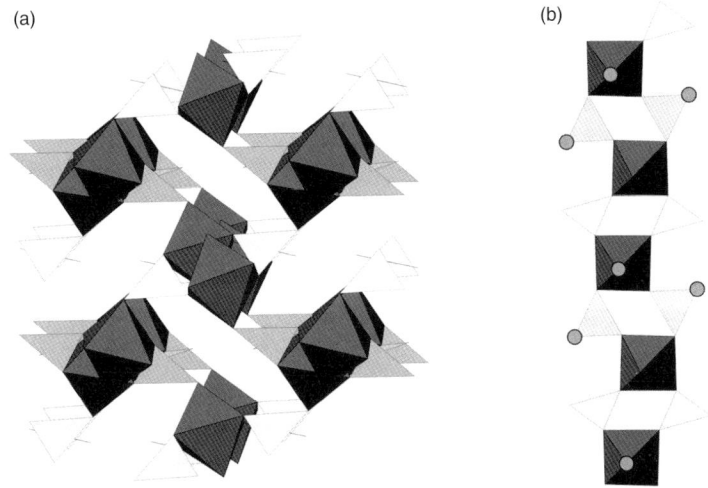

Fig. 3.21. Heteropolyhedral framework in $Ba[(VO)_2(SeO_3)_2(HSeO_3)_2]$ (a) and its fundamental chain (b). Circles indicate anions bridging between octahedra and tetrahedra.

(a)   (c)

(b)

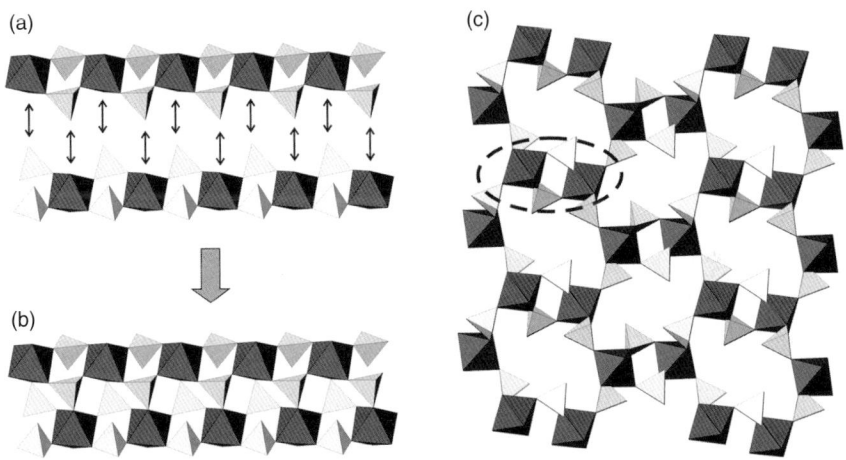

Fig. 3.22. Heteropolyhedral framework in $(H_3O)[Fe(HPO_4)_2]$: two kroehnkite-like chains (a) polymerize into a double chain (b) that are further condensed into a 3D framework (c).

trans-orientation with respect to other corners. Heteropolyhedral framework in $Ba[V_2(HPO_4)_4](H_2O)$ is based upon kroehkite chains with *cis*-orientation of terminal octahedral corners. As a result, the chains display a rather distorted configuration (Fig. 3.24).

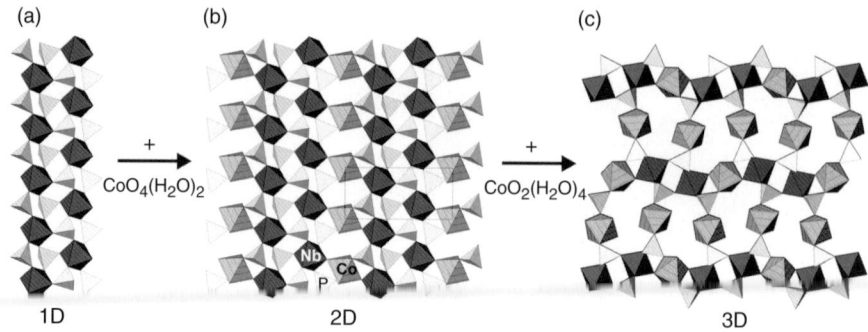

Fig. 3.23. Octahedral–tetrahedral framework in $Co_3(NbO)_2(PO_4)_4(H_2O)_{10}$ as built as a result of successive condensation of the $[(NbO)(PO_4)_2]^{3-}$ double kroehkite-type chains (a) via Co-centered octahedra ((b), (c)).

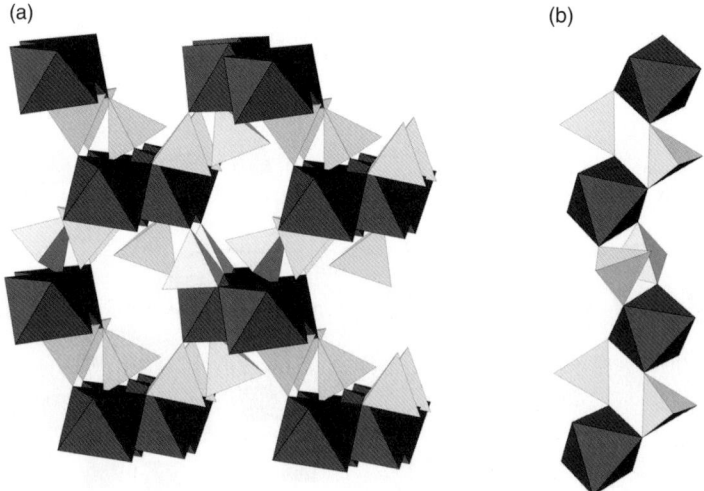

Fig. 3.24. Heteropolyhedral framework in $Ba[V_2(HPO_4)_4](H_2O)$ (a) is based upon kroehkite chains with *cis*-orientation of terminal octahedral corners (b).

3.5.3.2   *Other examples*

As has already been mentioned, autunite-type topology is parent to a number of inorganic oxysalt structure types. Figure 3.25(a) shows an octahedral–tetrahedral framework in the structures of $Cs[M_2(PO_4)(HPO_4)_2(H_2O)_2]$ ($M = V^{3+}$, In) (Haushalter *et al.* 1993; Dhingra and Haushalter, 1994). Its basic motif is the octahedral–tetrahedral chain $[MO_2(H_2O)(PO_4)]$ shown in Fig. 3.25(b). The chains are interlinked

(a)

(b)

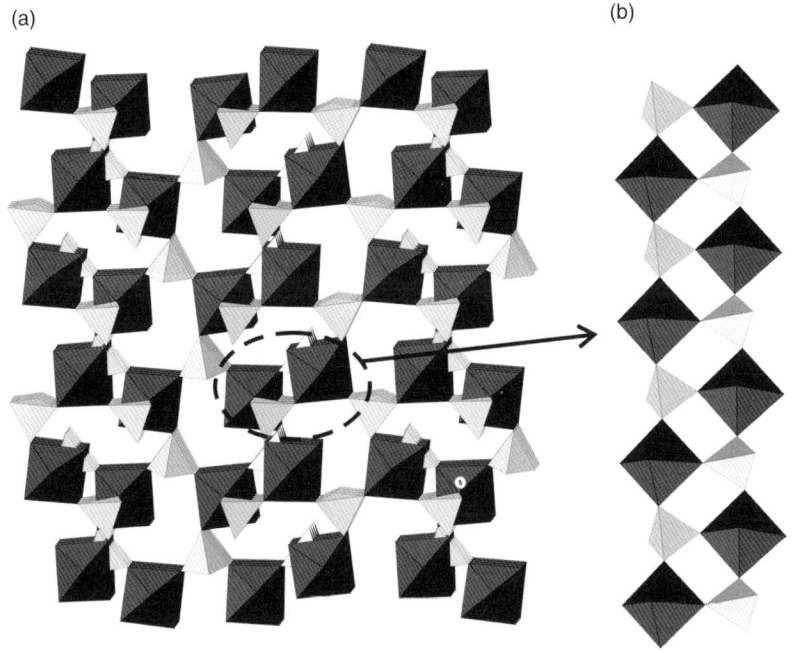

Fig. 3.25. Octahedral–tetrahedral framework in the structures of $Cs[M_2(PO_4)(HPO_4)_2(H_2O)_2]$ ($M = V^{3+}$, In) (a) and its fundamental chain (b) which is a derivative of the autunite topology.

by additional tetrahedral oxyanions so the M:P ratio of the resulting framework is equal to 2:3.

The structures of $Mn_2(SeO_3)_3(H_2O)_3$ (Koskenlinna and Valkonen, 1977) and mandarinoite, $Fe_2(Se_3O_9)(H_2O)_6$ (Hawthorne, 1984) are interesting examples of frameworks based upon two different types of fundamental chains (Fig. 3.26). Both structures contain kroehnkite-type chains (**II**) and chains with alternating 2-connected octahedra and SeO_3^{2-} anions (**I**). However, the isomeric variations of the I-type chains are different in the two structures. The structure of $Mn_2(SeO_3)_3(H_2O)_3$ (Fig. 3.26(a)) contains chains with *trans*-configuration of shared octahedral vertices (Fig. 3.26(b)), whereas, in the structure of mandarinoite, octahedra share their vertices with adjacent selenite anions in *cis*-configuration (Fig. 3.26(d)). As a consequence, the heteropolyhedral framework in mandarinoite is more open and accommodates two times more water molecules than that of $Mn_2(SeO_3)_3(H_2O)_3$.

Some interesting examples of frameworks with chains of interlinked tetrahedra and pentagonal bipyramids have been observed in the structures of uranyl compounds (Fig. 3.27).

The uranyl molybdate framework from the structure of α-$Cs_2[(UO_2)_2(MoO_4)_3]$ (Krivovichev *et al.* 2002a) (Fig. 3.27(a)) consists of corner-sharing UO_7 pentagonal

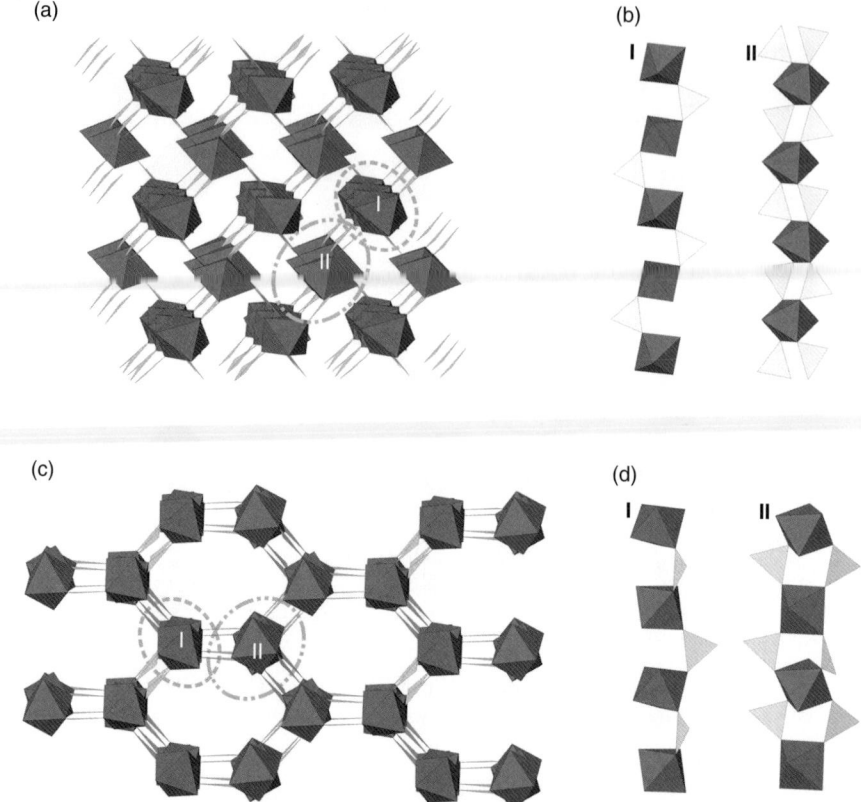

Fig. 3.26. Frameworks based upon fundamental chains of two different types: $Mn_2(SeO_3)_3$ $(H_2O)_3$ ((a), (b)) and mandarinoite, $Fe_2(Se_3O_9)(H_2O)_6$ ((c), (d)).

bipyramids and MoO_4 tetrahedra. The framework has one-dimensional channels running parallel to the a-axis that are occupied by Cs^+ cations. The framework can be derived from the uranyl molybdate sheet observed in β-$Cs_2[(UO_2)_2(MoO_4)_3]$ (see Chapter 2) by cutting it into 1D chains (Figs. 3.27(b) and (c)) and re-assembling them so that the 3D structure is formed. In fact, the $\alpha \rightarrow \beta$ high-temperature reconstructive phase transition in $Cs_2[(UO_2)_2(MoO_4)_3]$ has recently been observed by Nazarchuk *et al.* (2004). This phase transition involves a 3D $\rightarrow$ 2D transformation of the uranyl molybdate network.

The only known example of a uranyl selenate framework structure, $(H_3O)_2[(UO_2)(SeO_4)_3(H_2O)](H_2O)_4$ (Blatov *et al.* 1988), based upon a 3D framework of U and Se polyhedra is shown in Fig. 3.27(d). A description based upon polyhedra fails to capture the details of the framework. By application of the black-and-white graph theory technique, the architectural principles become transparent (Fig. 3.27(e)). The uranyl

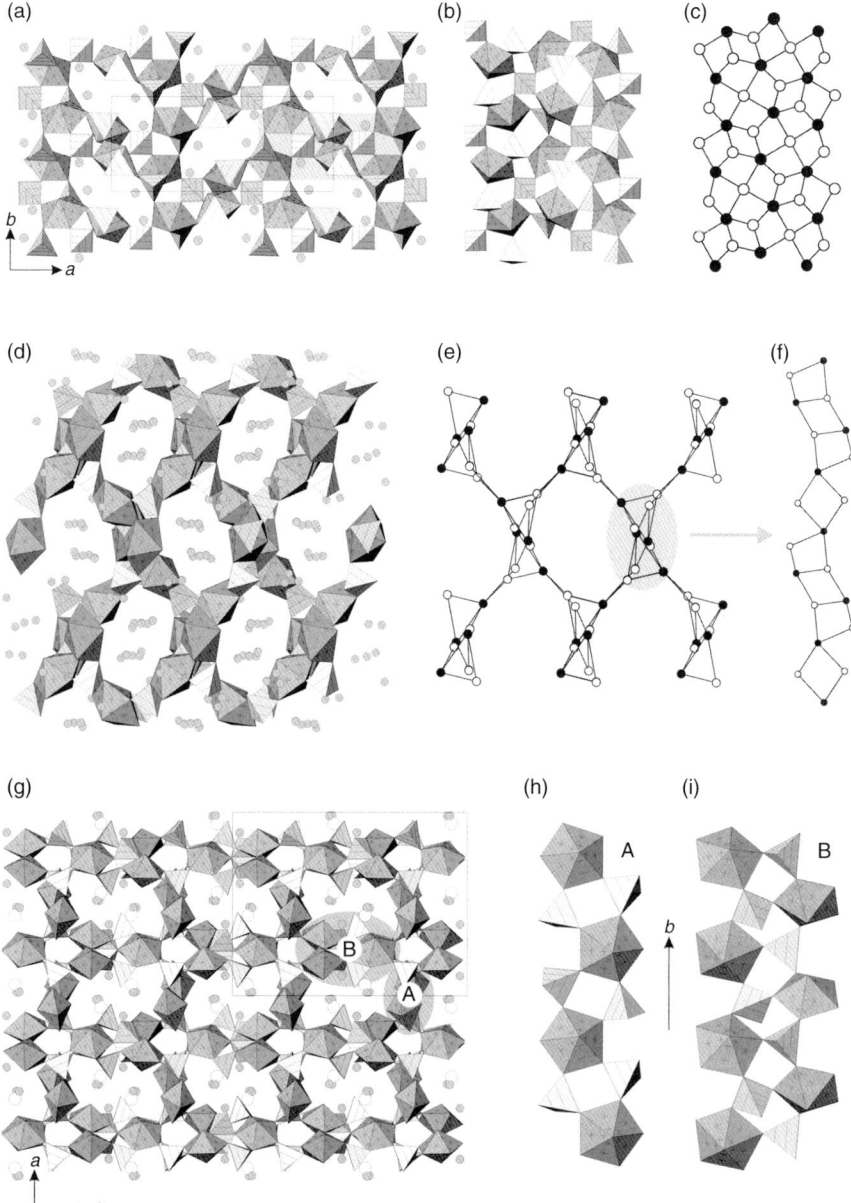

Fig. 3.27. Frameworks based upon fundamental chains in uranyl compounds: 3D uranyl molybdate framework in the structure of α-Cs$_2$[(UO$_2$)$_2$(MoO$_4$)$_3$] (a) the framework can be derived from the uranyl molybdate sheet observed in β-Cs$_2$[(UO$_2$)$_2$(MoO$_4$)$_3$] by cutting it into 1D chains ((b), (c)) and re-assembling them into a 3D structure; 3D framework of U and Se polyhedra in the structure of (H$_3$O)$_2$[(UO$_2$)(SeO$_4$)$_3$(H$_2$O)](H$_2$O)$_4$ (d) its black-and-white graph (e) can be described as being built by polymerization of 1D chains shown in (f); uranyl molybdate framework in the structure of Ba(UO$_2$)$_3$(MoO$_4$)$_4$(H$_2$O)$_4$ (g) can be described as being based upon two types of fundamental chains ((h), (i)).

selenate framework is built up by linkage of the U:Se = 2:3 chains of uranyl bipyramids and selenate tetrahedra whose graph is depicted in Fig. 3.27(f). The graph consists solely of 4-membered rings of black and white vertices.

The uranyl molybdate framework in the structure of $Ba(UO_2)_3(MoO_4)_4(H_2O)_4$ (Tabachenko *et al.* 1984) (Fig. 3.27(g)) can easily be described as based upon two types of fundamental chains, **A** and **B** (Figs. 3.27(h) and (i), respectively). Both chains are parallel to the *b* axis and cross-linked into the framework by sharing their corners.

3.5.4   *Frameworks with M–M and no T–T linkages*

3.5.4.1   *Frameworks based upon chains of corner-sharing M octahedra*

Chains of corner-sharing metal octahedra is one of the basic motifs in the structures of inorganic oxysalts. Here, we present some examples of heteropolyhedral frameworks consisting of such chains interlinked by tetrahedral oxyanions; corresponding compounds and references are given in Table 3.3.

The vanadium phosphate framework from the structure of $VO(H_2PO_4)_2$ (Linde *et al.* 1979) is shown in Fig. 3.28(a). It consists of linear octahedral chains linked into an open framework by doubly protonated bidentate phosphate tetrahedra. A titanite- or kieserite-type octahedral–tetrahedral framework (Fig. 3.28(b)) is common to a large number of inorganic oxysalts, including silicates, sulphates, selenates, phosphates, and arsenates. It can be described as being based upon the butlerite-type chains shown in Fig. 3.28(c). The chains are linked together by sharing terminal octahedral and tetrahedral corners. It can be seen that, due to the bridging tetrahedra, octahedral chains experience rather bent conformation that allowed some Russian crystallographers to call them "dancing chains". This framework has been observed, in particular, in the structure of pauflerite, a β-modification of $VOSO_4$. The structure of α-$VOSO_4$ also contains chains of corner-sharing octahedra (Fig. 3.28(d)). However, in this case, each tetrahedron connects four adjacent chains at the same time, whereas, in β-$VOSO_4$, tetrahedra share corners with three adjacent chains each. As a consequence, octahedral chains in α-$VOSO_4$ have linear conformation (Fig. 3.28(e)). The octahedral–tetrahedral framework in α-$VOSO_4$ can also be described as a fusion of autunite-type sheets (Fig. 3.28(f)) via octahedral corners that are unshared within the sheets (Fig. 3.28(g)). The structure related to α-$VOSO_4$ is that of vlodavetsite, $AlCa_2(SO_4)_2F_2Cl(H_2O)_4$ (Starova *et al.* 1995). In this mineral, octahedral–tetrahedral framework (Fig. 3.29(a)) contains chains of Al- and Ca-centered tetrahedra in the sequence ...–Al–Ca–Ca–Al–Ca–Ca–Al–... (Fig. 3.29(b)). By analogy with α-$VOSO_4$, vlodavetsite can also be described as based upon autunite-type sheets of Ca octahedra and SO_4^{2-} groups (Fig. 3.29(c)) but, in this case, interlinked by additional AlO_4F_2 octahedra (Fig. 3.29(d)).

Another example of an octahedral–tetrahedral framework with butlerite-type chains is shown in Fig. 3.30(a). Here, two butlerite-type chains are linked into a double chain (Fig. 3.30(b)), by analogy with kroehnkite-type chains. This framework has been observed in $Cs[(V_2O_3)(HPO_4)_2(H_2O)]$.

Table 3.3. Frameworks based upon chains of corner-sharing octahedra interlinked by isolated tetrahedra (examples)

$M{:}T$ ratio	Framework formula	Compound	Reference
1:2	$MX(TO_4)_2$	$VO(H_2PO_4)_2$	Linde *et al.* 1979
		$VO(H_2AsO_4)_2$	Wang and Lee 1991; Amoros *et al.* 1992
		$Li_4[VO(AsO_4)_2]$	Aranda *et al.* 1992
1:1	$MX(TO_4)$	kieserite $[Mg(SO_4)(H_2O)]$	Hawthorne *et al.* 1987
		dwornikite $[Ni(SO_4)(H_2O)]$	Wildner and Giester 1991
		szmikite $[Mn(SO_4)(H_2O)]$	Wildner and Giester 1991
		gunningite $[Zn(SO_4)(H_2O)]$	Wildner and Giester 1991
		szomolnokite $[Fe(SO_4)(H_2O)]$	Wildner and Giester 1991
		poitevnite $[(Cu,Fe)(SO_4)(H_2O)]$	Giester *et al.* 1994
		$[MSeO_4(H_2O)]$ M = Mg, Mn, Co, Ni, Zn	Giester and Wildner 1992
		titanite $Ca[TiO(SiO_4)]$	Zachariasen, 1930
		malayaite $Ca[SnO(SiO_4)]$	Higgins and Ribbe 1977
		vanadomalayaite $Ca[VO(SiO_4)]$	Basso *et al.* 1994
		$Mn[^{VI}SiO(SiO_4)]$	Arlt *et al.* 1998
		$Ca[AlFSiO_4]$	Troitzsch and Ellis 1999
		pauflerite β-$[VOSO_4]$	Boghosian *et al.* 1995; Krivovichev *et al.* 2007
		β-Li$[VOPO_4]$	Lii *et al.* 1991a
		$[VO(HPO_4)]$	Wilde *et al.* 1998
		Na$[VOPO_4]$	Lii *et al.* 1991b; Benhamada *et al.* 1992a
		tilasite $Ca[Mg(AsO_4)F]$	Bladh *et al.* 1972; Bermanec, 1994
		maxwellite $(Na,Ca)(Fe,Al)(AsO_4)(F,OH)$	Cooper and Hawthorne, 1995
		wilhelmkleinite $Zn[Fe_2(OH)_2(AsO_4)_2]$	Adiwidjaja *et al.* 2000
1:1	$MX(TO_4)$	α-Li$[VOPO_4]$	Dupre *et al.* 2004
		α-$[VOSO_4]$	Longo *et al.* 1970
		$[MoOPO_4]$	Kierkegaard and Longo 1970
		$NbOPO_4$	Amos *et al.* 1998
		$TaOPO_4$	Longo and Arnott 1970
		$VOMoO_4$	Eick and Kihlborg 1966
1:1	$MX(TO_4)$	$K[TiO(PO_4)]$	Tordjman *et al.* 1974
		α-Rb$(TiO)(PO_4)$	Lyakhov *et al.* 1993
		$Tl[TiO(PO_4)]$	Harrison *et al.* 1990
		$K[FeF(PO_4)]$	Belokoneva *et al.* 1990
		$(NH_4)[FeF(PO_4)]$	Loiseau *et al.* 1994
		$K[CrF(PO_4)]$	Slobodyanik *et al.* 1991
		$K[AlF(PO_4)]$	Slovokhotova *et al.* 1991; Kirkby *et al.* 1995
		β-Na$[SbO(GeO_4)]$	Belokoneva and Mill' 1994
		$Ag[SbO(SiO_4)]$	Belokoneva and Mill' 1994
		$K[VO(PO_4)]$	Phillips *et al.* 1990; Benhamada *et al.* 1991
		$(NH_4)[VO(PO_4)]$	Haushalter *et al.* 1994a; Schindler et al. 1997

Table 3.3. *Continued*

M:T ratio	Framework formula	Compound	Reference
		$(K,Li)[TiO(AsO_4)]$	Harrison *et al.* 1997
		$(NH_4)[GaF(PO_4)]$	Loiseau *et al.* 2000
		$(NH_4)[VF(PO_4)]$	Alda *et al.* 2003
1:1	$M_2X_3(TO_4)_2$	$Cs[(V_2O_3)(HPO_4)_2(H_2O)]$	Haushalter *et al.* 1994b
3:4	$M_3X_3(TO_4)_4$	$K_2(VO)_3(HPO_4)_4$	Lii and Tsai 1991
1:1	$MX(TO_4)$	$Rb[In(OH)PO_4]$	Lii 1996b
		$K[In(OH)PO_4]$	Hriljac *et al.* 1996

M = V, Ti, In, Sb, Fe, Mg, Mn, Zn, Ni, Al, Ga; X = O, OH, F; T = As, P, S, Se, Si, Ge.

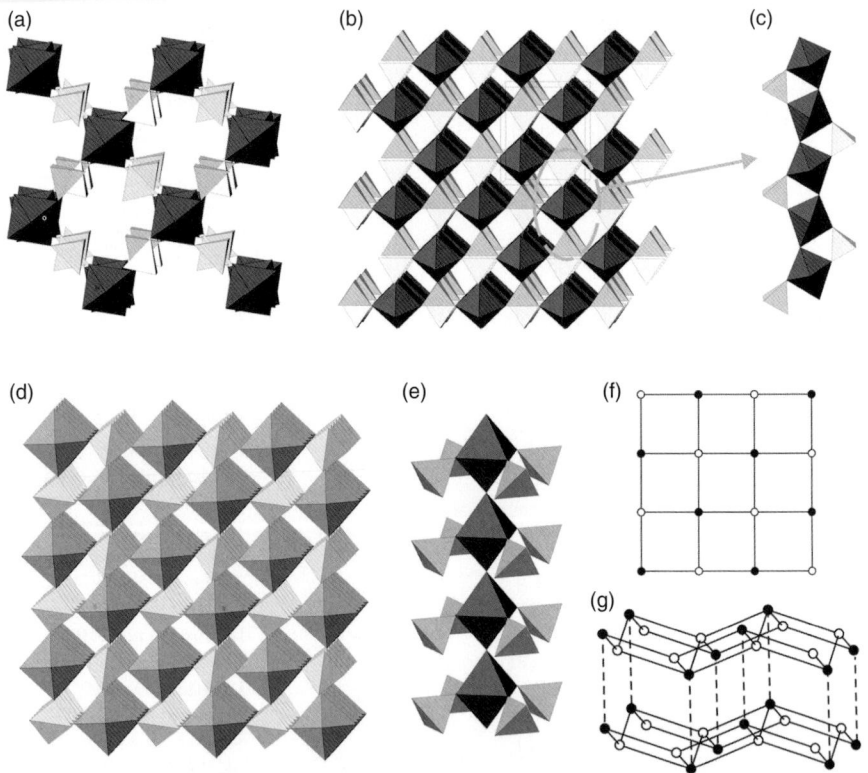

Fig. 3.28. Octahedral–tetrahedral frameworks based upon chains of corner-sharing octahedra: vanadium phosphate framework in $VO(H_2PO_4)_2$ (a); titanite- or kieserite-type octahedral–tetrahedral framework (b) based upon the butlerite-type chains (c); framework of α-$VOSO_4$ (d) and environment of its octahedral chain (e); the topology of this framework can also be described as consisting of autunite-type graphs (f) united by vertical links between black vertices (g).

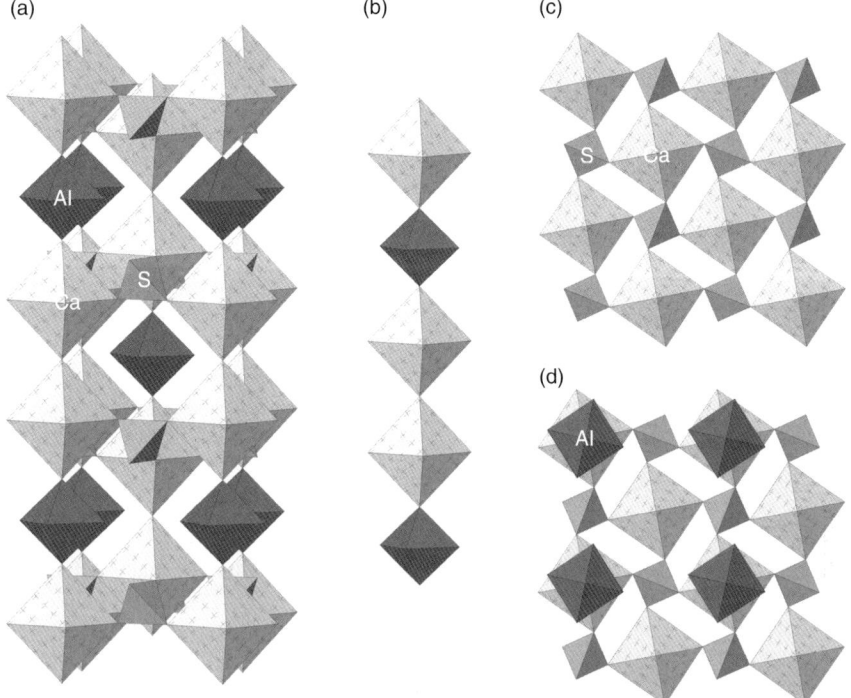

Fig. 3.29. The structure of vlodavetsite $AlCa_2(SO_4)_2F_2Cl(H_2O)_4$ (a) contains chains of Al- and Ca-centered tetrahedra in the sequence...–Al–Ca–Ca–Al–Ca–Ca–Al–...(b). Vlodavetsite framework can be described as being based upon autunite-type sheets of Ca octahedra and SO_4^{2-} groups (c) linked by additional AlO_4F_2 octahedra; figure (d) shows positions of Al-octahedra relative to the Ca-sulphate sheets.

The butlerite-type chain can be viewed as an octahedral chain incrustated on both sides by tetrahedral oxyanions. Another incrustation type can be seen in the funda-mental chain that occurs in the structure of $K_2(VO)_3(HPO_4)_4$ (Fig. 3.30(c)). Chains of corner-sharing octahedra have bridging tetrahedra on both sides in a symmetrical configuration (Fig. 3.30(d)) so that the chain has the $M{:}T$ ratio of 3:4 that is preserved during polymerization of the chains into a framework.

All octahedral chains described above in this section contain shared corners in a *trans*-configuration. The structures of $A[In(OH)PO_4]$ (A = K, Rb) possess chains of corner-sharing octahedra with a *cis*-configuration of shared corners (Figs. 3.31(a) and (b)). As a result, the chain has a spiral conformation with four octahedra within its identity period (Fig. 3.31(b)). In the structure of $K[TiO(PO_4)]$, a famous non-linear optical material KTP, the *trans*- and *cis*-configurations alternate within the octahe-dral chain (Figs. 3.31(c) and (d)). Due to the presence of *cis*-octahedra, the chains are, again, not linear and possess a wave-like conformation.

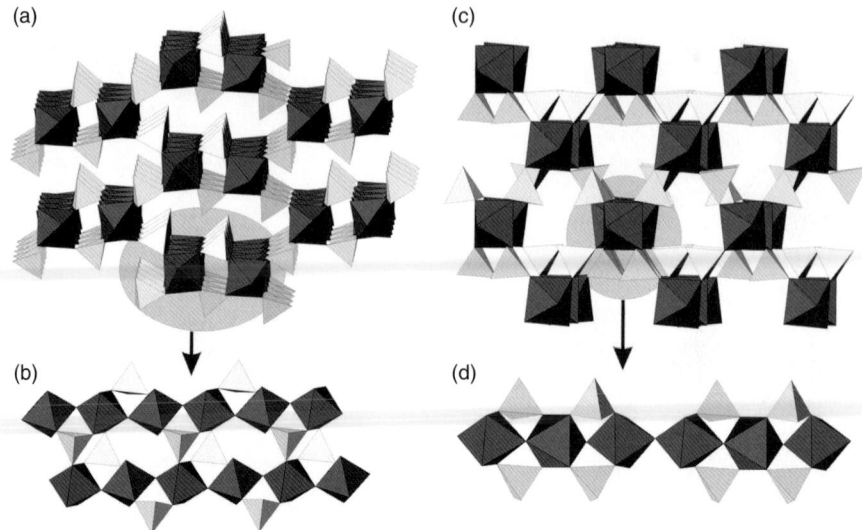

Fig. 3.30. Octahedral–tetrahedral framework with double butlerite-type chains in $Cs[(V_2O_3)(HPO_4)_2(H_2O)]$ (a) and its fundamental chain (b); framework in $K_2(VO)_3(HPO_4)_4$ (c) and its fundamental chain with the $M:T$ ratio of 3:4 (d).

3.5.4.2   Frameworks consisting of finite clusters of corner-sharing octahedra

Two frameworks based upon complex chains consisting of pairs of corner-sharing octahedra are shown in Fig. 3.32. Fundamental chains (Figs. 3.32(b) and (e)) of both frameworks are based upon double octahedra linked by bridging tetrahedra. However, the topology of linkage is different. In the structure of $Ni_{0.5}(VO)(PO_4)(H_2O)_{1.5}$ (Lii and Mao, 1992), all tetrahedra within the chain are tridentate (Figs. 3.32(b) and (c)), whereas, in the structure of $Ti_2O(PO_4)_2(H_2O)_2$ (Salvado *et al.* 1997; Poojary *et al.* 1997b), tetrahedra are bi- and tridentate within the chain (Figs. 3.32(e) and (f)). The chains are planar and orientations of their planes are different in the two structures. In $Ni_{0.5}(VO)(PO_4)(H_2O)_{1.5}$, the chains are inclined relative to each other, whereas, in $Ti_2O(PO_4)_2(H_2O)_2$, the chains are parallel. An interesting case of cyclic orientation of fundamental chains is shown in Fig. 3.33. The octahedral–tetrahedral framework of $(H_3O)_3(Ti_6(H_2O)_3)O_3(PO_4)_7(H_2O)$ (Serre and Feréy, 1999) consists of fundamental chains of pairs of corner-sharing octahedra linked by bidentate tetrahedra (Fig. 3.33(b)). The chains have three different orientations and are interlinked by an additional tetrahedron (Fig. 3.33(c)).

Figure 3.34 shows a three-dimensional framework based upon octahedral trimers interlinked by tetradentate tetrahedra. This kind of structural architecture has been observed in the structure of $A_2[Ti_3O_2(PO_4)_2(HPO_4)_2]$ (A = Rb, NH_4) (Harrison *et al.* 1994; Poojary *et al.* 1997b).

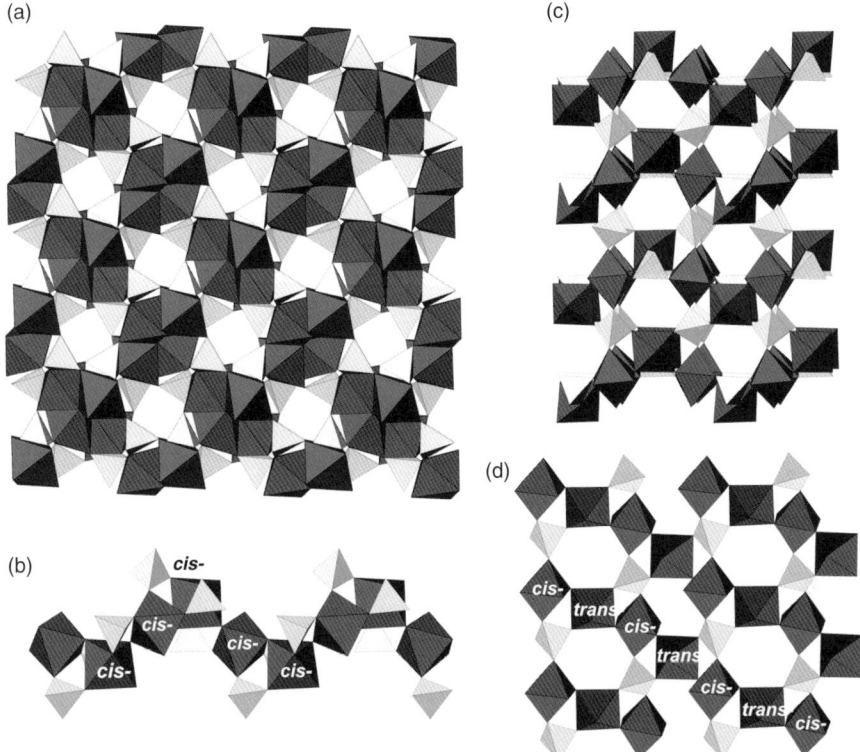

Fig. 3.31. Octahedral–tetrahedral frameworks based upon chains of corner-sharing octahedra with *cis*-orientations of shared octahedral corners: the structures of $A[In(OH)PO_4]$ ($A =$ K, Rb) possess spiral chains of *cis*-octahedra (b); the structure of $K[TiO(PO_4)]$ (c) has *trans*- and *cis*- configurations alternating within the octahedral chain (d).

3.5.5   *Frameworks with M=M and no T–T linkages*

There are numerous examples of octahedral–tetrahedral frameworks based upon chains of edge-sharing octahedra. Two of the simplest frameworks are shown in Fig. 3.35. The list of representative compounds is given in Table 3.4. The backbone of the framework shown in Fig. 3.35(a) is the chain of *trans*-edge-sharing octahedra incrustated by bidentate tetrahedra (Fig. 3.35(b)). This framework structure is common for a large number of oxysalts with simple formula $M(TO_4)$ (Table 3.4). Minerals of the descloizite group and related compounds are based upon a more open framework with composition $A[M_2(TO_4)(OH)]$ (Fig. 3.35(c)). The *M* cations are octahedrally coordinated by four O atoms of TO_4 groups and two OH anions. The resulting $MO_4(OH)_2$ octahedra share *trans*-O···OH edges to form chains (Fig. 3.35(d)) that are interlinked by tetrahedra into a three-dimensional framework. Comparison

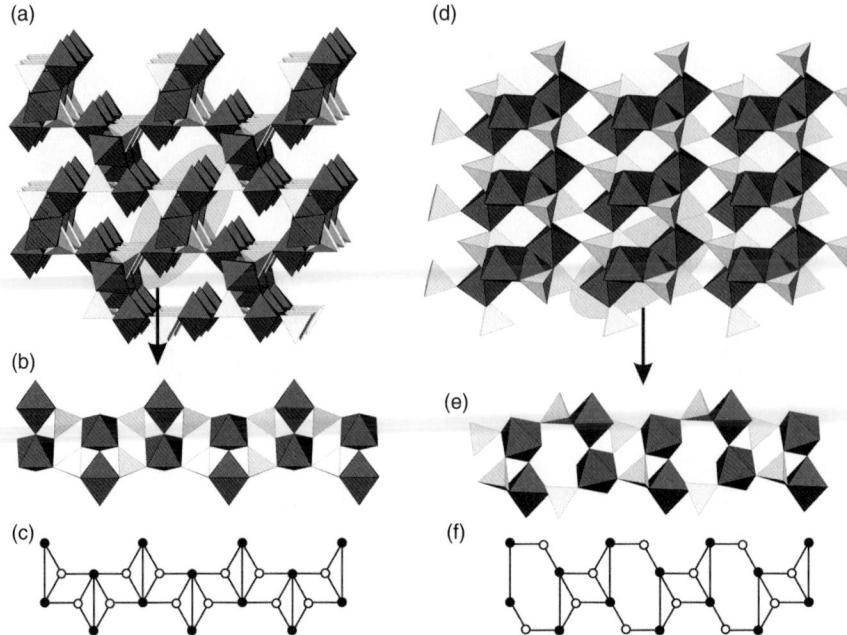

Fig. 3.32. Octahedral–tetrahedral frameworks with fundamental chains based upon octahedral dimers: framework in $Ni_{0.5}(VO)(PO_4)(H_2O)_{1.5}$ (a), its fundamental chain (b) and graph of the latter (c); framework in $Ti_2O(PO_4)_2(H_2O)_2$ (d), and its fundamental chain in polyhedral (e) and symbolic (f) description.

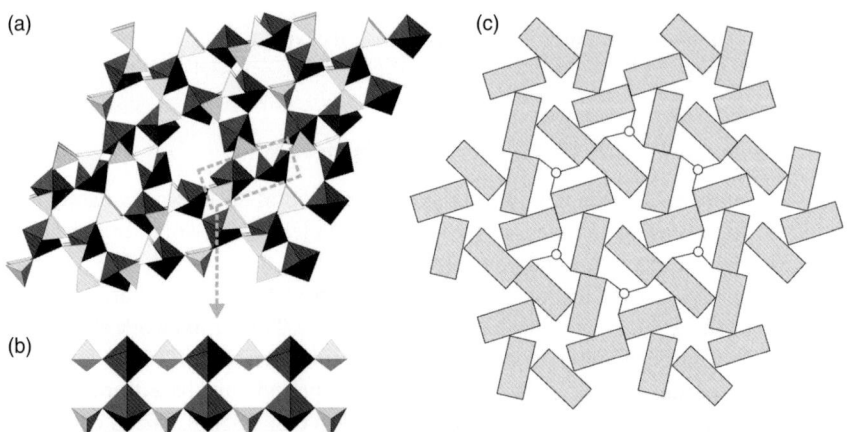

Fig. 3.33. Octahedral–tetrahedral framework in $(H_3O)_3(Ti_6(H_2O)_3)O_3(PO_4)_7(H_2O)$ (a), its fundamental chain (b) and the scheme of chain orientations showing an additional tetrahedron (white circle) providing three-dimensional framework integrity (c).

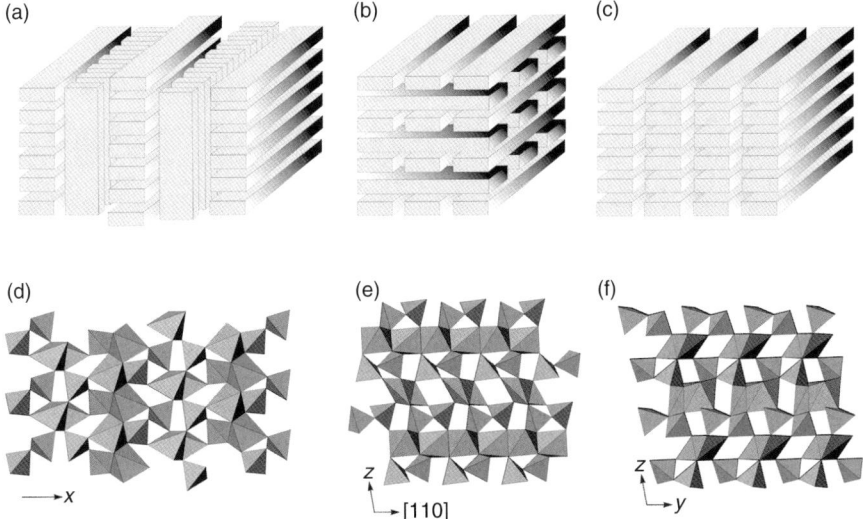

Fig. 3.38. Chains of Cu coordination polyhedra in the structures of $Cu_2V_2O_7$ polymorphs: blossite (a), ziesite (b) and γ-$Cu_2V_2O_7$ ((c), (d)).

1996; Braunbarth *et al.* 2000; Kuznicki *et al.* 2001; Nair *et al.* 2001), have received much attention in recent years because of their applications in catalysis, gas separation, optoelectronics, ion-exchange, etc. The structure of zorite was first reported by Sandomirskii and Belov (1979). It is based upon a complex octahedral–tetrahedral framework that consists of chains of corner-sharing octahedra cross-linked by double chains of disordered silicate tetrahedra (Fig. 3.40(a)). The octahedral chains run along [010] (Fig. 3.40(c)), whereas tetrahedral chains run along [001] (Fig. 3.40(b)). The tetrahedral chains are further linked by additional half-occupied Ti octahedra (or pentahedra, see discussion in Braunbarth *et al.* 2000). As a result, the 3D network corresponding to the zorite framework contains large polyhedral voids in the form of octangular prisms (Fig. 3.40(d)). These voids are arranged into columns so that large tunnels are formed along [001] (Fig. 3.40(e)).

3.5.7.2 *Benitoite net as based upon arrangement of polyhedral units and tubular units*

Figure 3.41(a) shows the structure of benitoite, $BaTiSi_3O_9$, projected along the *c*-axis (Zachariasen 1930b; Fischer 1969). The benitoite structure type is common for many minerals and inorganic compounds, including bazirite, $BaZrSi_3O_9$, pabstite, $BaSnSi_3O_9$ (Hawthorne 1987; Choisnet *et al.* 1972), $BaSi^{VI}Si_3^{IV}O_9$ (Finger *et al.* 1985), $KCaP_3O_9$ (Sandstroem and Bostroem 2004), etc. The structure represents a framework of isolated MO_6 octahedra and Si_3O_9 silicate rings. Nodal representation of the

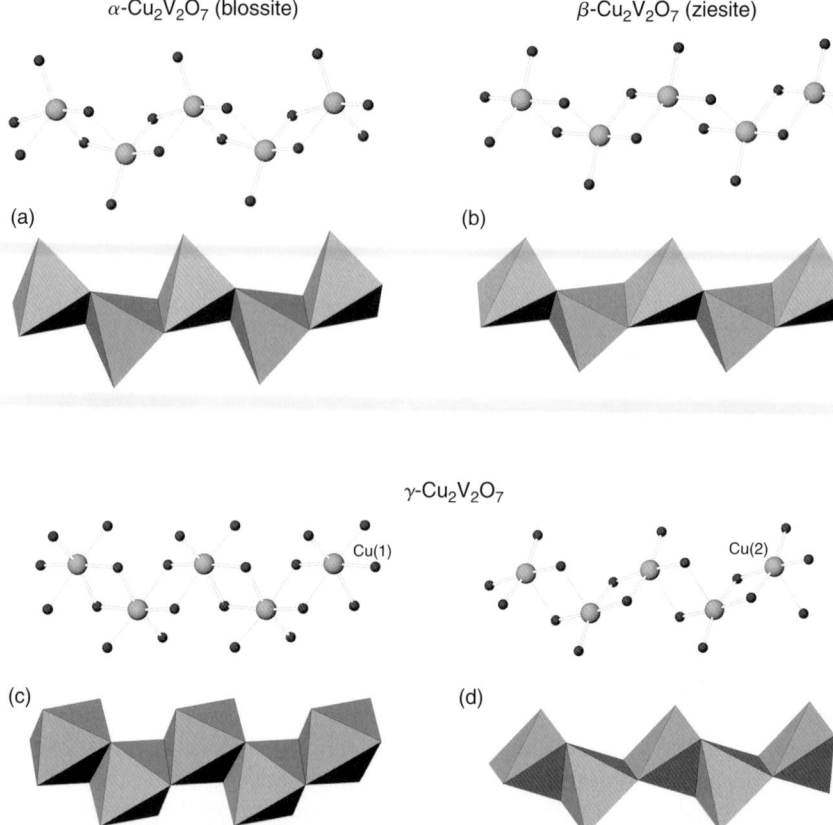

α-$Cu_2V_2O_7$ (blossite)

β-$Cu_2V_2O_7$ (ziesite)

(a)

(b)

γ-$Cu_2V_2O_7$

Cu(1)

Cu(2)

(c)

(d)

Fig. 3.39. Schemes illustrating orientation of chains of Cu^{2+} coordination polyhedra in the structures of blossite (a), ziesite (b), and γ-$Cu_2V_2O_7$ (c). Projections of the structures of blossite (d), ziesite (e) and γ-$Cu_2V_2O_7$ (f).

framework is shown in Fig. 3.41(b). Analysis of its topology allows to subdivide two polyhedral units, 4^3 and $6^3 3^2$, shown in Figs. 3.41(c) and (d), respectively. However, the benitoite 3D net cannot be constructed solely by these units as it contains the 1D tubular unit shown in Fig. 3.41(e). This unit cannot be unequivocally separated into polyhedral units and thus should be considered as an independent building unit of the benitoite net.

The tubular unit shown in Fig. 3.41(e) can be considered as a black-and-white graph on the surface of a cylinder. It consists of two types of rings: 6-membered rings **bw²bw²** and 4-membered rings **bwbw**. The tubular unit can be constructed using a procedure that is used to describe the topology of carbon nanotubes and that is known as *folding and gluing* (Kirby 1997). First, one constructs a tape-like black-and-white

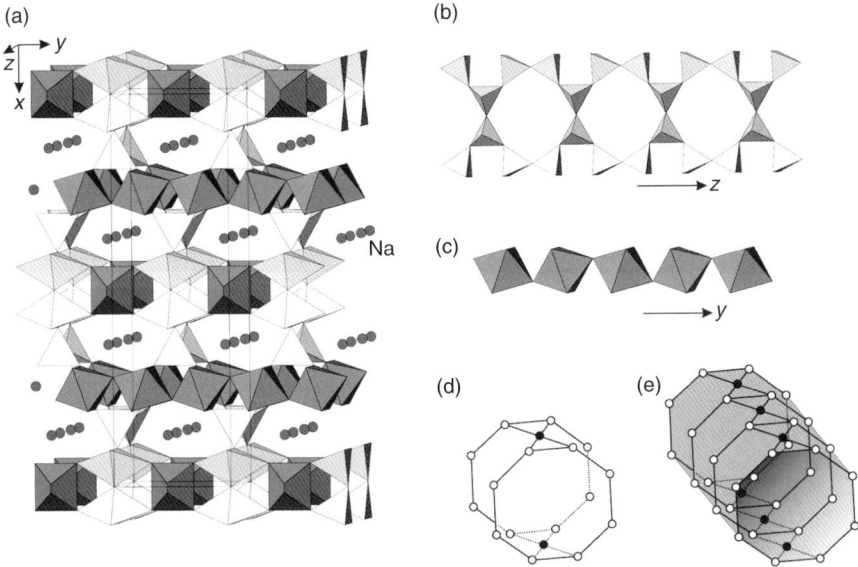

Fig. 3.40. Structure of zorite, $Na_6[Ti(Ti,Nb)_4(Si_6O_{17})_2(OH)_5](H_2O)_{10.5}$ (a) is based upon chains of corner-sharing octahedra (c) cross-linked by double chains of disordered silicate tetrahedra (b). The 3D network corresponding to the zorite framework contains large polyhedral voids in the form of octangular prism (d). These voids are arranged into columns so that large tunnels along [001] are formed (e).

graph that is a fragment of a tiling of a 2D plane. Equivalent points on the sides of the tape are identified by letters *a, b, c, d* .... In order to get the tubular unit, the tape is folded and opposite sides are glued by joining equivalent points to make a cylinder. The idealized model for a tubular unit in benitoite and its *prototape* are shown in Figs. 3.41(f) and (g), respectively.

3.5.7.3 *Tubular units, their topology, symmetry and classification*

Figure 3.42(a) shows the octahedral–tetrahedral framework observed in the structure of vlasovite, $Na_2ZrSi_4O_{11}$ (Voronkov and Pyatenko 1961; Voronkov *et al.* 1974; Vitins *et al.* 1996). This framework consists of chains of SiO_4 tetrahedra interlinked by ZrO_6 octahedra. The corresponding 3D net (Fig. 3.42(b)) is based upon tubular units (Fig. 3.42(c)) that can be constructed by folding and gluing the tape of regular hexagons as is shown in Figs. 3.42(d) and (e). The tubular unit is *achiral* as it has four mirror planes: three parallel and one perpendicular to the extension of the unit.

A different kind of tubular unit is observed in the octahedral–tetrahedral framework of hilairite-type $Na_2ZrSi_3O_9(H_2O)_3$ (Ilyushin *et al.* 1981a) (Fig. 3.43). This framework is built by corner sharing of isolated octahedra ZrO_6 and silicate tetrahedra of

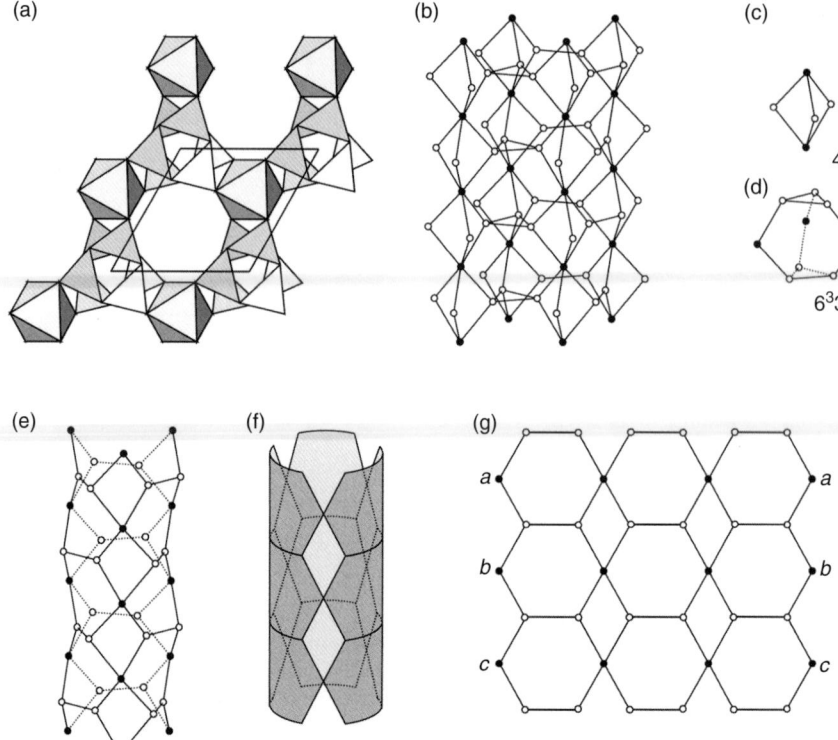

Fig. 3.41. Octahedral–tetrahedral framework in the structure of benitoite, BaTiSi$_3$O$_9$ projected along the c-axis (a) and its nodal representation (b). The benitoite net can be represented as consisting of polyhedral voids of two types ((c) and (d)) and tubular units ((e), (f)). The tubular unit can be constructed by folding and gluing the tape-like black-and-white graph shown in (g). See text for details.

the 6-er (or *sechster*) single Si$_6$O$_{18}$ chains. The hilairite net is assembled from tubular units composed of 8- and 3-membered rings. Unfolding of the tubular unit provides the tape shown in Fig. 3.43(d). To make the tubular unit, one has to fold the tape and to join the points identified by the same letters. In contrast to benitoite and vlasovite, equivalent points are not opposite to each other but, instead, are in diagonal orientation. Their joining produces a *chiral* unit that has a *helical* structure. It is noteworthy that the structures of the hilairite-group minerals (sazykinaite-(Y), hilairite, calciohilairite, komkovite and pyatenkoite-(Y)) have the same space group $R32$ that contains only rotational symmetry elements.

The concept of tubular units is useful not only from the point of structure description. It helps to recognize the internal structure of framework channels in terms of its

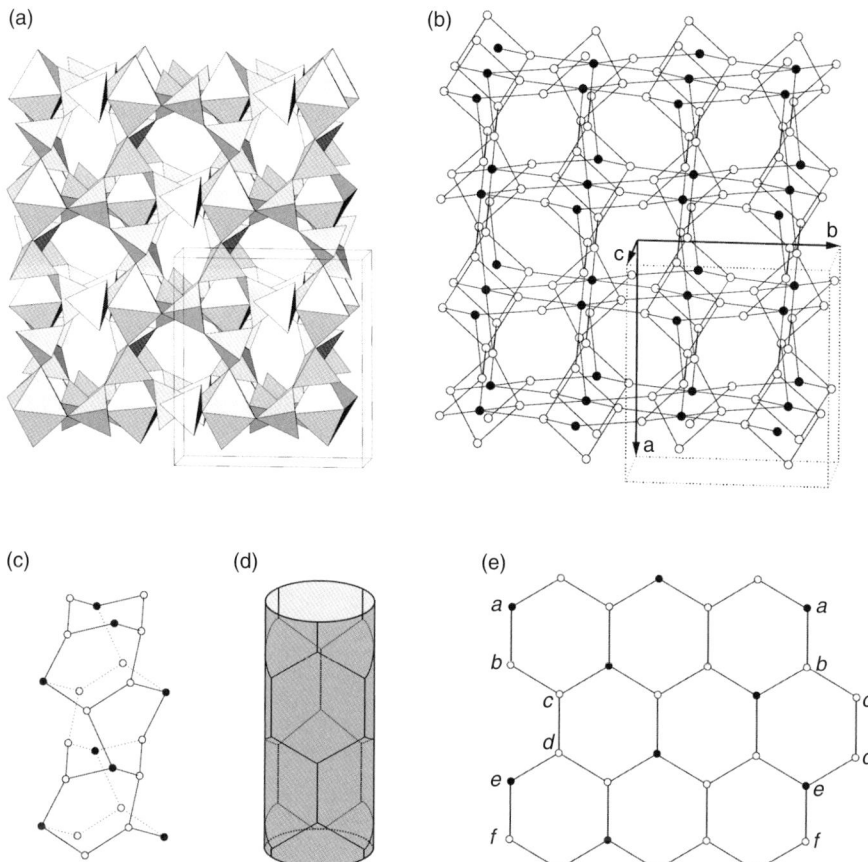

Fig. 3.42. Octahedral–tetrahedral framework in the structure of vlasovite, $Na_2ZrSi_4O_{11}$ (a) and its nodal representation (b). The 3D net (b) is based upon tubular units (c) that can be constructed by folding and gluing the tape of regular hexagons (d). Note that the tubular unit is achiral.

symmetry. For example, the 3D nets corresponding to the chiral 4:5 and 5:7 uranyl molybdate frameworks discussed above can also be described in terms of tubular units (Fig. 3.44). In both cases, tubular units have a chiral helical structure that is in agreement with the symmetry of the channels in the real structures ($6_5 22$ for 4:5 and 6_5 for 5:7 frameworks).

Thus, tubular units can be classified according to: (i) topology of their prototapes; (ii) their symmetry. The first feature allows classification of tubular units from the topological viewpoint and can be useful to identify windows and pores. The second helps to recognize such an important property of a framework channel as its chirality.

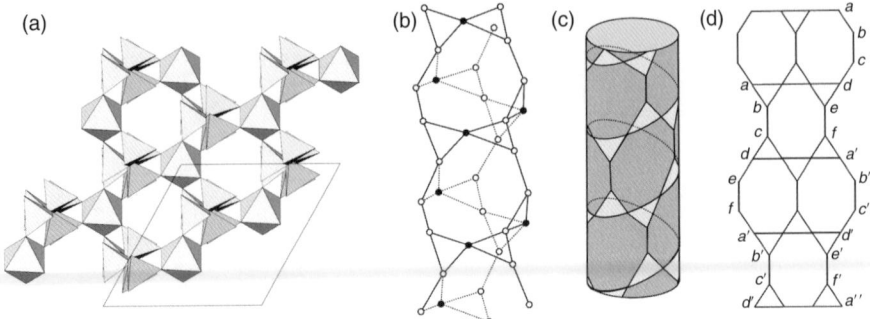

Fig. 3.43. Octahedral–tetrahedral framework in the structure of hilairite $Na_2ZrSi_3O_9(H_2O)_3$ (a) consists of tubular units (b,c) composed from 8- and 3-membered rings. Unfolding of the tubular unit provides the tape shown in (d). Note that the tubular unit is chiral.

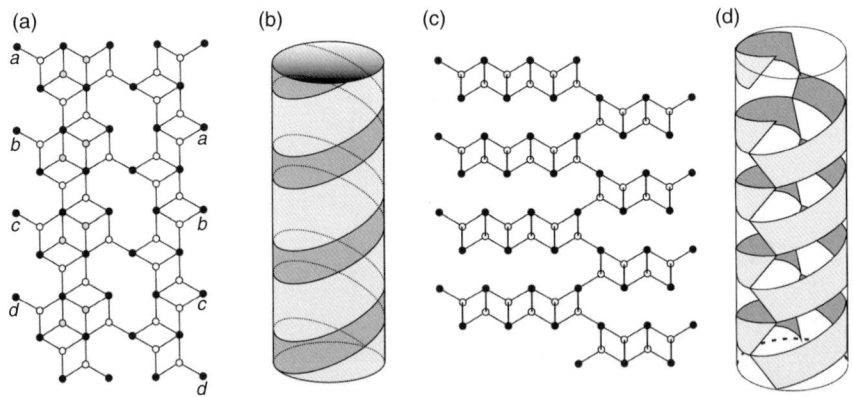

Fig. 3.44. Chiral tubular units in the structures of the U:Mo = 4:5 and 5:7 uranyl molybdate frameworks can be described in terms of tubular units ((b) and (d), respectively). The tubular units can be obtained by folding and gluing tapes shown in (a) and (c), respectively.

3.6   Frameworks based upon 2D units

3.6.1   *Frameworks based upon sheets with no T–T linkages*

The frameworks with no *T–T* linkages can be conveniently described as being based upon 2D topologies that have been reviewed in Chapter 2. For example, the octahedral–tetrahedral framework shown in Fig. 3.45(a) consists of 2D sheets with the topology corresponding to the graph *cc2–1:1–17* (Figs. 3.45(b) and (c)). Within the sheet, all tetrahedra are tridentate, thus leaving the fourth corner unshared. This corner can be oriented either up (**U**) or down (**D**) relative to the plane of the sheet. Octahedra within

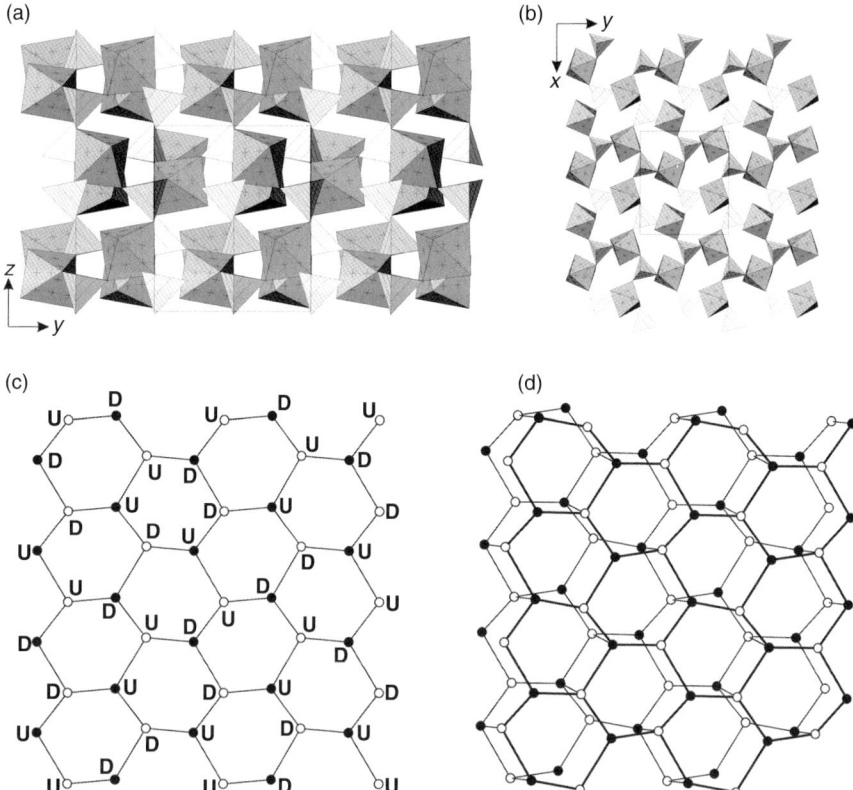

Fig. 3.45. Projection of the crystal structure of $Zn(SeO_4)(H_2O)_2$ onto the (100) plane (a), sheet of octahedra and tetrahedra parallel to the (001) plane (b) and its black-and-white graph (c), and scheme of connection of adjacent 2D graphs into a 3D graph (d).

the sheet are again tridentate, and each octahedron has two H_2O ligands occupying its corners so that these corners cannot act as bridging between adjacent polyhedra. The sixth corner is, however, again unshared within the sheet and can possess either **U** or **D** orientations. Figure 3.45(c) shows the 2D graph **cc2–1:1–17** with the letters **U** and **D** written near the vertices. The graph consists of 6-membered cycles with alternating black and white vertices and it can be seen that the orientational sequence within one cycle is **UUDDUD**. The whole framework might be depicted as a 3D graph consisting of the 2D **cc2–1:1–17** subgraphs linked through **D–U** linkage between octahedra and tetrahedra from adjacent sheets such that polyhedron from the sheet below has an **U**-orientation, whereas a polyhedron from the sheet above has a **D**-orientation. The fragment of the resulting graph is shown in Fig. 3.45(d). This type of framework structure is typical for a number of minerals and isostructural synthetic compounds of the variscite family, some examples of which are given in Table 3.5.

Table 3.5. Crystallographic data for compounds with general composition $M(TO_4)(H_2O)_2$ that belong to the variscite-type octahedral–tetrahedral framework

Mineral name	Chemical formula	Reference
scorodite	$Fe(AsO_4)(H_2O)_2$	Hawthorne 1976
yanomamite	$In(AsO_4)(H_2O)_2$	Tang *et al.* 2001
mansfeldite	$Al(AsO_4)(H_2O)_2$	Harrison 2000
variscite	$Al(PO_4)(H_2O)_2$	Kniep *et al.* 1977
strengite	$Fe(PO_4)(H_2O)_2$	Song *et al.* 2002
–	$Ga(PO_4)(H_2O)_2$	Loiseau *et al.* 1998
–	$Ga(AsO_4)(H_2O)_2$	Dick 1997
–	$In(PO_4)(H_2O)_2$	Xu *et al.* 1995
–	$Zn(SeO_4)(H_2O)_2$	Krivovichev 2006
–	$V(PO_4)(H_2O)_2$	Schindler *et al.* 1995
–	$In(PO_4)(H_2O)_2$	Sugiyama *et al.* 1999

The structure of $(VO)_3(PO_4)_2(H_2O)_9$ (Teller *et al.* 1992) is also based upon the octahedral–tetrahedral sheet with the ***cc2–1:1–17*** topology (Fig. 3.46). However, in this case, the sheets are not linked directly to each other, but through an additional V-centered octahedron located between the sheets. This type of framework is sometimes called pillared, implying that there is a kind of pillar linking adjacent sheets.

Two more examples of pillared frameworks are shown in Fig. 3.47. In both cases, the octahedral–tetrahedral sheets are interlinked by additional octahedra. However, the topology of the underlying sheets is different. In $(NH_3)[Ti_3(PO_4)_4(H_2O)_2]$ (Poojary *et al.* 1997b), the sheet has the ***cc2–1:2–1*** topology, whereas, in $Fe_3(HPO_4)_4(H_2O)_4$ (Vencato *et al.* 1986), the underying sheet's topology is that of the ***cc2–1:2–1*** graph (Chapter 2).

It is of interest that not all 2D topologies that exist as underlying graphs for 3D frameworks have been found in layered structures. Figure 3.48 shows two counterexamples. The structure of $CsIn_3H_2(SeO_3)_6(H_2O)_2$ (Rastsvetaeva *et al.* 1984) consists of linked sheets with the topology shown in Figs. 3.48(b) and (c). This topology has no analogs among known layered structures. The structure of α-$(NH_4)V(HPO_4)_2$ (Bircsak and Harrison 1998) is a pillared framework containing sheets with the elegant topology shown in Figs. 3.48(e) and (f). Again, we could not find any analogs of this topology in purely layered compounds.

In the examples of pillared frameworks described above, the role of pillars is usually played by octahedral species, i.e. polyhedra with higher number of vertices. In the structure of $(NH_4)_2Cd_2(SeO_4)_3(H_2O)_3$ (Martinez *et al.* 1990) (Fig. 3.49), autunite-type sheets are linked into an octahedral–tetrahedral framework by bidentate selenate tetrahedra, i.e. by species with lower coordination numbers.

Figures 3.50(a) and (c) show two frameworks that are based upon identical sheets of pairs of edge-sharing tetrahedra interlinked by selenite triangular pyramids (Fig. 3.50(b)). Both frameworks are characteristic for Fe(III) selenites: $FeH(SeO_3)_2$

(a) (b)

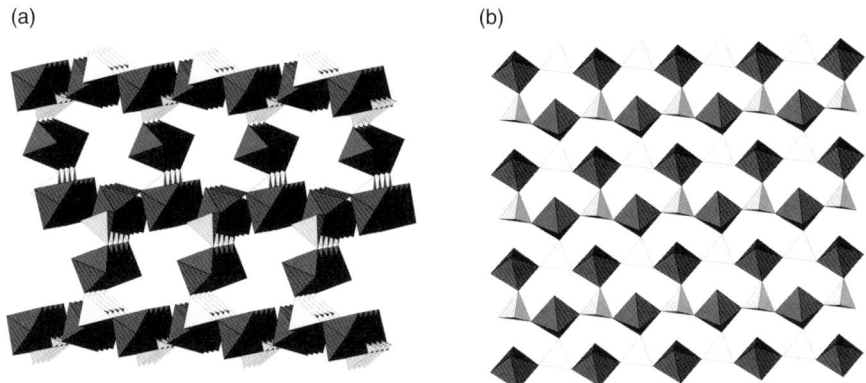

Fig. 3.46. Octahedral–tetrahedral framework in the structure of $(VO)_3(PO_4)_2(H_2O)_9$ (a) is also based upon the octahedral–tetrahedral sheet with the *cc2–1:1–17* topology (b).

(Fig. 3.50(a); Valkonen and Koskenlinna 1978) and $Fe_2(SeO_3)_3(H_2O)$ (Fig. 3.50(c); Giester 1993b). The sheet itself has the Fe:Se ratio equal to 1:1. Different chemical compositions result from different mode of linkages between the sheets. In $FeH(SeO_3)_2$, the sheets are linked by the bidentate SeO_2OH^- anions (protonation prevents one of the O^{2-} anions acting as a bridging ligand), whereas the SeO_3^{2-} anions in $Fe_2(SeO_3)_3(H_2O)$ are tridentate (no protonation is observed; instead, there are interstitial H_2O molecules in the framework cavities).

3.6.2 Frameworks based upon sheets with T–T linkages

3.6.2.1 Octahedral–tetrahedral frameworks with 2D tetrahedral anions

It is natural to describe octahedral–tetrahedral frameworks with $D_{tetr} = 2$ as being based upon 2D tetrahedral anions interlinked by arrays of MO_6 octahedra. Obviously, topological classification of such structures can be made using the topological properties of tetrahedral sheets as the classification criteria (Table 3.1).

Figure 3.51(a) shows the structure of cavansite, $Ca(VO)(Si_4O_{10})(H_2O)_4$ (Evans 1973; Rinaldi *et al.* 1975; Solov'ev *et al.* 1993). It consists of the $[VO(Si_4O_{10})]$ framework based upon Si_4O_{10} silicate sheets linked through isolated VO_5 tetragonal pyramids. Figure 3.51(b) shows an octahedral–tetrahedral framework present in the structure of KNAURSI [$=KNa_3(UO_2)_2(Si_4O_{10})_2(H_2O)_4$], the phase that formed during vapor hydration of an actinide-bearing borosilicate waste glass (Burns *et al.* 2000). The KNAURSI framework is based upon Si_4O_{10} silicate sheets interlinked via UO_6 distorted octahedra. The silicate sheets in the structures of cavansite and KNAURSI are shown in Figs. 3.51(c) and (e), respectively. Their idealized versions are shown in Figs. 3.51(d) and (f), respectively. Obviously, the overall topology of the sheets is identical: they consist of 4- and 8-membered rings of tetrahedra present in the 1:1 ratio. Thus, we can designate the sheet topology using its ring symbol as 8^14^1.

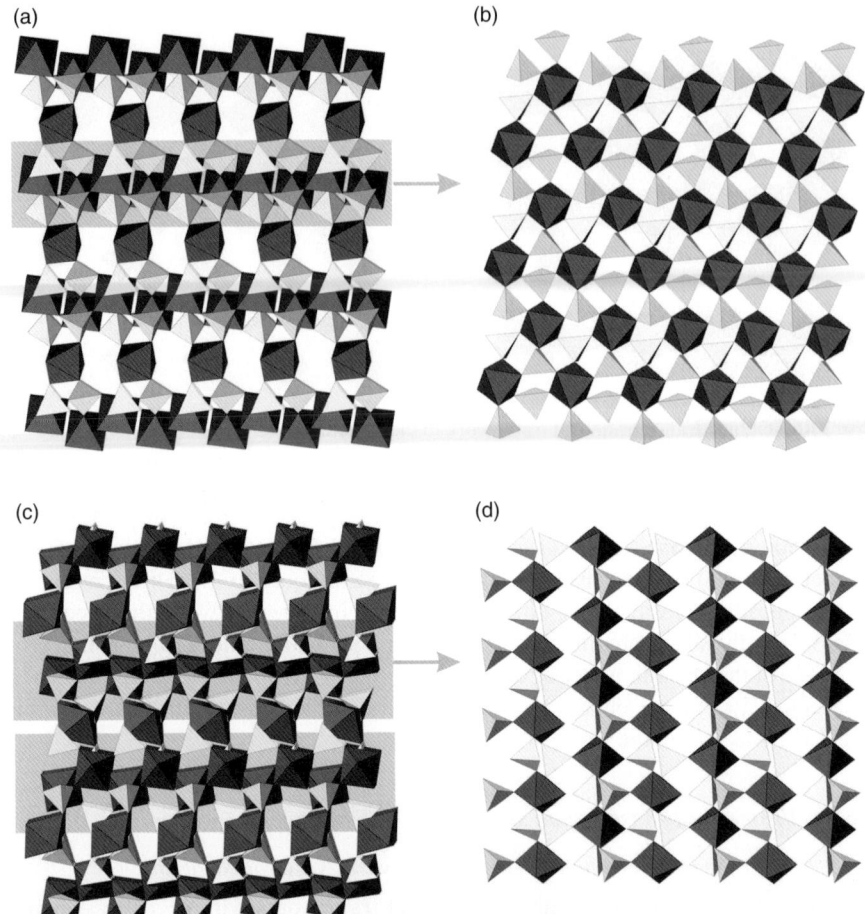

Fig. 3.47. Examples of pillared octahedral–tetrahedral frameworks. The topology of the underlying sheets of the framework in $(NH_3)[Ti_3(PO_4)_4(H_2O)_2]$ (a) is that of the *cc2–1:2–1* type (b), whereas, in $Fe_3(HPO_4)_4(H_2O)_4$ (c), the underying sheet's topology is that of the *cc2–1:2–1* type (d).

Inspection of the idealized versions of the Si_4O_{10} silicate sheets shown in Figs. 3.51(d) and (f) demonstrates that, despite identical ring symbols, the two sheets are topologically different: one cannot be transformed to the other without topological reconstructions. The point is that non-shared corners of tetrahedra have different patterns of "up" and "down" orientations relative to the plane of the sheet. Thus, the two sheets should be considered as *geometrical isomers*. A detailed search of octahedral–tetrahedral frameworks based upon Si_4O_{10} sheets with the 8^14^1 ring symbol allowed the four different geometrical isomers shown in Fig. 3.52 to be identified.

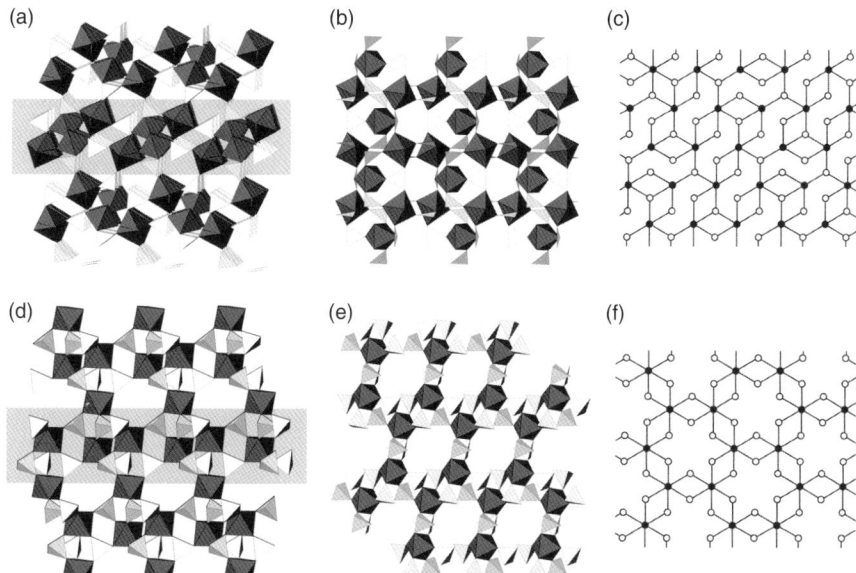

Fig. 3.48. 3D frameworks that are based upon sheets that have no analogs among purely layered compounds: framework in $CsIn_3H_2(SeO_3)_6(H_2O)_2$ (a), its basic sheet (b) and idealized graph of the latter (c); pillared framework in α-$(NH_4)V(HPO_4)_2$ (d), its basic sheet (e) and idealized graph of the latter (f).

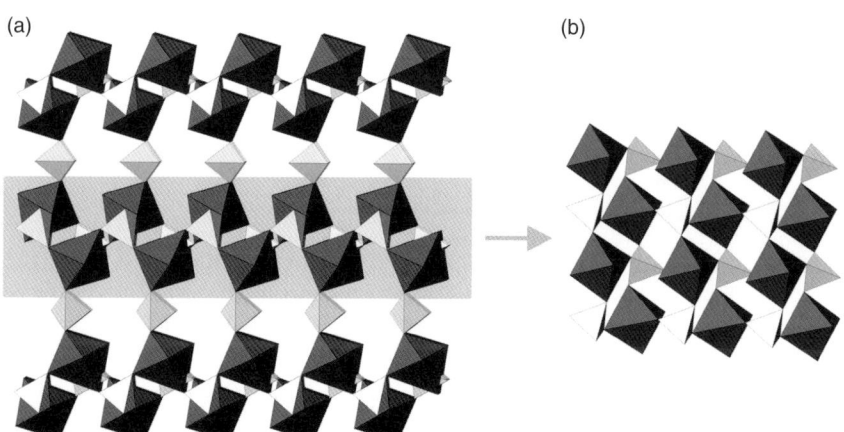

Fig. 3.49. Octahedral–tetrahedral framework in $(NH_4)_2Cd_2(SeO_4)_3(H_2O)_3$ (a) consists of autunite-type sheets (b) linked in third dimension by additional tetrahedra.

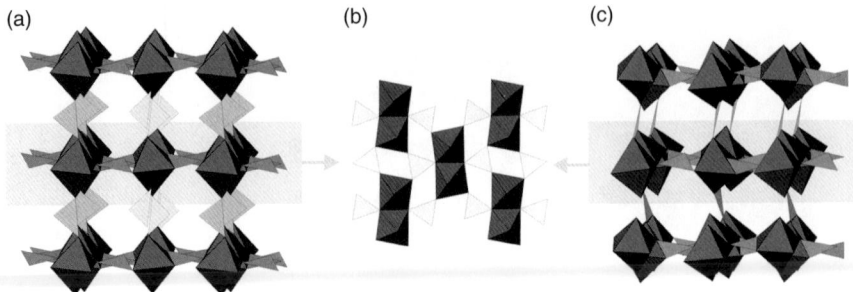

Fig. 3.50. Frameworks in the structures of $FeH(SeO_3)_2$ (a) and $Fe_2(SeO_3)_3(H_2O)$ (c) are based upon the same sheet of pairs of edge-sharing tetrahedra interlinked by selenite triangular pyramids (b).

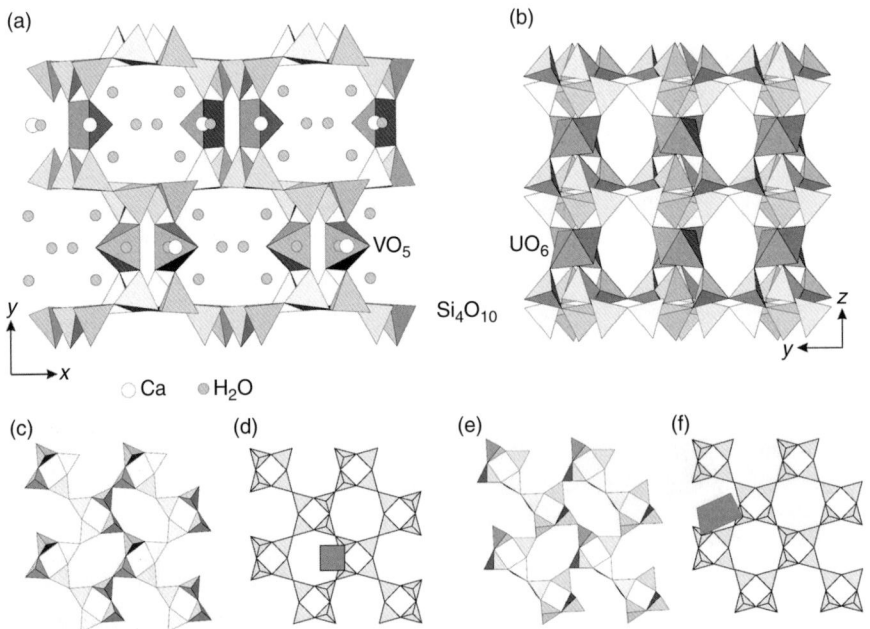

Fig. 3.51. Structure of cavansite, $Ca(VO)(Si_4O_{10})(H_2O)_4$ (a) and KNAURSI, $KNa_3(UO_2)_2$ $(Si_4O_{10})_2(H_2O)_4$ (b) are based upon Si_4O_{10} silicate sheets shown in ((c), (d)) and ((e), (f)), respectively. The sheets consist of 4- and 8-membered rings of tetrahedra and can be designated using their ring symbol as 8^14^1.

A 2D graph of the Si_4O_{10} sheet in KNAURSI corresponding to the 8^14^1 topology is given in Fig. 3.52(e) with the letters **u** and **d** written near the white nodes. The **u** and **d** designations indicate that the silicate tetrahedra symbolized by the white nodes have their non-shared corners oriented up or down relative to the plane of the sheet,

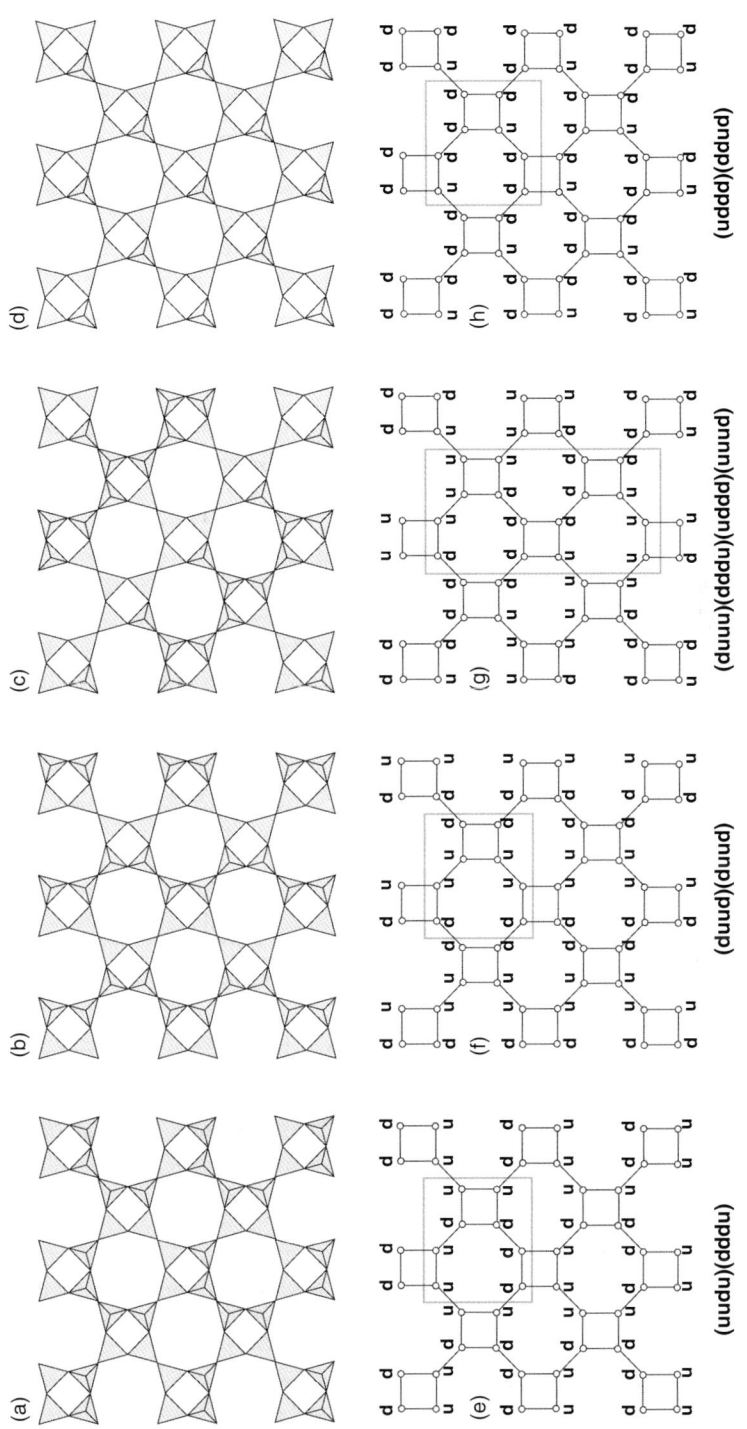

Fig. 3.52. Geometrical isomers of the Si_4O_{10} sheets with the 8^14^1 topology, their graphs and orientation matrices. See text for details.

respectively. The four geometrical isomers of the 8^14^1 sheets shown in Fig. 3.52 can be distinguished using different orientation matrices as described in Chapter 2. The list of respective minerals and compounds is given in Table 3.6.

It should be noted that the 8^14^1 sheet with the orientation matrix (**uddd**)(**ddud**) (Figs. 3.52(d) and (h)) is not observed as a single silicate sheet. Instead, it occurs as a half of a double silicate sheet Si_8O_{19} in the structures of montregianite, $Na_2[Y(Si_8O_{19})]$ $(H_2O)_5$ (Ghose *et al.* 1987), and its synthetic analogs (Rocha *et al.* 1997; Rocha *et al.* 2000).

It is noteworthy that the mixed frameworks based upon silicate tetrahedral sheets dominate structural topologies in natural and synthetic vanadium silicates. Recently, Wang *et al.* (2002a) reported a number of alkali-metal vanadium silicates, VSH-*n*, based upon sheets of different topology. Figure 3.53 shows three isomers of the Si_2O_5 silicate tetrahedral sheets of the mica-like topology. These sheets are composed solely of 6-membered rings of tetrahedra and therefore have the ring symbol 6^1. The isomers can be distinguished by their orientation matrices. The list of respective minerals and compounds is given in Table 3.6.

The structures of armstrongite, $CaZr[Si_6O_{15}](H_2O)_3$ (Kabalov *et al.* 2000), davanite $K_2Ti(Si_6O_{15})$ (Gebert *et al.* 1983), and dalyite, $K_2Zr(Si_6O_{15})$ (Fleet 1965), contain octahedral–tetrahedral frameworks based upon silicate sheets consisting of 4-, 6- and 8-membered rings of tetrahedra (ring symbol $4^16^18^1$). Figure 3.54(a) show three geometrical isomers of the $4^16^18^1$ sheets. It is interesting that the structure of sazhinite-(Ce), $Na_2Ce[Si_6O_{14}(OH)](H_2O)_{1.5}$ (Shumyatskaya *et al.* 1980) is based upon the $4^16^18^1$ sheet isomer different from that observed in armstrongite, probably, owing to the protonation of one of the O atoms at non-shared corners of silicate tetrahedra.

Figure 3.55 shows models of two silicate tetrahedral sheets that contain 5-membered rings of tetrahedra. The 5^28^1 sheet (Figs. 3.55(a) and (c)) are observed in the structures of the compounds with general formula $A_2M(Si_6O_{15})$ (A = K, Rb, Cs; M = Ti, Zr). It is noteworthy that the Cs framework titanosilicates have been extensively investigated owing to their application as ceramic waste materials. It was found that there are two modifications of $Cs_2Ti(Si_6O_{15})$: the *Cc* modification reported by Nyman *et al.* (2001) and the *C2/c* modification prepared by Grey *et al.* (1997). Both modifications are based upon silicate sheets of the same type linked into framework by isolated TiO_6 octahedra. However, the structures are different in the mutual orientations of adjacent silicate sheets. In the *C2/c* modification, chains of 5MRs of the adjacent sheets are parallel, whereas, in the structure of the *Cc* modification, these chains in the adjacent sheets are approximately perpendicular to each other. Thus, the octahedral–tetrahedral frameworks in *Cc*- and *C2/c*-$Cs_2Ti(Si_6O_{15})$ are topologically different, though closely related.

Figure 3.56 shows models of two porous silicate sheets that contain 12MRs. These sheets serve as a basis of very open heteropolyhedral frameworks observed in the structures of VSH-11 (the $4^66^28^312^1$ sheet) and VSH-4, -6 and -9 (the $4^36^212^1$ sheet) reported by Wang *et al.* (2002a).

The number of tetrahedra in a ring is not limited to 12. Figure 3.57(a) shows the structure of yakovenchukite, $K_3NaCaY_2[Si_{12}O_{30}](H_2O)_4$ (Krivovichev *et al.* 2007b),

Table 3.6. Minerals and inorganic compounds containing frameworks built upon tetrahedral sheets linked via isolated metal polyhedra or finite clusters of polyhedra ($D_{tetr} = 2$; $D_{nt} = 0$)

Net symbol	Sheet isomer	Mineral/name	Chemical formula	Reference
6^1	$(u^3d^3)(d^3u^3)$	VSH-3Rb	$Rb_2(VO)_2(Si_6O_{15})(H_2O)_{1.6}$	1
		VSH-3K	$K_2(VO)_2(Si_6O_{15})(H_2O)_{1.6}$	1
	$(u^2d^2)(d^2u^2)$	VSH-1K	$K_2(VO)(Si_4O_{10})(H_2O)$	2
		VSH-14Na	$Na_2(VO)(Si_4O_{10})(H_2O)_{1.4}$	1
	$(u^4d^4)(d^3u^4d)(u^2d^4u^2)\ (du^4d^3)$ $(d^4u^4)(u^3d^4u)\ (d^2u^4d^2)(ud^4u^3)$	pentagonite	$Ca(VO)(Si_4O_{10})(H_2O)_4$	3
4^18^1	$(u^2du)(d^3u)$	KNAURSI	$KNa_3(UO_2)_2(Si_4O_{10})_2(H_2O)_4$	4
		NAURSI	$Na_4(UO_2)_2(Si_4O_{10})_2(H_2O)_4$	5
		USH-1	$Na_4(UO_2)_2(Si_4O_{10})_2(H_2O)_4$	6
	$(du^2d)(du^2d)$	cavansite	$Ca(VO)(Si_4O_{10})(H_2O)_4$	7–9
		FDZG-1	$(C_4N_2H_{12})[ZrGe_4O_{10}F_2]$	10
		VSH-13Na	$Na_2(VO)(Si_4O_{10})(H_2O)_3$	1
	$(du^3)(d^3u)(ud^3)(u^3d)$	VSH-12Cs	$Cs_2(VO)(Si_4O_{10})(H_2O)_x$	1
		VSH-12LiX	$Li_2(VO)(Si_4O_{10})(H_2O)_x$	1
	$(ud^3)(d^2ud)$	montregianite	$Na_2[Y(Si_8O_{19})](H_2O)_5$	11
		AV-1	$Na_2[Y(Si_8O_{19})](H_2O)_5$	12
		AV-5	$Na_2[Ce(Si_8O_{19})](H_2O)_5$	13
		rhodesite	$HKCa_2(Si_8O_{19})(H_2O)_5$	14
		delhayelite	$Na_3K_7Ca_5(Al_2Si_{14}O_{38})F_4Cl_2$	15
		macdonaldite	$BaCa_4H_2(Si_{16}O_{38})(H_2O)_{10.4}$	16
$4^16^18^1$	$(u^3)(d^3)$	davanite	$K_2Ti(Si_6O_{15})$	17
		dalyite	$K_2Zr(Si_6O_{15})$	18
		armstrongite	$CaZr[Si_6O_{15}](H_2O)_3$	19, 20
	$(ud^2)(d^3)(du^2)(u^3)$	–	α-$K_2Ti(Si_6O_{15})$	21
		–	α-$K_3NdSi_6O_{15}(H_2O)_2$	22
	$(du^2)(ud^2)(ud^2)(du^2)$	sazhinite-(Ce)	$Na_2Ce[Si_6O_{14}(OH)](H_2O)_{1.5}$	23
		–	β-$K_3NdSi_6O_{15}$	24
5^28^1	–	–	$Cs_2Zr(Si_6O_{15})$	25
		SNL-A	$A_2Ti(Si_6O_{15})$ A = K, Rb, Cs	26, 27
		–	$Cs_2Ti(Si_6O_{15})$	28
$4^15^26^18^2$	–	–	$Na_3Nd(Si_6O_{15})(H_2O)_2$	29, 30
$4^66^28^312^1$	–	VSH-11RbNa	$(Rb,Na)_2(VO)(Si_4O_{10})\cdot(H_2O)_x$	1
$4^36^212^1$	–	VSH-4Cs	$Cs_2(VO)(Si_4O_{10})\cdot(H_2O)_{2.7}$	1
		VSH-4Rb	$Rb_2(VO)(Si_4O_{10})\cdot(H_2O)_3$	1
		VSH-6CsK	$(Cs,K)_2(VO)(Si_4O_{10})\cdot(H_2O)_3$	1
		VSH-6Rb	$Rb_2(VO)(Si_4O_{10})\cdot(H_2O)_3$	1
		VSH-9CsNa	$CsNa(VO)(Si_4O_{10})\cdot(H_2O)_4$	1
4^16^1	–	VSH-2Cs	$Cs_2(VO)(Si_6O_{14})\cdot(H_2O)_3$	2

References: (1) Wang *et al.* 2002a; (2) Wang *et al.* 2001; (3) Evans 1973; (4) Burns *et al.* 2000; (5) Li and Burns 2001; (6) Wang *et al.* 2002b; (7) Evans 1973; (8) Rinaldi *et al.* 1975; (9) Solov'ev *et al.* 1993; (10) Liu *et al.* 2003a; (11) Ghose *et al.* 1987; (12) Rocha *et al.* 1997; (13) Rocha *et al.* 2000; (14) Hesse *et al.* 1992; (15) Cannillo *et al.* 1970; (16) Cannillo *et al.* 1968; (17) Gebert *et al.* 1983; (18) Fleet 1965; (19) Canillo *et al.* 1973; (20) Kabalov *et al.* 2000; (21) Zou and Dadachov 1999; (22) Haile and Wuensch 2000a; (23) Shumyatskaya *et al.* 1980; (24) Haile and Wuensch 2000b; (25) Jolicart *et al.* 1996; (26) Nyman *et al.* 2000; (27) Nyman *et al.* 2001; (28) Grey *et al.* 1997; (29) Karpov *et al.* 1977; (30) Haile *et al.* 1997.

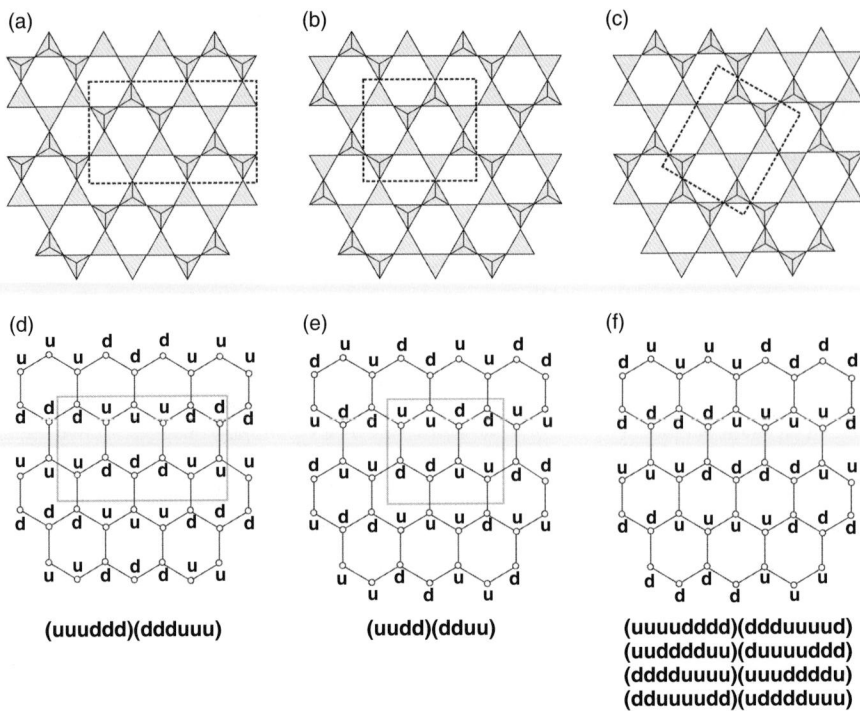

Fig. 3.53. Geometrical isomers of the Si_2O_5 sheets with the 6^1 topology, their graphs and orientation matrices. See text for details.

that contains an octahedral–tetrahedral framework consisting of $Si_{12}O_{30}$ sheets linked by isolated YO_6 octahedra. The $Si_{12}O_{30}$ sheet (Figs. 3.57(b)–(d)) contains 4-, 6- and 14-membered rings; its ring symbol is $14^16^14^3$.

All the structures based upon 2D silicate anions ($\boldsymbol{D}_{tetr} = 2$) described above have $\boldsymbol{D}_{nt} = 0$, i.e. their non-tetrahedral cations either form isolated octahedra or pentahedra or form finite clusters of polyhedra. Figure 3.58(a) demonstrates an example of structure for which $\boldsymbol{D}_{tetr} = 2$ and $\boldsymbol{D}_{nt} = 1$. This structure was recently reported by Wang *et al.* (2002c) for USH-8, $[(CH_3)_4N][(C_5H_5NH)_{0.8}((CH_3)_3NH)_{0.2}](UO_2)_2[Si_9O_{19}]F_4$. It is based upon complex Si_9O_{19} double sheets of silicate tetrahedra (Fig. 3.58(b)) [Wang *et al.* (2002c) pointed out that the sheet of this topology can be found in the structure of zeolite ferrierite, $Na_{1.5}Mg_2Si_{30.5}Al_{5.5}O_{72}(H_2O)_{18}$] and chains of edge-sharing UO_3F_4 pentagonal bipyramids (Fig. 3.58(c)).

3.6.2.2 Umbite-related frameworks

The structure of umbite, $K_2Zr(Si_3O_9)(H_2O)$ (Ilyushin 1981b, 1993), and its synthetic analog AM-2 (Lin *et al.* 1997) is based upon the framework of Si_3O_9 chains

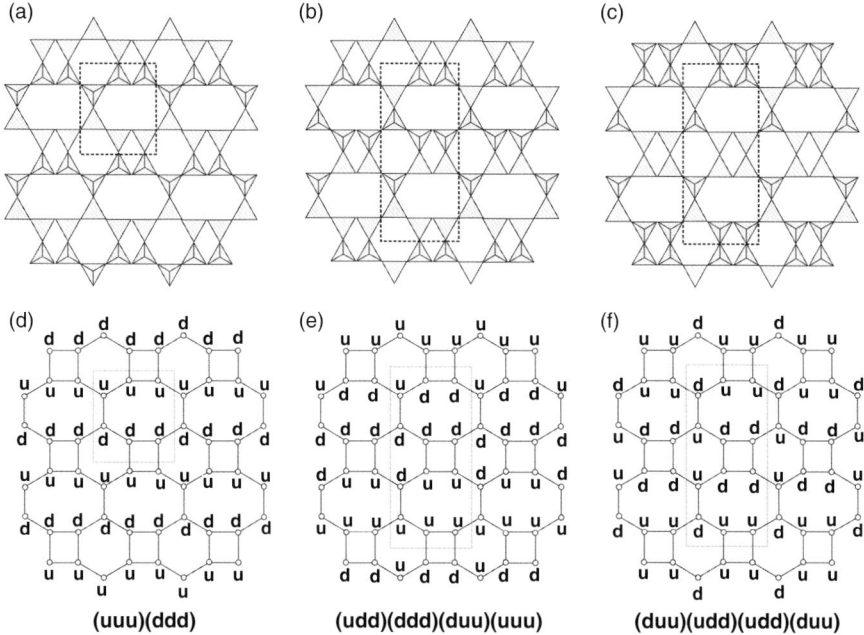

Fig. 3.54. Geometrical isomers of the Si_6O_{15} sheets with the $4^16^18^1$ topology, their graphs and orientation matrices. See text for details.

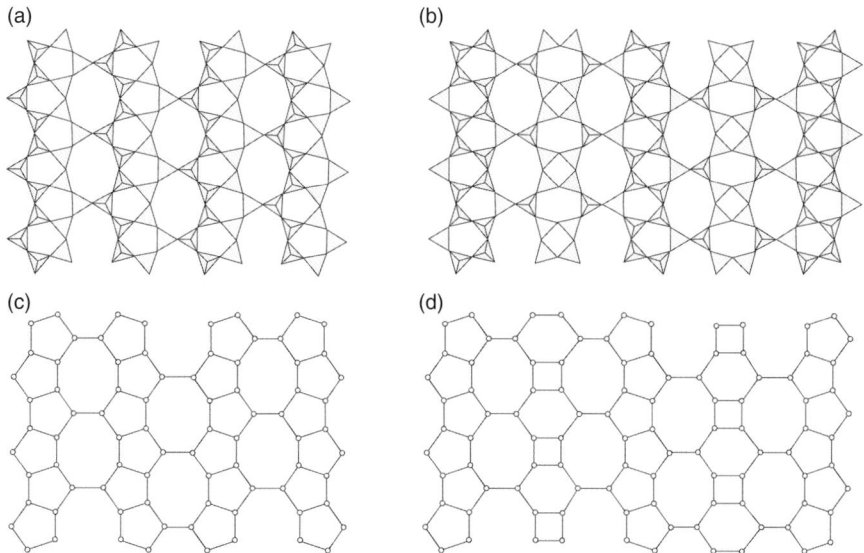

Fig. 3.55. Silicate tetrahedral sheets that contain 5-membered rings of tetrahedra: the 5^28^1 sheet ((a) and (c)) in the structures of $A_2M(Si_6O_{15})$ (A = K, Rb, Cs; M = Ti, Zr) and the $4^15^26^18^2$ sheet ((b) and (d)) in the structure of $Na_3Nd(Si_6O_{15})(H_2O)_2$.

(a)

(b)

(c)

(d)

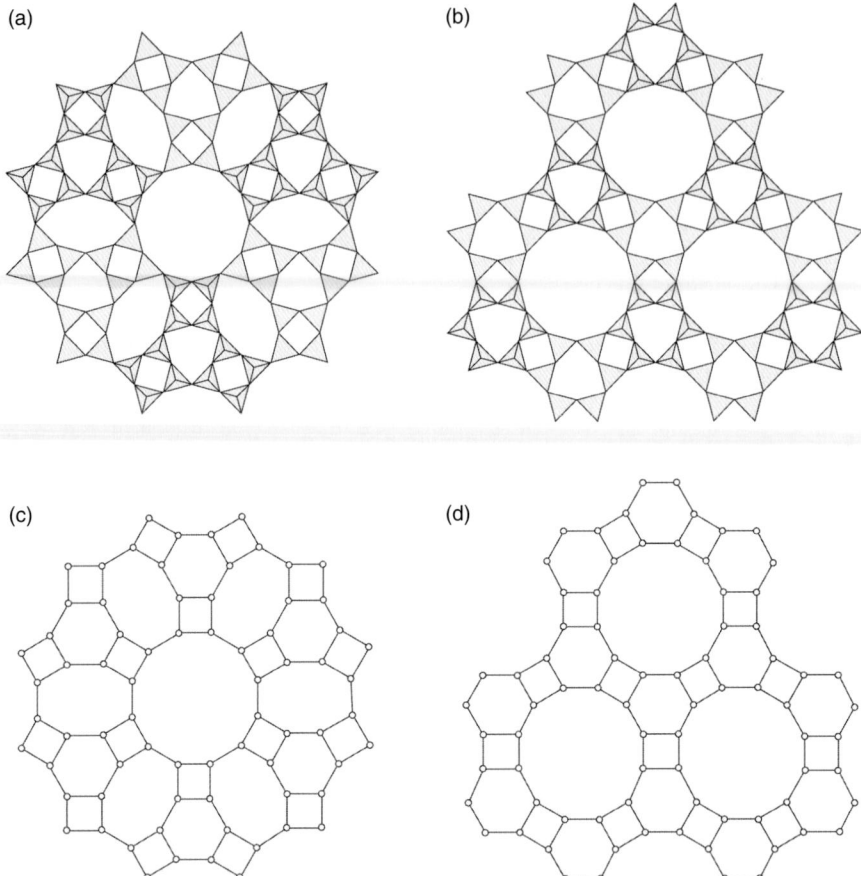

Fig. 3.56. Porous silicate sheets that contain 12MRs: the $4^6 6^2 8^3 12^1$ sheet from the structure of VSH-11 (a) and the $4^3 6^2 12^1$ sheet from the structures of VSH-4, -6 and -9 (b).

of corner-sharing SiO_4 tetrahedra and isolated ZrO_6 octahedra. The Ti analog of umbite, $K_2Ti(Si_3O_9)(H_2O)$ (Dadachov and Le Bail 1997; Bortun *et al.* 2000) possesses interesting ion-exchange properties. Plevert *et al.* (2003) reported a series of Zr germanates based upon umbite-related octahedral–tetrahedral frameworks. Despite the fact that the structure is based upon chains of corner-linked tetrahedra, the umbite-related frameworks are better described as based upon sheets of octahedra and tetrahedra shown in Figures 3.59(a) and (c) (Plevert *et al.* 2003). The sheets are geometrical isomers that differ by the orientation of one of the tetrahedra relative to the plane of the sheet. Nodal representations of the two sheets are shown in Figs. 3.59(e) and (f), respectively. The structures of umbite itself and ASU-25, $[C_3H_{12}N_2]$ $ZrGe_3O_9$, are based upon sheets with alternative **u** and **d** orientation of tetrahedra

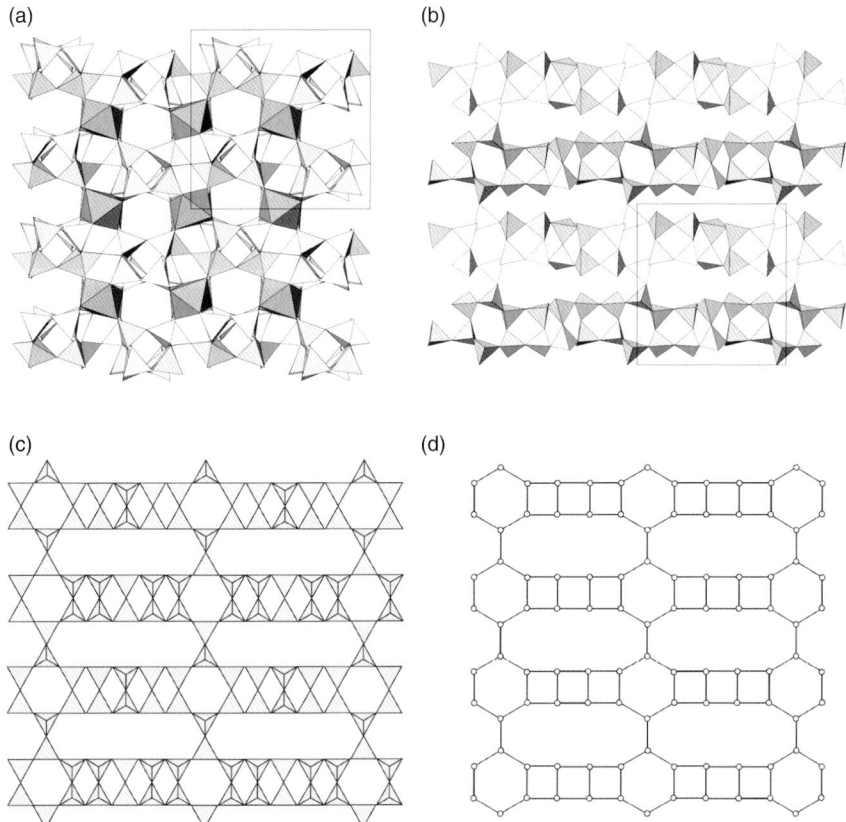

Fig. 3.57. Octahedral–tetrahedral framework in the structure of yakovenchukite, K_3NaCaY_2 $[Si_{12}O_{30}](H_2O)_4$ (a), its silicate sheet (b) and its ideal (c) and nodal (d) representations.

(Fig. 3.59(g)). The structure of ASU-26, $[C_2H_{10}N_2]ZrGe_3O_9$, is based upon sheets with one kind of orientation of tetrahedra (this accounts for the non-centrosymmetric space group, Pn, of this material) (Fig. 3.59(h)). In the structure of ASU-24, $[C_6H_{18}N_2]$ $[C_6H_{17}N_2]_2[Zr_3Ge_6O_{18}(OH_2,F)_4F_2](H_2O)_2$, sheets with alternative **u** and **d** orientation of tetrahedra are interlinked by additional $Zr(O,F)_6$ octahedra, thus forming a pillared layered structure (Fig. 3.59(i)).

3.6.2.3 The use of 2D nets to recognize structural relationships

Elpidite and $Ca_2ZrSi_4O_{12}$. The structures of elpidite, $Na_2ZrSi_6O_{15}(H_2O)_3$ (Neronova and Belov 1963, 1964; Cannillo *et al.* 1973; Sapozhnikov and Kashaev 1978), and $Ca_2ZrSi_4O_{12}$ (Colin *et al.* 1993) can easily be related to each other if described as being based upon 2D nets (Fig. 3.60). The 2D net in $Ca_2ZrSi_4O_{12}$ (Fig. 3.60(d)) can

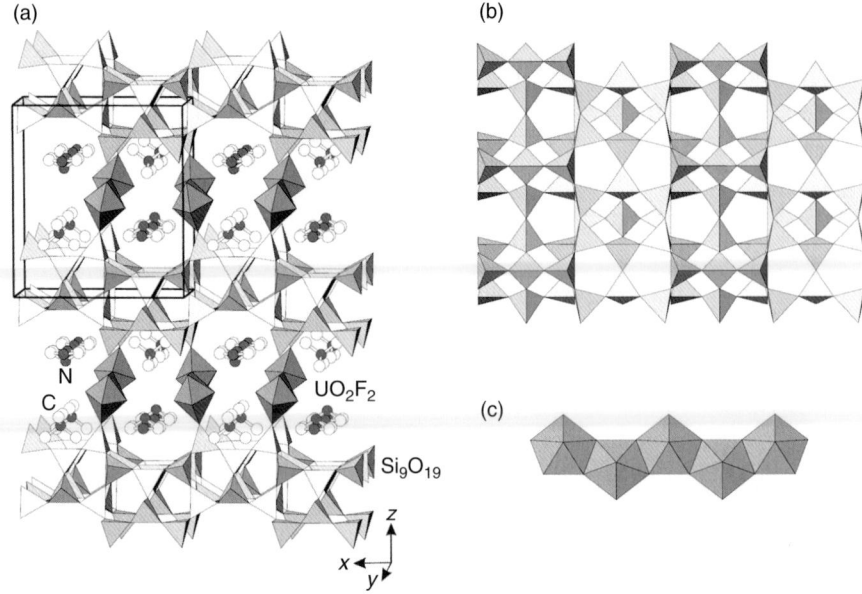

Fig. 3.58. The structure of USH-8, $[(CH_3)_4N][(C_5H_5NH)_{0.8}((CH_3)_3NH)_{0.2}](UO_2)_2[Si_9O_{19}]F_4$ (a), is based upon a heteropolyhedral framework consisting of Si_9O_{19} double sheets of silicate tetrahedra (b) and chains of edge-sharing UO_3F_4 pentagonal bipyramids (c).

be obtained from that in elpidite (Fig. 3.60(c)) by deleting pairs of white vertices and all edges incident upon those vertices. Thus, the $ZrSi_4O_{12}$ framework in $Ca_2ZrSi_4O_{12}$ can be obtained from the $ZrSi_6O_{15}$ framework in elpidite by extraction of two silicate tetrahedra and rearrangement of bonds between the sheets.

Gittinsite and SrZrSi₂O₇. The analysis of 3D nets in gittinsite, $CaZrSi_2O_7$ (Roelofsen-Ahl and Peterson 1989) and $SrZrSi_2O_7$ (Huntelaar *et al.* 1994) reveals that both nets are based upon the same 2D net shown in Fig. 3.61(a). It consists of heptagons **bwbw²bw** and triangles **bw²**. The structures differ in the mode the 2D nets are linked to each other. In gittinsite, the adjacent 2D nets have the same orientation of triangles (Fig. 3.61(b)), whereas, in $SrZrSi_2O_7$, the adjacent 2D nets are related to each other by rotation by 180° around an axis vertical to the plane of the nets (the triangles in the adjacent nets have opposite orientation; Fig. 3.61(c)).

3.6.3   *Structure description versus intuition*

For most cases discussed in this chapter so far, it was implicitly understood that the structure with *T–T* linkages should be described as being based upon the unit consisting of *T–T* links (e.g., tetrahedral anion), since this unit should be one of the strongest in the structure and its separation as such cannot be avoided. This point is in good

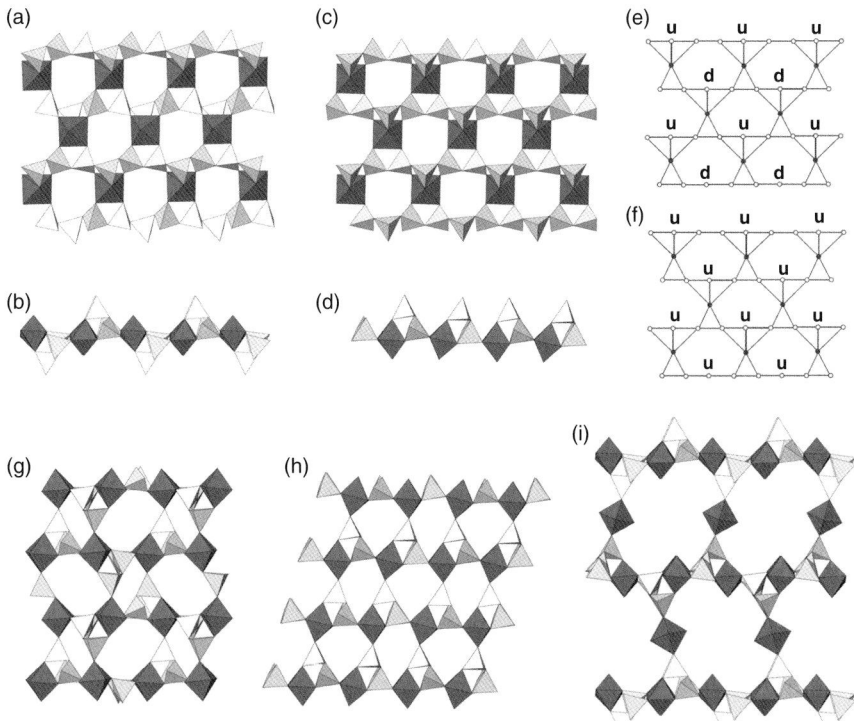

Fig. 3.59. The umbite-related octahedral–tetrahedral frameworks can be described as being based upon sheets of octahedra and tetrahedra (top view: (a) and (c); side view: (b) and (d), respectively), nodal representations of which are given in (e) and (f), respectively. Linkage of the (a) type produces frameworks in the structures of umbite and ASU-25 (g), whereas linkage of the sheets of the (c) type results in formation of the ASU-26 framework (h). Octahedral–tetrahedral framework in ASU-24 consists of the type (a) sheets interlinked by additional $Zr(O,F)_6$ octahedra, thus forming pillared layered structure (i).

agreement with general intuition that the simplest representation of structural topology must be the most reasonable from the viewpoint of common sense. However, in some but few cases, there is a need to override common sense in order to come to a clear and transparent description of a structure.

Recently, Alekseev *et al.* (2008b) described two modifications of a K uranyl polyphosphate, $K[(UO_2)(P_3O_9)]$, based upon complex 3D frameworks of polyphosphate chains, $[PO_3]^-$, and U-centered pentagonal bipyramids. In both structures, coordination polyhedra share corners only. The polyphosphate chains are different in the two structures, as shown in Fig. 3.62. The periodicity of the chains defined as the number of tetrahedra within the identity period of the chain is 12 for both

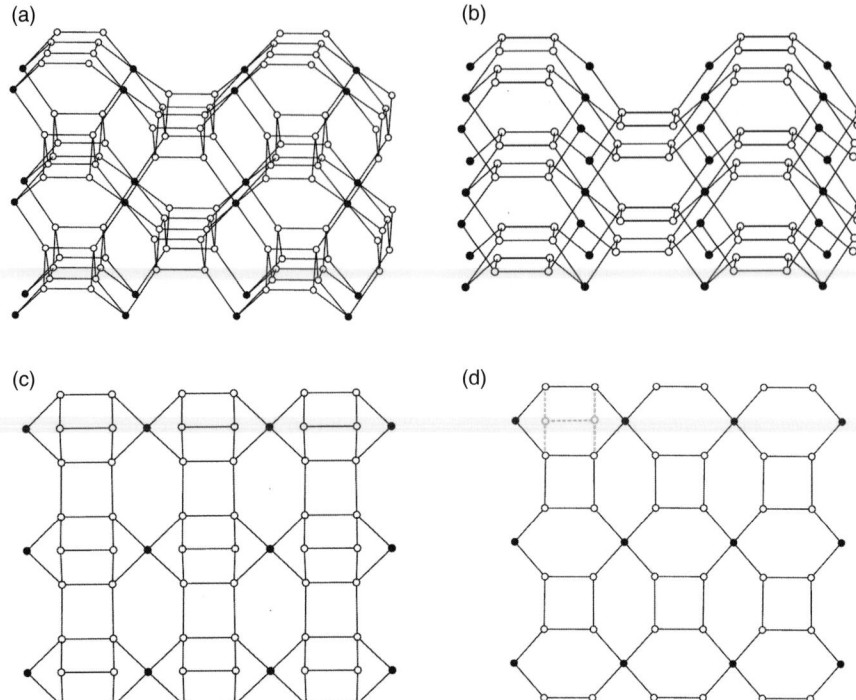

Fig. 3.60. 3D nets corresponding to the octahedral–tetrahedral frameworks in the structures of elpidite, $Na_2ZrSi_6O_{15}(H_2O)_3$ (a), and $Ca_2ZrSi_4O_{12}$ (b) can be described as being based upon 2D nets ((c) and (d), respectively).

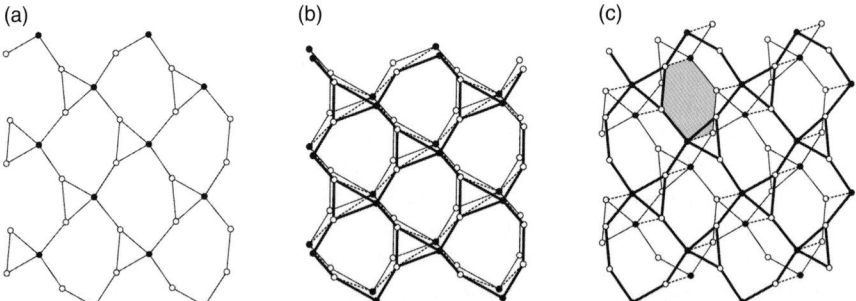

Fig. 3.61. 2D net consisting of heptagons **bwbw²bw** and triangles **bw²** (a) is a basis for the 3D nets in gittinsite, $CaZrSi_2O_7$ (b) and $SrZrSi_2O_7$ (c). In gittinsite, the adjacent 2D nets have the same orientation of triangles (b), whereas, in $SrZrSi_2O_7$, the adjacent 2D nets are related to each other by rotation by 180° around an axis vertical to the plane of the nets (c).

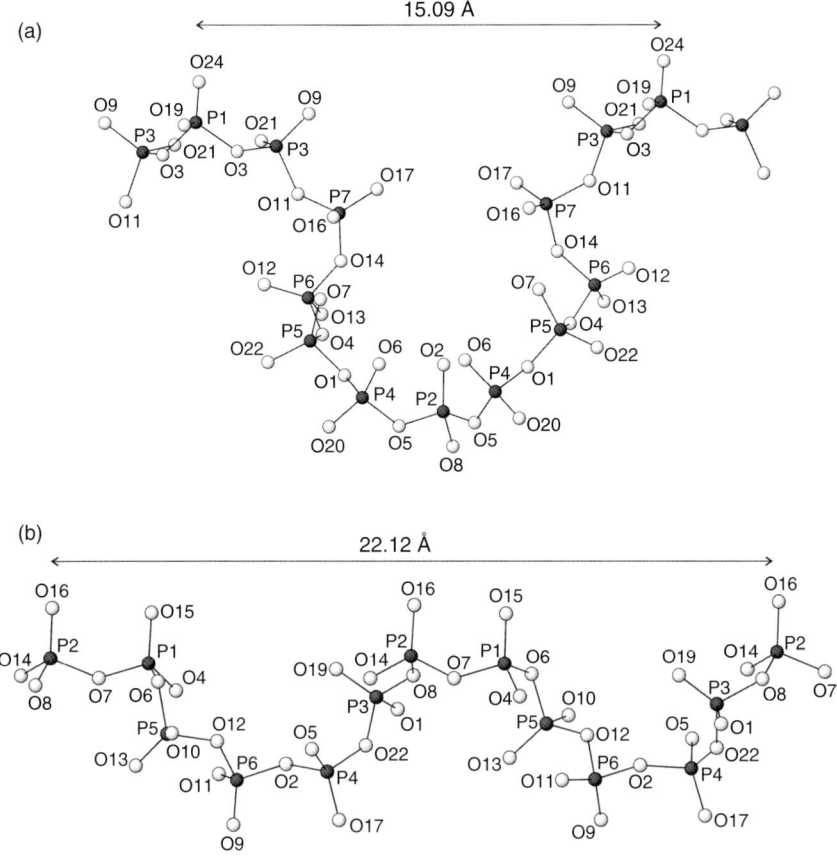

Fig. 3.62. Polyphosphate units in the structures of α- (a) and β- (b) K[(UO₂)(P₃O₉)]. Identity periods are given.

modifications. However, the identity periods are very different and are equal to 15.09 Å for the α-phase and 22.12 Å for the β-phase. This means that the chains in the α-phase are highly bent, whereas, in the β-phase, they are more linear. Visual analysis of the uranyl polyphosphate frameworks in both structures (Figs. 3.63(a) and 6.64(a)) does not provide any simple correlations between structural topologies. Thus, the method has been proposed that, counter-intuitively, includes imaginary breaking of the polyphosphate chains. First, the frameworks have been symbolized as graphs with white and black vertices corresponding to the P and U polyhedra, respectively. Then, the obtained graph was considered as being based upon 2D subgraphs that are selected such that all black U-vertices are *fully connected*, i.e. they do not form additional U–P linkages outside the subgraphs.

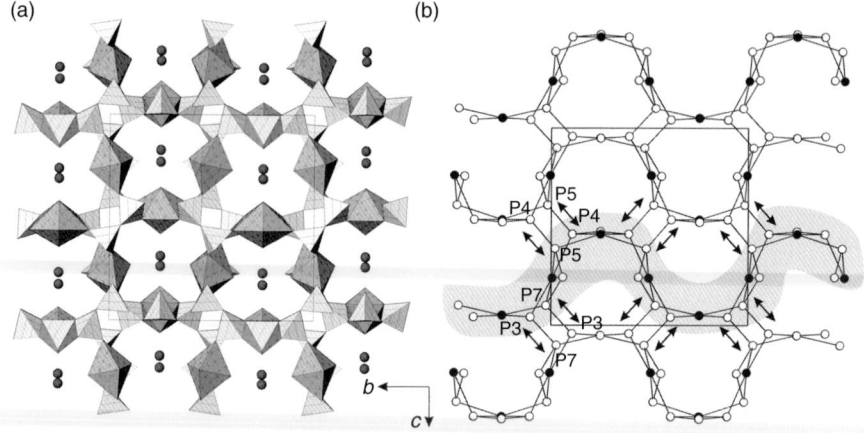

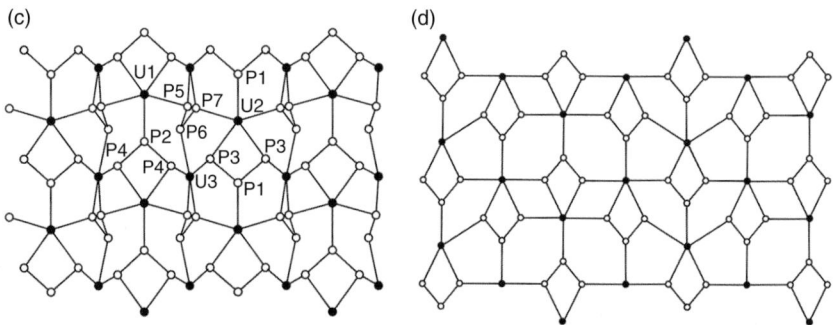

Fig. 3.63. Structure of α-K[(UO$_2$)(P$_3$O$_9$)] projected along the a-axis (a: U polyhedra = dark gray, P tetrahedra = gray; K atoms are shown as circles); the 3D graph corresponding to the topology of linkage of U and P polyhedra in the structure of α-K[(UO$_2$)(P$_3$O$_9$)] (gray area indicates the position of a 2D subgraph; double arrows show the P–P links between adjacent 2D subgraphs) (b); real (c) and idealized (d) topologies of the 2D subgraph.

It is noteworthy that the imaginary splitting of the 3D graphs into 2D subgraphs according to the last procedure involved breaking of some P–P links, i.e. edges between adjacent white vertices. This was obviously against the crystal chemical intuition that insists that the strongest bonds must be given priority over the weak bonds. However, this method is a more straightforward way to describe and to compare the structures of α- and β-K[(UO$_2$)(P$_3$O$_9$)].

Figure 3.63(b) shows a 3D black-and-white graph that corresponds to the topology of U–P and P–P linkages in the structure of α-K[(UO$_2$)(P$_3$O$_9$)]. The 3D graph can be subdivided into corrugated 2D subgraphs parallel to (001), as shown in Fig. 3.63(c). Note that this graph: (1) contains only 5-connected black vertices; that is, no edges

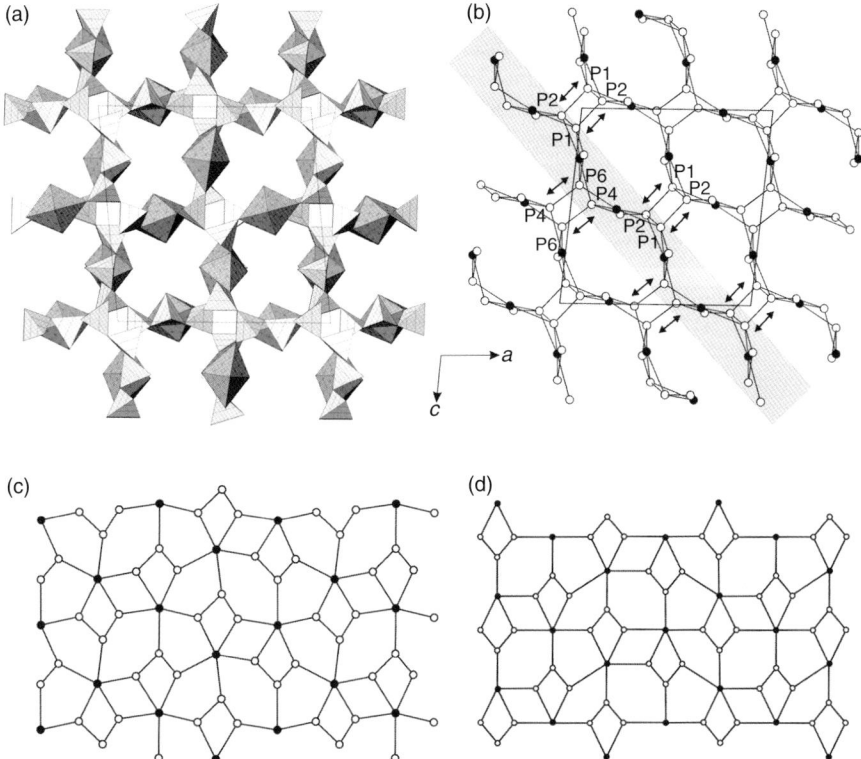

Fig. 3.64. Structure of β-K[(UO₂)(P₃O₉)] projected along the *b*-axis (a: U polyhedra = dark gray, P tetrahedra = gray; K atoms are shown as circles); the 3D graph corresponding to the topology of linkage of U and P polyhedra in the structure of β-K[(UO₂)(P₃O₉)] (gray area indicates position of a 2D subgraph; double arrows show the P–P links between adjacent 2D subgraphs) (b); real (c) and idealized (d) topologies of the 2D subgraph.

are emanating from these vertices outside the subgraph; (2) consists of 4-, 5-, and 6-membered rings of black and white vertices. Idealized representation of this graph is shown in Fig. 3.63(d).

Figure 3.64(b) shows a 3D black-and-white graph that corresponds to the topology of U–P and P–P linkages in the structure of β-K[(UO₂)(P₃O₉)]. This 3D graph can be subdivided into planar 2D subgraph shown in Figs. 3.64(c) and (d). It is obvious that this graph is closely related to the graph shown in Figs. 3.63(c) and (d). In particular, both graphs consist of 4-, 5-, and 6-membered rings. However, the arrangements of the rings are different.

In general, the graphs shown in Figs. 3.63(d) and 3.64(d) can be described as being built by condensation of chains of 4-membered rings depicted in Fig. 3.65(a). The

4-membered rings in these chains consist of three white and one black vertices, i.e. white vertices form an angular trimer. We distinguish between the **U** and **D** chains that have different orientations (Fig. 3.65(a)). The topologies of uranyl polyphosphate subgraphs shown in Figs. 3.63(d) and 3.64(d) can be obtained by combinations of the **U** and **D** chains shifted along their extensions and linked by additional edges (shown as dashed lines in Figs. 3.65(b) and (c)). Analysis of the resulting graphs unambiguously demonstrates that the two topologies are non-equivalent and cannot be transformed one into another without breaking of the edges, i.e. without breaking of the chemical bonds. Therefore, the structures of α- and β-K[(UO$_2$)(P$_3$O$_9$)] represent an interesting example of *combinatorial polymorphism* of heteropolyhedral frameworks.

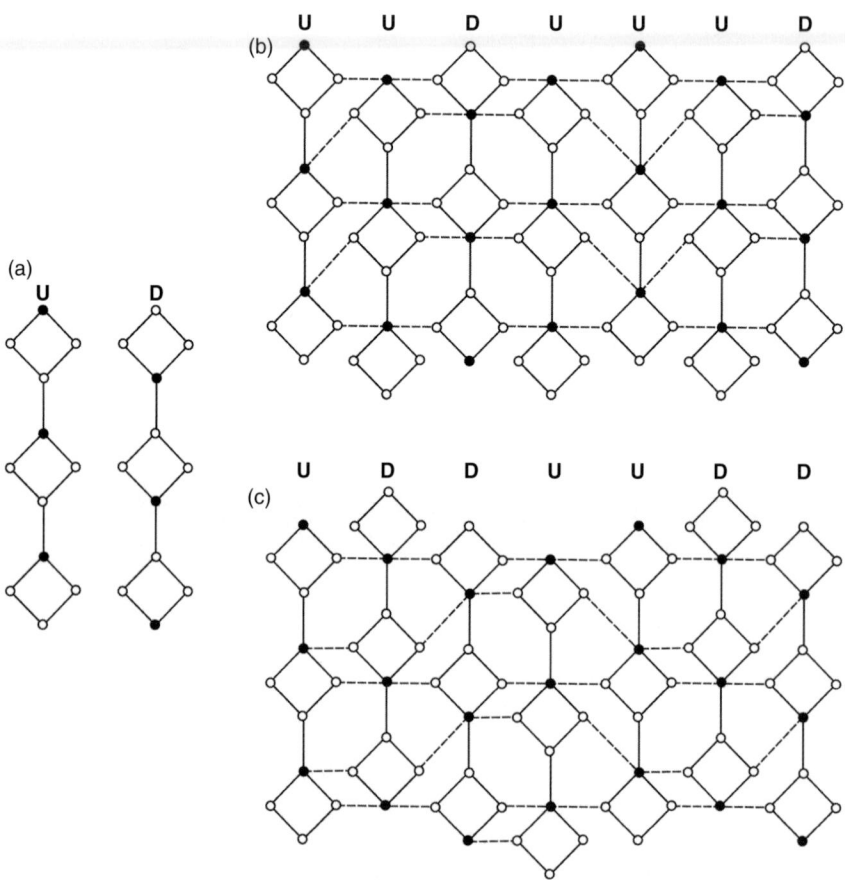

Fig. 3.65. Topologies of the 2D subgraphs observed in the structures of 1 and 2 as constructed from the chains of 4-membered rings: **U** and **D** chains (a); topology in α- (b) and β- (c) K[(UO$_2$)(P$_3$O$_9$)]. Dashed lines show links between chains. Note that the sequences of chains are …**UUUDUUUD**…in α- and …**UUDDUUDD**…in β-K[(UO$_2$)(P$_3$O$_9$)].

It is obvious that the framework in the α-phase is more topologically complex than that in the β-phase, since it is associated with a higher number of topologically non-equivalent vertices of the 3D graph.

The technique developed above has been applied to another uranyl polyphosphate structure, namely, that of $K[(UO_2)_2(P_3O_{10})]$. In contrast to α- and β-$K[(UO_2)(P_3O_9)]$, it consists of finite chains of three phosphate tetrahedra. Separation of the framework into 2D subgraphs with fully connected black vertices allows it to be described

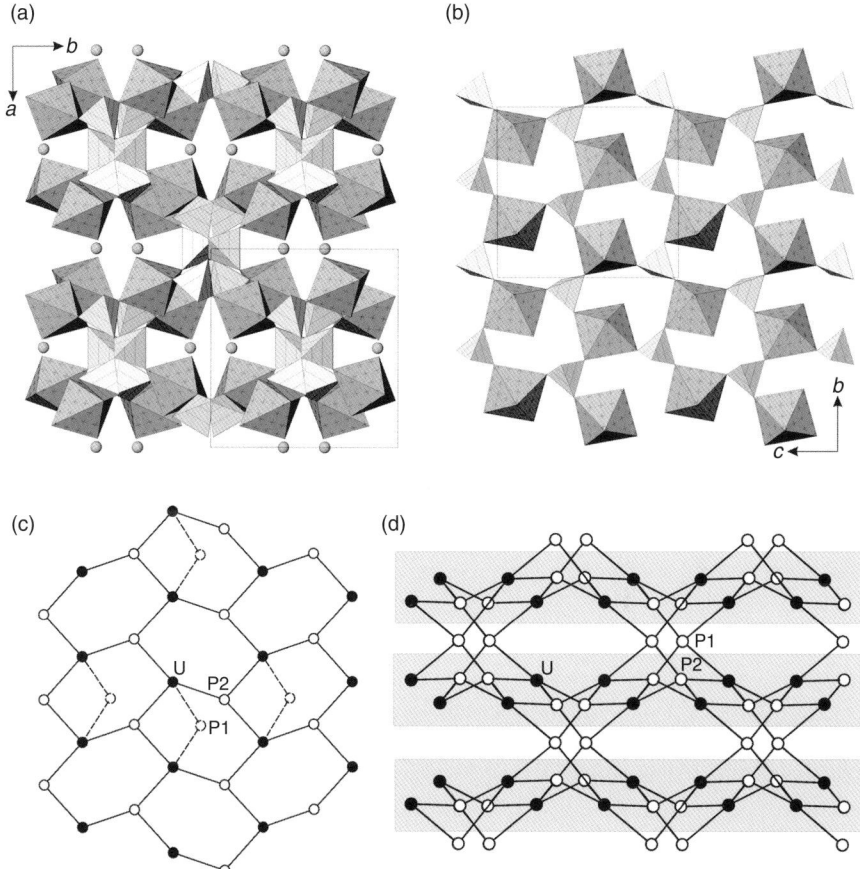

Fig. 3.66. The uranyl polyphosphate framework in the structure of $K[(UO_2)_2(P_3O_{10})]$ projected along the *c*-axis (a: U polyhedra = dark gray, Mo tetrahedra = gray); the octahedral–tetrahedral sheet of U and P polyhedra (b) and its graph (c) (dashed lines show links to the additional P1 vertices located in between the sheets); and the 3D graph corresponding to the topology of linkage of U and P polyhedra in the structure of $K[(UO_2)_2(P_3O_{10})]$ (gray area indicates position of a 2D subgraph).

as being based upon a hexagonal 2D net of alternating black-and-white vertices (Fig. 3.66(c)). These nets are linked by two P1–P2 links to an additional 2-connected white vertex (P1) that is located in the interlayer (Fig. 3.66(d)). It is of interest that the hexagonal 2D net shown in Fig. 3.66(c) and its corresponding octahedral–tetrahedral sheet (Fig. 3.66(b)) have been described above as a constituent of the variscite-type heteropolyhedral frameworks with the general formula $M(TO_4)(H_2O)_2$ (M = Fe, Al, Ga, In; T = P, As) (see Fig. 3.45).

Anion-topology approach

4.1 The concept of anion topology

The graphical (or nodal) representation of structure topologies developed in the two preceding chapters is most suitable for structures with relatively open architectures. For structures with dense structural units (e.g., those where edge-linkage dominates over corner-linkage), another technique might be more appropriate. In order to describe dense sheet structures in uranyl compounds, Burns *et al.* (1996) developed the method of anion topologies that can also be applied to some non-uranium compounds as well.

Let us consider the complex uranyl silicate sheet in the structure of uranophane (Fig. 4.1(a)). In this mineral, (UO_7) pentagonal bipyramids share their equatorial edges to form chains that are further linked by (SiO_4) tetrahedra. According to Burns *et al.* (1996), the anion topology of this sheet (Figs. 4.1(a) and (b)) can be constructed as follows: (1) each anion that is not bonded to at least two cations within the sheet, and that is not an equatorial anion of a bipyramid or pyramid within the sheet, is removed from further consideration (Fig. 4.1(c)); (2) cations are removed, along with all cation–anion bonds, leaving an array of unconnected anions (Fig. 4.1(d)); (3) anions are joined by lines, with only those anions that may be realistically considered as part of the same coordination polyhedron being connected (Fig. 4.1(e)); (4) anions are removed from further consideration, leaving only a series of lines that represent the anion topology.

The anion topology reflects the topology of arrangement of anions in the plane of the sheet and, as a consequence, is most suitable for planar 2D units. Uranyl oxysalts with dense sheets are especially appropriate for this method due to the strong tendency of uranyl-centered bipyramids to polymerize by sharing equatorial edges (Burns *et al.* 1997). However, some non-uranium oxysalts can also be treated with this approach. Figure 4.2(a) shows an octahedral–tetrahedral framework in the structure of $KMn_2O(PO_4)(HPO_4)$ (Lightfoot *et al.* 1988). The framework can be split into dense octahedral–tetrahedral sheets (Fig. 4.2(b)) consisting of chains of edge-sharing (MnO_6) octahedra interlinked by (PO_4) and (HPO_4) groups. The sheet is rather planar and thus can be described using the anion topology depicted in Fig. 4.2(c).

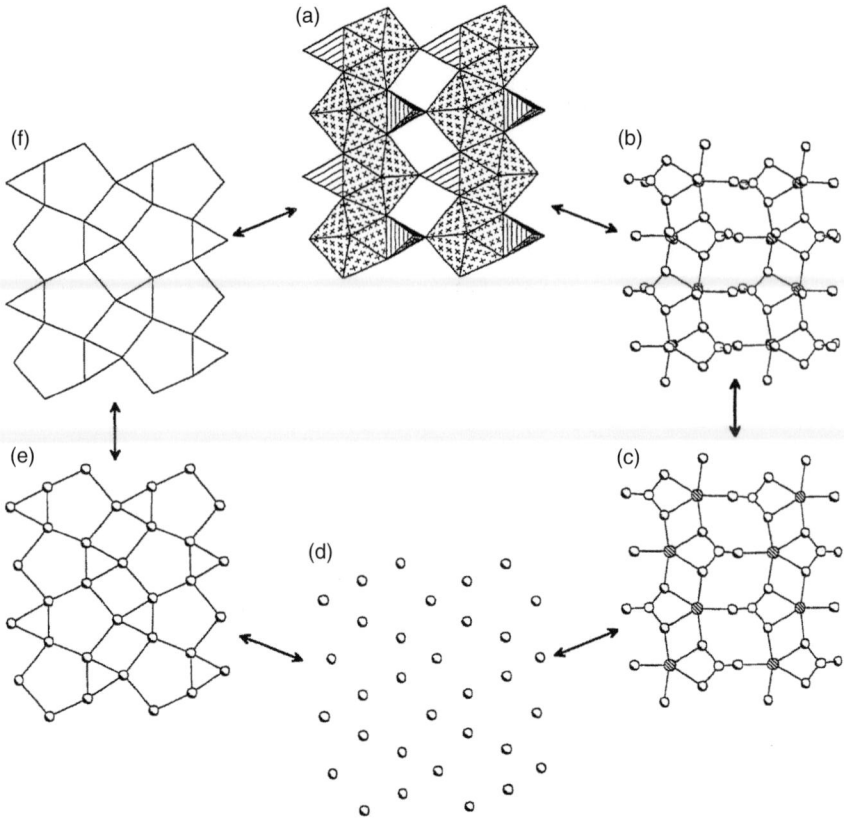

Fig. 4.1. Construction of anion topology (after Burns *et al.* (1996)). See text for details.

4.2 Classification of anion topologies

The anion topology is a tiling of the 2D Euclidean plane into convex polygons. The tiling is of the "edge-to-edge" type, which means that two polygons either do not have common points or share common corners or whole edges. In order to classify anion topologies, we use their cyclic symbol defined as $n_1^{m1}n_2^{m2}n_3^{m3}\ldots$, where n is the number of corners in a given polygon (3 for a triangle, 4 for a square, 5 for a pentagon, etc.), and m is the proportional number of these polygons in the anion topology. For example, the topology shown in Fig. 4.2(c) consists of squares and triangles in the proportion 3:2. Thus, the cyclic symbol for this anion topology is $4^3 3^2$.

Some representative examples of anion topologies, together with their parent sheets, are shown in Figs. 4.3, 4.4 and 4.5. Lists of the respective compounds are given in Tables 4.1, 4.2 and 4.3.

(a)

(b)

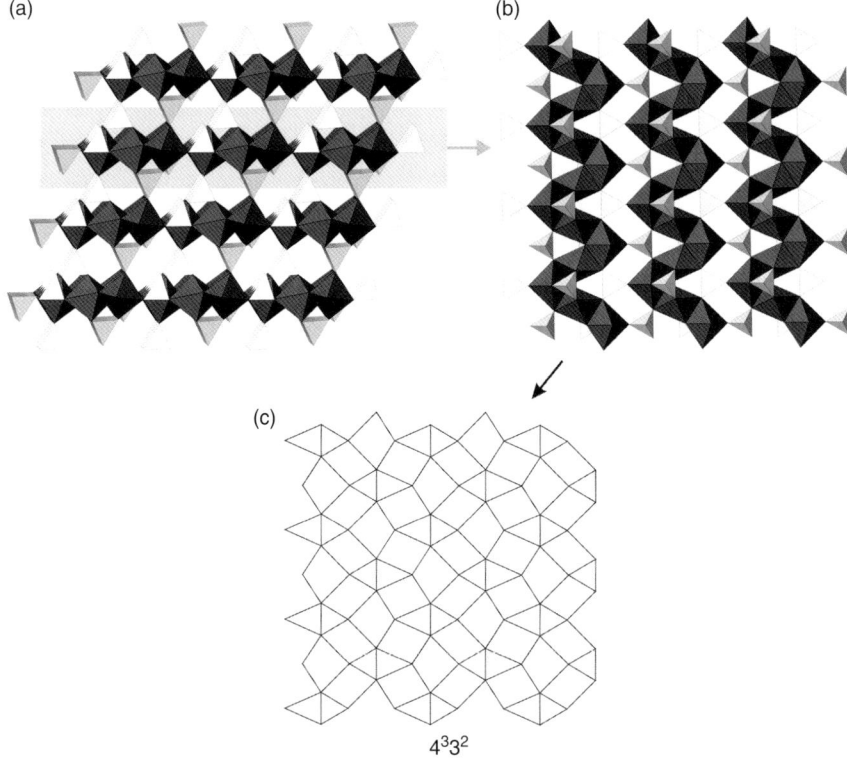

(c)

$4^3 3^2$

Fig. 4.2. Octahedral–tetrahedral framework in the structure of $KMn_2O(PO_4)(HPO_4)$ (a) can be split into dense octahedral–tetrahedral sheets (b) with anion topology shown in (c).

It is noteworthy that some anion topologies are remarkable in their ability to accommodate different cation populations (Burns *et al.* 1996; Burns 1999, 2005). For instance, phosphuranylite topology (Fig. 4.4(b)) consists of hexagons, pentagons, squares, and triangles (its cyclic symbol is $6^1 5^2 4^2 3^2$). The two sheets shown in Figs. 4.4(a) and (c) can both be described using this anion topology but its population by cations is different in the two structures. In phosphuranylite itself (Fig. 4.4(a)), hexagons and pentagons are populated by uranyl cations, whereas squares are empty. In the 2D unit shown in Fig. 4.4(c), only pentagons and triangles are occupied, whereas triangles and hexagons are empty. Another example is the uranophane anion topology that consists of pentagons, squares and triangles (Fig. 4.5(c)). This topology is extremely "popular" among uranyl oxides, hydroxides, and oxysalts (Burns *et al.* 1996), since it allows different populations by cations. Figure 4.5(a) shows a 2D unit from the structure of $Cs[NpO_2(CrO_4)](H_2O)$ (Grigor'ev *et al.* 1991c). Here, pentagons are occupied by the $[NpO_2]^+$ cations, triangles by chromate tetrahedra,

(a)   (b)

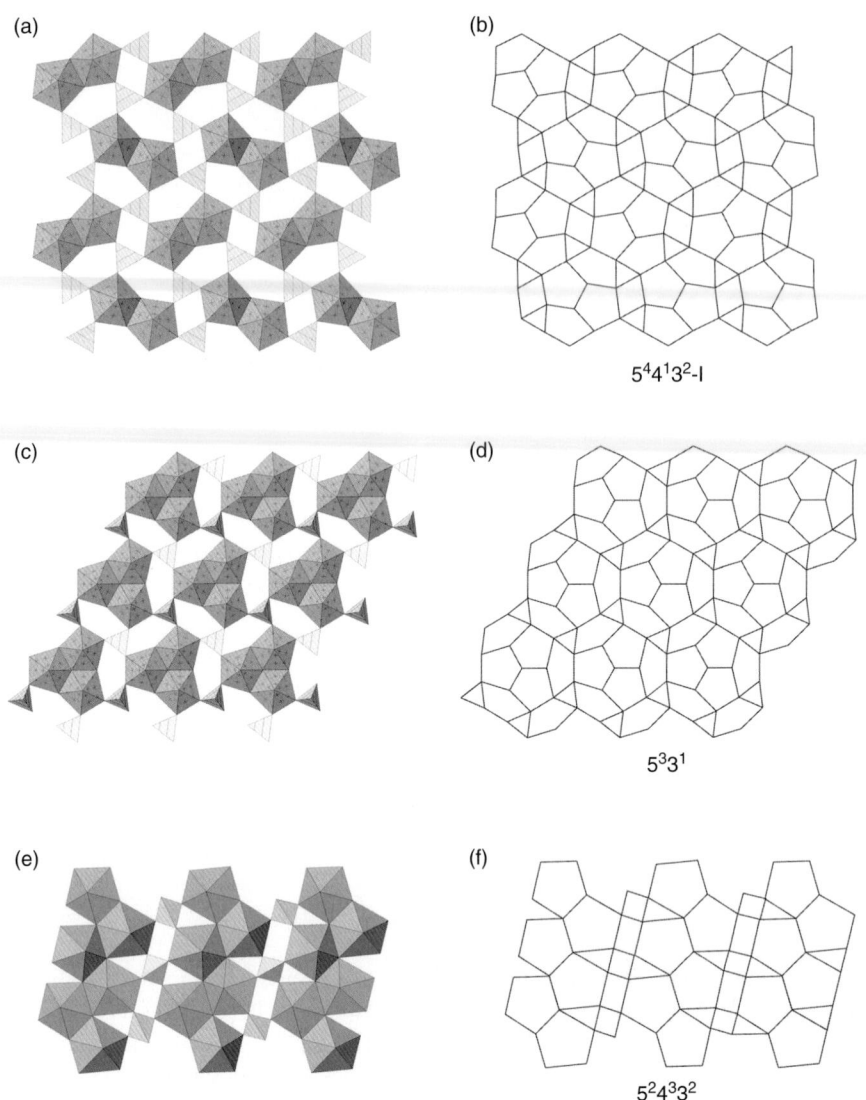

$5^4 4^1 3^2$-I

(c)   (d)

$5^3 3^1$

(e)   (f)

$5^2 4^3 3^2$

Fig. 4.3. Dense heteropolyhedral sheets in structures of uranyl oxysalts and their anion topologies (see Table 4.1).

and squares are vacant. A remarkable version of the uranophane-type sheet is the $[(NpO_2)_2(CrO_4)_3]^{4-}$ sheet in $(NH_4)_4[(NpO_2)_2(CrO_4)_3]$ (Grigor'ev *et al.* 1991c) (Figs. 4.5(b) and (d)). Only two thirds of the triangles are occupied by Np, which results in the formation of tetramers of edge-sharing (NpO_7) pentagonal bipyramids that are linked by the (CrO_4) tetrahedra.

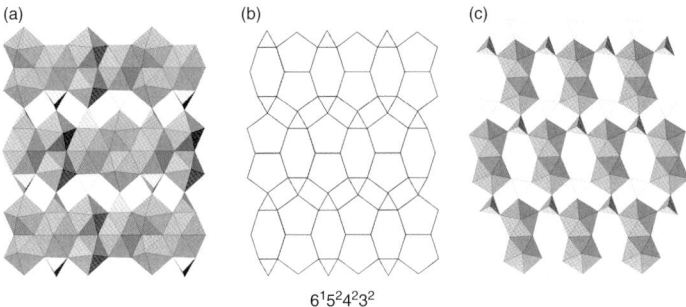

(a) (b) (c)

$6^1 5^2 4^2 3^2$

Fig. 4.4. Phosphuranylite anion topology (b) may have different cation populations ((a) and (c)).

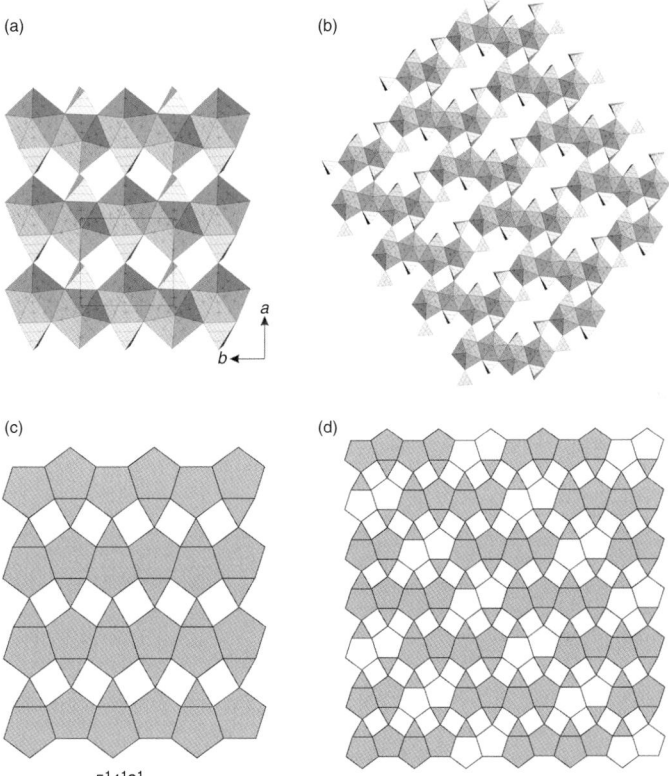

(a) (b)

a
b

(c) (d)

$5^1 4^1 3^1$

Fig. 4.5. Uranophane-type sheets in $Cs[NpO_2(CrO_4)](H_2O)$ (a) and $(NH_4)_4[(NpO_2)_2(CrO_4)_3]$ (b) have different populations by cations ((c) and (d): gray polygons correspond to the cation-occupied sites).

Table 4.1. Inorganic oxysalts based upon sheets with different anion-topologies.

Anion topology	Compound	Reference
$5^14^13^1$	$Cs[NpO_2(CrO_4)](H_2O)$	Grigor'ev et al. 1991c
	$(NH_4)_4[(NpO_2)_2(CrO_4)_3]$	Grigor'ev et al. 1991c
$6^15^24^23^{2*}$	$Cu[(UO_2)_2(OH)_2(SO_4)_2](H_2O)_8$ johannite	Mereiter 1982b
	$[C(NH_2)_3][(UO_2)(OH)(MoO_4)]$	Halasyamani et al. 1990
	$(UO_2)_3(MoO_4)_2(OH)_2(H_2O)_{10}$	Tali et al. 1994
	$(UO_2)_4(MoO_4)_3(OH)_2(H_2O)_4$	Tali et al. 1994
	$[N_2C_6H_{16}][UO_2F(SO_4)]_2$ USFO-3	Doran et al. 2005b
	$Sr[UO_2(OH)(CrO_4)](H_2O)_8$	Serezhkin et al. 1982
$5^44^13^2$	$K[(UO_2)(OH)(CrO_4)](H_2O)_{1.5}$	Serezhkina et al. 1990
$5^24^33^2$	$Zn(UO_2)_2(SO_4)O_2](H_2O)_{3.5}$	Burns et al. 2003
	$Co[(UO_2)_2(SO_4)O_2](H_2O)_{3.5}$	Burns et al. 2003
	$K_{2.71}[(UO_2)_4(SO_4)_2O_3(OH)] (H_2O)_3$	Vochten et al. 1995; Burns et al. 2003
	$Na_5(H_2O)_{12}[(UO_2)_8(SO_4)_4O_5 (OH)_3]$	Burns et al. 2003
	$(NH_4)_4[(UO_2)_2(SO_4)O_2]_2(H_2O)$	Burns et al. 2003
	$(NH_4)_2[(UO_2)_2(SO_4)O_2]$	Burns et al. 2003
	$Mg[(UO_2)_2(SO_4)O_2](H_2O)_{3.5}$	Burns et al. 2003
	$Mg_2[(UO_2)_2(SO_4)O_2]_2(H_2O)_{11}$	Burns et al. 2003
	$Mg_3(H_2O)_{18}[(UO_2)_4O_3(OH) (SO_4)_2]_2(H_2O)_{10}$ marecottite	Brugger et al. 2003
5^33^1	$(H_3O)_3[(UO_2)_3O(OH)_3(SeO_4)_2]$	Mit'kovskaya et al. 2003

* phosphuranylite topology with empty hexagons

Table 4.2. Inorganic oxysalts based upon uranophane anion topology ($5^14^13^1$) with occupied pentagons and triangles

Chemical formula	Mineral name	Reference
(ud) isomer (Fig. 4.6(a))		
$Ca[(UO_2)(SiO_3OH)]_2(H_2O)_5$	α-uranophane	Ginderow 1988
$K_2[(UO_2)(SiO_3OH)]_2(H_2O)_3$	boltwoodite	Burns 1998
$Mg[(UO_2)(SiO_3OH)]_2(H_2O)_6$	sklodowskite	Ryan and Rosenzweig 1977
$Cu[(UO_2)(SiO_3OH)]_2(H_2O)_6$	cuprosklodowskite	Rosenzweig and Ryan 1975
$Pb[(UO_2)(SiO_4)]_2(H_2O)$	kasolite	Rosenzweig and Ryan 1977
$Mg[(UO_2)(AsO_4)]_2(H_2O)_4$	seelite	Piret and Piret-Meunier 1994
$Cs[NpO_2(CrO_4)](H_2O)$	–	Grigor'ev et al. 1991c
(ud/du) isomer (Fig. 4.6(b))		
$(UO_2)[(UO_2)(VO_4)]_2(H_2O)_5$	–	Saadi et al. 2000
$(UO_2)[(UO_2)(PO_4)]_2(H_2O)_4$	–	Locock and Burns 2002a
$(UO_2)[(UO_2)(AsO_4)]_2(H_2O)_4$	–	Locock and Burns 2003e
$(UO_2)[(UO_2)(AsO_4)]_2(H_2O)_5$	–	Locock and Burns 2003e
(uudd) isomer (Fig. 4.6(c))		
$Ca[(UO_2)(SiO_3OH)]_2(H_2O)_5$	β-uranophane	Viswanathan and Harneit 1986
$A_2(UO_2)[(UO_2)(PO_4)]_4(H_2O)_2$ A = K, Rb, Cs	–	Locock and Burns 2002b
$[(C_2H_5)_3NH]_2[(UO_2)_2(PO_3OH)(PO_4)]$	–	Francis et al. 1998
$[(C_2H_5)_2NH_2]_2(UO_2)[(UO_2)(PO_4)]_4$	–	Danis et al. 2001
(uuuddd/uddduu/dduuud) isomer (Fig. 4.6(d))		
$[(C_3H_7)_4N][(UO_2)_3(PO_3OH)_2(PO_4)]$	–	Francis et al. 1998

Table 4.3. Inorganic oxysalts based upon phosphuranylite anion topology ($6^1 5^2 4^2 3^2$) with occupied hexagons, pentagons and triangles

Chemical formula	Mineral name	Reference
***(ud/ud)* isomer (Fig. 4.7(b))**		
$U^{6+}[(UO_2)_3(PO_4)_2(OH)_2](OH)_4(H_2O)_4$	vanmeersscheite	Piret and Deliens 1982
$Pb_2[(UO_2)_3(PO_4)_2O_2](H_2O)_5$	dumontite	Piret and Piret-Meunier 1988
$Pb_2[(UO_2)_3(AsO_4)_2O_2](H_2O)_5$	hügelite	Locock and Burns 2003c
$Cu[(UO_2)_3(SeO_3)_2O_2](H_2O)_8*$	marthozite	Cooper and Hawthorne 2001
$Ba[(UO_2)_3(SeO_3)_2O_2](H_2O)_3*$	guilleminite	Cooper and Hawthorne 1995c
***(uudd/uudd)* isomer (Fig. 4.7(c))**		
$Nd[(UO_2)_3(PO_4)O(OH)](H_2O)_6$	françoisite-(Nd)	Piret *et al.* (1988)
$Al[(UO_2)_3(PO_4)O(OH)](H_2O)_7$	upalite	Piret and Declercq (1983)
$CaK(H_3O)_3(UO_2)[(UO_2)_3(PO_4)_2O_2]_2(H_2O)_8$	phosphuranylite	Demartin *et al.* (1991), Piret and Piret-Meunier 1988
$Pb_3[(UO_2)_3(PO_4)_2O(OH)]_2(H_2O)_{12}$	dewindtite	Piret *et al.* (1990)
***(ud/du)* isomer (Fig. 4.7(d))**		
$Sr[(UO_2)_3(SeO_3)_2O_2]\cdot 4H_2O*$	–	Almond and Albrecht-Schmitt 2004
***(uddudd/uduudu)* isomer (Fig. 4.7(e))**		
$Ca_2Ba_4[(UO_2)_3(PO_4)_2O_2]_3(H_2O)_{16}$	bergenite	Locock and Burns 2003d
***(uudd/dduu)* isomer (Fig. 4.7(f))**		
$Al_2[(UO_2)_3(PO_4)_2(OH)_2](OH)_4(H_2O)_{10}$	phuralumite	Piret *et al.* (1979a)
$AlTh(UO_2)[(UO_2)_3(PO_4)_2O(OH)]_2(OH)_3(H_2O)_{15}$	althupite	Piret and Deliens (1987)
$Ca_2[(UO_2)_3(PO_4)_2O_2](H_2O)_7$	phurcalite	Atencio *et al.* (1991)

* In uranyl selenites, the "up" and "down" directions are determined by the orientation of Se^{4+} anions above or below the bases of (SeO_3) groups, respectively.

4.3   Anion topologies and isomerism

The uranophane and phosphuranylite topologies are some of the most commonly occurring in inorganic oxysalts that is likely to be the result of their high geometrical and energetical stability. Both topologies contain triangles that can be occupied by tetrahedral anions. In this case, a triangle constitutes a triangular basis of tetrahedron with three tetrahedral corners being within the plane of the sheet. The fourth corner does not participate in intra-sheet polymerization of polyhedra and may be oriented either up or down relative to the plane of the sheet. As with other cases of structural units containing tridentate tetrahedra (see Chapter 4), the possibility of "up" and "down" orientations results in the appearance of different orientational geometrical isomers. Locock and Burns (2003d, e) considered geometrical isomerism in phosphuranylite- and uranophane-related structures. Here, a slightly different approach to the description of isomers is employed by means of the concept of orientation matrix.

Figure 4.6 shows four different geometrical isomers of sheets with uranophane topology. The (**ud**) isomer has the shortest orientation matrices with the 1×2

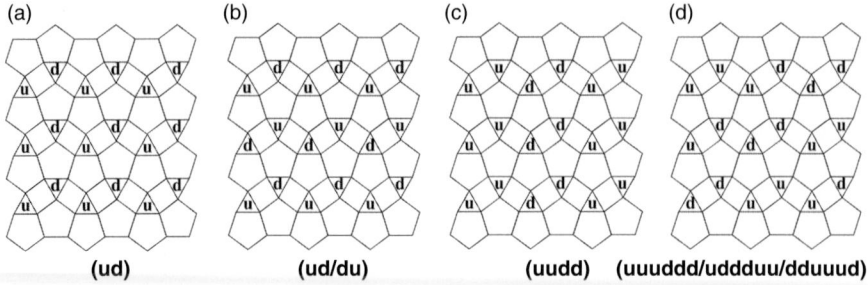

Fig. 4.6. Geometrical isomers of sheets with uranophane topology (see Table 4.2).

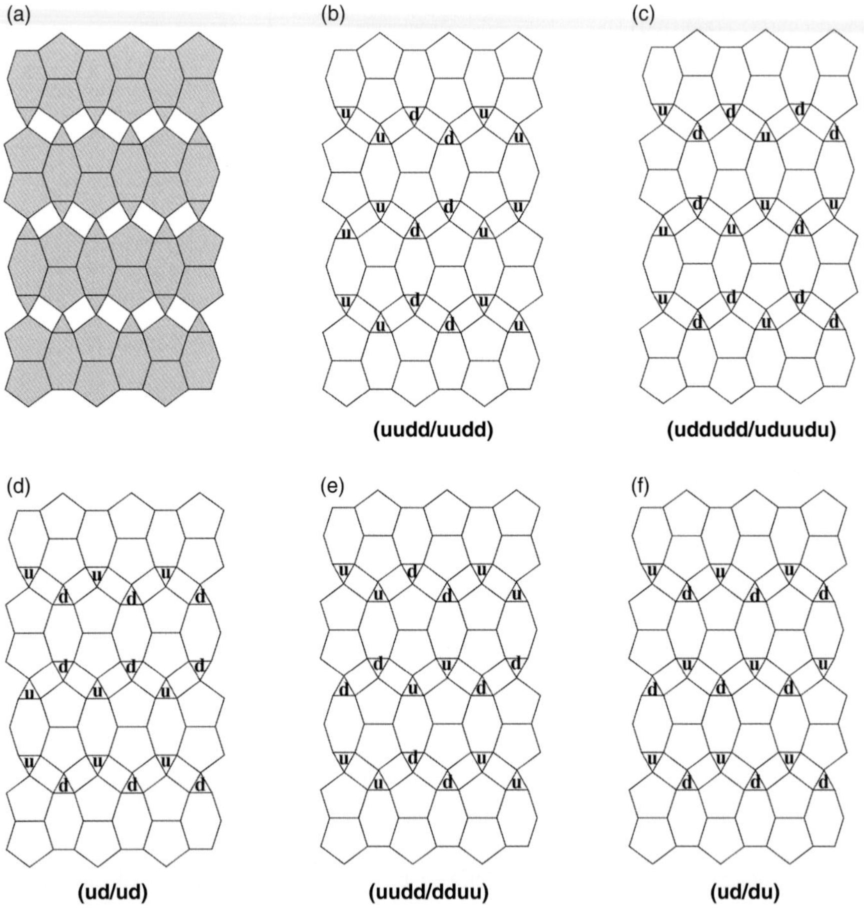

Fig. 4.7. Phosphuranylite sheet (a) and its geometrical isomers ((b), (c), (d), (e), (f)) (see Table 4.3 for details).

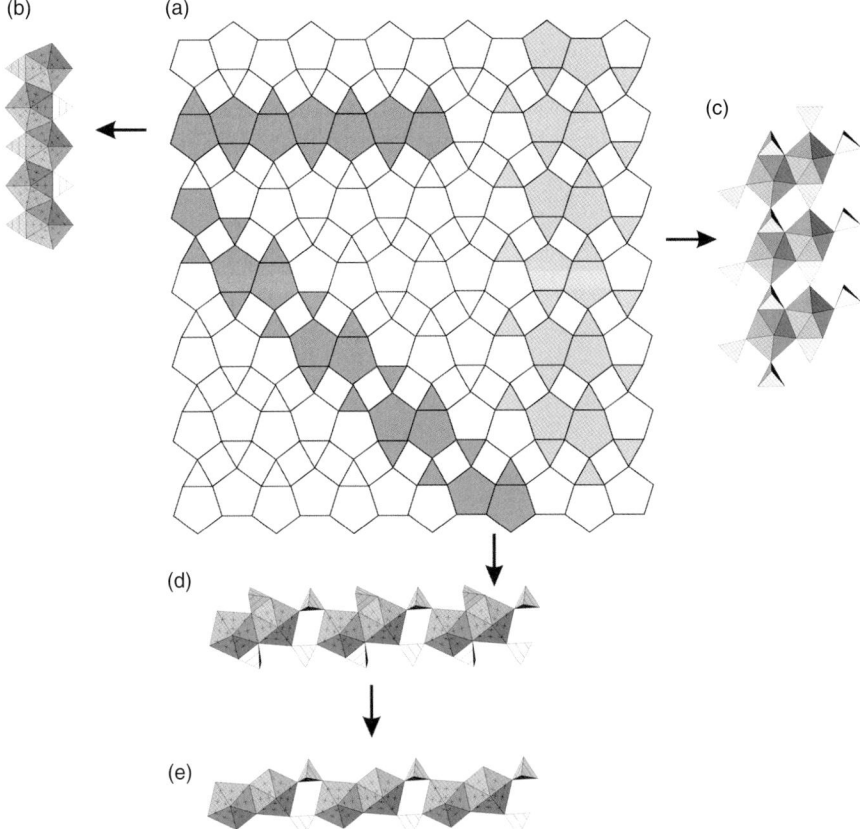

Fig. 4.8. Production of heteropolyhedral chains in inorganic oxysalts through excision of 1D regions from the uranophane anion topology. See text for details.

dimensions. The (**uuuddd/uddduu/dduuud**) isomer has the largest orientation matrix with the 3 × 6 dimensions (Fig. 4.6(d)). A list of the inorganic oxysalts containing different isomers of uranophane anion-topology sheets is given in Table 4.2.

Five isomers of the phosphuranylite-type sheets (i.e. sheets with occupied hexagons, pentagons and triangles) are shown in Fig. 4.7. A list of respective compounds is provided in Table 4.3.

4.4   1D derivatives of anion topologies

As is the case with 2D graphs and their derivatives, there are a number of 1D structural units that can be considered as being derived from corresponding anion topologies. Figure 4.8(a) shows how dense chains of edge-sharing polyhedra can be produced by cutting the uranophane anion-topology into polygonal chains. Cutting the topology parallel to the direction of the chains of edge-sharing pentagons generates the $[NpO_2(CrO_4)]$ chain in $Cs[NpO_2(CrO_4)](H_2O)_2$ (Fig. 4.8(b); Grigor'ev et al. 1991d). Excision of a chain in the perpendicular direction results in the $[(UO_2)(TO_4)_2]$ chain (T = P, As) (Fig. 4.8(c)) that has been observed in the structures of parsonsite, $Pb_2[(UO_2)(PO_4)_2](H_2O)_n$ (Burns 2000) and its As analog hallimondite (Locock et al. 2005). Finally, cutting the topology in the diagonal direction produces chains observed in $Cs_3[NpO_2(SO_4)_2](H_2O)_2$ (Grigor'ev et al. 1991e) (Fig. 4.8(d)), and in $K_2[(UO_2)F_2(SO_4)](H_2O)$ (Alcock et al. 1980) and $[N_2C_6H_{16}][UO_2F_2(SO_4)]$ (Doran et al. 2005b) (Fig. 4.8(e)).

Alternative approaches to structure description

5.1 Introductory remarks

The description of crystal structures of inorganic oxysalts in terms of cation-centered coordination polyhedra is the most popular and, in most cases, it provides adequate and efficient structural models. However, in some cases, this approach leads to a representation that lacks simplicity and clarity. This is an obvious indication that another structural principle is at work. In this chapter, we discuss two alternative approaches to the description of complex oxysalt structures. One is based upon anion-centered polyhedra and deals with the structures containing "additional" anions, i.e. anions not involved in strong polyhedra of high-valent cations. Another approach considers structure as a packing of cations (cation array) filled with anions. It is generally applicable to structures of large cations with a formal charge of 2+ or higher.

5.2 Anion-centered tetrahedra in inorganic oxysalts

5.2.1 Historical notes

A number of crystal structures can also be considered from the standpoint of the coordination of anions and, in particular, of oxygen atoms. Several researchers have used this approach for the description of crystal structures of inorganic compounds, identifying in the latter individual atomic groups formed by anions as the central atoms and by cations as the ligands. The development of this approach constitutes the subject of the crystal chemistry of compounds with oxocentered complexes, within the framework of which one considers "additional" oxygen atoms (i.e. atoms that do not participate in the formation of strong complexes of high-valent cations, $[T_mO_n]$ ($T =$ Si, Ge, B, S, P, V, As, Se, etc.) and water molecules) as the coordination centers. In most cases, the "additional" oxygen atoms are tetrahedrally coordinated by mono-, di- and trivalent metal atoms M to form oxocentered $[OM_4]$ tetrahedra. The M–O bonds formed within the oxocentered units are usually the strongest bonds formed by the M atoms in the structure. For this reason, $[OM_4]$ tetrahedra have been identified as basic structural units in many particular structures. For example, Sahl (1970) described the structure of lanarkite, $Pb_2O(SO_4)$, on the basis of the $[OPb_2]$ chains consisting of edge-linked $[OPb_4]$ tetrahedra and sulphate groups (SO_4). Apparently, the first systematic study on this topic was that of Bergerhoff and Paeslack (1968) where the authors specified only "additional" oxygen atoms as the centers for a series of tetrahedral complexes based on the $[OM_4]$ groups. The calculation of the stoichiometric cation/additional oxygen atom ratios, which are sometimes appreciably greater than unity, serves as the crystal chemical basis for this approach. Thus, in the structure of dolerophanite $Cu_2O(SO_4)$ (Effenberger 1985) the ratio is 2:1, which provides grounds for considering the "additional" oxygen atom as the coordination center. Bergerhoff and Paeslack (1968) identified six complexes based on $[OM_4]$ tetrahedra: a single tetrahedron, a chain, three types of layers, and one framework. Such a small structural diversity of complexes of $[OM_4]$ tetrahedra was due to the fact that at the time when the paper was written (1968) few structures with "additional" oxygen atoms were known. In the same year, Caro (1968) considered $[OM_4]$ tetrahedra and the cationic groups $(MO)^{n+}$ as structural units in rare-earth oxides and oxysalts. Later, Carre et al. (1984) analyzed a series of compounds with complexes based on oxocentered $[OLa_4]$ tetrahedra. They specified 20 five different types of complexes considered as derivatives of the oxocentred $[OLa]$ layer. Bengtsson and Holmberg (1990) published a review on the structural chemistry of crystalline and amorphous media containing $[OPb_4]$ tetrahedra. Oxocentered $[OPb_4]$ tetrahedra have been found in glasses of the $PbO–PbF_2$ system (Damodaran and Rao 1988), where they form a 3D framework made up of $O–Pb–O$ linkages and performing a function analogous to that of the $[SiO_4]$ tetrahedra in silicate glasses. Keller and co-workers (Keller, 1982, 1983; Keller and Langer 1994; Langecker and Keller 1994; Riebe and Keller 1988, 1989a, b, 1991) described a range of Pb oxyhalides containing $[OPb_4]$ tetrahedra as a basic structural unit. Schleid and co-workers elaborated in deep structural chemistry of rare-earth sulfides and halides with N- and O-centered metal tetrahedra (see Schleid 1996, 1999; Schleid and Lissner 2008 for reviews). In particular, Schleid and Wontcheu (2006) described a series of rare-earth oxyselenites based upon oxocentered tetrahedral units of variable dimensionality. Recently, Magarill, Borisov and others (Magarill et al. 2000) elaborated the structural classification of Hg^+ and Hg^{2+} oxysalts on the basis of the structural units consisting of $[OHg_4]$ tetrahedra.

The discovery of a large association of minerals based upon oxocentered $[OCu_4]$ tetrahedra in fumaroles of the Great Fissure Tolbachik eruption (GFTE, 1975–76, Kamchatka peninsula, Russia) (ponomarevite, melanothallite, piypite, fedotovite, kamchatkite, klyuchevskite, alumoklyuchevskite, ilinskite, averievite, georgbokiite, chloromenite, coparsite, burnsite, allochalcoselite, parageorgbokiite) by the research group of Filatov and Vergasova (1982–2007) prompted development of the crystal chemistry of minerals and inorganic compounds with anion-centered tetrahedra as a new branch of structural mineralogy and inorganic crystal chemistry (Filatov et al. 1992; Krivovichev et al. 1998a; Krivovichev and Filatov 2001). Filatov et al. (1992) identified ten types of complexes on the basis of $[OCu_4]$ complexes and noted their characteristic features: (1) the cationic nature of $[OM_x]$ tetrahedral complexes; (2) owing to the large size of the central atom and its relatively low charge (2–), the tetrahedra may link through edges as well as through vertices; (3) there is also a possibility of vertices being shared between more than two tetrahedra. All these features account for the exceptional diversity of the types of complexes in terms of their top-ology and geometry. These have been summarized in detail by Krivovichev *et al.* (1998a) and Krivovichev and Filatov (1999a, b, 2001). Below we shall provide a short overview of the most illustrative examples of inorganic oxysalt structures based upon units formed by $[OM_4]$ oxocentered tetrahedra. To distinguish topological types of tetrahedral units, we use the same notations as developed for units of cation-centered polyhedra in Chapter 2: ***acD–O:M–#*** where ***ac*** means "*anion-c*entered", ***D*** indicates dimensionality (0 – finite complexes, 1 – chains, 2 – sheets, 3 – frameworks), **O:M** ratio, **#** – registration number of the unit.

5.2.2 *0D units of oxocentered tetrahedra*

The simplest oxocentered unit is obviously an isolated $[OM_4]$ tetrahedron (Fig. 5.1). It has been encountered in a number of compounds, some of which are listed in Table 5.1. Table 5.2 lists some compounds that are based upon finite clusters consisting of linked oxocentered tetrahedra and we shall consider them in detail below.

Linkage of two $[OM_4]$ tetrahedra via one M atom results in formation of the $[O_2M_7]$ dimer (Fig. 5.1), an analog of the $[Si_2O_7]$ group in silicate chemistry. It has been found

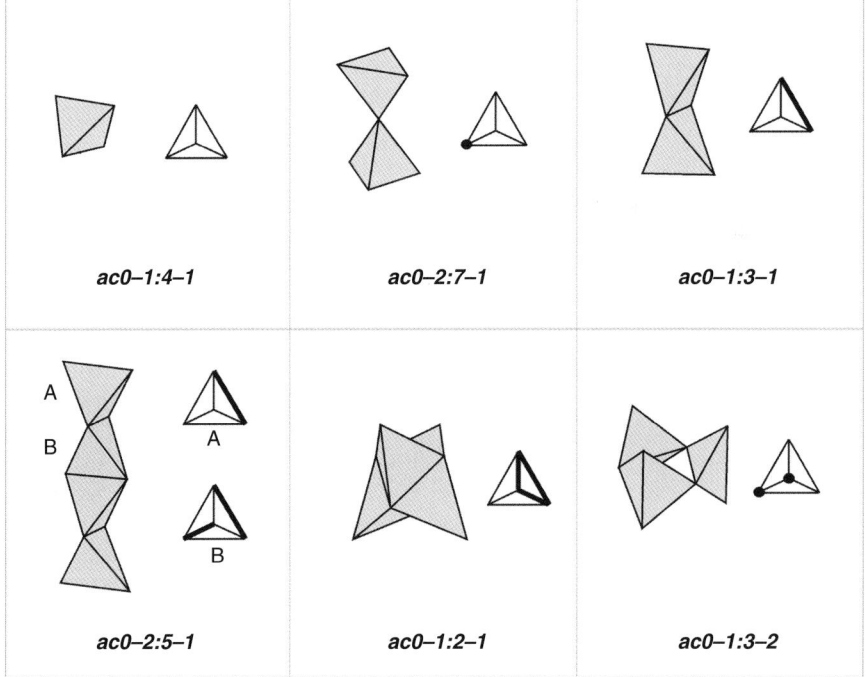

Fig. 5.1. 0D units of oxocentered (OM_4) tetrahedra in inorganic oxysalts.

Table 5.1.   Isolated oxocentered tetrahedra [OM_4] in structures of inorganic oxysalts

Compound	References
(Cu$_4$O)(PO$_4$)$_2$	Brunel-Laught *et al.* 1978; Anderson *et al.* 1978
(Cu$_4$O)(PO$_4$)$_2$ *Pnma*	Schwunck *et al.* 1998
(Cu$_4$O)(AsO$_4$)$_2$	Adams *et al.* 1995; Staack and Müller-Buschbaum 1996
(Cu(Cu,Mg)$_3$O)(As$_x$P$_{1-x}$O$_4$)$_2$ $x = 0.3$	Frerichs and Müller-Buschbaum 1996
KA′**(Cu$_4$O)**(VO$_4$)$_3$ A′ = Cu, Mg	von Postel and Müller-Buschbaum, 1993; Martin and Müller-Buschbaum 1994
BaMg$_2$**(Cu$_4$O)**$_2$(VO$_4$)$_6$	Vogt and Müller-Buschbaum 1991
A′(Cd,Cu)**(Cu$_4$O)**(VO$_4$)$_3$ A′ = Rb, Tl, K	Müller-Buschbaum and Mertens, 1997
TlCu**(Cu$_4$O)**(VO$_4$)$_3$	Moser and Jung 2000
Cu$_3$**(Cu$_4$O)**$_2$(VO$_4$)$_6$ fingerite	Finger 1985
(Fe$_4$O)(PO$_4$)$_2$	Bouchdoug *et al.* 1982
(Ca$_4$O)(PO$_4$)$_2$ hilgenstockite	Dickens *et al.* 1973
(M_4O) Cl$_6$ M = Ca, Sr, Ba	Meyer *et al.* 1991; Hagemann *et al.* 1996; Reckeweg and Meyer 1997; Frit *et al.* 1970
(PbCu$_3$O)(TeO$_6$)	Wedel and Müller-Buschbaum, 1996
(Pb$_4$O) Pb$_2$(BO$_3$)$_3$Cl	Behm 1983
(Pb$_4$O)$_2$Bi$_2$(PO$_4$)$_6$	Moore *et al.* 1982
(Pb$_4$O)(PO$_4$)$_2$	Krivovichev and Burns 2003i
Pb**(Pb$_4$O)**(OH)$_2$(CO$_3$)$_3$ "plumbonacrite"	Krivovichev and Burns 2000a
(Bi$_4$O)(AuO$_4$)$_2$	Geb and Jansen 1996
(M_4O)(AuO$_4$)$_2$ M = La, Nd, Sm, Eu	Ralle and Jansen 1994; Figulla-Kroschel and Jansen 2000
(M_4O)X$_6$ M = Yb, Sm, Eu; **X** = Cl, Br	Schleid and Meyer 1987a, b, c
(Be$_4$O)(NO$_3$)$_6$	Haley *et al.* 1997; Troyanov *et al.* 2000
Na**(Be$_4$O)**(SbO$_6$) swedenborgite	Huminicki and Hawthorne 2001
K$_6$**(Be$_4$O)**(CO$_3$)$_6$·7H$_2$O	Dahm and Adam 2002
K$_6$**(Be$_4$O)**(CO$_3$)$_6$]	Dahm and Adam 2002

Table 5.2.   Finite clusters of [OM_4] tetrahedra in structures of inorganic oxysalts

Type	Composition	Compound	References
ac0–2:7–1	[X$_2$A$_7$]	KCd**[Cu$_7$O$_2$]**(SeO$_3$)$_2$Cl$_9$ burnsite	Burns *et al.* 2000
ac0–1:3–1	[X$_2$A$_6$]	K$_2$**[Cu$_3$O]**(SO$_4$)$_3$ fedotovite	Starova *et al.* 1991
		NaK**[Cu$_3$O]**(SO$_4$)$_3$ euchlorine	Scordari and Stasi 1990b
		Zn**[Zn$_3$O]**(VO$_4$)$_2$	Waburg and Müller-Buschbaum 1986
		[Pb$_3$O](UO$_5$)	Sterns *et al.* 1986
		[Pr$_3$O](GeO$_4$)(PO$_4$)	Dzhurinskii *et al.* 1991
ac0–2:5–1	[X$_4$A$_{10}$]	**[Sr$_2$Bi$_3$O$_2$]**(VO$_4$)$_3$	Boje and Müller-Buschbaum 1992
ac0–1:2–1	[X$_4$A$_8$]	A′**[Pb$_8$O$_4$]**Br$_9$ A′ = Pb, Tl	Keller 1982, 1983
		[Sn$_8$O$_4$](SO$_4$)$_4$	Lundgren *et al.* 1982
		[Pb$_8$O$_4$]$_2$[Si$_{25}$Al$_{23}$O$_{96}$]	Yeom *et al.* 1997
		[(Pb^{2+}, Pb^{4+})$_4$Pb$_4$O$_4$] cluster in Pb$^{2+}_{44}$Pb$^{4+}_5$Tl$^+_{18}$O$_{17}$-Si$_{100}$Al$_{92}$O$_{384}$	Yeom *et al.* 1999
ac0–1:3–2	[X$_3$A$_9$]	**[Eu$_6$Ca$_3$O$_3$]**(BO$_3$)$_6$	Ilyukhin and Dzhurinskii 1993
ac0–8:13–1	[X$_8$A$_{13}$]	**[Pb$_{13}$O$_8$]**(OH)$_6$(NO$_3$)$_4$	Li *et al.* 2001; Kolitsch and Tillmans 2003

in the structure of burnsite, $KCd[\mathbf{Cu_7O_2}](SeO_3)_2Cl_9$, a fumarolic mineral. Two $[OCu_4]$ tetrahedra share a common Cu atom to form a dimer with opposite Cu–Cu–Cu faces parallel to each other (Fig. 5.2(a)). The O–Cu–O angle is 180°. The $(SeO_3)^{2-}$ groups attach to the triangular faces of the tetrahedra in a "face-to-face" fashion (Fig. 5.2(b)) that is typical for oxysalts with $[OCu_4]$ unit (Krivovichev *et al.* 1999a). The resulting $[\mathbf{Cu_7O_2}](SeO_3)_2{}^{6+}$ complexes link via Cu–O bonds into 3D microporous framework filled with K^+, Cd^{2+}, and Cl^- ions (Fig. 5.2(b)).

Linkage of two $[OM_4]$ tetrahedra via a common $M–M$ edge produces the $[O_2M_6]$ dimer. The local topology of linkage of $[OM_4]$ tetrahedra can be described using *connectivity diagrams* introduced in (Krivovichev 1997; Krivovichev *et al.* 1998a; see also Chapter 2). A connectivity diagram represents a view from above onto a regular tetrahedron resting on one of its triangular faces. The edges identified by semibold lines are common to two adjacent tetrahedra, whereas the corners designated by circles link the tetrahedron to another. If the number of tetrahedra sharing the same corner with the given one is more than 1, this number is written near the vertex of the diagram. Connectivity diagrams for the oxocentered tetrahedra in the finite clusters chain are shown in Fig. 5.1 near the polyhedral views of the clusters.

Linkage of the $[O_2M_6]$ dimers results in formation of a tetramer that occurs in the structure of $[\mathbf{Sr_2Bi_3O_2}](VO_4)_3$ (Fig. 5.1). The $\mathit{ac0–1:2–1}$ unit (Fig. 5.1) is also known as *stella quadrangula* or tetrahedral star. It represents an empty central tetrahedron that shares four of its faces with four $[OM_4]$ tetrahedra. The metallic sceleton of this unit consists of 8 M atoms and, in this form, represents one of the most common building blocks in the structures of intermetallic compounds (Nyman and Andersson

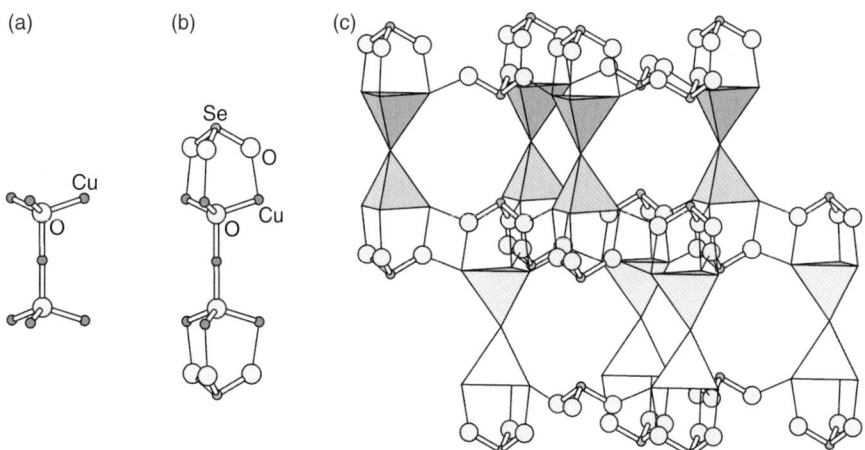

Fig. 5.2. Crystal structure of burnsite, $KCd[Cu_7O_2](SeO_3)_2Cl_9$: dimer of corner-sharing (OCu_4) tetrahedra (a), the dimer surrounded by two (SeO_3) groups in "face-to-face" positions (b), and arrangement of dimers into a 3D framework via Cu–O bonds.

1979; Häussermann *et al.* 1998). The [O_4M_8] *stella quadrangula* occurs as a separate moiety in [Sn_8O_4](SO_4)$_4$ (Fig. 5.3(a)). In the structure of [Bi_3O_2][$GaSb_2O_9$] (Sleight and Bouchard 1973), the [O_4Bi_8] clusters share peripheral corners to form a 3D network with the **dia** topology (Figs. 5.4(b) and (c); see Chapter 3). In this structure, two

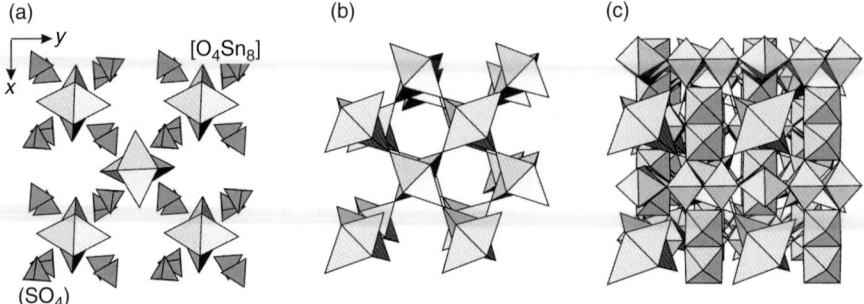

Fig. 5.3. The structure of $Sn_2O(SO_4)$ shown as an arrangement of [O_4Sn_8] stella quadrangula of four (OSn_4) tetrahedra and (SO_4) tetrahedra (a), **dia**-type framework of stella quadrangula of (OBi_4) tetrahedra in [Bi_3O_2][$GaSb_2O_9$] (b) and the structure of this compound as consisting of three interpenetrating frameworks (c).

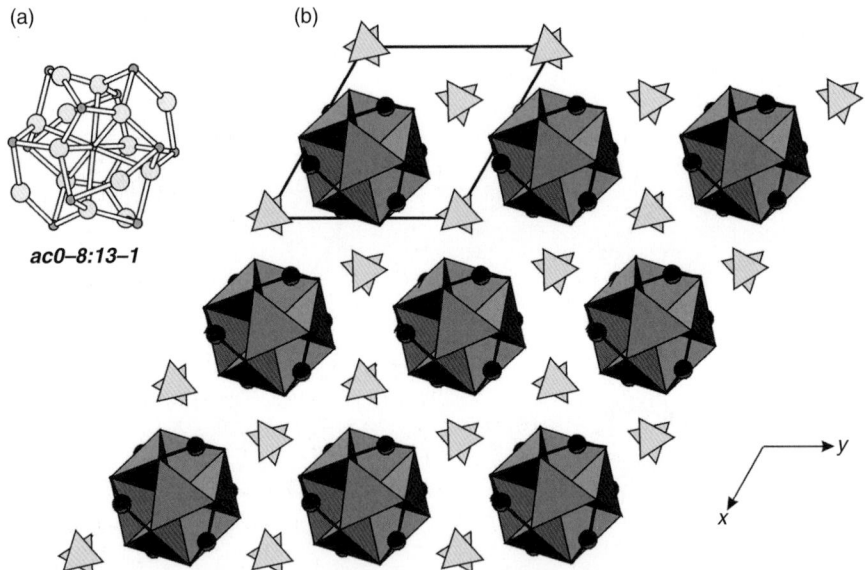

Fig. 5.4. [O_8Pb_{13}](OH)$_6$]$^{4+}$ cluster in [$Pb_{13}O_8$](OH)$_6$(NO_3)$_4$ (a) and the structure of this compound projected along the c-axis (b).

$[O_2Bi_3]$ networks interpenetrate each other and with the $[GaSb_2O_9]$ network of edge- and corner-sharing octahedra to form a rare example of inorganic compound with three interpenetrating networks.

The *ac0–1:3–2* unit is a simple 3-membered tetrahedral ring (Fig. 5.1) that has been observed in several Ca-*REE* borates (*REE* = rare earths) (Ilyukhin and Dzhurinskii 1993).

Figure 5.4(a) shows the $[[O_8Pb_{13}](OH)_6]^{4+}$ cluster from the structure of $[Pb_{13}O_8]$ $(OH)_6(NO_3)_4$ (Fig. 5.4(b)). In this unit, eight $[OPb_4]$ tetrahedra share one Pb^{2+} cation so that the latter has an eightfold cubic coordination. The coordination is highly symmetrical, which is not typical for Pb^{2+} cations in compounds with strong Lewis bases such as O^{2-} or OH^- anions. In the structure, the $[[O_8Pb_{13}](OH)_6]^{4+}$ clusters are packed in a close-packed array with NO_3^- anions in between.

5.2.3 *1D units of oxocentered tetrahedra*

The structural topology of the 1D units of oxocentered tetrahedra is remarkable in its analogy to topology of classical tetrahedral anions in silicate chemistry. Liebau (1985) introduced the concept of multiplicity in order to describe the structural diversity of silicate chains in terms of multiple structural units. Thus, pyroxenes and amphiboles contain single and double chains, respectively. A similar situation holds for chains of oxocentered tetrahedra.

Figure 5.5(a) shows the $[OM_3]$ chain of corner-sharing $[OM_4]$ tetrahedra with an identity period comprising of one tetrahedron. Merging two $[OM_3]$ chains by sharing common M atoms results in a double chain with the composition $[O_2M_4]$ (Fig. 5.5(b)). Addition of one more chain to the double chain produces a triple chain $[O_3M_5]$ (Fig. 5.5(c)). It is easy to deduce that the chains of this family will have a common structural formula $[O_nM_{n+2}]$. At the present time, only chains with $n \leq 3$ are known. Inorganic compounds containing these chains are listed in Table 5.3. It is noteworthy that the chains with $n = 1$ and 2 contain tetrahedra with one type of connectivity diagrams only, whereas the $n = 3$ chain consists of tetrahedra of two different connectivity types.

Figure 5.6(a) shows the $[OM_3]$ chain of corner-sharing $[OM_4]$ tetrahedra similar to that considered above but with two tetrahedra within its identity period (thus, it is an anolog of the pyroxene chains). This chain occurs, in particular, in the structure of chloromenite, $Cu_3[Cu_6O_2](SeO_3)_4Cl_6$ (Krivovichev *et al.* 1998b), another rare mineral from Kamchatka fumaroles. In its structure (Fig. 5.7), the chains are surrounded by SeO_3 groups (in the "face-to-face" fashion) and are linked by additional Cu^{2+} cations into a complex layered structure. Merging two *zweier* $[OM_3]$ chains by edge-sharing between tetrahedra results in the formation of stoiberite chain, $[O_2M_5]$, depicted in Fig. 5.6(b). Again, it can be shown that the general formula for chains of this family, should be written as $[O_nM_{2n+1}]$. Only chains with $n = 1$ and 2 are known.

Another family of 1D chains of oxocentered tetrahedra is shown in Fig. 5.8. A single chain is that of *trans*-edge-sharing tetrahedra. It had been first recognized by Sahl (1970) as a basic structure element in lanarkite, $Pb_2O(SO_4)$. It also occurs in a

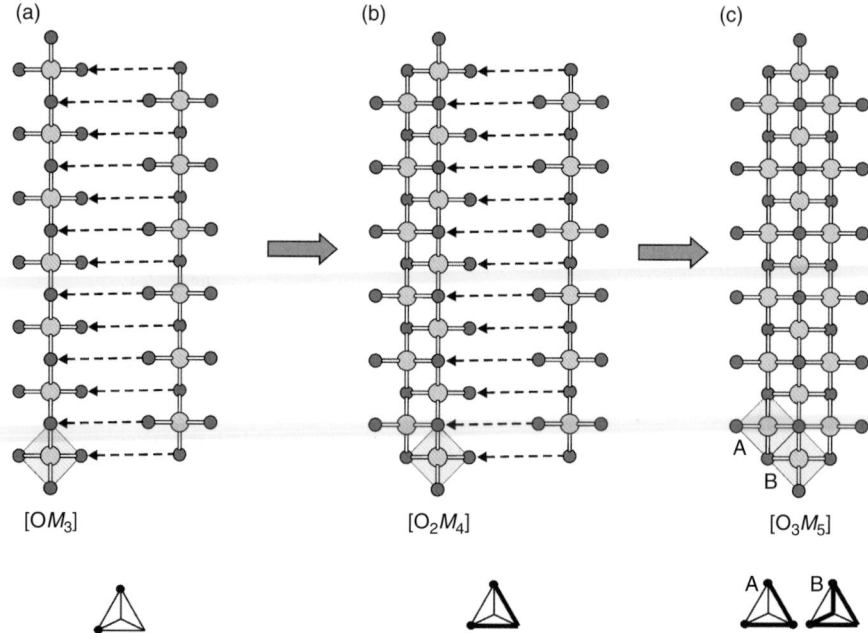

Fig. 5.5. The $[O_n M_{n+2}]$ family of chains of anion-centered tetrahedra: single chain (a), double chain (b), and triple chain (c) together with connectivity diagrams of their tetrahedra.

number of inorganic oxysalts listed in Table 5.4. Various schemes of conformation of these chains of anion-centered tetrahedra were considered by Krivovichev and Filatov (1998). Double chains of this family are also quite common (Table 5.4). In general, the chemical formula of the chains is $[O_n M_{n+1}]$. It is important that the topological complexity of the chains increases essentially for $n > 2$. The single and double chains consist of one topological type of tetrahedra: the connectivity diagrams are the same for all tetrahedra of the chain. In other words, all tetrahedra are *topologically* equivalent. For $n = 3$, the chains start to contain two topological types of tetrahedra (Fig. 5.8). As a consequence, triple chains are very rare. In the impressive series of papers, Mentré and co-workers described a whole class of Bi–M^{2+} oxophosphates (M = Cu, Co, Ni, Cd, Mn, etc.) consisting of chains of edge-sharing tetrahedra of various multiplicity (Abraham *et al.* 2002; Ketatni *et al.* 2003; Huvé *et al.* 2004, 2006; Mentré *et al.* 2006; Colmont *et al.* 2003, 2006). On the basis of X-ray diffraction structure analysis and high-resolution transmission electron microscopy (HRTEM) data, these authors identified several structure types with different arrangements and stacking sequences of chains. For the $[O_n M_{n+1}]$ family, the chains with $n = 3, 4, 5$, and 6 were discovered. It is interesting that, in the multiple chains, Bi^{3+} and M^{2+} cations occupy topologically and geometrically different positions (Fig. 5.8). There are two different metal sites in the multiple chains. The $M1$ atoms are on the periphery and

Table 5.3. Inorganic oxysalts containing single and multiple chains of corner-sharing [OM_4] tetrahedra

Type	Chemical formula	References
ac1–1:3–1	$Ca_2[MCa_2O](BO_3)_3$ M = Sm, Gd, Lu, Tb	Khamaganova et al. 1991; Norrestam et al. 1992
ac1–1:2–1	$[Pb_2O][SiO_3]$	Dent Glasser et al. 1981
	$[Pb_2O]_2(OH)_2(SO_3S)$ sidpietersite	Cooper and Hawthorne 1999
	$[Pb_2O]_2(OH)_2(SO_4)$	Steele et al. 1997
	$[Eu_4O_2][Al_2O_7]$	Brandle and Steinfink, 1969
	$[Pr_2O]_2[Ga_2O_7]$	Gesing et al. 1999
	$[Y_4O_2][Al_2O_7]$	Lehmann et al. 1987
ac1–3:5–1	$[La_5O_3]_2In_6S_{17}$	Gastaldi et al. 1987
ac1–1:3–2	$K[Cu_3O]Cl(SO_4)_2$ kamchatkite	Varaksina et al. 1990
	$Cu_3[Cu_6O_2](SeO_3)_4Cl_6$ chloromenite	Krivovichev et al. 1998b
	$[Cu_3O][(Mo,S)O_4SO_4]$ vergasovaite	Berlepsch et al. 1999
	$[Cu_3O](MoO_4)_2$	Steiner and Reichelt 1997
	$[Cu_3O][V_2O_7](H_2O)$	Leblanc and Férey 1990
	$Na_2Cu^I[Cu^{II}_3O](PO_4)_2Cl$	Etheredge and Hwu 1996
	$[Zn_3O](SO_4)_2$	Bald and Gruehn 1981
	$[Zn_3O](MoO_4)_2$	Söhnel et al. 1996
	$Cu^+Cu^{2+}[Cu^{2+}_3O](SeO_3)Cl_5$	Krivovichev et al. 2004b
	$Cu^+[Cu^{2+}_5PbO_2](SeO_3)_2Cl_5$ allochalcoselite	Krivovichev et al. 2006e
ac1–2:5–1	$[Cu_5O_2](VO_4)_2$ stoiberite	Shannon and Calvo 1973

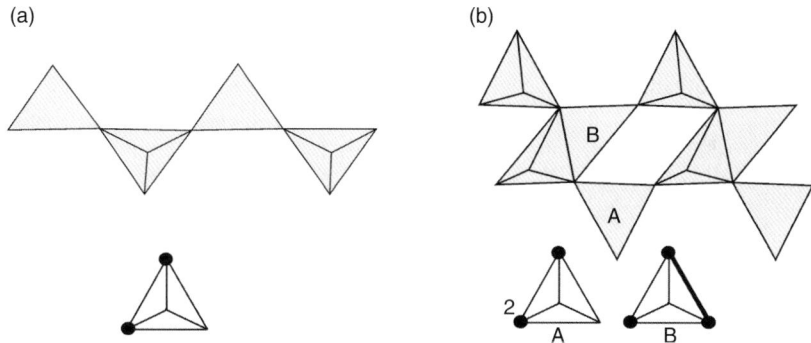

Fig. 5.6. Single and double chains of the [O_nM_{2n+1}] family.

are coordinated by two O atoms of the oxocentered tetrahedra. The $M2$ sites are in the middle parts of the chains and have fourfold coordination of four O atoms of the [OM_4] tetrahedra. This coordination has the form of a square pyramid with the $M2$ atom at its top. It is typical for cations with stereoactive lone electron pairs such as Bi^{3+} and Pb^{2+}. In the structures of the Bi–M^{2+} oxophosphates, the M^{2+} cations occupy the $M1$ sites, whereas the Bi^{3+} cations occupy the $M2$ sites. The higher the Bi

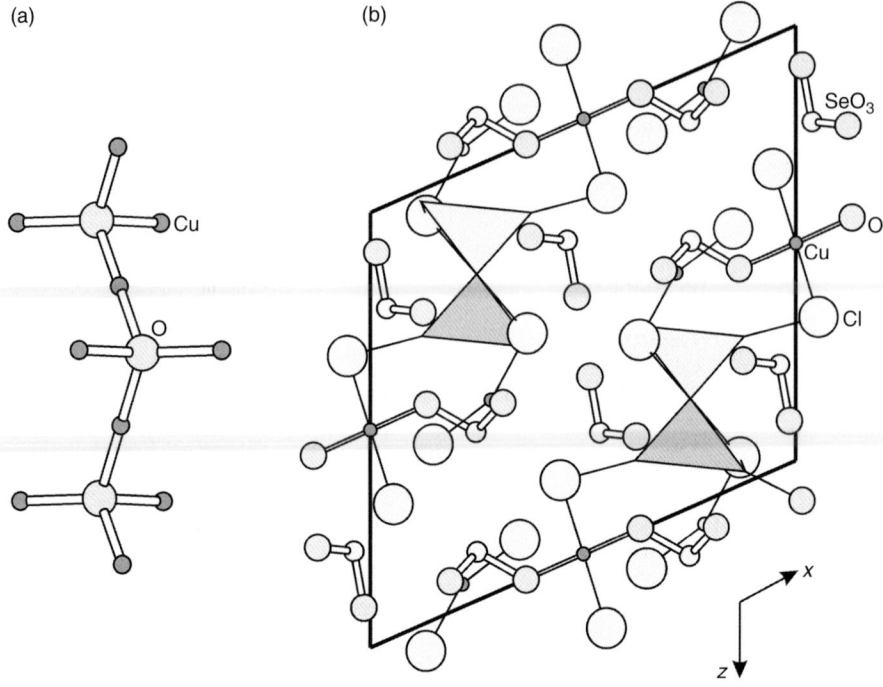

Fig. 5.7. The structure of chloromenite, $Cu_3[Cu_6O_2](SeO_3)_4Cl_6$: chain of corner-sharing (OCu_4) tetrahedra (a) and the structure projected along the b-axis (b).

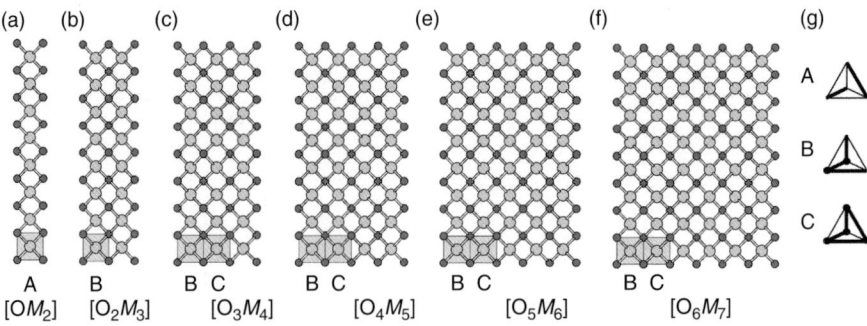

Fig. 5.8. The $[O_nM_{n+1}]$ family of chains with $n = 1$ (a), 2 (b), 3 (c), 4 (d), 5 (e), and 6 (f).

content in the structure, the wider the multiple chains, i.e. the larger the n value. With the absence of transitional metal atoms, continuous sheets of $[OBi_4]$ tetrahedra are observed, as in the Aurivillius phases. The role of phosphate groups in the Bi–M^{2+} oxophosphates is to link adjacent chains into a 3D structure via the M^{2+}–O_P bonds (O_P – atoms of the PO_4 tetrahedra).

Table 5.4.   Inorganic oxysalts containing chains of the $[O_nM_{n+1}]$ family

Type	Chemical formula	References
ac1–1:2–2	**[Pb$_2$O]**(SO$_4$) lanarkite	Sahl 1970
	[Pb$_2$O](WO$_4$)	Bosselet *et al.* 1985
	[Pb$_2$O](CrO$_4$) phoenicochroite	Williams *et al.* 1970; Ruckman *et al.* 1972; Morita and Toda 1984
	[Pb$_2$O](MoO$_4$)	Mentzen *et al.* 1984a
	Pb$_{2+x}$OCl$_{2+2x}$	Siidra *et al.* 2007a
	Pb$_2$**[Pb$_2$O]**(VO$_4$)$_3$	Krivovichev and Burns 2003j
	K$_3$**[Cu$_3$(Al, Fe)O$_2$]**(SO$_4$)$_2$ klyuchevskite	Gorskaya *et al.* 1992
	K$_2$**[Cu$_2$O]**(SO$_4$) · MeCl piypite	Effenberger and Zemann, 1984
	Na$_2$**[Cu$_2$O]**(SO$_4$) · MeCl	Kahlenberg *et al.* 2000
	(UO$_2$)**[Bi$_4$O$_2$]**O$_2$(AsO$_4$)$_2$ · 2H$_2$O walpurgite, orthowalpurgite	Krause *et al.* 1995; Mereiter 1982
	[La$_2$O](ReO$_4$)	Waltersson, 1976
	[Bi$_2$O](AuO$_4$)	Geb and Jansen 1996
	[Bi$_2$O]Cu(SeO$_3$)$_3$ · (H$_2$O)	Effenberger 1998
	[BiPbO](VO$_4$)	Wang and Li 1985
	[BiNiO](PO$_4$)	Abraham and Ketatni 1995
	[Er$_2$O][WO$_5$]	Tyulin *et al.* 1984
	[Pb$_2$O]$_6$[Mn(Mg,Mn)$_2$(Mn,Mg)$_4$ Cl$_4$(OH)$_{12}$ (SO$_4$)(CO$_3$)$_4$] philolithite	Moore *et al.* 2000
	[Yb$_2$O](SiO$_4$)	Smolin 1969
	[Dy$_2$O](GeO$_4$)	Brixner *et al.* 1985
	[Cu$_4$O$_2$]((As,V)O$_4$)Cl	Starova *et al.* 1998
	Lu$_2$**[Lu$_4$O$_2$]**(GeO$_4$)$_2$[Ge$_2$O$_7$]	Palkina *et al.* 1994
	Na**[Pr$_4$O$_2$]**Cl$_9$	Mattfeld and Meyer 1994
	K**[Pr$_4$O$_2$]**Cl$_9$	Mattfeld and Meyer 1994
	[Sc$_2$O](GeO$_4$)	Maksimov *et al.* 1974
	[Tb$_2$O](SeO$_3$)	Wontcheu and Schleid 2002
	[Sr$_2$O]I$_2$	Reckeweg and DiSalvo 2006
ac1–2:3–1	**[Pb$_3$O$_2$]**I$_2$	Kramer and Post 1985
	[Pb$_3$O$_2$](SO$_4$)	Sahl 1981; Mentzen *et al.* 1984b; Latrach *et al.* 1985a, b
	[Pb$_3$O$_2$]Cl$_2$ mendipite	Vincent and Perrault 1971; Krivovichev and Burns 2001c
	[Pb$_3$O$_2$](OH)Cl damaraite	Krivovichev and Burns 2001c
	[Pb$_3$O$_2$](OH)Br	Krivovichev and Burns 2001d
	Pb**[Pb$_3$O$_2$]**$_2$(OH)$_4$Cl$_2$	Krivovichev and Burns 2002c
	[Pb$_3$O$_2$](CO$_3$)	Krivovichev and Burns 2000b
	[Pb$_3$O$_2$](SeO$_3$)	Krivovichev *et al.* 2004c
	[Pb$_3$O$_2$]$_2$(OH)(CO$_3$)(NO$_3$)	Li *et al.* 2000
	[Pb$_3$O$_2$](OH)(NO$_3$)	Krivovichev *et al.* 2001
	[Pb$_3$O$_2$](NO$_3$)$_2$	Bataille *et al.* 2004
	Cu**[Pb$_3$O$_2$]**(OH)$_2$Cl$_2$ chloroxiphite	Finney *et al.* 1977
	[BiMg$_2$O$_2$](*T*O$_4$) *T* = V, P, As	Huang *et al.* 1993
	[BiCu$_2$O$_2$](PO$_4$)	Abraham *et al.* 1994
	[BiCu$_2$O$_2$]((P,V)O$_4$)	Mentre *et al.* 2006
	[Bi$_2$CdO$_2$](GeO$_4$)	Dinnebier *et al.* 1996
	[Pb$_2$BiO$_2$](PO$_4$)	Mizrahi *et al.* 1997

Table 5.4. *Continued*

Type	Chemical formula	References
	[BiCa$_2$O$_2$](VO$_4$)	Radoslavljevic *et al.* 1998
	[La$_3$O$_2$](ReO$_6$) *P*2$_1$/*m*	Rae-Smith *et al.* 1984
	[Y$_3$O$_2$](ReO$_6$)	Baud *et al.* 1981
	[Sm$_3$O$_2$](ReO$_6$)	Besse *et al.* 1976
	Gd**[Gd$_3$O$_2$]**(WO$_5$)$_2$	Tyulin and Efremov 1987
	Gd**[Gd$_3$O$_2$]**(MoO$_5$)$_2$	Tyulin and Efremov 1987
	[Gd$_3$O$_2$](GaO$_4$)	Yamane *et al.* 1999
	Sm**[Sm$_3$O$_2$]**(MoO$_5$)$_2$	Klevtsov *et al.* 1975
	Tb**[Tb$_3$O$_2$]**(MoO$_5$)$_2$	Xue *et al.* 1995
	Nd**[Nd$_3$O$_2$]**(WO$_5$)$_2$	Polyanskaya *et al.* 1970
acl–3:4–1	**[La$_4$O$_3$]**[AsS$_3$]$_2$	Palazzi and Jaulmes 1981
	[Ho$_4$O$_3$][Mo$_4$O$_8$]	Gougeon *et al.* 1990
	Bi$_{1.2}$$M_{1.2}PO_{5.5}$ (*M* = Mn, Co, Zn)	Abraham *et al.* 2002
	Bi$_{6.2}$Cu$_{6.2}$O$_8$(PO$_4$)$_5$	Ketatni *et al.* 2003
acl–4:5–1	Bi$_{5.625}$Cu$_{2.062}$(PO$_4$)$_3$O$_6$	Colmont *et al.* 2006
acl–5:6–1	Bi$_{20}$Cd$_{7.42}$Cu$_{0.58}$O$_{24}$(PO$_4$)$_{12}$	Colmont *et al.* 2006
acl–6:7–1	Bi$_{15.32}$Cd$_{10}$O$_{18}$(PO$_4$)$_{10}$	Huvé *et al.* 2006

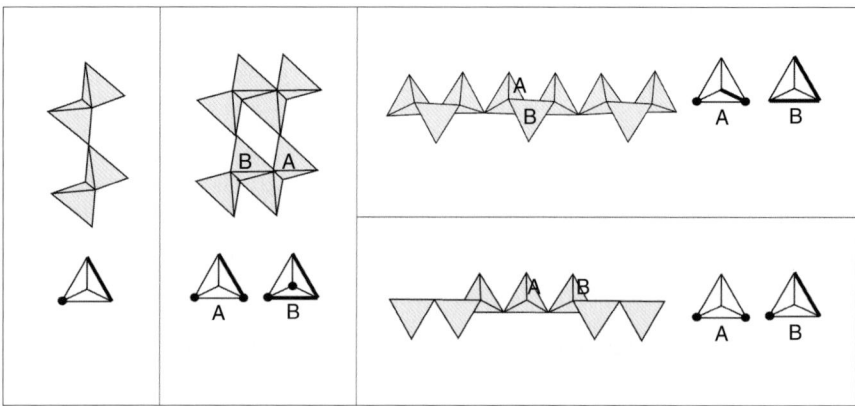

Fig. 5.9. Chains of (*OM$_4$*) tetrahedra with alternating edge and corner sharing between tetrahedra.

In addition to single chains where either corner- or edge-sharing tetrahedra occur, there are chains where corner and edge linkages alternate. Some examples of these chains are shown in Fig. 5.9. The structures of georgbokiite and parageorgbokiite, Cu$_5$O$_2$(SeO$_3$)$_2$Cl$_2$ (Krivovichev *et al.* 1999b, 2007d), contain [O$_2$Cu$_5$] chains of [OCu$_4$] tetrahedra (Fig. 5.10(a)). Each tetrahedron is covered in the "face-to-face" fashion by a SeO$_3$ pyramid, so that the 1D units $|$[O$_2$Cu$_5$](SeO$_3$)$_2|$ are formed (Fig. 5.10(b)). They

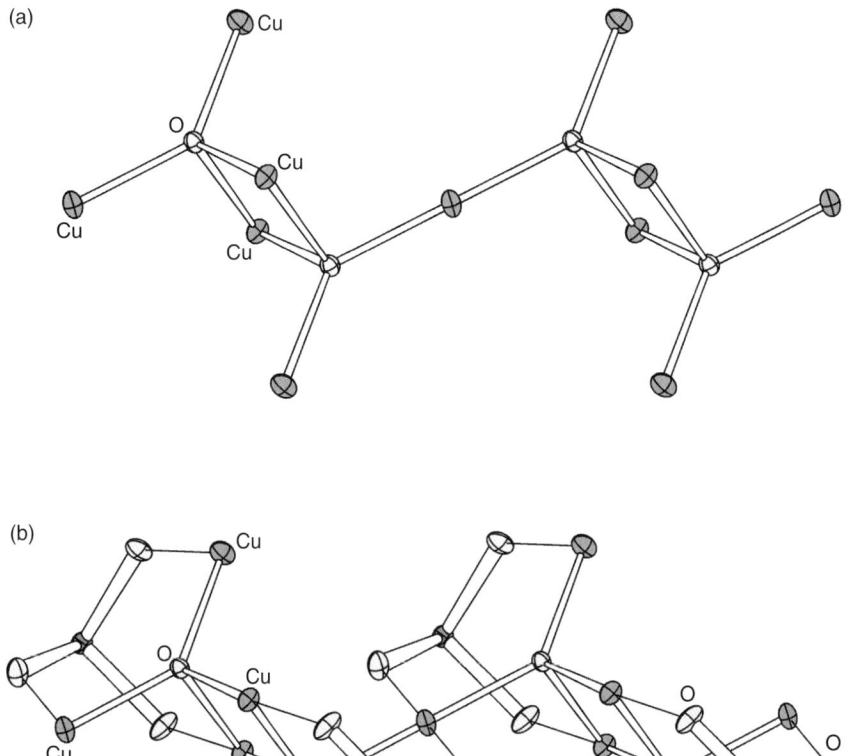

Fig. 5.10. Chain of (OCu$_4$) tetrahedra in georgbokiite and parageorgbokiite (a) and complex chain formed by a "face-to-face" attachment of (SeO$_3$) groups (b).

pack differently in geogbokiite and parageorgbokiite (Fig. 5.11), which represents an interesting case of polymorphism in compounds with oxocentered tetrahedra. On a purely geometrical basis, double georgbokiite chains may be subdivided in kyanite, Al$_2$O(SiO$_4$), which was first shown by Bergerhoff and Paeslack (1968).

The chains shown in Figs. 5.12(a) and (b) are interesting examples of topological complexity. First, both chains can be constructed by successive growth of an initial fragment by addition of new tetrahedra via edge sharing. There is no "pure" corner linkage in the chains. Both chains have been observed in Pb oxysalts (Table 5.5) and can be considered as excisions from the [OPb] layer of [OPb$_4$] tetrahedra in tetragonal PbO (see below). The [O$_3$Pb$_5$] chain (Fig. 5.12(a)) contains tetrahedra of three

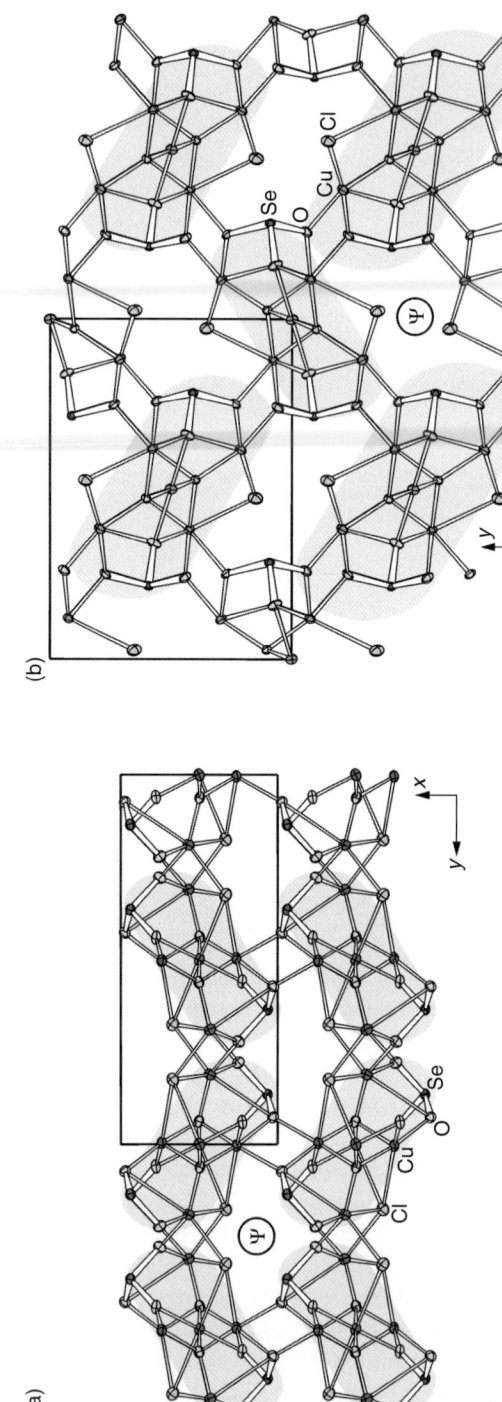

Fig. 5.11. Arrangement of complex $|[O_2Cu_5](SeO_3)_2]|$ in georgbokiite (a) and parageorgbokiite (b). The chains run perpendicular to the plane of the figure; their sections are highlighted.

connectivity types, whereas the $[O_7Pb_{10}]$ chain (Fig. 5.12(b)) contains five topological types of tetrahedra. However, it is easy to note that, although B and C tetrahedra are topologically equivalent, their configurations in the chain are different. To distinguish topological tetrahedra with different configurations, the concept of *corona* can be used (Krivovichev *et al.* 1998a). The first corona of a tetrahedron consists of the tetrahedron itself and all tetrahedra that have at least one common corner with the given one. The second corona consists of the first corona plus all tetrahedra that share with tetrahedra from the first corona at least one common corner. Figure 5.12(c) shows that the first coronas of both the B and C tetrahedra consist of six tetrahedra arranged in the same way. In contrast, the second coronas for B and C consist of 12 and 14 tetrahedra, respectively. Thus, the B and C tetrahedra are *configurationally* non-equivalent. Similarly, the D and G tetrahedra are configurationally non-equivalent, although they are equivalent in the topological sense. In summary, in the $[O_7Pb_{10}]$ chain, there are five classes of topological equivalence and seven classes of configurational equivalence. As has been noted by Krivovichev and Filatov (2001), the $[O_7Pb_{10}]$ chain in the structure of $Pb_{10}O_7(OH)_2F_2(SO_4)$ is one of the most complex structural units based on anion-centered tetrahedra in inorganic oxysalts.

5.2.4 2D units of anion-centered tetrahedra

The 2D units consisting of anion-centered tetrahedra are much more diverse than 0D and 1D units. Here we shall describe the most representative types only.

The structure of averievite, $[Cu_5O_2](VO_4)_2 \cdot nMCl$ (M = K, Rb, Cs) (Starova *et al.* 1997) (Fig. 5.13) is based upon a 2D sheet of corner-sharing $[OCu_4]$ tetrahedra. This sheet is a topological analog of the silicate sheets in micas but with another arrangement of non-shared vertices of tetrahedra. In averievite, (VO_4) tetrahedra are in "face-to-face" positions relative to the $[OCu_4]$ tetrahedra that forms a complex electroneutral 2D layer $|[O_2Cu_5](VO_4)_2|$. The layers are stacked one under another so that the large channels are formed, which are occupied by alkali metal and chlorine ions. The same type of tetrahedral sheets is observed in $[Pb_2Cu_3O_2](NO_3)_2(SeO_3)_2$ (Effenberger 1986). However, in this case, the sheet is formed by heterometallic $[OCu_3Pb]$ tetrahedra, where Pb atoms occupied non-shared vertices of tetrahedra.

The $[O_nM_{n+1}]$ family of chains shown in Fig. 5.8 generates an interesting family of sheets shown in Fig. 5.14. In these sheets, adjacent chains are linked by sharing common M atoms. The stoichiometry of the resulting 2D units is $[O_nM_{n+0.5}]$ or $[O_{2n}M_{2n+1}]$. A list of oxysalts consisting of sheets with $n = 1, 2,$ and 3 is given in Table 5.6. It is noteworthy that, in Bi–Cu oxysalts, the role of Bi^{3+} and Cu^{2+} cations is again different and identical to that mentioned for the mixed Bi–M^{2+} chains. Observations of this kind led Krivovichev and Filatov (1999a) to the following *size classification* of M cations that form anion-centered tetrahedra: (1) small tetrahedra formed by cations with $r_M = 0.5–0.7$ Å: e.g., Cu^{2+}, Zn^{2+}, rarely Co^{2+}, Ni^{2+}, Fe^{2+}, Fe^{3+}, and Al^{3+} in octahedral coordination and (2) large cations with $r_M \sim 1.0$ Å: e.g. Pb^{2+}, Sn^{2+}, *REE*$^{3+}$, and Bi^{3+}. Tetrahedra formed by cations from both classes simultaneously are designated as heterometallic. Owing to the greater distances between the centers of large

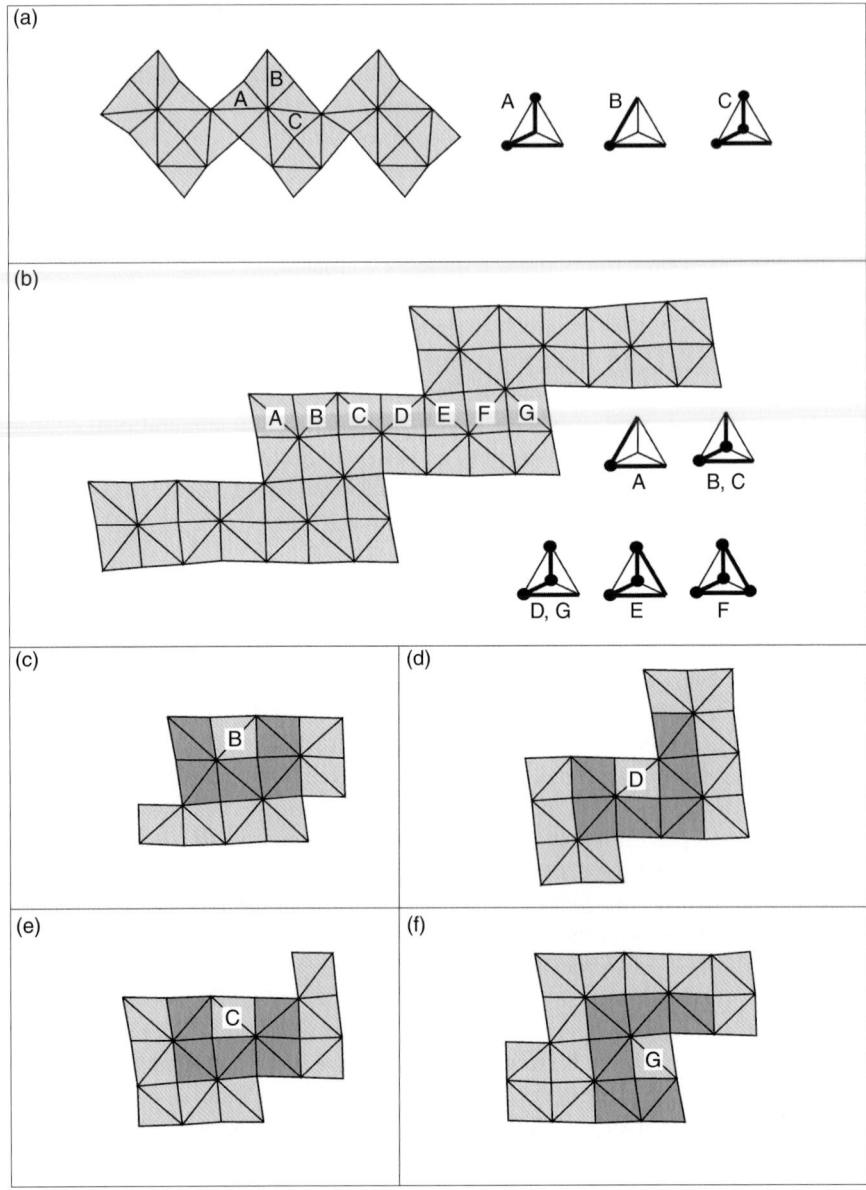

Fig. 5.12. Chains of (OPb$_4$) tetrahedra in Pb$_5$O$_4$(TO$_4$) (T = Cr, S) (a) and Pb$_{10}$O$_7$(OH)$_2$F$_2$(SO$_4$) (b) and topological configurations of tetrahedra in the latter (c).

Table 5.5. Inorganic oxysalts containing chains of oxocentered (OM_4) tetrahedra with edge- and corner-linkage between tetrahedra

Type	Chemical formula	Reference
acl–2:5–2	α-$[\mathbf{Cu_5O_2}](SeO_3)_2Cl_2$ georgbokiite	Galy et al. 1979; Krivovichev et al. 1999b
	β-$[\mathbf{Cu_5O_2}](SeO_3)_2Cl_2$ parageorgbokiite	Krivovichev et al. 2007d
	$[\mathbf{Cu_5O_2}](PO_4)_2$	Brunel-Lauegt and Guitel 1977
acl–1:2–3	$[\mathbf{Al_2O}](SiO_4)$ kyanite	Winter and Ghose 1979
acl–3:7–1	$[Pb_2O]_2[\mathbf{Pb_7O_3}]O(GeO_4)[Ge_2O_7]$	Kato et al. 1995
	$[Pb_2O]_2[\mathbf{Pb_7O_3}]O(SiO_4)\,[Si_2O_7]$	Kato 1982
acl–3:5–2	$[\mathbf{Pb_5O_3}]O^{[3n]}(MoO_4)_2$	Vassilev and Nihtianova 1998
	$[\mathbf{Pb_5O_3}]O^{[3n]}(SO_4)_2$	Steele and Pluth 1998
	$[\mathbf{Pb_5O_3}]O^{[3n]}(CrO_4)_2$	Krivovichev et al. 2004a
acl–3:8–1	$[\mathbf{Pb_8O_3}]Cu(AsO_3)_2Cl_5$ freedite	Pertlik 1987
acl–7:10–1	$[\mathbf{Pb_{10}O_7}](OH)_2F_2(SO_4)$	Krivovichev and Burns 2001e

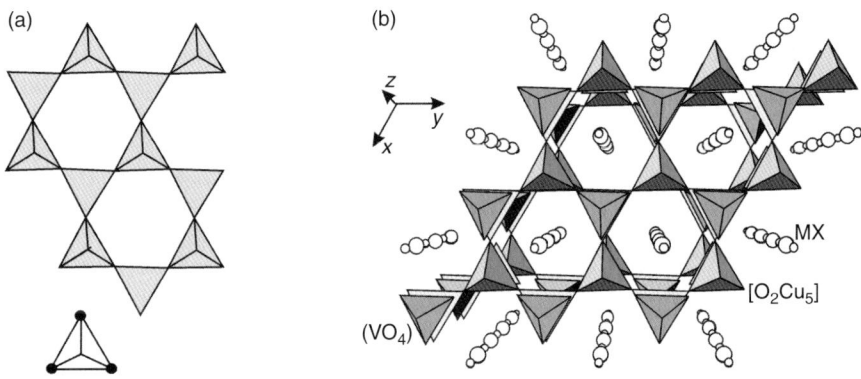

Fig. 5.13. Sheet of (OCu_4) tetrahedra in averievite, $[\mathbf{Cu_5O_2}](VO_4)_2 \cdot nMCl$ (M = K, Rb, Cs) (a) and the structure of averievite projected along the c-axis (b).

tetrahedra, repulsive forces between them are considerably lower than those between the centers of small tetrahedra. Thus, the following statement is valid: if a structural unit is composed of heterometallic (XM_4) tetrahedra, small cations prefer to link a small number of tetrahedra than large cations and to link tetrahedra via corners than via edges.

Another family of 2D units based upon oxocentered tetrahedra can be constructed by combination of the chains shown in Fig. 5.6. One of such units has been observed in $[Yb_5O_4]Li(BO_3)_3$ (Jubera et al. 2001) and is depicted in Fig. 5.15. Six $[OYb_3]$ chains of $[OYb_4]$ tetrahedra are linked by edge sharing between tetrahedra to form a complex sixfold chain $[O_3Yb_4]$. The sixfold chains are further corner-linked via double $[OYb_2]$ chains of edge-sharing tetrahedra (Fig. 5.6(b)).

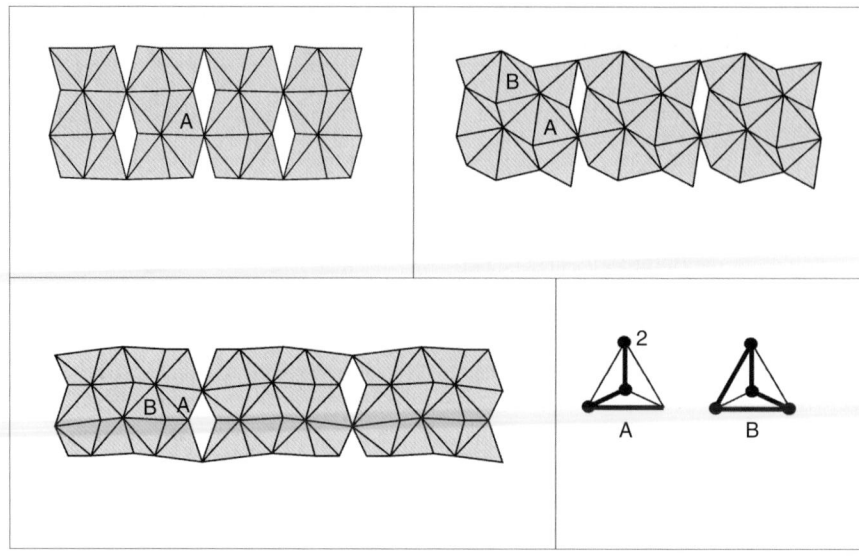

Fig. 5.14. A family of layers of oxocentered tetrahedra obtained by condensation of the chains of the $[O_nM_{n+1}]$ family.

Table 5.6. Inorganic oxysalts containing 2D units of oxocentered (OM_4) tetrahedra

Type	Chemical formula	Reference
ac2–2:5–1	$[Cu_5O_2](VO_4)_2 \cdot nMCl$ (M = K,Rb,Cs) averievite	Starova *et al*. 1997
	$[Pb_2Cu_3O_2](NO_3)_2(SeO_3)_2$	Effenberger 1986
ac2–4:5–1	$A'_2[Bi_2Cu_3O_4](AsO_4)_2 \cdot H_2O$ (A′ = Na, K)	Effenberger and Miletich 1995
	$[M_5O_4][Re_2O_8]$ M = Tm, Dy	Baud et al. 1983; Ehrenberg *et al*. 1999
	$[M_5O_4][Mo_2O_8]$ M = Y, Gd	Torardi *et al*. 1985
ac2–6:7–1	$[Cu_3Bi_4O_6](VO_4)_2$	Deacon *et al*. 1994
ac2–4:5–2	$[Bi_{34.7}O_{32}]O^{[3n]}{}_4(SO_4)_{16}$*	Aurivillius 1987
ac2–1:2–1	$[Gd_2O](SiO_4)$	Smolin and Tkachev 1969
	$\{[Ho_2O](GeO_4)\}_2\{Na(OH)\}$	Christensen 1972
	$[A_2O](GeO_4)$ A = Eu, Nd, Gd	Kato *et al*. 1979; Vigdorchik *et al*. 1986; Brixner *et al*. 1985
ac2–1:2–2	$[Cu_2O](SO_4)$	Effenberger 1985b
	$[Cu_3BiO_2](SeO_3)_2Cl$ francisite	Pring *et al*. 1990
	$[Cu_3ErO_2](SeO_3)_2Cl$	Berrigan and Gatehouse 1996
	$[Hg_2O](NO_3)$	Brodersen *et al*. 1985; Kamenar *et al*. 1986

Linkage of dimers of edge-sharing (OM_4) tetrahedra may provide different topologies and two of them occur in inorganic oxysalts. In the structures of doleroph-anite, $Cu_2O(SO_4)$, and francisite, $[Cu_3BiO_2](SeO_3)_2Cl$, $[O_2M_6]$ dimers share M atoms not involved in the edge-linkage to form a 2D topology (Fig. 5.16(a)). Nazarchuk

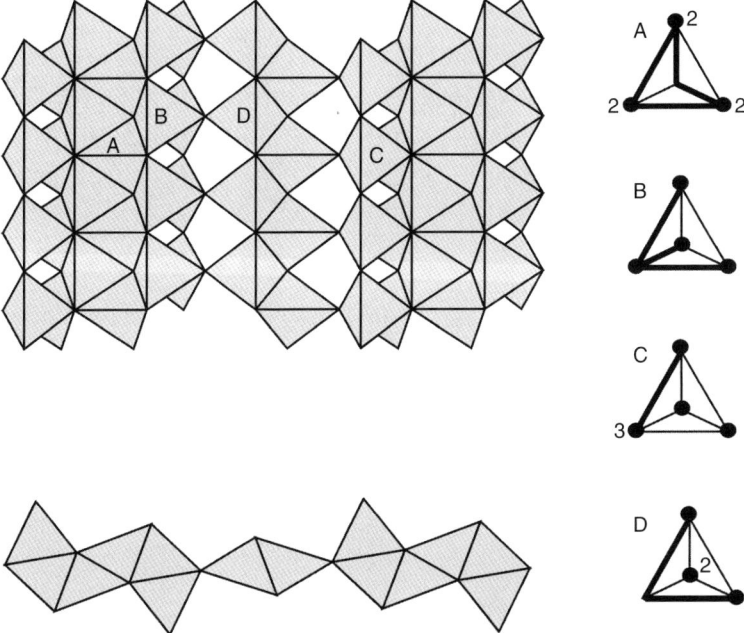

Fig. 5.15. 2D unit of (OYb$_4$) tetrahedra in the structure of [Yb$_5$O$_4$]Li(BO$_3$)$_3$.

et al. (2000) investigated the thermal expansion of francisite and observed its highly anisotropic character – the α_c coefficient (17.0 × 10^{-6} K^{-1}) is about two times larger than α_a (9.0 × 10^{-6} K^{-1}) and more than three times larger than α_b (4.7 × 10^{-6} K^{-1}). The explanation of the observed anisotropy is straightforward but only in terms of anion-centered tetrahedra. The additional O atoms in francisite are central for oxocentered heterometallic [OCu$_3$Bi] tetrahedra. Two tetrahedra form a dimer through the common Cu–Bi edge and then dimers are united via Cu corners into a 2D sheet perpendicular to the (001) plane. The oxygen base of the (SeO$_3$)$^{2-}$ groups are in "face-to-face" positions relative to the oxocentered tetrahedra. The explanation of thermal-expansion anisotropy in terms of oxocentered tetrahedra is straightforward. 1. The [O$_2$Cu$_5$Bi] layers of [OCu$_3$Bi] tetrahedra are parallel to the (001) plane – this explains why the thermal expansion along [001] is the largest. 2. The internal expansion of the layers is also anisotropic: α_a is about two times larger than α_b (Fig. 5.17). This is clearly explained by the mode of linkage of tetrahedra within the layer (Fig. 5.17). The [OCu$_3$Bi] tetrahedra are linked along [100] via Cu corners (i.e. via *two* O–Cu bonds) but along [010] via Cu–Bi edges (i.e. via *two* O–Cu and *two* O–Bi bonds). Consequently, the thermal expansion along [100] is twice that along [010]. This example shows that the structure description in terms of oxocentered tetrahedra not only provides a convenient description of structural topology but also helps to understand the anisotropy of physical properties of the compound.

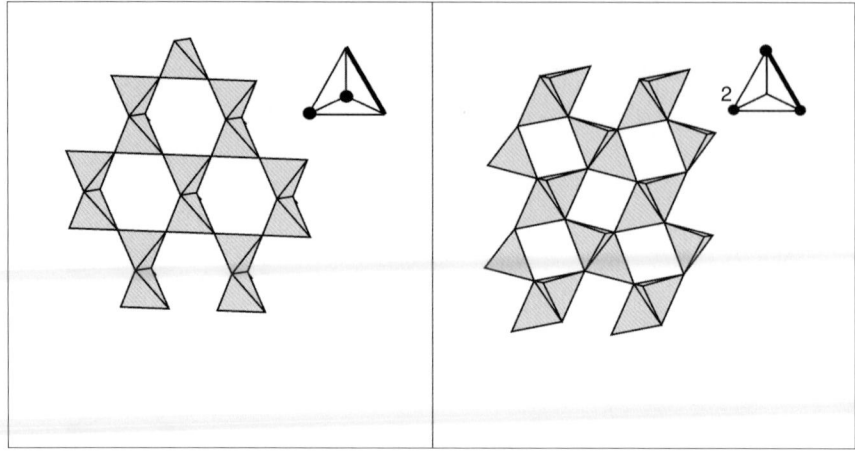

Fig. 5.16. 2D units produced by linkage of dimers of edge-sharing (OM_4) tetrahedra.

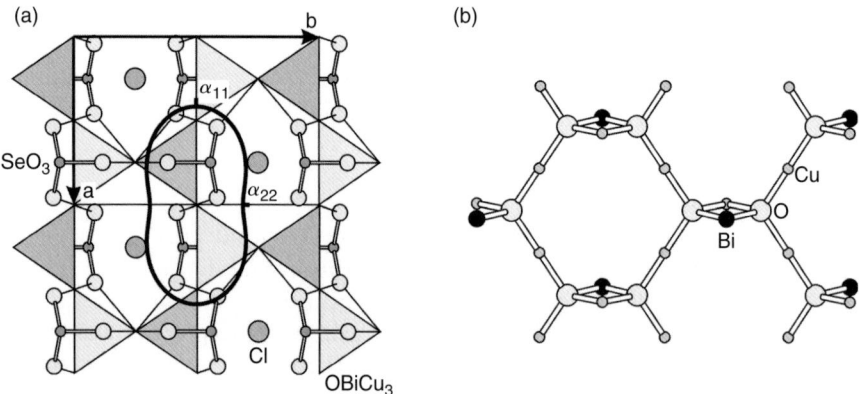

Fig. 5.17. (a) Projection of the crystal structure of francisite on (001) plane together with a section of the figure of thermal expansion coefficients; (b) structure of layer of oxocentered copper-bismuth tetrahedra.

Figure 5.16(b) shows another type of 2D sheet based upon linked dimers of edge-sharing tetrahedra. In this case, a "free" vertex of one dimer is linked to adjacent dimer through its M atom involved in linkage of tetrahedra via edges. As a consequence, the fourth vertex of each tetrahedron does not participate in any linkage at all. This type of 2D unit is typical for *REE* oxysalts (Table 5.6).

Figures 5.18(a)–(c) shows tetrahedral sheet, defect tetrahedral sheet, and chain observed in the structures of tetragonal PbO, $Pb_8O_5(TO_4)_2$ (T = P, As), and $Pb_5O_4(TO_4)$ (T = S, Cr), respectively. It is evident that the latter two units may be considered as

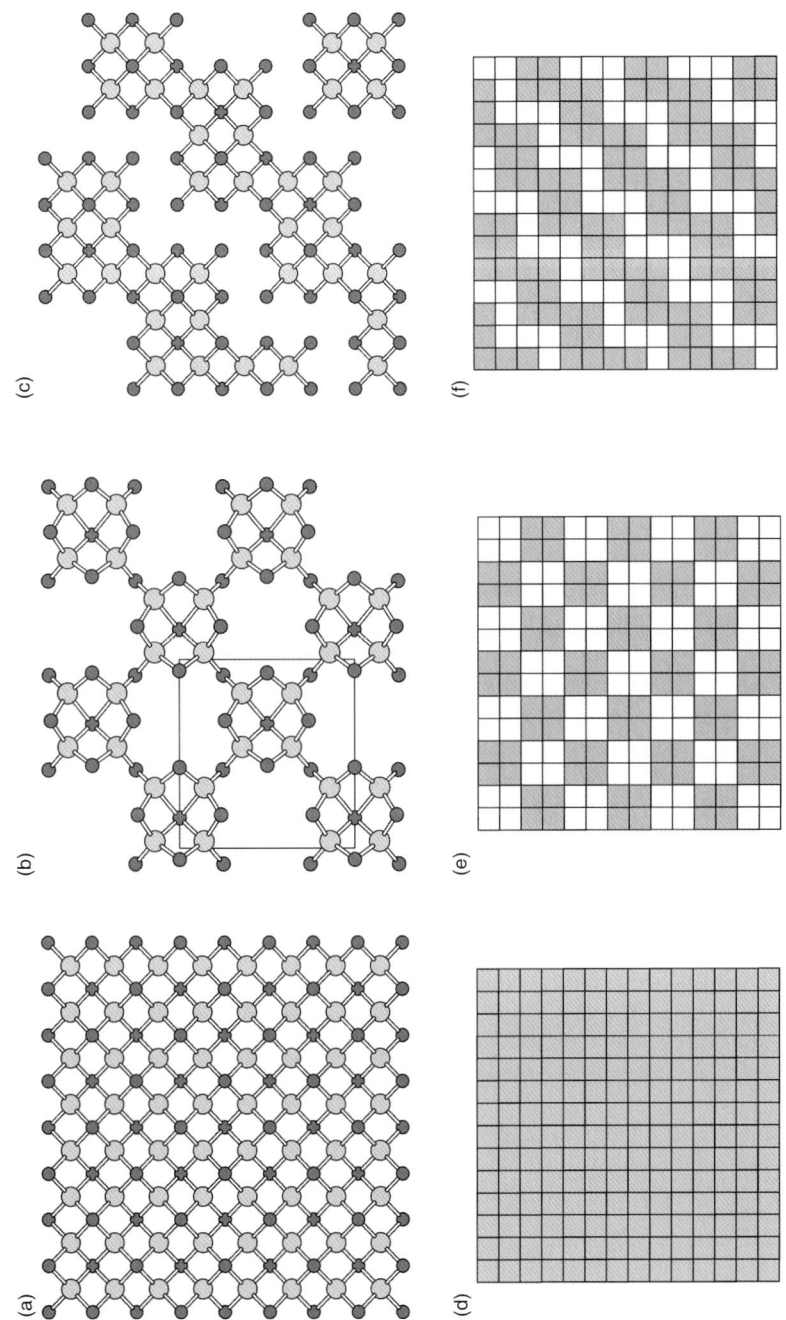

Fig. 5.18. Continuous sheet of (OPb$_4$) tetrahedra in the structure of tetragonal PbO (a), [O$_4$Pb$_7$] sheet in the structure of Pb$_8$O$_5$(AsO$_4$)$_2$ (b), and [O$_3$Pb$_5$] chain in the structure of Pb$_5$O$_4$(CrO$_4$) (c), and their lattice representation ((d), (e), and (f), respectively). Each black square symbolizes a (OPb$_4$) tetrahedron and each white square corresponds to a vacancy.

derivatives of the first, i.e. of the continuous [OPb] sheet of OPb_4 tetrahedra from the tetragonal modification of PbO. Figures 5.18(b) and (c) demonstrate that the $[O_4Pb_7]$ and $[O_3Pb_5]$ units can be obtained from the [OPb] sheet by deletion of some of the O and Pb atoms. This operation can also be thought of as a removal of a certain portion of OPb_4 tetrahedra. If a continuous sheet of OPb_4 tetrahedra is represented as a lattice of black squares (Fig. 5.18(d); each black square corresponds to a OPb_4 tetrahedron), then the $[O_4Pb_7]$ and $[O_3Pb_5]$ units may be modelled as an arrangement of black and white squares (where white squares symbolize "vacant" tetrahedra, i.e. vacancies left after removal of tetrahedra from the continuous [OPb] sheet; Figures 5.18(e) and (f)).

Representation of the PbO derivatives using a lattice of black and white squares offers a simple and clear interpretation of a number of different structure types. These represent not only structures based upon anion-centered tetrahedra (e.g. $[OPb_4]$, $[OBi_4]$, $[OLa_4]$, etc.) but also some structures consisting of layers of cation-centered tetrahedra (Krivovichev et al. 2004a). Figure 5.19 provides lattice representations of eight tetrahedra layers that can be considered as derivatives of a [OPb] sheet in tetragonal PbO. Note that the vacancies (=white squares) are either isolated (Figs. 5.19(c) and (g)) or are grouped into chains (Figs. 5.19(d), (g) and (h)) or islands of various form (usually rectangular) (Figs. 5.19(b), (e), (f) and (i)). Table 5.7 gives a list of respective inorganic compounds.

Obviously, the lattice representation may be extended to one-dimensional tetrahedral units as well. Figure 5.20 gives lattice representations of eleven types of chains based upon edge-sharing anion-centered tetrahedra (units based upon cation-centered tetrahedra are not considered here). Figure 5.20(h) depicts a scheme of a $[O_5Pb_7]$ complex chain that occurs in the structures of $[Pb_{13}O_{10}]Br_6$ (Riebe and Keller 1990) and $[Pb_{13}O_{10}]Cl_6$ (Siidra et al. 2007). However, in these structures, the $[O_5Pb_7]$ chains are not isolated but are linked into a complex $[O_{10}Pb_{13}]$ framework with channels occupied by halide anions (Fig. 5.21).

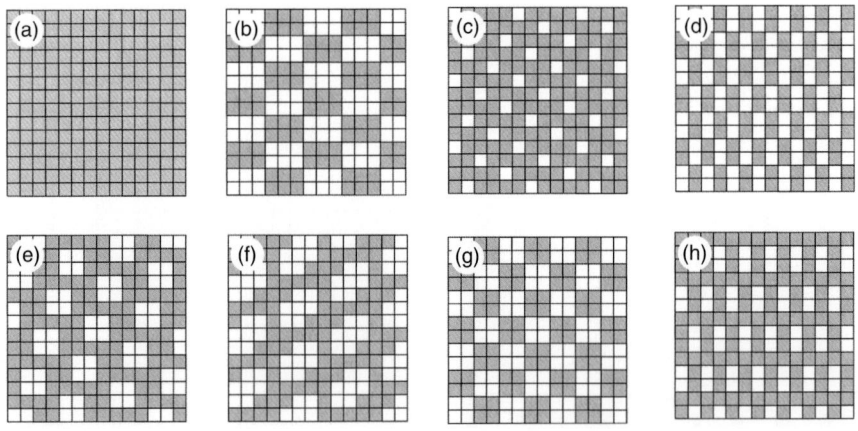

Fig. 5.19. Lattice representations of 2D PbO-related units in inorganic compounds.

A remarkably complex PbO-derivative sheet has been observed in $Pb_{31}O_{22}Br_{10}Cl_8$ (Krivovichev *et al.* 2006d). This structure contains 31 symmetrically independent Pb^{2+} cations, 18 halide sites statistically occupied by Br^- and Cl^- ions, and 22 oxygen positions (Fig. 5.22(a)). All 22 O^{2-} anions are tetrahedrally coordinated by Pb^{2+} cations, thus forming oxocentered OPb_4 tetrahedra. The basis of the structure is a defect PbO-derived sheet with the composition $[O_{22}Pb_{30}]^{16+}$ (Fig. 5.22(b)). It has two types of vacancies: single tetrahedra and 4-membered blocks of tetrahedra with the 2×2 dimensions. The $[O_{22}Pb_{30}]^{16+}$ unit is remarkable in its exceptional topological complexity that has no analogs among the known PbO derivatives. The 22 symmetrically independent OPb_4 tetrahedra play different roles in the topological organization of the layer. Figure 5.23(a) provides a description of the 2D sheet using the model of square lattices mentioned above. Each black square is labelled by a number that corresponds to the designation of the O site at the center of the OPb_4 tetrahedron. The topological

Table 5.7. Inorganic oxysalts containing PbO-related sheets of (OM_4) and (FM_4) tetrahedra

Type	Chemical formula	Reference
ac2–1:1–1	$[Pb_2F_2](SO_4)$ grandreefite	Kampf 1991
	$[(Pb,Si)_7O_8]Cl_2$ asisite	Rouse *et al.* 1998
	$[PbBiO_2]Cl$ perite	Ketterer and Kraemer 1985
	$[PbSbO_2]Cl$ nadorite	Giuseppetti and Tadini 1973
	$[Pb_3Sb_{0.6}As_{0.4}O_3(OH)]Cl_2$ thorikosite	Rouse and Dunn 1985
	$[(Pb,Mo,\square)_8O_8]Cl_2$ parkinsonite	Symes *et al.* 1994
	$[Bi_2O_2](CO_3)$ bismuthite	Greaves and Blower 1988
	$Ca[Bi_2O_2](CO_3)_2$ beyerite	Lagercrantz and Sillen 1948
	$[BiO]F$ zavaritskite	Aurivillius 1964
	$[BiO]Cl$ bismoclite	Keramidas *et al.* 1993
	$[Bi_2O_2](MoO_4)$ koechlinite	Zemann 1956; Pertlik and Zemann 1982
	$[Bi_2O_2](WO_4)$ russelite	Wolfe *et al.* 1969
	$Ca[BiO]F(CO_3)$ kettnerite	Grice *et al.* 1999
	$[BiO]Br$	Ketterer and Kraemer 1986b
	$[LaO]_2SO_4$	Zhukov *et al.* 1997
	$[NdO]_3(PO_4)$	Palkina *et al.* 1995
	$[LaO]_2(MoO_4)$	Efremov *et al.* 1987; Xue *et al.* 1995
	$[BiO]_2[GeO_3]$	Aurivillius *et al.* 1964; Shashkov *et al.* 1986
	$[BiO]_2[SiO_3]$	Ketterer and Kraemer 1986a
	$[YO](NO_3)$	Pelloquin *et al.* 1994
ac2–4:5–3	$[AgPb_4O_4]Cl$	Riebe and Keller 1988
ac2–3:5–1	$[Pb_5O_3](GeO_4)$	Kato 1979
ac2–9:14–1	$[Pb_{14}O_9](VO_4)_2Cl_4$ kombatite	Cooper and Hawthorne 1994c
	$[Pb_{14}O_9](AsO_4)_2Cl_4$ sahlinite	Bonaccorsi and Pasero 2003
ac2–5:8–1	$[Pb_8O_5](OH)_2Cl_4$ blixite	Krivovichev and Burns 2006
ac2–4:7–1	$Pb[Pb_7O_4]O(PO_4)_2$	Krivovichev *et al.* 2003i
	$Pb[Pb_7O_4]O(AsO_4)_2$	Krivovichev *et al.* 2004a
ac2–4:7–2	$[MBi_6O_4](PO_4)_4$ $M = Bi_{0.67}$, Pb	Ketatni *et al.* 1998
ac2–7:10–1	$[Pb_{10}O_7](SO_4)Cl_4(H_2O)$ symesite	Welch *et al.* 2000
ac2–11:15–1	$Pb[Pb_{30}O_{22}]Br_{10}Cl_8$	Krivovichev *et al.* 2006d

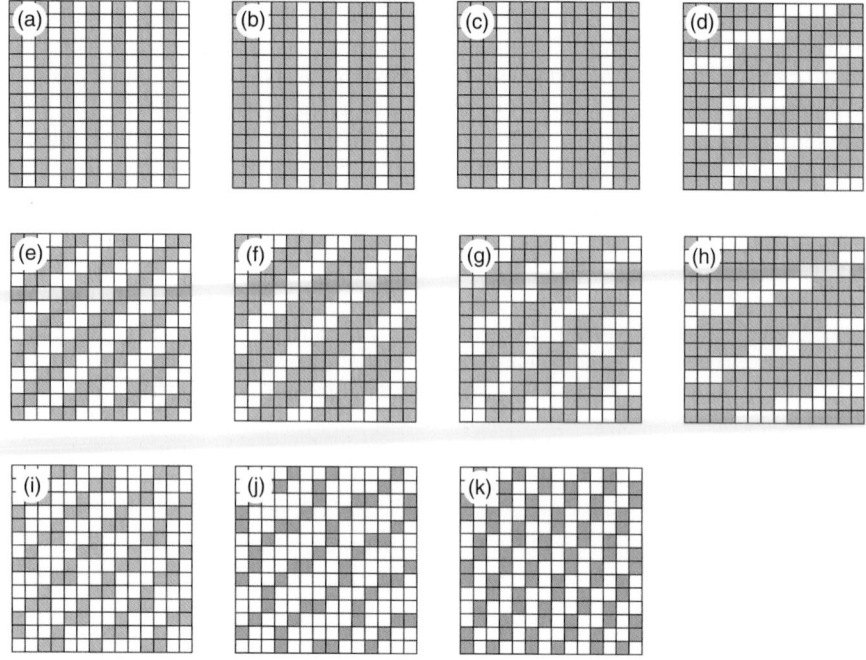

Fig. 5.20. Lattice representations of 1D PbO-related units of oxocentered tetrahedra in inorganic compounds.

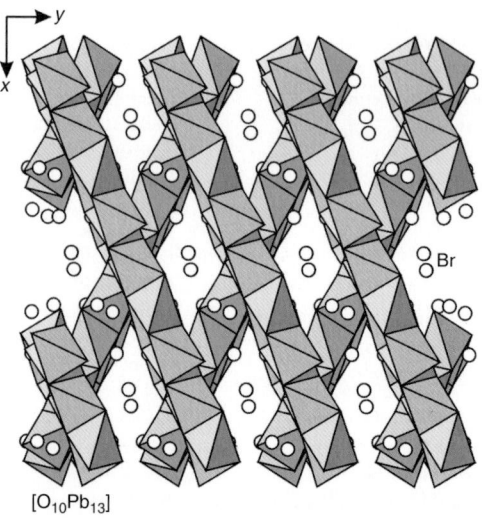

Fig. 5.21. 3D framework of (OPb$_4$) tetrahedra in Pb$_{13}$O$_{10}$X$_6$ (X = Cl, Br).

(a)

(b)

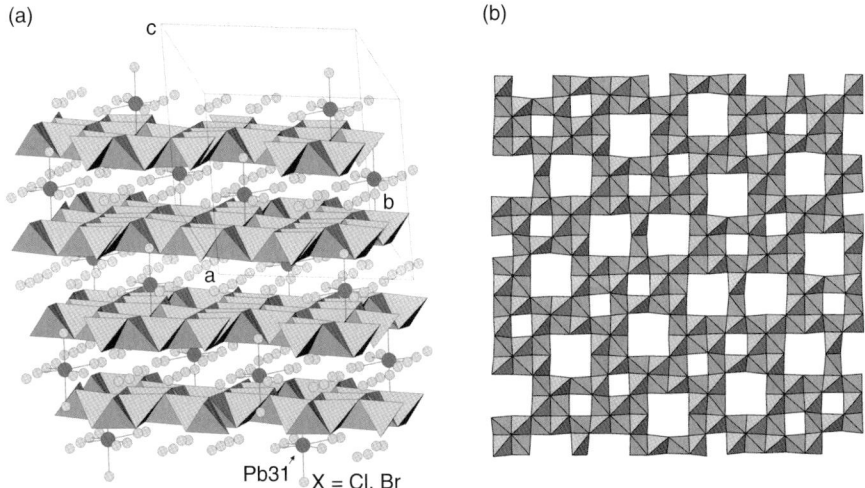

Pb31 X = Cl, Br

Fig. 5.22. Crystal structure of $Pb_{31}O_{22}Br_{10}Cl_8$ (a) and the $[O_{22}Pb_{30}]^{16+}$ 2D layer composed from edge-sharing (OPb_4) tetrahedra (c).

function of a tetrahedron within the layer can be visualized by investigation of the local coordination of a given square by the adjacent squares, i.e. by configuration of its coronas. Figure 5.23(b) provides the schemes of the first coronas for all the 22 tetrahedra present in the $[O_{22}Pb_{30}]^{16+}$ blocks. There are some coronas that are common for several tetrahedra. For instance, the $O(1)Pb_4$, $O(7)Pb_4$, $O(8)Pb_4$, and $O(10)$ Pb_4 tetrahedra have the same coronas consisting of 6 tetrahedra arranged around the central one in the same way. In order to further investigate whether topological functions of the tetrahedra are different, one has to examine their second coronas. Figure 5.24 demonstrates that, despite the fact that the first coronas of some tetrahedra are identical, their second coronas are different and therefore the topological functions of the tetrahedra are different. For instance, the $O(9)Pb_4$ and $O(19)Pb_4$ tetrahedra have identical first but different second coronas (Fig. 5.24(a)). The tetrahedra centered by the $O(5)$, $O(11)$, $O(13)$, $O(15)$, $O(16)$, $O(18)$, $O(20)$, and $O(21)$ atoms are distiguished by their second coronas, whereas their first coronas are the same (Fig. 5.24(c)). The situation is more complicated for the $O(1)Pb_4$ and $O(8)Pb_4$ tetrahedra as they have identical first and second coronas (Fig. 5.25). However, their third coronas are different and therefore their topological functions within the sheet are inequivalent. To our knowledge, this is the first example of a structure with edge-sharing tetrahedra where third coronas are necessary to reveal topological differences between single tetrahedra. All 22 symmetrically independent tetrahedra in the $[O_{22}Pb_{30}]^{16+}$ block have unique functions in the topology of this unit. This topological complexity is exceptional and, as far as we know, has not been observed in any ordered tetrahedral structure. From the chemical viewpoint, the appearance of such complexity should be

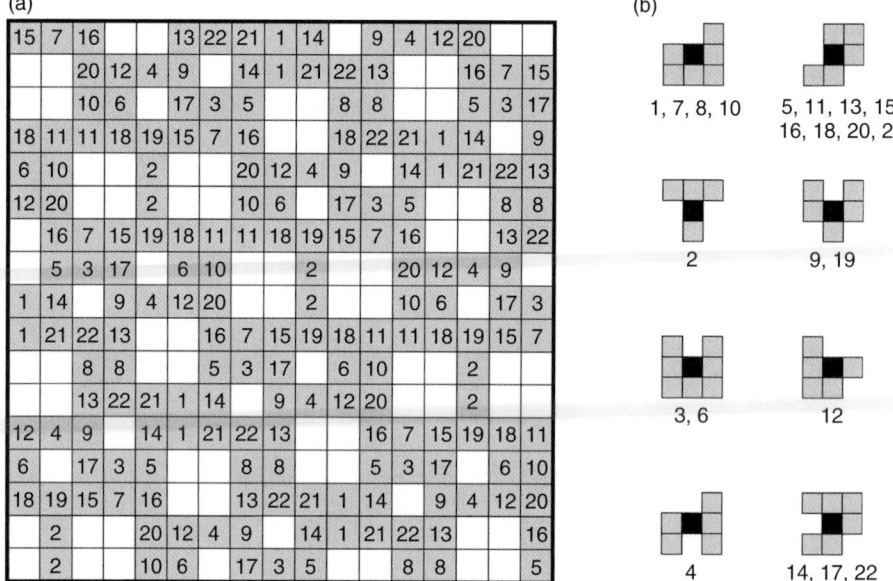

Fig. 5.23. Topological structure of the $[O_{22}Pb_{30}]^{16+}$ 2D layer in the structure of $Pb_{31}O_{22}Br_{10}Cl_8$ (a) and first coronas (local coordinations) of central (OPb$_4$) tetrahedra (shown as black squares) (b).

ascribed to the incorporation of octahedral halide clusters into the metal oxide matrix that induces modification of the latter in a complex way. However, the model of black and white squares proposed to describe this level of complexity is rather simple.

It is noteworthy that the method of derivation of various structural units from the tetragonal PbO structure was previously applied by Carré *et al.* (1984) to obtain chains of [OLa$_4$] tetrahedra observed in the structures of lanthanum oxysulphides. In fact, the [OPb] sheet shown in Fig. 5.18(a) is just a slice of the fluorite structure if the latter is described as an arrangement of FCa$_4$ tetrahedra. A majority of anion-centered tetrahedral units, including finite clusters, chains, sheets and frameworks, may be considered as fluorite derivatives (Krivovichev and Filatov 1999b).

5.2.5 *3D units of anion-centered tetrahedra*

A number of the structures of oxides and fluorides can be described in terms of anion-centered tetrahedra. This is especially true for fluorite-derivative frameworks since the structure of fluorite, CaF$_2$, can itself be considered as a dense framework of (FCa$_4$) tetrahedra. Most of the defect fluorite 3D frameworks do not occur in oxysalts and will not be considered here. Kang and Eyring (1997, 1998) elaborated a modular description of fluorite derivatives that was further extended in (Krivovichev 1999a, b).

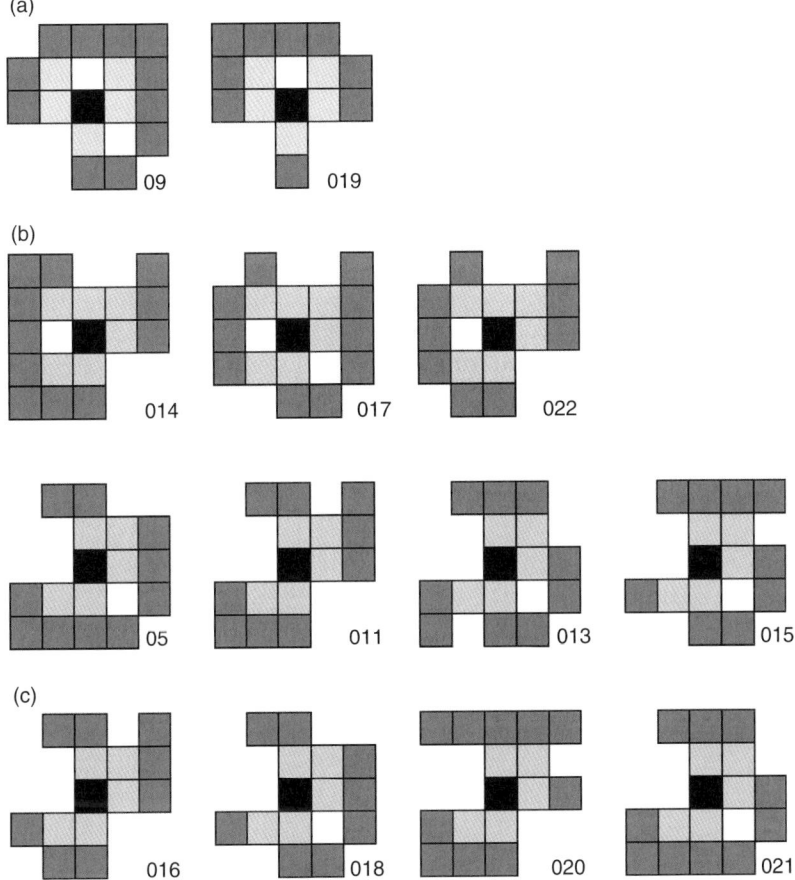

Fig. 5.24. Local coordinations of (OPb$_4$) tetrahedra within the [O$_{22}$Pb$_{30}$]$^{16+}$ 2D layer in the structure of Pb$_{31}$O$_{22}$Br$_{10}$Cl$_8$.

As for frameworks of cation-centered polyhedra (see preceding chapter), regular 3D nets are of importance. For instance, the structure of [Nd$_5$O$_4$](MoO$_4$)$_3$ (Hubert *et al.* 1973) can be considered as the **bcu** network of fluorite clusters with channels occupied by isolated MoO$_4$ tetrahedra (Fig. 5.26). The structure of melanothallite, Cu$_2$OCl$_2$ (Arpe and Muller-Buschbaum 1977; Krivovichev *et al.* 2002d) and moses-ite, [Hg$_4$N$_2$](SO$_4$)(H$_2$O) (Airoldi and Magnano 1967), are based upon the **dia** networks of (OCu$_4$) and (NHg$_4$) tetrahedra, respectively. A tridimite-type framework of (NHg$_4$) tetrahedra has been observed in the structure of kleinite, **[Hg$_2$N](Cl,H$_2$O,SO$_4$)** (Giester *et al.* 1996).

Again, similar to the frameworks of cation-centered polyhedra, frameworks of anion-centered tetrahedra can also be classified into different groups with simple

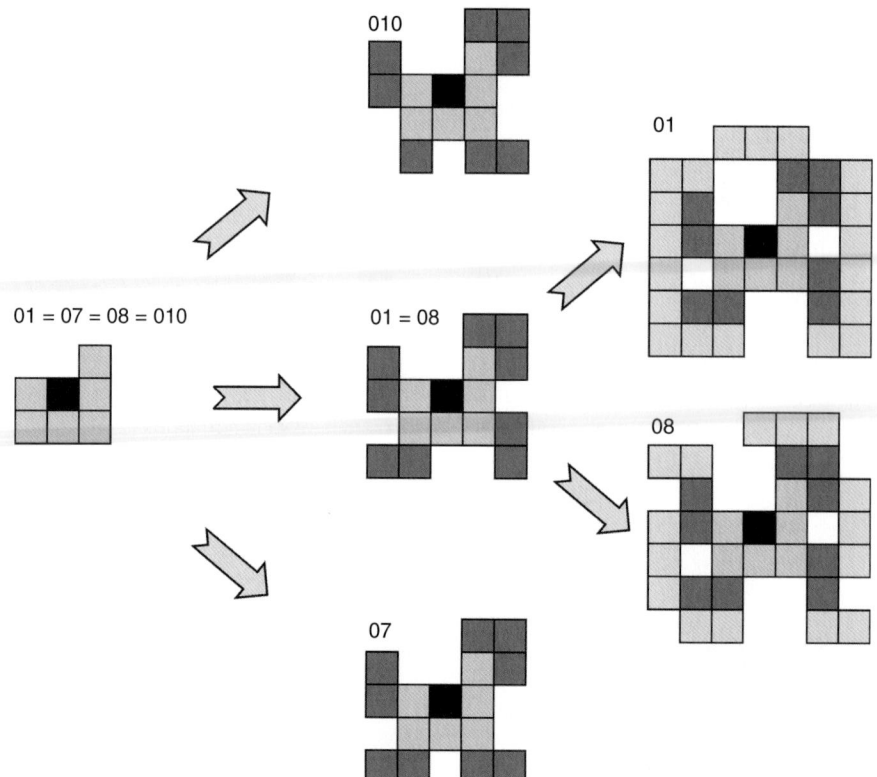

Fig. 5.25. Description of topology of the $[O_{22}Pb_{30}]^{16+}$ 2D layer in the structure of $Pb_{31}O_{22}Br_{10}Cl_8$. OPb_4 tetrahedra centered by the O(1), O(7), O(8), and O(10) atoms have the same first coronas. The second coronas are different for the O(10)- and O(7)-centered tetrahedra; however, they are the same for the O(1)Pb_4 and O(8)Pb_4 tetrahedra. The O(1)Pb_4 and O(8)Pb_4 tetrahedra have different third coronas.

organization principles. Thus, the framework shown in Fig. 5.4(b) is based upon *stella quadrangula* as a fundamental building block. Figure 5.27 shows framework of (strongly distorted) (OBi$_4$) tetrahedra in the structures of Bi oxyhalides. In the construction principle, these structures are very similar to the structures of tunnel oxides. In the latter, chains of edge-sharing cation-centered octahedra share edges to form multiple chains that are further arranged perpendicularly to form frameworks with variable dimensions. The channels have rectangular sections and are filled by cations and water molecules. In the structures of Bi oxyhalides, the [OBi$_3$] single chains form multiple [O$_n$Bi$_{n+2}$] chains of various widths that are further condensed to form 3D frameworks with channels that have triangular sections. The framework in (Cu$_5$Cl)Bi$_{48}$O$_{59}$Cl$_{30}$ (Aurivillius 1990) is built from ninefold chains ($n = 9$), whereas $n = 4$ and 10 for the structure of Bi$_6$O$_7$FCl$_3$ (Hopfgarten 1975), and $n = 4$ and 16 for the structure of Bi$_{12}$O$_{15}$Cl$_6$ (Hopfgarten 1976). The channels are filled by chloride ions

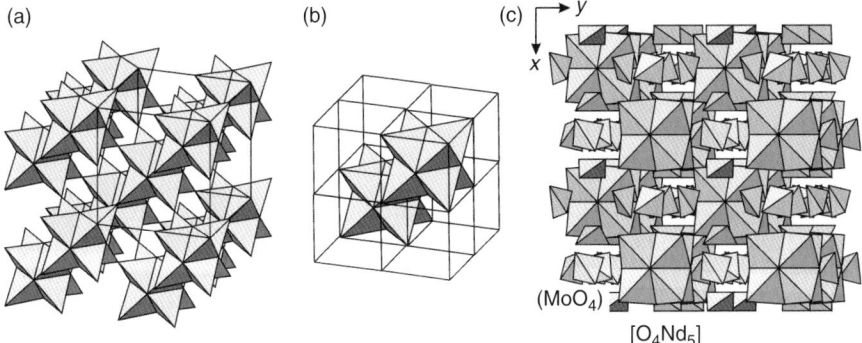

Fig. 5.26. Schematic representations of the framework of fluorite clusters in $[Nd_5O_4](MoO_4)_3$ (a), their arrangement in the unit cell (b) and crystal structure shown in terms of (ONd_4) and (MoO_4) tetrahedra (c).

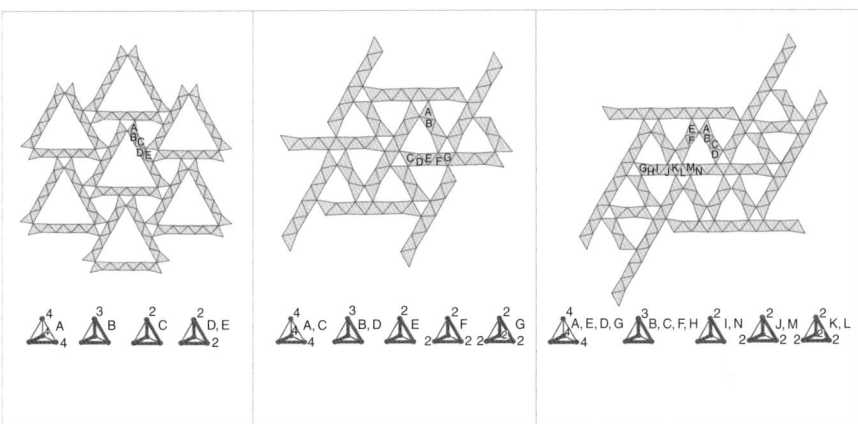

Fig. 5.27. Oxocentered tetrahedral frameworks built by condensation of (OBi_4) tetrahedra: (a) $[O_9Bi_8]$ framework in $(Cu_5Cl)Bi_{48}O_{59}Cl_{30}$ built from ninefold chains; (b) $[(O_6F)Bi_6]$ framework from $Bi_6O_7FCl_3$ built from tenfold and fourfold chains; (c) $[O_7Bi_6]$ framework from $Bi_{12}O_{15}Cl_6$ built from 16-fold and fourfold chains.

except for the structure of $(Cu_5Cl)Bi_{48}O_{59}Cl_{30}$, where the channels host Cu^{2+} cations as well.

5.2.6 Which cations may form anion-centered tetrahedra?

As has been mentioned above, the description of structures of inorganic oxysalts in terms of anion-centered tetrahedra is secured for selected compounds only, namely, those containing additional O^{2-} and N^{3-} anions. However, one would be interested to know which *cations* may form anion-centered tetrahedra. Or, taking into account accumulated structure data, why certain cations do form and others do not form anion-centered tetrahedra? Another closely related question is when description in terms of anion-centered tetrahedra should be preferred over description in terms of cation-centered polyhedra? These problems have been extensively discussed by Krivovichev and Filatov (2001), who suggested two general principles.

Principle 1. Anion-centered tetrahedra $(X^{n-}M_4)$ may exist if cation M may form an $M–X$ bond with the bond valence of $n/4$. The more typical the $M–X$ bonds with $s_{MX} = n/4$, the more common are the $(X^{n-}M_4)$ tetrahedra.

For $X = O^{2-}$ (the most common case), this principle can be re-formulated as follows.

Principle 1′. Oxocentered tetrahedra (OM_4) may exist if cation M may form an $M–O$ bond with the bond valence of 0.5 valence units.

In a certain sense, this principle is a variation of the bond-valence matching principle. Krivovichev and Filatov (2001) analyzed all stable cations of the common elements of the Periodic Table and determined those that are able to form oxocentered tetrahedra. However, in order for the description in terms of anion-centered tetrahedra to be justified, another principle has been proposed.

Principle 2. Structure description in terms of $(X^{n-}M_4)$ anion-centered tetrahedra is justified, if the $M–X$ bonds are the strongest bonds in the MX_m coordination polyhedra.

Combination of the two principles provides a simple and efficient framework for explanation of the experimental data.

Example 1. Si^{4+} cations usually have tetrahedral coordination with the average $Si^{4+}–O$ bond valence of 1.0 v.u. This is much larger than the value of 0.5 v.u. required for oxocentered tetrahedra. As a consequence, (OSi^{4+}_4) tetrahedra have never been observed.

Example 2. Al^{3+} cations frequently occur in octahedral coordination. The average $^{VI}Al–O$ bond-valence in the $(Al^{3+}O_6)$ octahedra is 0.5 v.u. Thus, (OAl^{3+}_4) tetrahedra can be recognized in a number of structures including Al_2O_3 and kyanite, $Al_2O(SiO_4)$. However, the $Al^{3+}–O$ bonds to additional O atoms in the $(Al^{3+}O_6)$ octahedra will not be the strongest and therefore description in terms of oxocentered tetrahedra is not justified in this case.

Example 3. Trivalent REE^{3+} cations (e.g. La^{3+}) usually have a coordination number greater than 6. However, their coordination is rather flexible due to the large size of the cations (Bandurkin and Dzurinskii 1998) and formation of $REE^{3+}–O$ bonds with bond-valences of 0.5 v.u. is possible. As a consequence, the $REE^{3+}–O$ bonds to additional O atoms are the shortest $REE^{3+}–O$ bonds in the structure and its description in terms of $(OREE_4)$ tetrahedra is crystal chemically justified (see Caro 1968; Carre *et al.* 1984; Schleid 1996).

Example 4. In compounds with strong Lewis bases such as additional O atoms, Pb^{2+} and Bi^{3+} cations have highly irregular coordinations due to the stereochemical activity of the $6s^2$ lone electron pairs. The typical bond valences of the Pb^{2+}–O and Bi^{3+}–O bonds for additional O atoms are 0.501 and 0.596 v.u., respectively (Krivovichev and Filatov 2001). As a consequence, both (OPb_4) and (OBi_4) tetrahedra are common and their subdivision in inorganic structures is totally justified. The (OPb_4) tetrahedra are particularly widespread, since the typical Pb^{2+}–O bond-valence for additional O atoms is exactly what is required by Principle 1. The respective Pb^{2+}–O bonds in (OPb_4) tetrahedra are the strongest bonds formed by Pb^{2+} cations in oxysalts with additional O atoms. However, the *typical* Bi^{3+}–O bond-valence is 0.596 v.u. and therefore (OBi_4) tetrahedra are usually distorted and (OBi_3) configurations are common as well.

As a concusion, analysis performed in (Krivovichev and Filatov 2001) demonstrated that description of structures of inorganic oxysalts in terms of (OM_4) oxocentered tetrahedra is most efficient for $M = Cu^{2+}$, Pb^{2+} and REE^{3+}, sometimes for Bi^{3+}. As can be seen, compounds with these cations constitute a major part of data listed in tables given in this section.

5.3   Anion-centered octahedra in inorganic oxysalts

5.3.1   *Some notes on antiperovskites*

Description of structures of inorganic oxysalts in terms of anion-centered octahedra is not so common due to the fact that anions that are able to form octahedral units are usually of low charge (no more than –3). Thus, the (XA_6) octahedra are not as strong as (XA_4) tetrahedra. However, in many cases, a description based upon anion-centered octahedra provides a clear and elegant representation of the structure. Moreover, it indicates further parallels between cation- and anion-centered approaches, since most of the structures of this kind can be considered as inverted analogs of perovskite, which are sometimes called antiperovskites.

The classical perovskite has a formula ABX_3 (A, B = cations; X = anions) and might be regarded as a cubic close-packing of the A and X ions with octahedral interstices occupied by the B cations. The A and X atoms form close-packed layers with the AX_3 stoichiometry that alternate along the body diagonal of a cubic unit cell in the *ccc* sequence. Hexagonal perovskites differ from cubic perovskites in that they contain hexagonal *h* sequences of the AX_3 layers (Mitchell 2002). This leads to the occurrence of face-sharing BX_6 octahedra that are usually stabilized by the formation of the B–B metal–metal bonds. A number of polytypes of hexagonal perovskites are known with the $2H$ polytype being a "pure" hexagonal *hh* close packing of the AX_3 layers.

Antiperovskites or inverse perovskites are inorganic compounds with a perovskite structure but with cations replaced by anions and vice versa. The ideal antiperovskite structure is adopted by some ternary carbides and nitrides with a general formula A_3BX. The examples of the simplest antiperovskites with *Pm-3m* symmetry are Mg_3NAs, Mg_3NSb and the new superconductor Ni_3CMg (Mitchell 2002). Another example of a cubic antiperovskite is the structure of Na_3OCl (Hippler *et al.* 1990), which is based upon a framework of corner-sharing [ONa$_6$] oxocentered octahedra with cavities occupied by large Cl$^-$ anions. Recently, a number of variations of antiperovskite structures have been described with structural units composed from corner- and face-sharing N-centered octahedra (Gaebler et al. 2004, 2005, 2007; Gaebler and Niewa 2007). Among them, there are examples of cubic and orthorhombically distorted framework antiperovskites (Gaebler et al. 2005; Chern et al. 1992; Ming et al. 1992), inverse Ruddlesden–Popper phases (Gaebler et al. 2007), and hexagonal antiperovskites (Gaebler and Niewa 2007). The latter structures are "inverse" analogs of hexagonal perovskites that can be considered as polytypic variations of cubic perovskite structure. Gaebler and Niewa (2007) recently reported inverse hexagonal perovskite structures $(Sr_{3-x}Ba_xN)E$ (E = Sb, Bi) based upon anion-centered [N^{3-}A$_6$] octahedra (A = Sr, Ba) and corresponding to the 2H, 4H, and 9R polytypes. The occurrence of a certain polytype is controlled by the x compositional parameter.

Below, we shall demonstrate that there exist a number of inorganic oxysalts that can be considered as antiperovskite derivatives consisting of anion-centered [XA$_6$] octahedra. In most cases, this approach provides a simple and unique desciption of otherwise complicated inorganic structures.

5.3.2 Chemical considerations

As has been noted in the preceding section, description in terms of anion-centered units is usually justified for minerals and compounds containing so-called "additional" anions X, i.e. anions that are not bonded to a high-valent cation. A simplified formula for these compounds can be written as $A_mX_n(MO_l)_k$ with $n \leq m$. X is usually N^{3-}, O^{2-}, F$^-$, rarely Cl$^-$. As examples, consider lanarkite, Pb$_2$O(SO$_4$), and nacaphite, Na$_2$CaF(PO$_4$), that can both be described in terms of anion-centered polyhedra. Several hundred structures with tetrahedrally coordinated X anions have been reported so far (Krivovichev et al. 1998a; Krivovichev, Filatov, 2001). The number of structures with O^{2-}- and N^{3-}-centered metal octahedra is also remarkable, though they are not so diverse as those with anion-centered tetrahedra.

Here we consider silicates, sulphates and phosphates with a general formula $A_mX_n(TO_4)_k$; A = Na$^+$, K$^+$, Ca^{2+}; X = O^{2-}, F$^-$, Cl$^-$; T = Si^{4+}, S^{6+}, P^{5+}. The minerals and related inorganic compounds under consideration are listed in Table 5.8 according to the X:A ratio. In all these compounds, anions X are octahedrally coordinated by the A cations that results in the formation of [AX$_6$] octahedra. Linkage of these octahedra via corners and faces generates complex structural units that can be directly derived from the known perovskite and antiperovskite structures.

5.3.3 Structural diversity

The simplest antiperovskite unit is a cubic framework of corner-sharing octahedra. It is realized in the structure of sulphohalite, Na$_6$FCl(SO$_4$)$_2$ (Pabst 1934), where the framework consists of alternating [FNa$_6$] and [ClNa$_6$] octahedra (Fig. 5.28). Tetrahedral (SO$_4$) groups reside in the framework cavities. Due to the F–Cl ordering,

Table 5.8. Antiperovskite units in minerals and related inorganic compounds*

X:M	X	M	Mineral name	Chemical formula	Reference
Framework antiperovskites					
1:3–3C	F, Cl	Na	sulphohalite	$[\textbf{FClNa}_6](SO_4)_2$	Pabst 1934
	F	K	–	$[\textbf{FK}_3](SO_4)$	Skakle *et al.* 1996
1:3–2H	F	Na, Ca	nacaphite	$[\textbf{FNa}_2\textbf{Ca}](PO_4)$	Krivovichev *et al.* 2007
			–	$[\textbf{FNaCa}_2]GeO_4$	Schneemeyer *et al.* 2001
			–	$[\textbf{FNaCa}_2]SiO_4$	Andac *et al.* 1997
	O, F	Ca	–	$[(\textbf{O}_{1-2x}\textbf{F}_{2x})\textbf{Ca}_{3-x}](GeO_4)$	Nishi and Takeuchi 1984
1:3–5H	F, Cl	Na	galeite	$[\textbf{F}_4\textbf{ClNa}_{15}](SO_4)_5$	Fanfani *et al.* 1975a
1:3–7H	F, Cl	Na	schairerite	$[\textbf{F}_6\textbf{ClNa}_{21}](SO_4)_7$	Fanfani *et al.* 1975b
1:3–9R	F	Na	kogarkoite	$[\textbf{FNa}_3](SO_4)$	Fanfani *et al.* 1980
	O	Ca	hatrurite	$[\textbf{OCa}_3](SiO_4)$	Nishi *et al.* 1985; Mumme, 1995
			–	$[\textbf{OCa}_3](GeO_4)$	Nishi and Takeuchi 1986
1:3–15R	F	Na,Ca	–	$[\textbf{FNa}_2\textbf{Ca}](PO_4)$	Sokolova *et al.* 1999
Broken antiperovskites					
1:4	F,O	Ca	–	$K[\textbf{FO}_2\textbf{Ca}_{12}](SO_4)_2(SiO_4)_4$	Fayos *et al.* 1985
	F	Na,Ca	arctite	$Ba[\textbf{F}_3\textbf{Na}_5\textbf{Ca}_7](PO_4)_6$	Sokolova *et al.* 1984
			polyphite	$[\textbf{F}_6\textbf{Ca}_3\textbf{Na}_{15}]Na_2Mg(Ti,Mn)_4$ $O_2(Si_2O_7)_2(PO_4)_3$	Sokolova *et al.* 1987

* Composition of anion-centered units is given in bold.

(a)   (b)

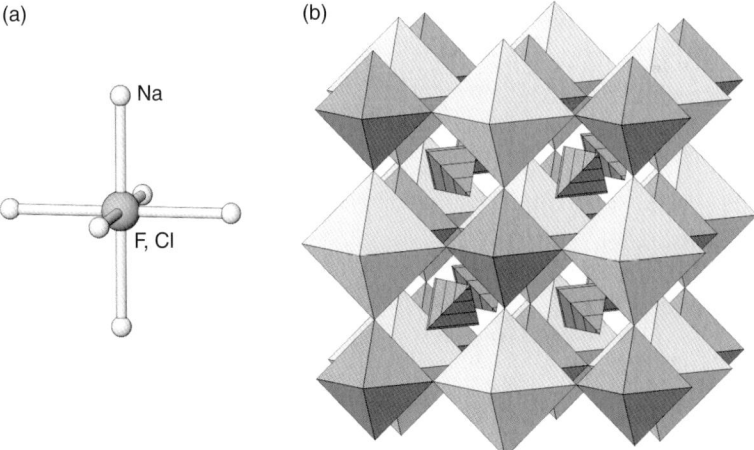

Fig. 5.28. Anion-centered octahedron (a) in the structure of sulphohalite (b). Note that the structure (b) is shown as a combination of anion- (octahedra) and cation- (tetrahedra) centered coordination polyhedra. Anion-centered $[FNa_6]$ and $[ClNa_6]$ octahedra are dark and light, respectively; sulphate tetrahedra are lined.

the *a* parameter is doubled in comparison with the 'usual' cubic perovskite unit cell. Thus, sulphohalite is actually an ordered double antiperovskite with the anti-elpasolite structure and the $Fm\bar{3}m$ space group.

The structure of nacaphite, $Na_2CaF(PO_4)$ (Sokolova *et al.* 1989a; Sokolova and Hawthorne, 2001; Krivovichev *et al.* 2007d) (Fig. 5.29(a)), is a hexagonal antiperovskite with the 2*H*-polytype arrangement of the $[FNa_4Ca_2]^{7+}$ octahedra that share faces to produce $[FNa_2Ca]^{3-}$ chains parallel to the *c*-axis. Due to the complete Ca–Na ordering recently established by Krivovichev *et al.* (2007e), the chains are distorted and the whole structure is monoclinic rather than hexagonal. However, a pseudo-hexagonal character of the atomic arrangement is reflected in pseudo-merohedral

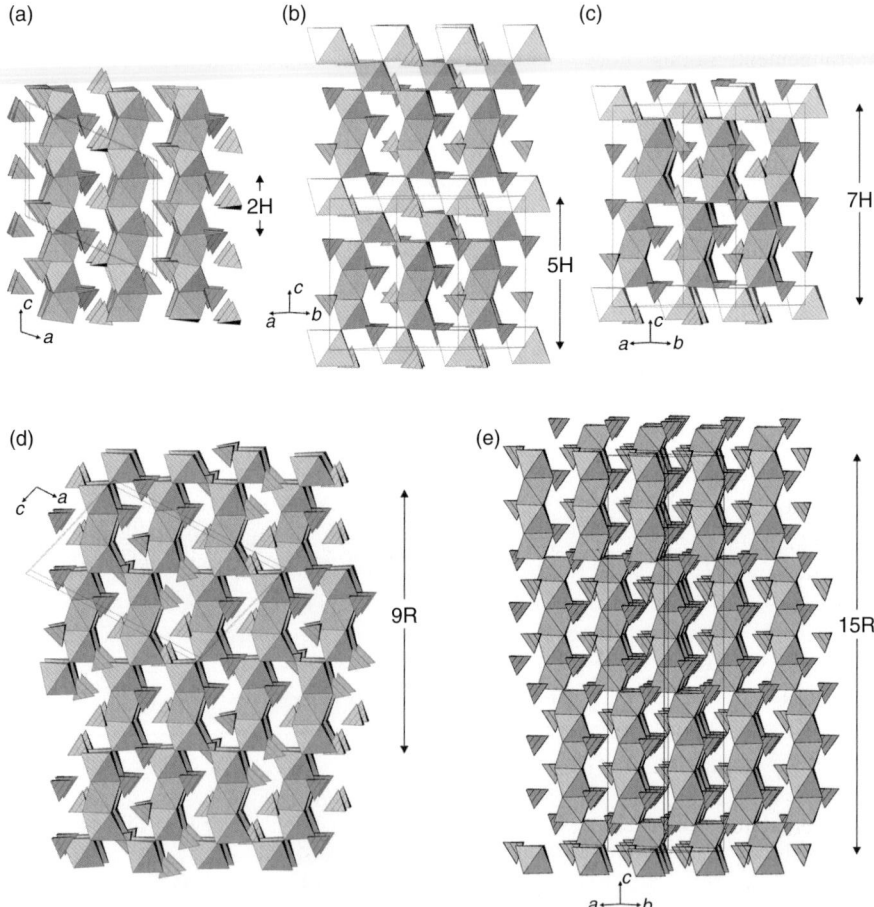

Fig. 5.29. Crystal structures of minerals with antiperovskite structure shown as a combination of anion-centered octahedra and cation-centered tetrahedra: (a) nacaphite; (b) galeite; (c) schairerite; (d) kogarkoite; (e) rhombohedral polymorph of nacaphite. $[FNa_6]$ and $[ClNa_6]$ octahedra are dark and light, respectively; TO_4 tetrahedra ($T = P, S$) are lined.

polysynthetic twinning that is typical for nacaphite (Khomyakov *et al.* 1980). It is of interest that the structure of nacaphite is closely related to isoformular silicate and germanate compounds listed in Table 5.8. Among the N-centered antiperovskites, the 2*H* polytype has been observed for [NBa$_3$]*T* *T* = Sb, Bi (Gaebler *et al.* 2004) and [NBa$_3$]Na (Rauch and Simon 1992), where [NBa$_6$] octahedra share faces to produce extended chains. In normal perovskites, this structure type is observed for Ba[NiO$_3$].

The 5*H* and 7*H* polytypes are observed in the structures of galeite, Na$_{15}$F$_4$Cl(SO$_4$)$_5$ (Fanfani *et al.* 1975a) (Fig. 5.29(b)), and schairerite, Na$_{21}$F$_6$Cl(SO$_4$)$_7$ (Fanfani *et al.* 1975b) (Fig. 5.29(c)), respectively. Again, the arrangement of Cl$^-$ and F$^-$ anions is completely ordered, and the frameworks consist of [FNa$_6$]$^{5+}$ and [ClNa$_6$]$^{5+}$ octahedra. The fluorine-centered octahedra usually share faces, whereas the chlorine-centered ones do not. As far as we know, there are no analogs of the 5*H* and 7*H* polytypic arrangements among antiperovskite and perovskite compounds.

Figure 5.29(d) shows the structure of kogarkoite, Na$_3$F(SO$_4$) (Fanfani *et al.* 1980), that is similar to the structure of hatrurite, Ca$_3$O(SiO$_4$) (Golovastikov *et al.* 1975; Nishi *et al.* 1985; Mumme, 1995; Dunstetter *et al.* 2006), an important phase known in the cement industry as a "tricalcium silicate". Here, [XA$_6$] octahedra share faces to form triplets further linked into a three-dimensional framework by sharing corners. This arrangement corresponds to a 9*R* hexagonal antiperovskite also known for (Sr$_{3-x}$Ba$_x$N)E (E = Sb, Bi) (2.50 ≤ *x* ≤ 2.55) (Gaebler and Niewa 2007). The cation-centered perovskite analog is that of the Ba[MnO$_3$] structure type.

Closely related to the 9*R* polytype is the 15*R* polytype shown in Fig. 5.29(e). Here, the octahedral framework consists of pentaplets of the [XA$_6$] octahedra. This structure is observed for the rhombohedral polymorph of nacaphite, Na$_2$CaF(PO$_4$), reported by Sokolova *et al.* (1999).

The occurence of hexagonal perovskites with the ABX$_3$ composition is usually explained by the large size of the A cation, i.e. hexagonal perovskite crystallizes when the A cation is too large for the ideal cubic perovskite structure. A similar situation is also observed for the hexagonal antiperovskites. For instance, Ba$_3$O(SiO$_4$) (Tillmans and Grosse 1978) and Cd$_3$O(SiO$_4$) (Eysel and Breuer 1983) contain a perovskite framework of corner-sharing [OA$_6$] octahedra, whereas harturite, Ca$_3$O(SiO$_4$), has a 9*R*-antiperovskite structure. K$_3$F(SO$_4$) (Skakle *et al.* 1996) adopts a cubic perovskite framework, whereas kogarkoite, Na$_3$F(SO$_4$), belongs to the 9*R*-antiperovskite structure. Thus, by analogy with perovskites, antiperovskite structure types in minerals are largely controlled by the dimensional factors such as size of the framework-forming anions and cations.

Intercalated antiperovskite structures have also been observed. Figure 5.30(a) shows the structure of K[Ca$_{12}$FO$_2$](SO$_4$)(SiO$_4$)$_4$, a fluorine-containing phase encountered in cement clinker (Fayos *et al.* 1985), whereas Fig. 5.30(c) shows the structure of arctite, Ba[Na$_5$Ca$_7$F$_3$](PO$_4$)$_6$ (Sokolova *et al.* 1984). Both structures contain isolated triplets of oxygen- and/or fluorine-centered octahedra shown in Figs. 5.30(b) and (d), respectively. These triplets might be regarded as parts of either 2*H* or 9*R* polytypes with intercalated layers of K$^+$ (or Ba^{2+}) cations and (SO$_4$) (or (PO$_4$)) tetrahedra.

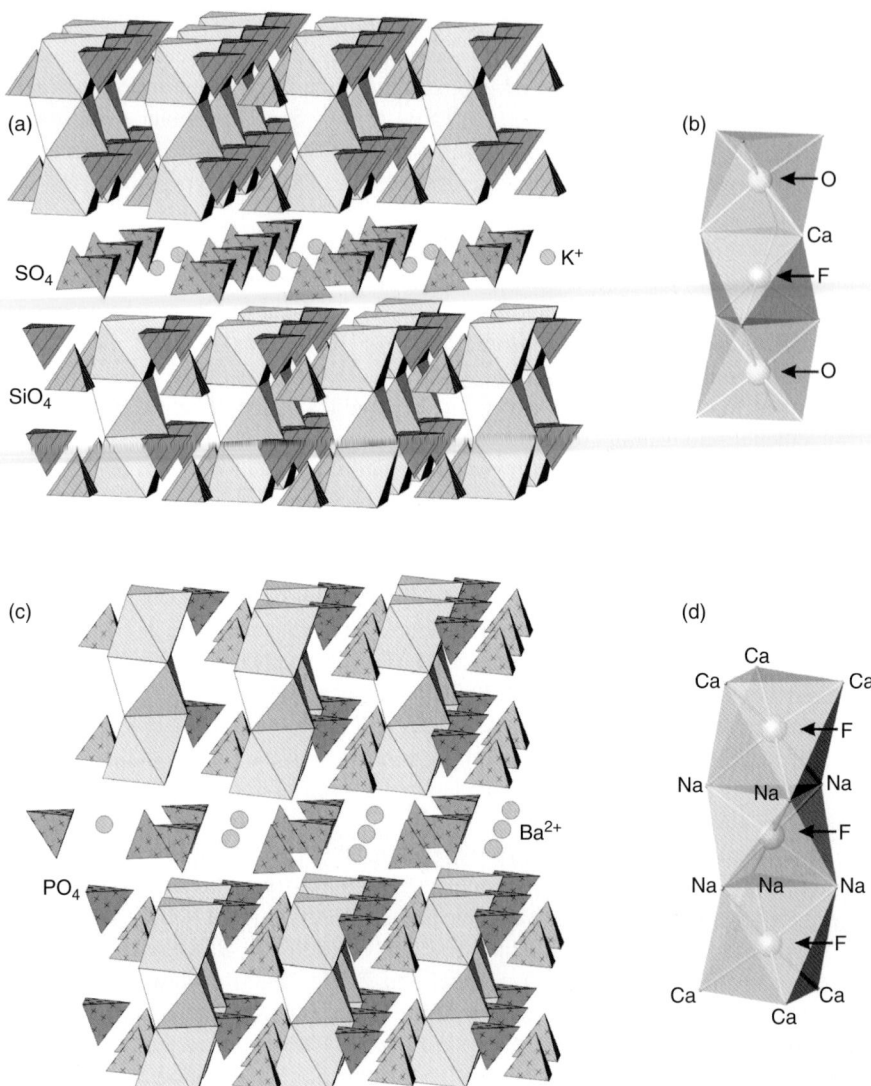

Fig. 5.30. Crystal structures $K[Ca_{12}FO_2](SO_4)(SiO_4)_4$ (a) and arctite, $Ba[Na_5Ca_7F_3](PO_4)_6$ (c) shown as a combination of anion-centered octahedra and cation-centered tetrahedra, and structures of face-sharing octahedral triplets ((b), (d), respectively).

An interesting case of a composite structure is that of polyphite, $[Ca_3Na_{15}F_6]$ $Na_2Mg(Ti,Mn)_4O_2(Si_2O_7)_2(PO_4)_3$, a complex silicate–phosphate (Sokolova *et al.* 1987) (Fig. 5.31). It has a modular structure and is composed of alternating *2H* anti-perovskite (nacaphite-like) and *HOH* heterophyllosilicate modules.

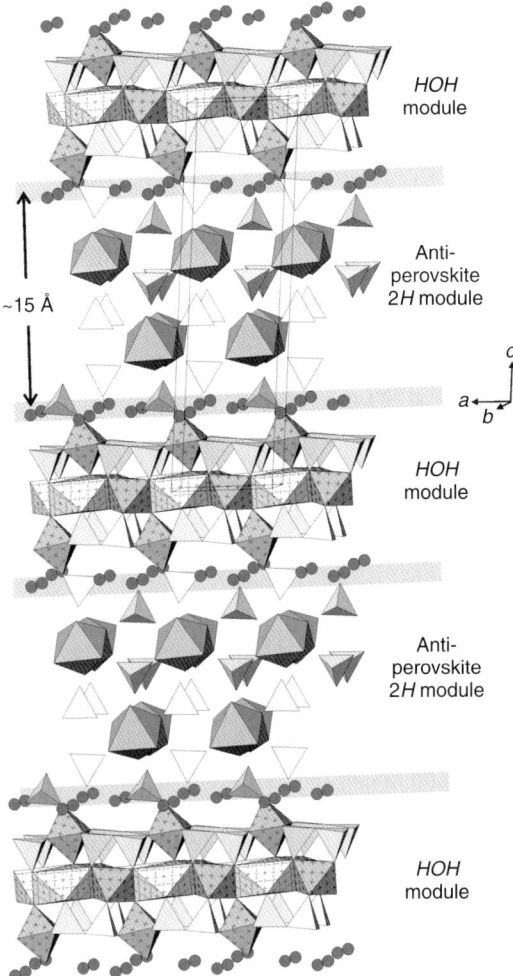

Fig. 5.31. Polyphite, $[Ca_3Na_{15}F_6]Na_2Mg(Ti,Mn)_4O_2(Si_2O_7)_2(PO_4)_3$, has a modular structure composed from alternating $2H$ antiperovskite (nacaphite-like) and *HOH* heterophyllosilicate modules. Note that the *HOH* modules consist solely of cation-centered polyhedra, whereas the antiperovskite modules are combinations of F-centered octahedra and cation-centered tetrahedra. The octahedra share faces to form one-dimensional chains running approximately perpendicular to the plane of the figure.

5.4 Cation arrays in inorganic oxysalts

Another alternative approach to the structures of inorganic oxysalts is that based upon their description as an arrangement of cations filled by anions. It has been frequently observed that, in some simple oxysalts, the arrangement of cations corresponds to the arrangement of atoms in alloys, sulphides, phosphides, silicides or other binary compounds. For instance, the cation array in Ag_2SO_3 is identical to that observed in Ag_2S, whereas the cation array in β-Ca_2SiO_4 is the structure of Ca_2Si. This approach has been systematically developed by O'Keeffe and Hyde (1985) and Vegas and Jansen (2002) who provided numerous examples of geometrical relationships between cation arrays in simple oxysalts and structures of binary compounds. A slightly different approach has been exploited by Borisov and co-workers (Borisov and Podberezskaya 1984; Bliznyuk and Borisov 1992). They analyzed a large number of structures of oxides, fluorides and oxysalts of heavy elements such as *REE* and actinides in terms of close-packed arrays of cations filled by anions. Cations of these elements are large and have a formal charge of 3+ and more that forces them to arrange in a symmetrical fashion corresponding to arrangement of spheres in close and closest packings.

Detailed discussion of this approach is beyond the scope of this book, and we provide only illustrative examples of description of complex inorganic oxysalts in terms of cation arrays.

Structures of *REE* and Ba fluorocabonates are usually described using a modular approach (Ferraris *et al.* 2004) in terms of stacking of 2D modules of cation-centered polyhedra. Another description may be based upon cation arrays formed by REE^{3+} and Ba^{2+} cations (Krivovichev *et al.* 2003b). Table 5.9 provides crystallographic data for some *REE* and Ba fluorocabonates, including both minerals and synthetic compounds. In the structures of all these compounds, REE^{3+} and Ba^{2+} cations are arranged according to the cubic closest packing. The latter has an ideal cubic symmetry with the space group $Fm\bar{3}m$ and four systems of close-packed layers (oriented perpendicular to threefold axes). Figure 5.32 shows a cation array in the structures of kukharenkoite-(Ce) and -(La), with cations linked by imaginary bonds not longer than 6 Å. One of the four systems of close-packed layers is parallel to (401). It is interesting that, in the structures, triangular (CO_3) groups are parallel to this plane as well. Thus, orientation of triangular ions might be chosen as a basis for selection of close-packed layers in cation arrays. The structure of the (401) close-packed layer in kukharenkoite-(La) is shown in Fig. 5.33(a). According to the formula $Ba_2(La,Ce)(CO_3)_3F$, the Ba:La ratio is 2:1. The Ba^{2+} and La^{3+} cations form rows parallel to the *b*-axis according to the sequence... BaBaLa... According to Zaitsev *et al.* (1996), the pseudosymmetrical cation arrangement in kukharenkoite-(Ce) results in strong pseudosymmetry of diffraction pattern. It is noteworthy that zhonghuacerite (which is probably identical to kukharenkoite-(Ce)) was described by Zhan and Tao (1981) as trigonal. Zaitsev *et al.* (1996) noted the presence in kukharenkoite-(Ce) of a pseudo-rhombohedral cell with hexagonal axes $a' = 5.076(1)$ and $c' = 9.821(9)$ Å. Transition between pseudo and real cells is depicted in Fig. 5.32(b) and can be described by the following transformation: $a' = (a - 3b + 4c)/6; b' = b; c' = (2a - c)/3$.

As in the structures of kukharenkoite-(Ce) and -(La), in cebaite-(Ce) and huang-hoite-(Ce), heavy cations form a cubic closest-packed arrangement. The close-packed cation layers that are parallel to the planes of the (CO_3) groups are parallel to (102)

Table 5.9. Crystallographic data for natural Ba and REE fluorocarbonates and isostructural compounds

Mineral name	Chemical formula	Sp. gr.	a (Å)	b (Å)	c (Å)	β (°)	V(Å^3)	V_R(Å^3)	References
kukharenkoite-(Ce)	$Ba_2Ce(CO_3)_3F$	$P2_1/m$	13.374	5.1011	6.653	106.56	435.06	217.53	Krivovichev *et al.* 1998c
kukharenkoite-(La)	$Ba_2(La,Ce)(CO_3)_3F$	$P2_1/m$	13.396	5.111	6.672	106.628	437.7	218.85	Krivovichev *et al.* 2004b
cebaite-(Ce)	$Ba_3Ce_2(CO_3)_5F_2$	$C2/m$	21.42	5.078	13.30	94.8	1441.58	216.24	Yang 1995
–	$Ba_3La_2(CO_3)_5F_2$	$C2/m$	21.472	5.098	13.325	94.96	1453.14	217.97	Mercier and Leblanc 1993
huanghoite-(Ce)	$BaCe(CO_3)_2F$	$R\bar{3}m$	5.072	–	38.460	–	856.84	214.21	Yang and Pertlik 1993
–	$BaSm(CO_3)_2F$	$R\bar{3}m$	5.016	–	37.944	–	826.78	206.70	Mercier and Leblanc 1993

V_R – volume of rhombohedral subcell.

and (001) in cebaite-(Ce) and huanghoite-(Ce), respectively. The close-packed cation layer in cebaite-(Ce) is shown in Fig. 5.33(b). According to the Ba:Ce ratio of 3:2, triple rows of Ba^{2+} cations alternate with double rows of Ce^{3+} cations. In huanghoite-(Ce), close-packed layers of Ba^{2+} and Ce^{3+} cations alternate along the direction

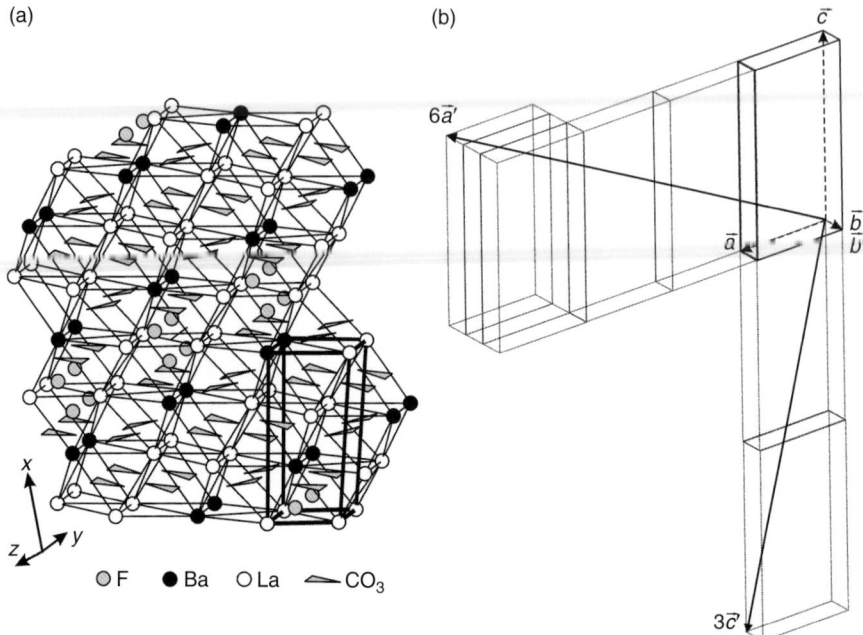

Fig. 5.32. Structure of kukharenkoite-(La) viewed as a framework of heavy cations with F^- and CO_3^{2-} anions in interstices (a) and geometrical relationships between the true cell and the pseudo-rhombohedral subcell (b) (see text for details).

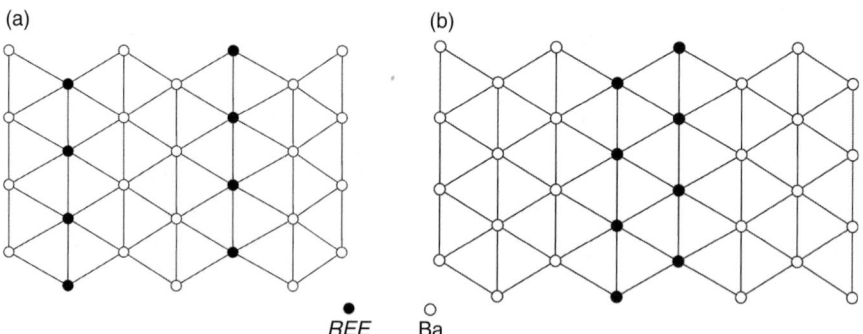

Fig. 5.33. Structure of closely packed cation layers in kukharenkoite-(La) (a) and cebaite-(Ce) (b).

perpendicular to (001). Rhombohedral subcells (in hexagonal axes) may be obtained through the following transformations.

For cebaite-(Ce): $a' = (2a + 5b + c)/10$; $b' = b$; $c' = (a - 7c)/10$, $|a'| = 5.061$ Å, $|b'| = 5.078$ Å, $(|a'| + |b'|)/2 = 5.070$ Å, $|c'| = 9.726$ Å.

For huanghoite-(Ce): $a' = a$; $b' = b$; $c' = c/4$. $|a'| = |b'| = 5.072$ Å, $|c'| = 9.615$ Å.

Table 5.9 provides volumes of rhombohedral subcells (V_R) in all structures analyzed. It can be seen that these volumes are in the range of 206–219 Å³.

It is of interest to analyze the positions of anions relative to the Ba-*REE* cation arrays. The close-packed arrays have octahedral and tetrahedral cavities. In Ba-*REE*

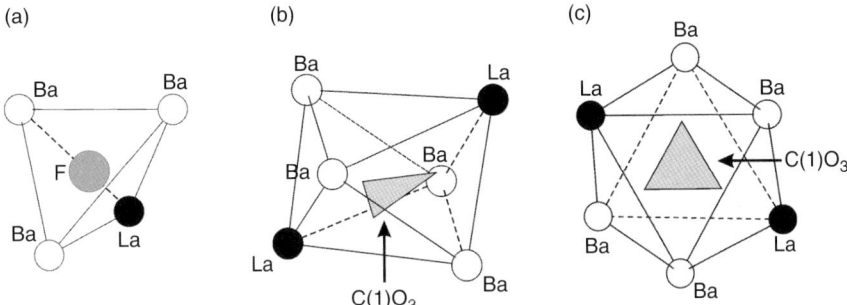

Fig. 5.34. Coordination of anions F⁻ (a) and CO_3^{2-} ((b) and (c)) in the structure of kukharenkoite-(La).

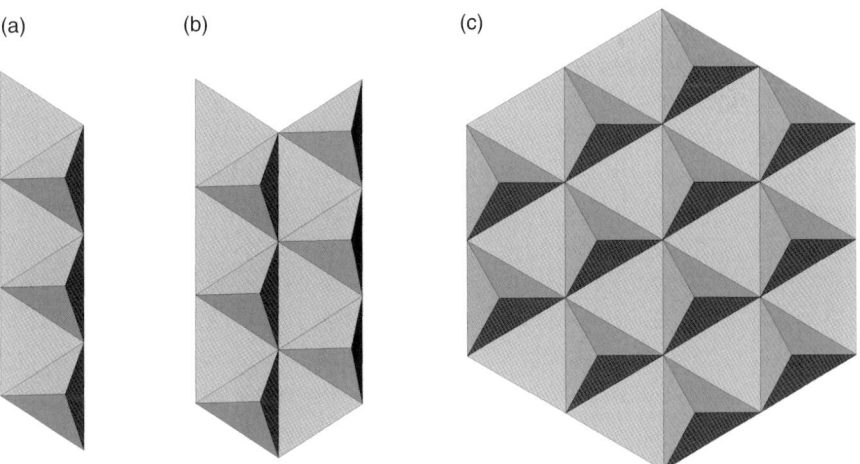

Fig. 5.35. Complexes of F-centered tetrahedra FM_4 (M = Ba, *REE*) in kukharenkoite-(La) (a), cebaite-(Ce) (b) and huanghoite-(Ce) (c).

fluorocarbonates, (CO_3) groups and F^- anions are in octahedral and tetrahedral cavities, respectively (Fig. 5.34). We recall that, for n spheres forming a closest packing, there are n octahedral and $2n$ tetrahedral cavities. Looking at the formulas given in Table 5.9, one may observe that the number of (CO_3) groups is equal to the number of cations, (Ba+*REE*). This means that all octahedral cavities are occupied by carbonate groups. The number of "additional" F^- anions varies from structure to structure and equals 1/6 for kukharenkoite-(Ce) and -(La), 1/5 for cebaite-(Ce) and 1/4 for huanghoite-(Ce). Thus, formally these structures can also be described in terms of F-centered tetrahedra (FM_4) (M = Ba, REE) (Krivovichev *et al.* 1998c). In kukharenkoite-(Ce) and -(La), cebaite-(Ce) and huanghoite-(Ce), (FM_4) tetrahedra share edges to form double chains, quadruple chains, and sheets, respectively (Fig. 5.35). This approach allows to describe structures of Ba–*REE* fluorocarbonates as derivatives of fluorite, CaF_2, where all tetrahedral cavities in a cubic close-packed cation array are filled by F^- anions, whereas all octahedral cavities are empty. Transition from the structure of fluorite to the structures of Ba-*REE* fluorocarbonates includes ordered subsitutions of cations by Ba and *REE*, exclusion of some F^- anions and filling of octahedral voids by (CO_3) groups.

Description of complex structures in terms of cation arrays is sometimes the only way to present transparent and constructive structure model. It has been especially effective for some Pb oxysalts with hexagonal arrays of Pb^{2+} cations (Steele *et al.* 1998, 1999; Krivovichev and Burns 2000b; Weil 2002).

6

Dimensional reduction in inorganic oxysalts

6.1 Introduction

As has been shown in previous chapters, inorganic oxysalts are characterized by a large diversity of structures and chemical compositions. The composition–structure relationships in this class of inorganic compounds can be rationalized using the principle of dimensional reduction initially proposed by Long *et al.* (1996) for the description of decreasing dimensionality of chalcogenide structural units in Re sulphides and selenides. Later, this idea was generalized for a wide class of ternary compounds (Tulsky and Long 2001) and was applied to various materials, including organic–inorganic composites (Haddad *et al.* 2003). The principle of dimensional reduction states that incorporation of an ionic reagent A_aX into the framework of a parent compound, MX_n, according to the reaction $kA_aX + MX_n \rightarrow A_{ka}MX_{n+k}$, results in the decreased dimensionality of the MX_{n+k} structural unit. The higher the k value, the lower the dimensionality of the MX_{n+k} unit. The example of dimensional reduction is the row of V fluorides (Tulsky and Long 2001):

$$VF_3 \ (3D) + KF \rightarrow KVF_4 \ (2D) + KF \rightarrow K_2VF_5 \ (1D) + KF \rightarrow K_3VF_6 \ (0D).$$

Along this row, dimensionality of V fluoride structural unit changes from 3 (perovskite-type octahedral framework in VF_3) to 0 upon the addition of the ionic component KF.

Below, we shall demonstrate how the principle of dimensional reduction can be applied to inorganic oxysalts, using quaternary systems as examples.

6.2 Alkali-metal uranyl molybdates

Alekseev *et al.* (2007) considered structures of alkali-metal uranyl molybdates from the viewpoint of dimensional reduction and demonstrated that, at least for the part of the system with U ≤ Mo, this principle provides a reasonable description of the relationships between structure and chemical composition.

The general formula of an alkali-metal uranyl molybdate can be written as $A_xU_nMo_mO_y$, where A is an alkali metal. The analysis of the "structure–composition" correlation involves the following steps:

(1) to represent the $A_xU_nMo_mO_y$ formula ($n \le m$) as a sum of $UMoO_6$ ($= UO_2MoO_4$), A_2O, and A_2MoO_4 (A = alkali metal): $A_xU_nMo_mO_y = kA_2O + nUMoO_6 + (m - n)A_2MoO_4$, where $x = 2(k + m - n)$ and $y = k + 2n + 4m$;

(2) to analyze dimensionality of uranyl molybdate structural units and its changes upon incorporation of ionic reagents A_2O and A_2MoO_4 into a parent framework of UO_2MoO_4 that consists of a 3D array of U and Mo coordination polyhedra.

The row of relationships between different compositions of the alkali-metal uranyl molybdates with respect to the dimensionality of the uranyl molybdate structural units is shown in Fig. 6.1. In this figure, the horizontal line corresponds to the addition of the A_2MoO_4 component, whereas the vertical line corresponds to the addition of A_2O. For some compositions, such as $A_6[(UO_2)_3O(MoO_4)_5]$, that have not yet been described for alkali metals, information on their Ag and Tl counterparts was added. There is no structural information available on the $A_4[(UO_2)(MoO_4)_3]$ compounds and the structure of $Na_4[(UO_2)(CrO_4)_3]$ (Krivovichev and Burns 2003h) was used as a counterpart, suggesting structural similarities between chromates and molybdates. Table 6.1 summarizes the basic features of the structure of respective compounds and provides references. The analysis included anhydrous compounds only since incorporation of water or hydroxyl may modify the structural integrity (see below).

Table 6.1. Structural data on monovalent cation uranyl molybdates

Formula type	A	Uranyl molybdate structural unit	Reference
UO_2MoO_4	–	3D framework of UO_7 and MoO_4 polyhedra	Serezhkin *et al.* 1972
$A_2(UO_2)_3(MoO_4)_4$	Na, K	3D framework of UO_7 and MoO_4 polyhedra	Krivovichev and Burns 2007
$A_2(UO_2)_2(MoO_4)_3$	Cs, Rb	3D framework of UO_7 and MoO_4 polyhedra	Krivovichev *et al.* 2002c
	Cs	2D layer of UO_7 and MoO_4 polyhedra	Krivovichev *et al.* 2002c
$A_2(UO_2)(MoO_4)_2$	Na	2D layer of UO_7 and MoO_4 polyhedra	Krivovichev *et al.* 2002a
	K, Rb, Cs	2D layer of UO_7 and MoO_4 polyhedra	Krivovichev *et al.* 2002a; Krivovichev and Burns 2002a,b
$A_6(UO_2)(MoO_4)_4$	Na, Tl	0D unit $[(UO_2)_2(MoO_4)_8]^{12+}$	Krivovichev and Burns 2001b, 2003b
	Rb, Cs	0D unit $[(UO_2)(MoO_4)_4]^{6+}$	Krivovichev and Burns 2002a,b
$A_4(UO_2)_3Mo_3O_{14}$	Cs	2D layer of UO_7 and MoO_n ($n = 4,5$) polyhedra	Krivovichev and Burns 2002b
$A_2(UO_2)MoO_5$	Cs	1D chain of UO_6 and MoO_4 polyhedra	Alekseev *et al.* 2007
$A_6(UO_2)_2Mo_4O_{17}$	Na, K, Rb	1D chain of UO_6 and MoO_5 polyhedra	Krivovichev and Burns 2001b, 2002a
$A_6(UO_2)_3Mo_5O_{21}$	Ag	2D layer of UO_7 and MoO_n ($n = 4,5$) polyhedra	Krivovichev and Burns 2002b

It is clear from Fig. 6.1 that dimensional reduction is valid for alkali-metal uranyl molybdates with U ≤ Mo. Moreover, the relationships between different compositions and structures may be visualized using the $UO_2MoO_4 - A_2O - A_2MoO_4$ compositional diagram (Fig. 6.2). The observed trends of decreasing dimensionality of the uranyl molybdate units allow fields of dimensionality to be specified on this diagram. Though the borders between different fields are not clearly and unambiguously defined, some relationships are noteworthy. For instance, compounds with the $A_2[(UO_2)_2(MoO_4)_3]$ formula are known to crystallize in two modifications, 2D layered tetragonal and 3D framework orthorhombic (Krivovichev *et al.* 2002c). Thus, the point **3** in Fig. 6.2 is located on the border between the fields of dimensionality 2 and 3. The diagram helps to explain the finest structural pecularities of the compounds. For example, the structure $Cs_2[(UO_2)O(MoO_4)]$ contains planar chains of UO_6 octahedra and MoO_4 tetrahedra. The chains are arranged within the (010) plane so the structure has a pseudo-2-dimensional character (Fig. 6.3). The formula $Cs_2[(UO_2)$ $O(MoO_4)]$ can be obtained from the formula $Cs_4(UO_2)_3Mo_3O_{14}$ formula by incorporation of Cs_2O:

$$Cs_4(UO_2)_3Mo_3O_{14} + Cs_2O = 3Cs_2(UO_2)MoO_5.$$

The structure of $Cs_4(UO_2)_3Mo_3O_{14}$ is based upon uranyl molybdate layers (Krivovichev and Burns 2002b), i.e. units with dimensionality 2. Incorporation of the ionic reagent Cs_2O results in a decrease of the dimensionality from 2 to 1 but a 2-dimensional character of the structure is retained to some extent.

6.3   Inorganic oxysalts in the Ca$_3$(TO_4)$_2$ – H$_3$(TO_4) – (H$_2$O) system

The $Ca_3(TO_4)_2 - H_3(TO_4) - (H_2O)$ system (T = P, As) includes 16 different phases, and many are known as minerals in Nature (Table 6.2). In order to analyze their structures from the viewpoint of dimensional reduction, their compositions should be represented as a sum of their components:

$$Ca_nH_m(TO_4)_k(H_2O)_l = n/3Ca_3(TO_4)_2 + kH_3(TO_4) + l(H_2O)$$

Here, we consider the $Ca_3(TO_4)_2$ component as being responsible for the structural organization, i.e. the basis of the structure is a structural unit formed by linkage of Ca-centered polyhedra and (TO_4) tetrahedra. $H_3(TO_4)$ is considered as the ionic component, whereas the H_2O molecules are considered as modifiers of the structure dimensionality (see Chapter 1). One may expect that incorporation of $H_3(TO_4)$ and H_2O into the structure would result in the decreasing dimensionality of the Ca-(TO_4) unit.

Indeed, the diagram in Fig. 6.4 demonstrates that dimensionality of the heteropolyhedral unit changes upon addition of $H_3(TO_4)$ and H_2O from 3 (frameworks) to 2 (sheets). No 1D or 0D structural units have been observed in this system to date, most likely due to the relatively large size of the Ca^{2+} ions that favors coordination numbers of 7 and 8.

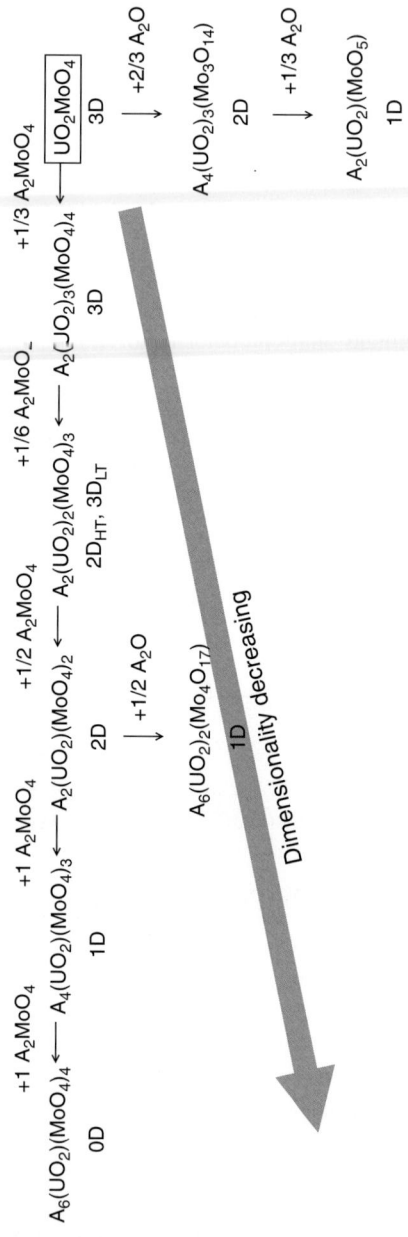

Fig. 6.1. Relationships between chemical compositions in the system $UO_2MoO_4 - A_2MoO_4 - A_2O$ (A = Na, K, Rb, Cs, Tl^I, Ag^I). Dimensionalities of uranyl molybdate units are given below the formulas.

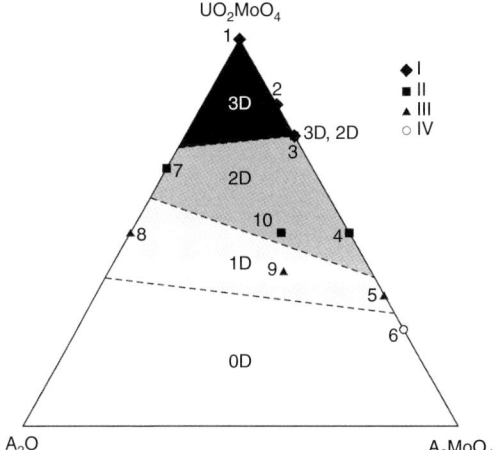

Fig. 6.2. Dimensionality fields on the compositional diagram of the UO$_2$MoO$_4$ – A$_2$MoO$_4$ – A$_2$O (A = Na, K, Rb, Cs, Tl^I, Ag^I) system. Points correspond to the following compounds: **1** – UO$_2$MoO$_4$, **2** – A$_2$(UO$_2$)$_3$(MoO$_4$)$_4$, **3** – A$_2$(UO$_2$)$_2$(MoO$_4$)$_3$, **4** – A$_2$(UO$_2$)(MoO$_4$)$_2$, **5** – A$_4$(UO$_2$) (MoO$_4$)$_3$, **6** – A$_6$(UO$_2$)(MoO$_4$)$_4$, **7** – A$_4$(UO$_2$)$_3$Mo$_3$O$_{14}$, **8** – A$_2$(UO$_2$)MoO$_5$, **9** – A$_6$(UO$_2$)$_2$Mo$_4$O$_{17}$, **10** – A$_6$(UO$_2$)$_3$Mo$_5$O$_{21}$. Numbers in the fields indicate the dimensionality of uranyl molybdate units. See Table 6.1 for references.

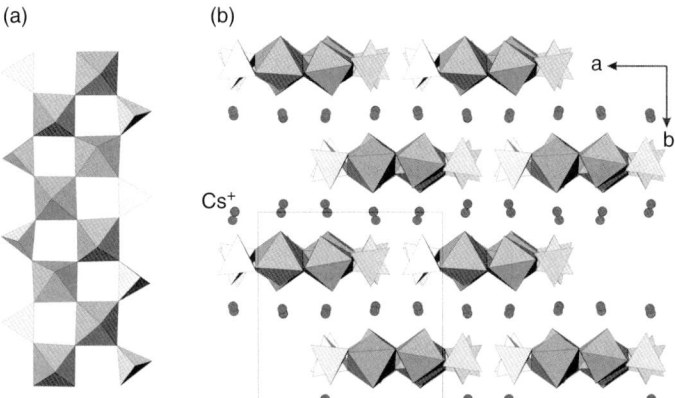

Fig. 6.3. Uranyl molybdate chain in the structure of Cs$_2$[(UO$_2$)O(MoO$_4$)] (a) and the structure projected along the c-axis (b). Legend: U polyhedra = dark gray, Mo tetrahedra = gray; Cs atoms are shown as circles.

The point **6** is on the border between compositional regions of 2- and 3-dimensional units. It corresponds to the composition 5Ca$_3$(TO$_4$)$_2$ · 2H$_3$(TO$_4$) · 27(H$_2$O) = Ca$_5$(HAsO$_4$)$_2$(AsO$_4$)$_2$(H$_2$O)$_9$. There are at least two minerals that have this chemical formula, guerinite and ferrarisite. The structure of guerinite (Fig. 6.5(a)) consists of

Table 6.2. Phases in the $Ca_3(TO4)_2 - H_3(TO_4) - (H_2O)$ system (T = P, As): structural information

Point*	Chemical formula	D**	Reference
1	$Ca(H_2PO_4)_2(H_2O)$	2	Schroeder *et al.* 1975
2	$CaHPO_4$ monetite	3	Catti *et al.* 1980b
	$CaHAsO_4$ weilite	3	Ferraris and Chiari 1970
3	$Ca(H_2PO_4)_2$	2	Dickens *et al.* 1973b
	$Ca(H_2AsO_4)_2$	2	Ferraris *et al.* 1972a
4	$Ca(HPO_4)_2(H_2O)_2$ brushite	2	Curry and Jones 1971
	$Ca(HPO_4)_2(H_2O)_2$ pharmacolite	2	Ferraris *et al.* 1971
5	$Ca_8H_2(PO_4)_6(H_2O)_5$	3	Mathew *et al.* 1988
6	$Ca_5(HAsO_4)_2(AsO_4)_2(H_2O)_9$ guerinite	2	Catti and Ferraris 1974
	$Ca_5(HAsO_4)_2(AsO_4)_2(H_2O)_9$ ferrarisite	3	Catti *et al.* 1980a
7	$Ca(HAsO_4)(H_2O)_3$	2	Catti and Ferraris 1973
8	$Ca_5(HAsO_4)_2(AsO_4)_2(H_2O)_4$ sainfeldite	3	Ferraris and Abbona 1972
9	$Ca_3(AsO_4)_2(H_2O)_{11}$ phaunouxite	2	Catti and Ivaldi 1983
10	$Ca_3(AsO_4)_2(H_2O)_{10}$ rauenthalite	2	Catti and Ivaldi 1983
11	$Ca_5(HAsO_4)_2(AsO_4)_2(H_2O)_5$ vladimirite	3	Catti and Ivaldi 1981
12	$Ca(HAsO_4)(H_2O)$ haidingerite	2	Ferraris *et al.* 1972b

  *   as given in Fig. 6.4

 ** D – dimensionality of structural unit of Ca-centered polyhedra and (TO_4) tetrahedra

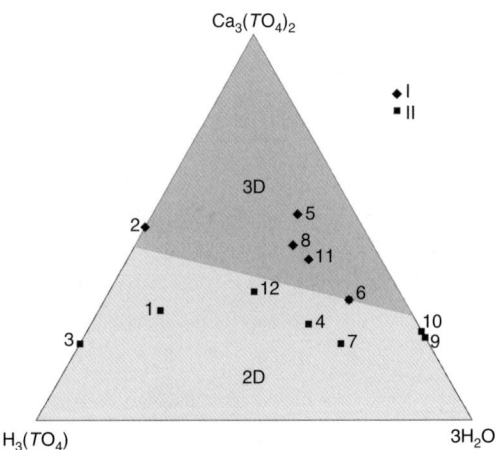

Fig. 6.4. Compositional diagram of the $Ca_3(TO_4)_2 - H_3(TO_4) - (H_2O)$ system (T = P, As). Legend: **I** – 3D frameworks, **II** – 2D sheets. Dimensionality fields are shown. See Table 6.2 for the list of compounds.

sheets of Ca polyhedra and (AsO_4) tetrahedra, whereas the structure of ferrarisite (Fig. 6.5(b)) is based upon dense sheets of Ca polyhedra and (AsO_4) tetrahedra linked by additional Ca-octahedra into a 3D framework. Thus, the framework in ferrarisite has a pronounced pseudo-2D character and can also be characterized as having a

(a)   (b)

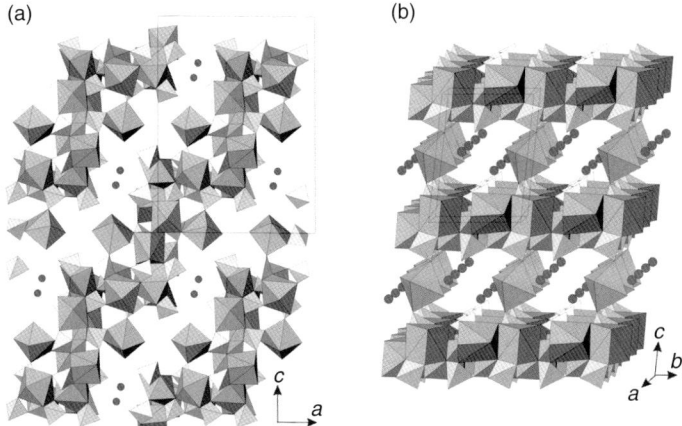

Fig. 6.5. Structures of guerinite (a) and ferrarisite (b). Ca polyhedra are dark gray, (TO_4) tetrahedra are light gray.

pillared architecture. The composition 5:2:27 is therefore transitional between the structures with 2D and 3D structural units.

6.4   Inorganic oxysalts in the $M_2(TO_4)_3$ – $H_2(TO_4)$ – (H_2O) system

Figure 6.6 shows the compositional diagram of the $M_2(TO_4)_3$ – $A_2(TO_4)$ – $4(H_2O)$ system, where T = S, Se, A = Na, K, Rb, Cs, NH_4, H_3O, and M is an octahedrally coordinated trivalent cation (Al^{3+}, Fe^{3+}, In^{3+}, Sc^{3+}). The list of example compounds is given in Table 6.3. From the diagram, it is clear that the tendency is similar to that observed for the $Ca_3(TO_4)_2$ – $H_3(TO_4)$ – (H_2O) system. However, it is remarkable that, except for the compounds with the $M_2(TO_4)_3$ stoichiometry, frameworks are absent in the structures of this system. Also, there are no such well-defined regions of dimensionality as those in alkali-metal uranyl molybdates and Ca phosphate and arsenate hydrates. For instance, consider the line in the diagram originating from the right corner ($4H_2O$)) and ending at the point **6** (composition $M_2(TO_4)_3 \cdot A_2(TO_4) = AM(TO_4)_2$). This line describes compounds with the composition $AM(TO_4)_2(H_2O)_n$. The points **8** and **9** correspond to the structures with no connections between octahedra and tetrahedra. The points **2** and **5** symbolize compositions of compounds containing sheets with the **cc2–1:2–3** topology (Fig. 6.7). Thus, the dimensionality of the structural unit is increasing from the points **9** to **2**. However, the dimensionality corresponding to the next composition along the line (point **4**) is 1 (chain structure with the **cc1–1:2–2** topology), which is in obvious contradiction to the principle of dimensional reduction. In order to explain this contradiction, one has to remember that the dimensionality of a structural unit is the result of connectivity of its subunits, i.e. in

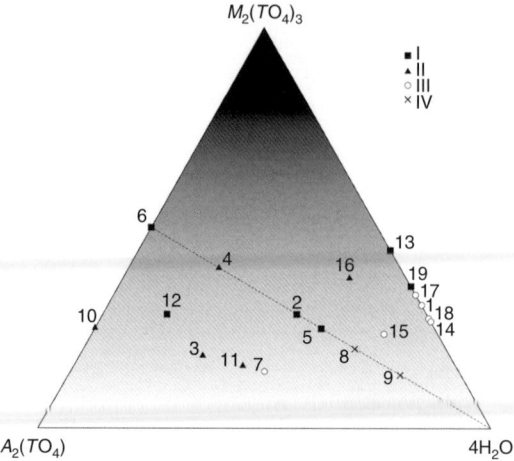

Fig. 6.6. Compositional diagram of the $M_2(TO_4)_3 - A_2(TO_4) - 4(H_2O)$ system, where $T = S$, Se, $A = $ Na, K, Rb, Cs, NH$_4$, H$_3$O, and M is an octahedrally coordinated trivalent cation (Al^{3+}, Fe^{3+}, In^{3+}, Sc^{3+}). Legend: **I** – 2D sheets, **II** – 1D chains, **III** – finite clusters, **IV** – no connections between tetrahedra and octahedra. See Table 6.3 for the list of example compounds.

Table 6.3.   Phases in the $M_2(TO_4)_3 - A_2(TO_4) - 4(H_2O)$ system* (examples of compositions)

Point**	Chemical formula	D***	Reference
1	$Fe_2(SO_4)_3(H_2O)_9$ coquimbite	0	Fang and Robinson 1970
2	$NH_4Fe(SO_4)_2(H_2O)_3$	2	Palmer et al. 1972
3	$Na_3Fe(SO_4)_3(H_2O)_3$ ferrinatrite	1	Scordari 1977
4	$KFe(SO_4)_2(H_2O)$ krausite	1	Effenberger et al. 1986
5	$KFe(SO_4)_2(H_2O)_4$ goldichite	2	Graeber and Rosenzweig 1971
6	$KFe(SO_4)_2$ yavapaiite	2	Graeber and Rosenzweig 1971
7	$NaFe(SO_4)_2(H_2O)_6$ amarillite	0	Li et al. 1990
8	$NaAl(SO_4)_2(H_2O)_6$ tamarugite	Ø****	Robinson and Fang 1969
9	$NaAl(SO_4)_2(H_2O)_{11}$ mendozite	Ø****	Fang and Robinson 1972
10	$(NH_4)In(SO_4)_3$	1	Jolibois et al. 1980, 1981
11	$Na_3(Sc(SO_4)_3)(H_2O)_5$	1	Sizova et al. 1974
12	$ScH(SeO_4)_2(H_2O)_2$	2	Valkonen 1978
13	$Al_2(SO_4)_3(H_2O)_5$	2	Fisher et al. 1996a
14	$Fe_2(SO_4)_3(H_2O)_{11}$	0	Thomas et al. 1974
15	$(H_3O)_2(Al_2(SO_4)_3)_2(SO_4)(H_2O)_{22}$	0	Fisher et al. 1996g
16	$Al_3(HSO_4)(SO_4)_4(H_2O)_9$	1	Fisher et al. 1996e
17	$Al_2(SO_4)_3(H_2O)_8$	0	Fisher et al. 1996f
18	$Al_2(SO_4)_3(H_2O)_{10.5}$	0	Fisher et al. 1996d
19	$In_2(SeO_4)_3(H_2O)_5$	2	Kadoshnikova et al. 1978

 * $T = $ S, Se; $A = $ Na, K, Rb, Cs, NH$_4$, H$_3$O; $M = $ octahedrally coordinated trivalent cation (Al^{3+}, Fe^{3+}, In^{3+}, Sc^{3+})

 ** as given in Fig. 6.6

 *** D – dimensionality of structural unit of M-centered polyhedra and (TO_4) tetrahedra

 **** no connections between octahedra and tetrahedra

our case, of coordination polyhedra. We recall that the *connectedness* of a polyhe-dron, s, is the number of polyhedra with which it shares common corners (Liebau 1985; Krivovichev *et al.* 1997a). Intuitively, the higher the connectedness of polyhe-dra, the higher the dimensionality of a structural unit composed from polyhedra. For the unit consisting of n MO_6 octahedra and m TO_4 tetrahedra, we may define separ-ately average connectedness values for octahedra, s_M, and tetrahedra, s_T. The average connectedness of a polyhedron in the unit can then be calculated as $\hat{s} = (ns_M + ms_T)/(n + m)$. In terms of the graph theory, average connectedness is equal to the average valency of vertices in a graph. Figure 6.7 shows that the average connectedness of the chains with the ***ccl–1:2–2*** topology is 3.33, whereas the average connectedness of the sheets with the ***cc2–1:2–3*** topology is 2.67. Thus, despite the fact that dimensionality of the ***ccl–1:2–2*** chain is lower, its connectedness is higher. This feature is due to the higher complexity of the ***ccl–1:2–2*** graph in comparison to ***cc2–1:2–3***. The former has 2- and 3-connected white vertices and 5-connected black vertices, whereas the latter has 2-connected white and 4-connected black vertices. Thus, the ***ccl–1:2–2*** graph has three topologically different vertices, whereas the ***cc2–1:2–3*** graph has only two. It is interesting that both ***ccl–1:2–2*** and ***cc2–1:2–3*** topologies are deriva-tives of the ***ccl–1:2–2*** graph that describes the topology of polyhedral linkage in the compound with the stoichiometry **6**.

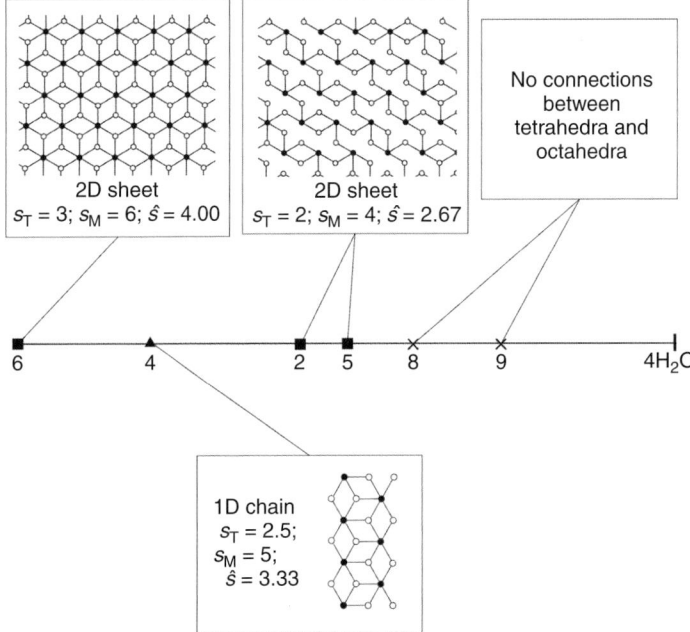

Fig. 6.7. Section of the $M_2(TO_4)_3 - A_2(TO_4) - 4(H_2O)$ system along the line $AM(TO_4)_2 - 4H_2O$ with comments concerning topology and connectedness of structural units. See text for details.

6.5   Inorganic oxysalts in the $M(TO_4) - H_2(TO_4) - (H_2O)$ system

The compositional diagram of the $M(TO_4) - A_2(TO_4) - 4(H_2O)$ system is shown in Fig. 6.8. Here, T = S, Se, A = Na, K, Rb, Cs, NH$_4$, H$_3$O, and M is an octahedrally coordinated divalent cation (Mg^{2+}, Fe^{2+}, Cd^{2+}, Zn^{2+}, etc.). Table 6.4 provides the list of example compounds. The system is characterized by the widespread occurence of framework structures, which is in stark contrast with the system containing tri-valent metal cations. The change of dimensionality versus composition is similar to that observed in the systems considered above and is in agreement with dimensional reduction. For example, the dimensionality increases when going from point **1** (no connections) through **2** (0D cluster of octahedron sharing two corners with two adja-cent tetrahedra) and **3** (1D krohnkite-type chain) to **6** (3D framework).

An interesting change of dimensionality is observed along the $M(TO_4) - 4(H_2O)$ line. The points **4, 16, 14**, and **11** correspond to framework structures. With addition of H$_2$O, the framework density is generally decreasing (framework density, FD, is defined as a number of M and T atoms per 1 nm^3). For the points **14** and **11**, a devi-ation is observed that can be explained by the fact that these points correspond to the structures of (Cd(SO$_4$))$_3$(H$_2$O)$_8$ and Cu(SO$_4$)(H$_2$O)$_3$, respectively. Cd^{2+} cations are larger than Cu^{2+} cations and thus the framework density for the Cd sulphate is higher than that for the Cu sulphate.

The points **7** and **8** symbolize compositions of compounds containing 0D and 1D units, respectively. The chain structure is more hydrated than the finite-cluster

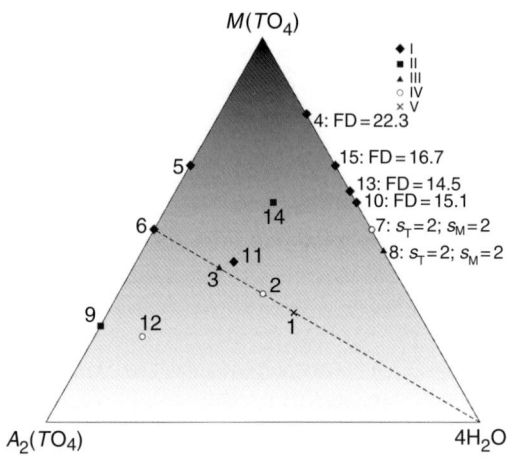

Fig. 6.8. Compositional diagram of the $M(TO_4) - A_2(TO_4) - 4(H_2O)$ system, where T = S, Se, A = Na, K, Rb, Cs, NH$_4$, H$_3$O, and M is an octahedrally coordinated divalent cation (Mg^{2+}, Fe^{2+}, Cd^{2+}, Zn^{2+}, etc.). Legend: **I** – 3D frameworks, **II** – 2D sheets, **III** – 1D chains, **IV** – finite clusters, **V** – no connections between tetrahedra and octahedra. See Table 6.4 for the list of example compounds.

Table 6.4.   Phases in the $M(TO_4) - A_2(TO_4) - 4(H_2O)$ system* (examples of compositions)

Point**	Chemical formula	D***	Reference
1	$(NH_4)_2Fe(SO_4)_2(H_2O)_6$ mohrite	Ø****	Montgomery *et al.* 1967
2	$K_2[Mg(H_2O)_4(SO_4)_2]$ leonite	0	Jarosch 1985
3	$K_2Fe(SO_4)_2(H_2O)_2$	1	Ishigami *et al.* 1999
4	$Fe(SO_4)(H_2O)$ poitevnite	3	Giester *et al.* 1994
5	$K_2Zn_2(SO_4)_3$	3	Moriyoshi and Itoh 1996
6	$Na_2Zn(SO_4)_2$	3	Berg and Thorup 2005
7	$Zn(SO_4)(H_2O)_4$	0	Blake *et al.* 2001
8	$Cu(SO_4)(H_2O)_5$ chalcanthite	1	Bacon and Titterton 1975
9	$Na_6[Mg(SO_4)_4]$ vanthoffite	2	Fischer and Hellner 1964
10	$[Cu(H_2O)_3(SO_4)]$ bonattite	3	Zahrobsky and Baur 1968
11	$Na_{12}[Mg_7(H_2O)_{12}(SO_4)_9](SO_4)_4(H_2O)_3$ löweite	3	Fang and Robinson 1970b
12	$Na_6(Zn(SO_4)_4(H_2O)_2)$	0	Heeg *et al.* 1986
13	$(CdSO_4)_3(H_2O)_8$	3	Caminiti and Johansson 1981
14	$K_2Cd_3(SO_4)_4(H_2O)_5$	2	Fleck and Giester 2003
15	$Zn(SeO_4)(H_2O)_2$	3	Krivovichev 2006

* T = S, Se; A = Na, K, Rb, Cs, NH_4, H_3O; M = octahedrally coordinated divalent cation (Mg^{2+}, Fe^{2+}, Cd^{2+}, Zn^{2+}, etc.)

** as given in Fig. 6.8

*** D – dimensionality of structural unit of M-centered polyhedra and (TO_4) tetrahedra

**** no connections between octahedra and tetrahedra

structure, which is seemingly in contradiction with dimensional reduction. However, analysis of the connectedness values shows that both structural units have $\hat{s} = 2$. Both units consist of 2-connected octahedra and tetrahedra. In the structure of **8**, they form linear 1D chains of the ***cc1–1:1–1*** topology, whereas, in the structure of 7, they are closed in a cyclic structure (4-membered ring with the ***cc0–1:1–2*** graph).

6.6   Concluding remarks

The systems considered above demonstrate that dimensional reduction indeed takes place in inorganic oxysalts and may be used for analysis and prediction of structural connectivities. In some cases, however, deviations may occur due to the possible formation of structural isomers with different dimensionality and complexity. However, the general trend is in agreement with dimensional reduction: incorporation of ionic components and water into parent salts results in child compounds with decreasing dimensionality of the structural unit.

References

Abbona, F., Boistelle, R., and Haser, R. (1979). Hydrogen bonding in $MgHPO_4(H_2O)_3$ (newberyite). *Acta Crystallogr.* **B35**, 2514–2518.

Abraham, F. and Ketatni, E. M. (1995). Crystal structure of a new bismuth nickel oxophosphate: $BiNiOPO_4$. *Eur. J. Solid State Inorg. Chem.* **32**, 429–437

Abraham, F., Ketatni, E. M., Mairesse, G., and Mernari, B. (1994). Crystal structure of a new bismuth copper oxyphosphate: $BiCu_2PO_6$. *Eur. J. Solid State Inorg. Chem.* **31**, 313–323.

Abraham, F., Cousin, O., Mentré, O. and Ketatni, E. M. (2002). Crystal structure approach of the disordered new compounds $Bi_{-1.2}M_{-1.2}PO_{5.5}$ (M = Mn, Co, Zn): the role of oxygen-centered tetrahedra linkage in the structure of bismuth–transition metal oxy-phosphates. *J. Solid State Chem.* **167**, 168–181.

Adams, R. D., Layland, R., and Payen, C. (1995). $Cu_4(AsO_4)_2(O)$: a new copper arsenate with unusual low temperature magnetic properties. *Inorg. Chem.* **34**, 5397–5398.

Adiwidjaja, G., Friese, K., Klaska, K. H., and Schlueter, J. (1999). The crystal structure of kastningite $(Fe_{0.5}Mn_{0.5})(H_2O)_4(Al_2(OH)_2(H_2O)_2(PO_4)_2)(H_2O)_2$ a new hydroxyl aquated orthophosphate hydrate mineral. *Z. Kristallogr.* **214**, 465–468.

Adiwidjaja, G., Friese, K., Klaska, K.H., Moore, P.B., and Schlueter, J. (2000). The crystal structure of the new mineral wilhelmkleinite $ZnFe^{3+}_2(OH)_2(AsO_4)_2$. *Z. Kristallogr.* **215**, 96–101.

Ahmed Farag, I. S., El Kordy, M. A., and Ahmed, N. A. (1981). Crystal structure of praseodymium sulfate octahydrate. *Z. Kristallogr.* **155**, 165–171.

Airoldi, R. and Magnano, G. (1967). Sulla struttura del solfato (di)mercurioammonico. *Rassegna Chimica* **5**, 181–189

Alcock, N. W., Roberts, M. M., and Brown, D. (1982). Actinide structural studies. Part 3. The crystal and molecular structures of $(UO_2SO_4)(H_2SO_4)(H_2O)_5$ and $(NpO_2SO_4)_2(H_2SO_4)(H_2O)_4$. *Dalton Trans.* **1982**, 869–873.

Alda, E., Bazab, B., Mesa, J. L., Pizarro, J. L., Arriortua, M. I., and Rojo, T. (2003). A new vanadium(III) fluorophosphate with ferromagnetic interactions, $(NH_4)(V(PO_4)F)$. *J. Solid State Chem.* **173**, 101–108.

Alekseev, E. V., Krivovichev, S. V., Armbruster, T., Depmeier, W., Suleimanov, E. V., Chuprunov, E. V., and Golubev, A. V. (2007). Dimensional reduction in alkali metal uranyl molybdates: synthesis and structure of $Cs_2[(UO_2)O(MoO_4)]$. *Z. Anorg. Allg. Chem.* **633**, 1979–1984.

Alekseev, E. V., Krivovichev, S. V., and Depmeier, W. (2008a). The crystal structure of $[CH_3NH_3][(UO_2)(H_2AsO_4)_3]$. *Radiochem.* **50**, 445–449.

Alekseev, E. V., Krivovichev, S. V., Depmeier, W., and Knorr, K. (2008b). Complex topology of uranyl polyphosphate frameworks: crystal structures of α, β–$K[(UO_2)(P_3O_9)]$ and $K[(UO_2)_2(P_3O_{10})]$. *Z. Anorg. Allg. Chem.* **634**, 1527–1532.

Alekseev, E. V., Krivovichev, S. V., and Depmeier, W. (2008c). A crown ether as template for microporous and nanostructured actinide compounds. *Angew. Chem. Int. Ed.* **47**, 549–551.

Alkemper, J. and Fuess, H. (1998). The crystal structures of $NaMgPO_4$, $Na_2CaMg(PO_4)_2$ and $Na_{18}Ca_{13}Mg_5(PO_4)_{18}$: new examples for glaserite related structures. *Z. Kristallogr.* **213**, 282–287.

Almond, P. M. and Albrecht-Schmitt, T. E. (2002a). Hydrothermal syntheses, structures, and properties of the new uranyl selenites $Ag_2(UO_2)(SeO_3)_2$, $M[(UO_2)(HSeO_3)(SeO_3)]$ (M = K, Rb, Cs, Tl), and $Pb(UO_2)(SeO_3)_2$. *Inorg. Chem.* **41**, 1177–1183.

Almond, P. M. and Albrecht-Schmitt, T. E. (2002b). Expanding the remarkable structural diversity of uranyl tellurites: hydrothermal preparation and structures of $K[UO_2Te_2O_5(OH)]$, $Tl_3\{(UO_2)_2[Te_2O_5(OH)](Te_2O_6)\} \cdot 2H_2O$, β-$Tl_2[UO_2(TeO_3)_2]$, and $Sr_3[UO_2(TeO_3)_2](TeO_3)_2$. *Inorg. Chem.* **41**, 5495–5501.

Almond, P. M. and Albrecht-Schmitt, T. E. (2003). Do secondary and tertiary ammonium cations act as structure-directing agents in the formation of layered uranyl selenites? *Inorg. Chem.* **42**, 5693–5698.

Almond, P. M. and Albrecht-Schmitt, T. E. (2004). Hydrothermal synthesis and crystal chemistry of the new strontium uranyl selenites, $Sr[(UO_2)_3(SeO_3)_2O_2] \cdot 4H_2O$ and $Sr[(UO_2)(SeO_3)_2] \cdot 2H_2O$. *Am. Mineral.* **89**, 976–980.

Almond, P. M., Peper, S. M., Bakker, E., and Albrecht-Schmitt, T. E. (2002). Variable dimensionality and new uranium oxide topologies in the alkaline-earth metal uranyl selenites $AE[(UO_2)(SeO_3)_2]$ (AE=Ca, Ba) and $Sr[(UO_2)(SeO_3)_2] \cdot 2H_2O$. *J. Solid State Chem.* **168**, 358–366.

Amoros, P. and LeBail, A. (1992). Synthesis and crystal structure of α-$NH_4(VO_2)(HPO_4)$. *J. Solid State Chem.* **97**, 283–291.

Amoros, P., Beltran-Porter, A., Villeneuve, G., and Beltran-Porter, D. (1992). Vanadyl dihydrogenoarsenate, $VO(H_2AsO_4)_2$: crystal structure and superexchange pathways. *Eur. J. Solid State Inorg. Chem.* **29**, 257–272.

Amoros, P., Beltran-Porter, D., LeBail, A., Férey, G., and Villeneuve, G. (1988). Crystal structure of $A(VO_2)(HPO_4)(A = NH_4^+, K^+, Rb^+)$ solved from X-ray powder diffraction. *Eur. J. Solid State Inorg. Chem.* **25**, 599–607.

Amos, T. G., Yokochi, A., and Sleight, A. W. (1998). Phase transition and negative thermal expansion in tetragonal $NbOPO_4$. *J. Solid State Chem.* **141**, 303–307.

Andac, O., Glasser, F. P., and Howie, R. A. (1997). Dicalcium sodium fluoride silicate. *Acta Crystallogr.* **C53**, 831–833.

Anderson, J. B., Shoemaker, G. L., and Kostiner, E. (1978). The crystal structure of $Cu_4(PO_4)_2O$. *J. Solid State Chem.* **25**, 49–57.

Andreev, G. B., Antipin, M. Yu., Fedoseev, A. M., and Budantseva, N. A. (2001). Synthesis and crystal structure of ammonium monohydrate dimolybdatouranylate, $(NH_4)_2[UO_2(MoO_4)_4] \cdot H_2O$. *Russ. J. Coord. Chem.* **27,** 208–210.

Anisimova, N., Schuelke, U., Bork, M., Hoppe, R., and Serafin, M. (1996). Preparation and crystal structure of a new acentric caesium iron hydrogen phosphate, $Cs_3Fe_3H_{15}(PO_4)_9$. *Z. Anorg. Allg. Chem.* **622**, 1920–1926.

Anisimova, N., Chudinova, N., and Serafin, M. (1997). Preparation and crystal structure of potassium iron hydrogen phosphate $KFe_3(HPO_4)_2(H_2PO_4)_6 \cdot 4(H_2O)$. *Z. Anorg. Allg. Chem.* **623**, 1708–1714.

Anisimova, N., Ilyukhin, A., and Chudinova, N. (1998). Preparation and crystal structure of sodium iron hydrogen phosphate, $NaFe(HPO_4)(H_2PO_4)_2 \cdot H_2O$. *Z. Anorg. Allg. Chem.* **624**, 1509–1513.

Anthony, J. W., McLean, W. J., and Laughon, R. B. (1972). The crystal structure of yavapaiite: a discussion. *Am. Mineral.* **57**, 1546–1549.

Aranda, M. A. G., Attfield, J. P., Bruque, S., and Martinez-Lara, M. (1992). Order and disorder of vanadyl chains: crystal structures of vanadyl dihydrogen arsenate $(VO(H_2AsO_4)_2)$ and the lithium derivative $Li_4VO(AsO_4)_2$. *Inorg. Chem.* **31**, 1045–1049.

Arlt, J., Jansen, M., Klassen, H., Schimmel, G., and Heymer, G. (1987). $Na_5AlF_2(PO_4)_2$: Darstellung, Kristallstruktur und Ionenleitfaehigkeit. *Z. Anorg. Allg. Chem.* **547**, 179–187.

Arlt, T., Armbruster, T., Ulmer, P., and Peters, T. (1998). $MnSi_2O_5$ with the titanite structure: a new high-pressure phase in the $MnO–SiO_2$ binary system. *Am. Mineral.* **83**, 657–660.

Armbruster, T., Krivovichev, S. V., Weber, T., Gnos, E., Organova, N. I., Yakovenchuk, V. N. (2004). Origin of diffuse superstructure reflections in labuntsovite-group minerals. *Am. Mineral.* **89**, 1655–1666.

Arpe, R. and Müller-Buschbaum, H. (1977). Ueber Oxocuprate, XXII. Zur Kristallchemie von Kupferoxichlorid: Cu_2Cl_2O. *Z. Naturforsch.* **32b**, 380–382.

Atencio, D., Neumann, R., Silva, A. J. G. C., and Mascarenhas, Y. P. (1991). Phurcalite from Perus, São Paulo, Brazil, and redetermination of its crystal structure. *Can. Mineral.* **29**, 95–105.

Attfield, J.-P., Battle, P. D., and Cheetham, A. K. (1985). The spiral magnetic structure of beta-chromium(III) orthophosphate. *J. Solid State Chem.* **57**, 357–361.

Attfield, M. P., Morris, R. E., Burshtein, I., Campana, C. F., and Cheetham, A. K. (1995). The synthesis and characterization of a one-dimensional aluminophosphate: $Na_4Al(PO_4)_2(OH)$. *J. Solid State Chem.* **118**, 412–416.

Aurivillius, K. (1964). The crystal structure of bismuth oxide fluoride. *Acta Chem. Scand.* **18**, 1823–1830.

Aurivillius, B. (1987). Pyrolysis products of $Bi_2(SO_4)_3$. Crystal structures of $Bi_{26}O_{27}(SO_4)_{12}$ and $Bi_{14}O_{16}(SO_4)_5$. *Acta Chem. Scand.* **A41**, 415–422.

Aurivillius, B. (1990). Crystal structures of $(MI_5Cl)(Bi_{48}O_{59}Cl_{30})$, MI = Cu, Ag. *Acta Chem. Scand.* **44**, 111–122.

Aurivillius, B., Lindblom, C. I., and Stenson, P. (1964). The crystal structure of Bi_2GeO_5. *Acta Chem. Scand.* **18**, 1555–1557.

Ayi, A. A., Choudhury, A., Natarajan, S., Neeraj, S., and Rao, C. N. R. (2001). Transformations of low-dimensional zinc phosphates to complex open-framework structures. Part 1: zero-dimensional to one-, two- and three-dimensional structures. *J. Mater. Chem.* **2001**, 1181–1191.

Ayyappan, S., Bu, X., Cheetham, A.K., Natarajan, S., and Rao, C. N. R. (1998). A simple ladder tin phosphate and its layer derivative. *Chem. Commun.* **1998**, 2181–2182.

Bachmann, H. G. (1953). Die Kristallstruktur des Descloizit. *N. Jb. Mineral. Mh.* **1953**, 193–208.

Bacon, G. E. and Titterton, D. H. (1975). Neutron diffraction studies of $CuSO_4(H_2O)_5$ und $CuSO_4(D_2O)_5$. *Z. Kristallogr.* **141**, 330–341.

Baerlocher, Ch., Meier, W. M., and Olson, D. H. (2001). *Atlas of zeolite framework types.* Elsevier, Amsterdam London New York Oxford Paris Shannon Tokyo.

Baggio, R. F., de Benyacar, M. A. R., Perazzo, B. O., and de Perazzo, P. K. (1977). Crystal structure of ferroelectric guanidinium uranyl sulphate. *Acta Crystallogr.* **B33**, 3495–3499.

Bald, L. and Gruehn, R. (1981). Die Kristallstruktur von einem Sulfat-reichen Oxidsulfat des Zinks. *Naturwiss.* **68**, 39.

Balic Zunic, T., Moelo, Y., Loncar, Z., and Micheelsen, H. (1994). Dorallcharite, $Tl_{0.8}K_{0.2}Fe_3(SO_4)_2(OH)_6$, a new member of the jarosite-alunite family. *Eur. J. Mineral.* **6**, 255–263.

Bandurkin, G. A. and Dzhurinskii, B. F. (1998). Polyhedra $[LnO_n]$: coordination numbers, shape, interpolyhedral bonds, Ln-O distances. *Zh. Neorg. Khim.* **43**, 709–711.

Barbier, J. (1988). $LiMgVO_4$: powder neutron refinement and crystal chemistry. *Eur. J. Solid State Inorg. Chem.* **25**, 609–619.

Bars, O., le Marouille, J. Y., and Grandjean, D. (1977). Etude de chromates, molybdates et tungstates hydrates, 1. Etude structurale de $MgMoO_4(H_2O)_5$. *Acta Crystallogr.* **B33**, 1155–1157.

Bartl, H. (1989). Water of crystallization and its hydrogen-bonded cross linking in vivianite $Fe_3(PO_4)_2 \cdot 8(H_2O)$; a neutron diffraction investigation. *Fres. Z. Analyt. Chem.* **333**, 401–403.

Bartl, H. and Rodek, E. (1983). Untersuchung der Kristallstruktur von $Nd_2(SO_4)_3(H_2O)_8$. *Z. Kristallogr.* **162**, 13–14.

Bartl, H., Catti, M., Joswig, W., and Ferraris, G. (1983). Investigation of the crystal structure of newberyite, $MgHPO_4(H_2O)_3$, by single crystal neutron diffraction. *Tscherm. Mineral. Petrogr. Mitt.* **32**, 187–194.

Basso, R., Palenzona, A., and Zefiro, L. (1989). Crystal structure refinement of a Sr-bearing term related to copper vanadates and arsenates of adelite and descloizite groups. *N. Jb. Mineral. Mh.* **1989**, 300–308.

Basso, R., Lucchetti, G., Zefiro, L., and Palenzona, A. (1993). Mozartite, $CaMn(OH)SiO_4$, a new mineral species from the Cerchiara mine, northern Apennines, Italy. *Can. Mineral.* **31**, 331–336.

Basso, R., Lucchetti, G., Zefiro, L., and Palenzona, A. (1994). Vanadomalayaite, $CaVOSiO_4$, a new mineral vanadium analog of titanite and malayaite. *N. Jb. Mineral. Mh.* **1994**, 489–498

Bataille, T., Audebrand, N., Boultif, A., and Louër, D. (2004). Structure determination of thermal decomposition products from laboratory X-ray powder diffraction. *Z. Kristallogr.* **219**, 881–891.

Baud, G., Besse, J. P., Chevalier, R., and Gasperin, M. (1981). Les differentes formes cristallines de Y_3ReO_8: Relations avec la structure fluorine. *J. Solid State Chem.* **38**, 186–191.

Baud, G., Besse, J. P., Chevalier, R., and Gasperin, M. (1983). Synthese et etude structurale de l'oxyde double $Dy_5Re_2O_{12}$. *Mater. Chem. Phys.* **8**, 93–99.

Baur, W. H. (1960). Die Kristallstruktur von $FeSO_4(H_2O)_4$. *Naturwiss.* **47**, 467.

Baur, W. H. (1962). Zur Kristallchemie der Salzhydrate. Die Kristallstrukturen von $MgSO_4(H_2O)_4$ (Leonhardtit) und $FeSO_4(H_2O)_4$ (Rozenit). *Acta Crystallogr.* **15**, 815–826.

Baur, W. H. (1969a). A comparison of the crystal structures of pseudolaueite and laueite. *Am. Mineral.* **54**, 1312–1322.

Baur, W. H. (1969b). The crystal structure of paravauxite, $FeAl_2(PO_4)_2(OH)_2$ $(OH_2)_6(H_2O)_2$. *N. Jb. Mineral. Mh.* **1969**, 430–433.

Baur, W. H. (1974). The geometry of polyhedral distortions. Predictive relationships for the phosphate groups. *Acta Crystallogr.* **B30**, 1195–1215.

Baur, W. H. (1981). Interatomic distance predictions for computer simulation of crystal structures. In: O'Keeffe, M. and Navrotsky, A., eds. *Structure and bonding in crystals*, Vol. 2. Academic Press: New York, pp. 31–52.

Baur, W. and Fischer, R. X. (2000). *Zeolite structure codes from ABW to CZP.* Landolt-Bernstein: Numerical Data and Functional Relationships in Science and Technology. Vol. 14. Subvol. B. Springer, Berlin.

Baur, W. and Fischer, R. X. (2002). *Zeolite-type crystal structures and their chemistry. Framework type codes DAC to LOV.* Landolt-Bernstein: Numerical Data and Functional Relationships in Science and Technology. Vol. 14. Subvol. C. Springer, Berlin.

Baur, W. H. and Rao, B. R. (1967). The crystal structure of metavauxite. *Naturwiss.* **54**, 561–561.

Baur, W. H. and Rolin, J. L. (1972). Salt hydrates. IX. The comparison of the crystal structure of magnesium sulfate pentahydrate with copper sulfate pentahydrate and magnesium chromate pentahydrate. *Acta Crystallogr.* **B28**, 1448–1455.

Bayliss, P. and Atencio, D. (1985). X-ray powder-diffraction data and cell parameters for copiapite–group minerals. *Can. Mineral.* **23**, 53–56.

Bazán, B., Mesa, J. L., Pizarro, J. L., Lezama, L., Arriortua, M. I., and Rojo, T. (2000). A new inorganic-organic hybrid iron(III) arsenate: $(C_2H_{10}N_2)[Fe(HAsO_4)_2(H_2AsO_4)]$ (H_2O). hydrothermal synthesis, crystal structure, and spectroscopic and magnetic properties. *Inorg. Chem.* **39**, 6056–6060.

Bean, A. C., Peper, S. M., and Albrecht-Scmitt, T. E. (2001). Structural relationships, interconversion, and optical properties of the uranyl iodates, $UO_2(IO_3)_2$ and $UO_2(IO_3)_2(H_2O)$: a comparison of reactions under mild and supercritical conditions. *Chem. Mater.* **13**, 1266–1272.

Bean, A. C., Scott, B. L., Albrecht-Schmitt, T. E., and Runde, W. (2003). Structural and spectroscopic trends in actinyl iodates of uranium, neptunium, and plutonium. *Inorg. Chem.* **42**, 5632–5636.

Behm, H. (1983). Hexalead Chloride Triorthoborate Oxide, $Pb_4O(Pb_2(BO_3)_3Cl)$. *Acta Crystallogr.* **C39**, 1317–1319.

Behrens, E. A. and Clearfield, A. (1997). Titanium silicates, $M_3HTi_4O_4(SiO_4)_3 \cdot 4H_2O$ (M = Na⁺, K⁺), with three-dimensional tunnel structures for the selective removal of strontium and cesium from wastewater solutions. *Micropor. Mater.* **11**, 65–75.

Behrens, E. A., Poojary, D. M., and Clearfield, A. (1996). Syntheses, crystal structures, and ion-exchange properties of porous titanosilicates, $HM_3Ti_4O_4(SiO_4)_3 \cdot 4(H_2O)$ (M = H⁺, K⁺, Cs⁺), structural analogues of the mineral pharmacosiderite. *Chem. Mater.* **8**, 1236–1244.

Behrens, E. A., Poojary, D. M., and Clearfield, A. (1998). Syntheses, X-ray powder structures, and preliminary ion-exchange properties of germanium-substituted titanosilicate pharmacosiderites: $HM_3(AO)_4(BO_4)_3 \cdot 4(H_2O)$ (M = K, Rb, Cs; A = Ti, Ge; B = Si, Ge). *Chem. Mater.* **10**, 959–967.

Belkhiria, M. S., Laaribi, S., Ben Hadj Amara, A., and Ben Amara, M. (1998). Structure of $Na_3Fe(PO_4)_2$ from powder X-ray data. *Annal. Chim.* **23**, 117–120.

Belokoneva, E. L. and Mill', B. V. (1994). Crystal structures of β-$NaSbOGeO_4$ and $AgSbOSiO_4$ and migration paths of ions in structures of $KTiOPO_4$ type. *Zh. Neorg. Khim.* **39**, 355–362

Belokoneva, E. L., Yakubovich, O. V., Tsirel'son, V. G., and Urusov, V. S. (1990). The corrected crystal structure and electron structure of the non-linear crystal $KFeFPO_4$ – a structural analog of $KTiOPO_4$. *Izv. AN SSSR, Neorg. Mater.* **26**, 595–601.

Bengtsson, L. and Holmberg, B. (1990). Cationic lead(II) hydroxide complexes in molten alkali-metal nitrate. *J. Chem. Soc., Faraday Trans.* **86**, 351–359.

Benhamada, L., Grandin, A., Borel, M. M., Leclaire, A., and Raveau, B. (1991a). Structure of KVP_2O_7. *Acta Crystallogr.* **C47**, 424–425.

Benhamada, L., Grandin, A., Borel, M.M., Leclaire, A., and Raveau, B. (1991b). Structure of barium vanadium(III) diphosphate. *Acta Crystallogr.* **47**, 2437–2438.

Benhamada, L., Grandin, A., Borel, M. M., Leclaire, A., and Raveau, B. (1991c). $KVPO_5$, an intersecting tunnel structure closely related to the hexagonal tungsten bronze. *Acta Crystallogr.* **C47**, 1138–1141.

Benhamada, L., Grandin, A., Borel, M. M., Leclaire, A., and Raveau, B. (1992a). Synthesis and crystal structure of a new vanadium(IV) phosphate: $NaVPO_5$. *C. R. Hebd. S. Acad. Sci. Ser. 2*, **314**, 585–589.

Benhamada, L., Grandin, A., Borel, M. M., Leclaire, A., Leblanc, M., and Raveau, B. (1992b). $Na_5V_2P_3O_{14} \cdot H_2O$, a tetravalent vanadium phosphate with a layer structure and a pure pyramidal coordination of V(IV). *J. Solid State Chem.* **96**, 390–396.

Berg, R. W. and Thorup, N. (2005). The reaction between ZnO and molten $Na_2S_2O_7$ or $K_2S_2O_7$ forming $Na_2Zn(SO_4)_2$ or $K_2Zn(SO_4)_2$, studied by Raman spectroscopy and X-ray diffraction. *Inorg. Chem.* **44**, 3485–3493.

Berg, R.W., Boghosian, S., Bjerrum, N.J., Fehrmann, R., Krebs, B., Straeter, N., Mortensen, O.S., Papatheodorou, G.N. (1993) Crystal structure and spectroscopic characterization of $CsV(SO_4)_2$. Evidence for an electronic Raman transition. *Inorg. Chem.*, **32**, 4714–4720.

Bergerhoff, G. and Paeslack, J. (1968). Sauerstoff als Koordinationszentrum in Kristallstrukturen. Z. *Kristallogr.* **126**, 112–123.

Berlepsch, P., Armbruster, Th., Brugger, J., Bykova, E. Y., and Kartashov, P. M. (1999). The crystal structure of vergasovaite $Cu_3O[(Mo,S)O_4SO_4]$, and its relation to synthetic $Cu_3O[MoO_4]_2$. *Eur. J. Mineral.* **11**, 101–110.

Bermanec, V. (1994). Centro-symmetric tilasite from Nezilovo, Macedonia: a crystal structure refinement. *N. Jb. Mineral. Mh.* **1994**, 289–294.

Berrigan, R. and Gatehouse, B. M., 1996: $Cu_3Er(SeO_3)_2O_2Cl$, the erbium analogue of francisite. *Acta Crystallogr.*, **C52**, 496–497.

Besse, J. P., Bolte, M., Baud, G., and Chevalier, R. (1976). Structure cristalline d'oxydes doubles de rhenium. I. Sm_3ReO_8. *Acta Crystallogr.* **B32**, 3045–3048.

Bialek, R. and Gramlich, V. (1992). The superstructure of $K_3HGe_7O_{16} \cdot 4H_2O$. Z. *Kristallogr.* **198**, 67–77.

Bino, A. (1980). Triangular triniobium cluster in aqueous solution. *J. Am. Chem. Soc.* **102**, 7990–7991.

Bino, A. (1982). A trinuclear niobium cluster with six bridging sulfates and metal to metal bond order of 2/3. *Inorg. Chem.* **21**, 1917–1920.

Bircsak, Z. and Harrison W. T. A. (1998a). Template cooperation effect leading to the new layered aluminophosphate $CN_3H_6 \cdot Al(HPO_4)_2 \cdot 2H_2O$. *Chem. Mater.* **10**, 3016–3019.

Bircsak, Z. and Harrison, W. T. A. (1998b). $Ba_2(VO_2)(PO_4)(HPO_4) \cdot H_2O$, a new barium vanadium(V) phosphate hydrate containing trigonal bipyramidal VO_5 groups. *J. Solid State Chem.* **140**, 272–277.

Bircsak, Z. and Harrison, W. T. A. (1998c). α-ammonium vanadium hydrogen phosphate, α-$(NH_4)V(HPO_4)_2$. *Acta Crystallogr.* **C54**, 1195–1197.

Bircsak, Z., Hall, A. K., Harrison, W. T. A. (1999). Synthesis, structure, and magnetism of $CN_3H_6 \cdot VO(H_2O)(HPO_4)(H_2PO_4) \cdot H_2O$, a new guanidinium vanadium(IV) phosphate. *J. Solid State Chem.* **142**, 168–173.

Bladh, K. W., Corbett, R. K., and McLean, W. J. (1972). The crystal structure of tilasite. *Am. Mineral.* **57**, 1880–1884.

Blake, A. J., Cooke, P. A., Hubberstey, P., and Sampson, C. L. (2001). Zinc(II) sulfate tetrahydrate. *Acta Crystallogr.* **E57**, 109–111.

Blatov, V. A., Serezhkina, L. B., Serezhkin, V. N., and Trunov, V. K. (1988). Crystal structure of uranyl selenate, $2UO_2SeO_4 \cdot H_2SeO_4 \cdot 8H_2O$. *Koord. Khim.* **14**, 1705–1708.

Blount, A. M. (1974). The crystal structure of crandallite. *Am. Mineral.* **59**, 41–47.

Bliznyuk, N. A. and Borisov, S. V. (1992). Development of methods of geometrical analysis of structures of inorganic compounds. *Zh. Strukt. Khim.* **33**, 145–165.

Boghosian, S., Fehrmann, R., and Nielsen, K. (1994). Synthesis and crystal structure of $Na_3V(SO_4)_3$. Spectroscopic characterization of $Na_3V(SO_4)_3$ and $NaV(SO_4)_2$. *Acta Chem. Scand.* **48**, 724–731.

Boghosian, S., Eriksen, K. M., Fehrmann, R., and Nielsen, K. (1995). Synthesis, crystal structure redetermination and vibrational spectra of β-$VOSO_4$. *Acta Chem. Scand.* **49**, 703–708.

Boje, J. and Müller-Buschbaum, H. (1992). Zur Kenntnis von $Sr_2Bi_3V_3O_{14}$. *Z. Anorg. Allg. Chem.* **618**, 39–42.

Bokii, G. B. and Gorogotskaya, L. I. (1969). Crystal chemical classification of sulfates. *Zh. Strukt. Khim.* **10**, 624–632.

Boldt, K., Engelen, B., and Unterderweide, K. (1997). A new acid selenite: $Mg(HSeO_3)_2$. *Acta Crystallogr.* **53**, 666–668.

Bonaccorsi, E. and Pasero, M. (2003). Crystal structure refinement of sahlinite, $Pb_{14}(AsO_4)_2O_9Cl_4$. *Mineral. Mag.* **67**, 15–21.

Bonhomme, F., Thoma, S. G., and Nenoff, T. M. (2001). A linear DABCO templated fluorogallophosphate: synthesis and structure determination from powder diffraction data of $Ga(PO_4H)_2F \cdot [N_2C_6H_{14}]$. *J. Mater. Chem.* **11**, 2559–2563.

Borel, M. M., Leclaire, A., Chardon, J., Daturi, M., and Raveau, B. (2000). Dimorphism of the vanadium(V) monophosphate $PbVO_2PO_4$: alpha-layered and beta-tunnel structures. *J. Solid State Chem.* **149**, 149–154.

Borene, J. (1970). Structure cristalline de la parabutlerite. *Bull. Soc. Franc. Mineral. Crist.* **93**, 185–189.

Borene, J. and Solery, J.P. (1972). Structure cristalline de sulfates doubles hydrates de Wyrouboff. *Acta Crystallogr.* **B28**, 2687–2694.

Borisov, S. V. and Podberezskaya, N. V. (1984). *Stable cationic frameworks in the structures of fluorides and oxides.* Novosibirsk, Nauka.

Bortun, A. I., Bortun, L. N., Poojary, D. M., Xiang, O., and Clearfield, A. (2000). Synthesis, characterization, and ion exchange behaviour of a framework potassium titanium trisilicate $K_2TiSi_3O_9 \cdot H_2O$ and its protonated phases. *Chem. Mater.* **12**, 294–305.

Borup, F., Berg, R. W., and Nielson, K. (1990). The crystal structure of $K_7Nb(SO_4)_6$ and $K_7Ta(SO_4)_6$. *Acta Chem. Scand.* **44**, 328–331.

Bosman, W. P., Beurskens, P. T., Smits, J. M. M., Behm, H., Mintjens, J., Meisel, W., and Fuggle, J. C. (1986). Structure of an oxonium iron(II) orthophosphate hydrate. *Acta Crystallogr.* **C42**, 525–528.

Bosselet, F., Mentzen, B. F., and Bouix, J. (1985). Etude de la solution $Pb_2O(S_{1-x}W_xO_4)$ et Structure Cristalline de la Variete alpha du Monooxodiplomb(II) Tetraoxotungstate(VI) $Pb_2O(WO_4)$ par diffraction X sur poudres. *Mater. Res. Bull.* **20**, 1329–1337.

Bouchdoug, M., Courtois, A., Gerardin, R., Steinmetz, J., and Gleitzer, C. (1982). Preparation et etude d'un oxyphosphate $Fe_4(PO_4)_2O$. *J. Solid State Chem.* **42**, 149–157.

Boudjada, A. and Guitel, J.C. (1981). Structure cristalline d'un orthoarseniate acide de fer(III) pentahydrate: $Fe(H_2AsO_4)_3(H_2O)_5$. *Acta Crystallogr.* **B37**, 1402–1405.

Boutfessi, A., Boukhari, A., and Holt, E. M. (1996). Lead(II) diiron(III) pyrophosphate and barium diiron(III) pyrophosphate. *Acta Crystallogr.* **C52**, 1594–1597.

Bragg, W. L. (1930). The structure of silicates. *Z. Kristallogr.* **74**, 237–305.

Brandenburg, N. P. and Loopstra, B. O. (1973). Uranyl sulphate hydrate, $UO_2SO_4(H_2O)_{3.5}$. *Cryst. Struct. Commun.* **2**, 243–246.

Brandle, C. D. and Steinfink, H. (1969). The crystal structure of $Eu_4Al_2O_9$. *Inorg. Chem.* **8**, 1320–1324.

Braunbarth, C., Hillhouse, H. W., Nair, S., Tsapatis, M., Burton, A., Lobo, R. F., Jacubinas, R. M., and Kuznicki, S. M. (2000). Structure of strontium ion-exchanged ETS-4 microporous molecular sieves. *Chem. Mater.* **12**, 1857–1865.

Breidenstein, B., Schlüter, J., and Gebhard, G. (1992). On beaverite: new occurrence, chemical data and crystal structure. *N. Jb. Mineral. Mh.* **1992**, 213–220.

Bridson, J. N., Quinlan, S. E., and Tremaine, P. R. (1998). Synthesis and crystal structure of maricite and sodium iron(III) hydroxyphosphate. *Chem. Mater.* **10**, 763–768.

Brixner, L., Calabrese, J., and Chen, H. Y. (1985). Structure and luminescence of Gd_2GeO_5 and Dy_2GeO_5. *J. Less-Common Met.* **110**, 397–410.

Brodalla, D. and Kniep, R. (1980). $(H_3O)(Al_3(H_2PO_4)_6(HPO_4)_2)\cdot(H_2O)_4$: Ein Al-O-P vernetztes Phosphat mit Oxoniumionen enthaltenden Hohlraeumen. *Z. Naturforsch.* **35b**, 403–404.

Brodersen, K., Liehr, G., and Schottner, G. (1985). Kristallstruktur des $Hg_4O_2(NO_3)_2$. *Z. Anorg. Allg. Chem.* **531**, 158–166

Brotherton, P. D., Maslen, E. N., Pryce, M. W., and White, A. H. (1974). Crystal structure of collinsite. *Austr. J. Chem.* **27**, 653–656.

Brown, I. D. (2002). *The chemical bond in inorganic chemistry. The bond valence model.* Oxford University Press, Oxford.

Brugger, J., Meisser, N., Schenk, K., Berlepsch, P., Bonin, M., Armbruster T., Nyfeler, D., and Schmidt, S. (2000). Description and crystal structure of cabalzarite, $Ca(Mg,Al,Fe)_2(AsO_4)_2(H_2O,OH)_2$, a new mineral of the tsumcorite group. *Am. Mineral.* **85**, 1307–1314.

Brugger, J., Armbruster, T., Criddle, A., Berlepsch, P., Graeser, S., and Reeves, S. (2001). Description, crystal structure, and paragenesis of krettnichite, $PbMn^{3+}_2(VO_4)_2(OH)_2$, the Mn^{3+} analogue of mounanaite. *Eur. J. Mineral.* **13**, 145–158.

Brugger, J., Krivovichev, S. V., Kolitsch, U., Meisser, N., Andrut, M., Ansermet, S., and Burns, P. C. (2002). Description and crystal structure of manganolotharmeyerite, $Ca(Mn^{3+},Mg^{2+},)_2(AsO_4)_2(OH,H_2O)_2$, from the Starlera Mn-deposit, Swiss Alps, and a redefinition of lotharmeyerite. *Can. Mineral.* **40**, 1597–1608.

Brugger, J., Burns, P. C., and Meisser, N. (2003). Contribution to the mineralogy of acid drainage of uranium minerals: Marecottite and the zippeite-group. *Am. Mineral.* **88**, 676–685.

Brunel-Laugt, M. and Guitel, J. C. (1977). Structure cristalline de $Cu_5O_2(PO_4)_2$. *Acta Crystallogr.* **B33**, 3465–3468.

Brunel-Laugt, M., Durif, A., and Guitel, J. (1978). Structure cristalline de $Cu_4(PO_4)_2O$. *J. Solid State Chem.* **25**, 39–47.

Bruque, S., Aranda, M. A. G., Losilla, E. R., Olivera-Pastor, P., and Maireles-Torres, P. (1995). Synthesis optimization and crystal structures of layered metal(IV) hydrogen phosphates, α-$M(HPO_4)_2\cdot(H_2O)$ (M = Ti, Sn, Pb). *Inorg. Chem.* **34**, 893–899.

Budantseva, N. A., Andreev, G. B., Fedoseev, A. M., and Antipin, M. Yu. (2003). Neptunium(VI)chromate complex with 1,2-diammoniopropane {$H_3NCH_2CH(NH_3)CH_3$} [$(NpO_2)_2(CrO_4)_3(H_2O)$]·$3H_2O$: synthesis, crystal structure and properties. *Russ. J. Coord. Chem.* **29**, 653–657.

Bukovec, P. and Golic, L. (1975). The salts and double salts of rare earths. II. Crystal structure of cesium bis-sulfato tri-aquo praseodimate(III) monohydrate. *Documenta Chemica Yugoslavica. Vestnik Slovenskega Kemijskega Drustva* **22**, 19–25.

Bukovec, N., Golic, L., and Siftar, J. (1979). The salts and double salts of rare earths. Structural study of dehydration differences between $Cs(Pr(SO_4)_2(H_2O)_3)(H_2O)$ and $Cs(Lu(SO_4)_2(H_2O)_3)(H_2O)$. *Documenta Chemica Yugoslavica. Vestnik Slovenskega Kemijskega Drustva*, **26**, 377–385.

Buerger, M. J., Dollase, W. A., and Garaycochea-Wittke, I. (1967). The structure and composition of the mineral pharmacosiderite. *Z. Kristallogr.* **125**, 92–108.

Burns, P. C., Grice, J. D., and Hawthorne, F. C. (1995). Borate minerals I: polyhedral clusters and fundamental building blocks. *Can. Mineral.* **33**, 1131–1151.

Burns, P. C. (1998). The structure of boltwoodite and implications of solid solution toward sodium boltwoodite. *Can. Mineral.* **36**, 1069–1075.

Burns, P. C. (1999). The crystal chemistry of uranium. *Rev. Mineral.* **38**, 23–90.

Burns, P. C. (2000). A new uranyl phosphate chain in the structure of parsonsite. *Am. Mineral.* **85**, 801–805.

Burns, P. C. (2005). U^{6+} minerals and inorganic compounds: insights into an expanded structural hierarchy of crystal structures. *Can. Mineral.* **43**, 1839–1894.

Burns, P. C. and Hawthorne, F. C. (1994). The crystal structure of humberstonite, a mixed sulfate-nitrate mineral. *Can. Mineral.* **32**, 381–385.

Burns, P. C. and Hayden, L. A. (2002). A uranyl sulfate cluster in $Na_{10}((UO_2)(SO_4)_4)(SO_4)_2$·$3(H_2O)$. *Acta Crystallogr.* **58**, i121–i123.

Burns, P. C., Miller, M. L., and Ewing, R. C. (1996). U^{6+} minerals and inorganic phases: a comparison and hierarchy of structures. *Can. Mineral.* **34**, 845–880.

Burns, P. C., Ewing, R. C., and Hawthorne, F. C. (1997). The crystal chemistry of hexavalent uranium: Polyhedral geometries, bond-valence parameters, and polymerization of polyhedra. *Can. Mineral.* **35**, 1551–1570.

Burns, P. C., Olson, R. A., Finch, R. J., Hanchar, J. M., and Thibault, Y. (2000). $KNa_3(UO_2)_2(Si_4O_{10})_2(H_2O)_4$, a new compound formed during vapor hydration of an actinide-bearing borosilicate waste glass. *J. Nucl. Mater.* **278**, 290–300.

Burns, P. C., Krivovichev, S. V., and Filatov, S. K. (2002). New Cu^{2+} coordination polyhedra in the crystal structure of burnsite, $KCdCu_7O_2(SeO_3)_2Cl_9$. *Can. Mineral.* **40**, 1587–1595.

Burns, P. C., Deely, K. M., Hayden, L. A. (2003). The crystal chemistry of the zippeite group. *Can. Mineral.* **41**, 687–706.

Burns, P. C., Alexopoulos, C. M., Hotchkiss, P. J., and Locock, A. J. (2004). An unprecedented uranyl phosphate framework in the structure of [$(UO_2)_3(PO_4)O(OH)(H_2O)_2$]($H_2O$). *Inorg. Chem.* **43**, 1816–1818.

Bykov, A. B., Chirkin, A. P., Dem'yanets, L. N., Doronin, S. N., Genkina, E. A., Ivanov-Shits, A. K., Kondratyuk, I. P., Maksimov, B. A., Mel'nikov, O. K.,

Muradyan, L. N., Simonov, V. I., and Timofeeva, V. A. (1990). Superionic conductors $Li_3M_2(PO_4)_3$ (M = Fe, Sc, Cr): synthesis, structure and electrophysical properties. *Solid State Ion.* **38**, 31–52.

Cahill, C. L., Krivovichev, S. V., Burns, P. C., Bekenova, G. K., Shabanova, T. A. (2001). The crystal structure of mitryaevaite, $Al_5(PO_4)_2[(P,S)O_3(O,OH)]_2F_2(OH)_2$ $(H_2O)_8 \cdot 6.48H_2O$, determined from microcrystal by means of synchrotron radiation. *Can. Mineral.* **39**, 179–186.

Caminiti, R. and Johansson, G. (1981). A refinement of the crystal structure of the cadmium sulfate $(CdSO_4)_3(H_2O)_8$. *Acta Chem. Scand.* **A35**, 451–455.

Caminiti, R., Marongiu, G., and Paschina, G. (1982a). Oxonium diaquadisulphatoindium(III) dihydrate $(H_3O)^+(In(H_2O)_2(SO_4)_2)^- \cdot 2H_2O$. *Cryst. Struct. Commun.* **11**, 955–958.

Caminiti, R., Marongiu, G., and Paschina, G. (1982b). A comparative X-ray diffraction study of aqueous $MnSO_4$ and crystals of $MnSO_4 \cdot 5(H_2O)$. *Z. Naturforsch.* **B37**, 581–586.

Calvo, C. and Faggiani, R. (1975). α cupric vanadate. *Acta Crystallogr.* **B31**, 603–605.

Cannillo, E., Rossi, G., Ungaretti, L., and Carobbi, S. G. (1968). The crystal structure of macdonaldite. *Atti della Accad. Naz. Lincei, Scienze Fis. Mat. Natur. Rend., Ser 8*, **45**, 399–414.

Cannillo, E., Rossi, G., and Ungaretti, L. (1970). The crystal structure of delhayelite. *Rend. Soc. Ital. Mineral. Petrol.* **26**, 63–75.

Cannillo, E., Rossi, G., and Ungaretti, L. (1973). The crystal structure of elpidite. *Am. Mineral.* **58**, 106–109.

Caro, P. E. (1968). OM_4 tetrahedra and the cationic groups $(MO)^{n+}$ in rare earth oxides and oxysalts. *J. Less-Common Met.* **16**, 367–377.

Carré, D., Guittard, M., Jaulmes, S., Mazurier, A., Palazzi, M., Pardo, M. P., Laurelle, P., and Flahaut J. (1984). Oxysulfides formed by a rare earth and a second metal. II. Shear structures. *J. Solid State Chem.* **55**, 287–292.

Catti, M. and Ferraris, G. (1983). Hydrogen bonding in the crystalline state. Crystal structure of $CaHAsO_4(H_2O)_3$. *Acta Crystallogr.* **29**, 90–96.

Catti, M. and Ferraris, G. (1974). Crystal structure of $Ca_5(HAsO_4)_2(AsO_4)_2(H_2O)_9$ (guerinite). *Acta Crystallogr.* **30**, 1789–1794.

Catti, M. and Franchini-Angela, M. (1976). Hydrogen bonding in the crystalline state. Structure of $Mg_3(NH_4)_2(HPO_4)_4(H_2O)_8$ (hannayite), and crystal–chemical relationships with schertelite and struvite. *Acta Crystallogr.* **B32**, 2842–2848.

Catti, M. and Ivaldi, G. (1981). Mechanism of the reaction $Ca_5H_2(AsO_4)_4(H_2O)_9$ (ferrarisite) – $Ca_5H_2(AsO_4)_4(H_2O)_5$ (dimorph of vladimirite), and structure of the latter phase. *Z. Kristallogr.* **157**, 119–130.

Catti, M. and Ivaldi, G. (1983). On the topotactic dehydration $Ca_3(AsO_4)_2(H_2O)_{11}$ (phaunouxite)–$Ca_3(AsO_4)_2(H_2O)_{10}$ (rauenthalite), and the structure of both minerals. *Acta Crystallogr.* **B39**, 4–10.

Catti, M., Ferraris, G., and Ivaldi, G. (1977). Hydrogen bonding in the crystalline state. Structure of talmessite, $Ca_2(Mg,Co)(AsO_4)_2 \cdot (H_2O)_2$, and crystal chemistry of related minerals. *Bull. Soc. Franc. Mineral. Crystallogr.* **100**, 230–236.

Catti, M., Ferraris, G., and Ivaldi, G. (1979). Refinement of the crystal structure of anapaite, $Ca_2Fe(PO_4)_2(H_2O)_4$. Hydrogen bonding and relationships with the bihydrated phase. *Bull. Mineral.* **102**, 314–318.

Catti, M., Chiari, G., and Ferraris, G. (1980a). The structure of ferrarisite, $Ca_5(HAsO_4)_2(AsO_4)_2(H_2O)_9$. Disorder, hydrogen bonding, and polymorphism with guerinite. *Bull. Mineral.* **103**, 541–546.

Catti, M., Ferraris, G., and Mason, S. A. (1980b). Low-temperature ordering of hydrogen atoms in $CaHPO_4$ (monetite): X-ray and neutron diffraction study at 145K. *Acta Crystallogr.* **B36**, 254–259.

Cavellec, M., Riou, D., and Férey, G. (1994). Oxyfluorinated microporous compounds. XI. Synthesis and crystal structure of ULM-10: the first bidimensional mixed-valence iron fluorophosphate with intercalated ethylenediamine. *J. Solid State Chem.* **112**, 441–447.

Ceshron, F., Ginderow, D., Giraud, R., Pelisson, P., and Pillard, F. (1987). La nickelaustinite, $Ca(Ni,Zn)(AsO_4)(OH)$: Nouvelle espece minerale du district Cobalto-Nickelifere de Bou-Azzer, Maroc. *Can. Mineral.* **25**, 401–407.

Chakrabarti, S., Pati, S. K., Green, M. A., and Natarajan, S. (2003). Synthesis, structure and magnetic properties of a new iron arsenate, $[C_{10}N_4H_{28}][\{FeF(OH)(HAsO_4)\}_4]$, with a layer structure *Eur. J. Inorg. Chem.* **2003**, 3820–3825.

Chakrabarti, S., Pati, S. K., Green, M. A., and Natarajan, S. (2004). Hydrothermal synthesis, structure and magnetic properties of a one-dimensional iron arsenate, $^1_\infty[NH_3(CH_2)_2NH(CH_2)_2NH_3]$ $[Fe_2F_4(HAsO_4)_2]$. *Eur. J. Inorg. Chem.* **2004**, 3846–3851.

Chang, F. M., Jansen, M., and Schmitz, D. (1983). Structure of pentahydrogendioxygen(1+) diaquadisulfatomanganate(III), $(H_5O_2)^+(Mn(H_2O)_2(SO_4)_2)^-$. *Acta Crystallogr.* **C39**, 1497–1498.

Chapman, D. M. and Roe, A. L. (1990). Synthesis, characterization and crystal chemistry of microporous titanium-silicate materials. *Zeolites* **10**, 730–737.

Chavez, A. V., Nenoff, T. M., Hannooman, L., and Harrison, W. T. A. (1999). Tetrahedral-network organo-zincophosphates: syntheses and structures of $(N_2C_6H_{14}) \cdot Zn(HPO_4)_2 \cdot H_2O$, $H_3N(CH_2)_3NH_3 \cdot Zn_2(HPO_4)_3$, and $(N_2C_6H_{14}) \cdot Zn_3(HPO_4)_4$. *J. Solid State Chem.* **147**, 584–591.

Chen, C., Yang, Y.-L., Huang, K.-L., Sun, Z.-H., Wang, W., Yi, Z., Liu, Y.L., and Pang, W.-Q. (2004). Hydrothermal synthesis and intercalation behavior of a layered titanium phosphate $Ti_2(H_2PO_4)(HPO_4)(PO_4)_2 \cdot 0.5C_6N_2H_{16}$, with an extended γ-phase intercalated into organic amine. *Polyhedron* **23**, 3033–3042.

Cheng, C.-Y. and Wang, S.-L. (1992). Hydrothermal synthesis and structural characterization of a new vanadyl(IV) arsenate: $BaVO(AsO_4)(H_2AsO_4) \cdot H_2O$. *Dalton Trans.* **1992**, 2395–2397.

Chern, M. Y., DiSalvo, F. J., Parise, J. B., and Goldstone, J. A. (1992). The structural distorsion of the anti-perovskite nitride Ca_3AsN. *J. Solid State Chem.* **96**, 426–435.

Chibiskova, N. T., Maksimova, N. V., Storozhenko, D. A., Palkina, K. K., Skorikov, V. M., and Shevchuk, V. G. (1984). Structure of crystals of the bisulfate $(Cs_2SO_4)_3Gd_2(SO_4)_3$. *Izv. Akad. Nauk SSSR, Neorg. Mater.* **20**, 1944–1946.

Chippindale, A. M. and Walton, R. I. (1999). $[C_9H_{20}N][Al_2(HPO_4)_2(PO_4)]$: an aluminum phosphate with a new layer topology. *J. Solid State Chem.* **145**, 731–738.

Chippindale, A. M., Powell, A. W., Bull, L. M., Jones, R. H., Cheetham, A. K., Thomas, J. M., and Xu, R. (1992). Synthesis and characterization of two layered aluminophosphates. *J. Solid State Chem.* **96**, 199.

Chippindale, A. M., Brech, S. J., Cowley, A. R., and Simpson, W. M. (1996). Novel pillared layer structure of the organically templated indium phosphate $[In_8(HPO_4)_{14}(H_2O)_6](H_2O)_5(H_3O)(C_3N_2H_5)_3$. *Chem. Mater.* **8**, 2259–2264.

Chippindale, A. M., Cowley, A. R., Huo, Q., Jones, R. H., and Law, A. D. (1997). Synthesis and structure of a new layered aluminium phosphate $[BuNH_3]_3[Al_3P_4O_{16}]$. *Dalton Trans.* **1997**, 2639–2643.

Choisnet, J., Deschanvres, A., and Raveau, B. (1972). Sur de nouveaux germanates et silicates de type benitoite. *J. Solid State Chem.* **4**, 209–218.

Choudhury, A. and Rao, C. N. R. (2002). Hydrothermal synthesis and structure of organically templated one-dimensional and three-dimensional iron fluorophosphate. *J. Solid State Chem.* **43**, 632–642.

Choudhury, A., Natarajan, S., and Rao, C. N. R. (1999). A hybrid open-framework iron phosphate-oxalate with a large unidimensional channel, showing reversible hydration. *Chem. Mater.* **11**, 2316–2318.

Choudhury, A., Neeraj, S., Natarajan, S., and Rao, C. N. R. (2001). Transformations of the low-dimensional zinc phosphates to complex open-framework structures. Part 2: one-dimensional ladder to two- and three-dimensional structures. *J. Mater. Chem.* **11**, 1537–1546.

Choudhury, A., Neeraj, S., Natarajan, S., and Rao, C. N. R. (2002). Transformations of two-dimensional layered zinc phosphates to three-dimensional and one-dimensional structures. *J. Mater. Chem.* **12**, 1044–1052.

Christ, C. L. and Clark, J. R. (1977). A crystal-chemical ciassification of borate structures with emphasis on hydrated borates. *Phys. Chem. Miner.* **2**, 59–87.

Christensen, A. N. (1972). Hydrothermal Preparation and Crystal Structure of $NaHo_4(GeO_4)_2O_2OH$. *Acta Chem. Scand.* **26**, 1955–1960.

Chukanov, N. V., Pekov, I. V., and Khomyakov, A. P. (2002). Recommended nomenclature for labuntsovite group minerals. *Eur. J. Mineral.* **14**, 165–173.

Chukanov, N. V., Pekov, I. V., Zadov, A. E., Voloshin, A. V., Subbotin, V. V., Sorokhtina, N. V., Rastsvetaeva, R. K., and Krivovichev, S. V. (2003). *Minerals of the Labuntsovite group.* Nauka: Moscow (in Russian).

Clark, A. M., Cooper, A. G., Embrey, P. G., and Fejer, E. E. (1986). Waylandite: new data from an occurence in Cornwall, with a note on "agnesite". *Mineral. Mag.* **50**, 731–733.

Clark, L. A., Pluth, J. J., Steele, I., Smith, J. V., and Sutton, S. R. (1997). Crystal structure of austinite, $CaZn(AsO_4)OH$. *Mineral. Mag.* **61**, 677–683.

Clearfield, A. (2001). Structure and ion exchange properties of tunnel type titanium silicates. *Solid State Sci.* **3**, 103–112.

Clearfield, A. and Duax, W.L. (1969). The crystal structure of the ion exchanger zirconium bis(monohydrogen orthoarsenate) monohydrate. *Acta Crystallogr.* **B25**, 2658–2662.

Clearfield, A. and Smith, D. (1969). The crystallography and structure of alpha-zirconium bis(monohydrogen orthophosphate) monohydrate. *Inorg. Chem.* **8**, 431–436.

Clearfield, A. and Troup, J. M. (1973). On the mechanism of ion exchange in crystalline zirconium phosphates. VII. The crystal structure of alpha-zirconium bis(ammonium orthophosphate) monohydrate. *J. Phys. Chem.* **77**, 243–247.

Clearfield, A., Bortun, L. N., and Bortun, A. I. (2000). Alkali metal ion exchange by the framework titanium silicate $M_2Ti_2O_3SiO_4 \cdot nH_2O$ (M = H, Na). *React. Funct. Polymers* **43**, 85–95.

Colin, S., Dupre, B., Venturini, G., Malaman, B., and Gleitzer, C. (1993). Crystal structure and infrared spectrum of the cyclosilicate $Ca_2ZrSi_4O_{12}$. *J. Solid State Chem.* **102**, 242–249.

Colmont, M., Huvé, M., Ketatni, E. M., Abraham, F., and Mentré, O. (2003). Double (*n* = 2) and triple (*n* = 3) $[M_4Bi_{2n-2}O_{2n}]^{x+}$ polycationic ribbons in the new $Bi_{.9}Cd_{.3.72}M_{.1.28}O_3(PO_4)_3$ oxyphosphate (*M* = Co, Cu, Zn). *J. Solid State Chem.* **176**, 221–233.

Colmont, M., Huvé, M., and Mentré, O. (2006). Bi^{3+}/M^{2+} oxyphosphate: a continuous series of polycationic species from the 1D single chain to the 2D planes. Part 2: crystal structure of three original structural types showing a combination of new ribbonlike polycations. *Inorg. Chem.* **45**, 6612–6621.

Cooper, M. A. and Hawthorne, F. C. (1994a). The crystal structure of curetonite, a complex heteropolyhedral sheet mineral. *Am. Miner.* **79**, 545–549.

Cooper, M. A. and Hawthorne, F. C. (1994b). The crystal structure of wherryite, $Pb_7Cu_2(SO_4)_4(SiO_4)_2(OH)_2$, a mixed sulfate-silicate with $(M(TO_4)_2\varphi)$ chains. *Can. Mineral.* **32**, 373–380.

Cooper, M. A. and Hawthorne, F. C. (1994c). The crystal structure of kombatite, $Pb_{14}(VO_4)_2O_9Cl_4$, a complex heteropolyhedral sheet mineral. *Am. Mineral.* **79**, 550–554.

Cooper, M. A. and Hawthorne, F. C. (1995a). The crystal structure of maxwellite. *N. Jb. Mineral. Mh.* **1995**, 97–104.

Cooper, M. A. and Hawthorne, F. C. (1995b). The crystal structure of mottramite, and the nature of Cu-Zn solid solution in the mottramite-descloizite series. *Can. Mineral.* **33**, 1119–1124.

Cooper, M. A. and Hawthorne, F. C. (1995c). The crystal structure of guilleminite, a hydrated Ba-U-Se sheet structure. *Can. Mineral.* **33**, 1103–1109.

Cooper, M. A. and Hawthorne, F. C. (1999). The structure topology of sidpietersite, $Pb^{2+}_4(S^{6+}O_3S^{2-})O_2(OH)_2$, a novel thiosulfate structure. *Can. Mineral.* **37**, 1275–1282.

Cooper, M. A. and Hawthorne, F. C. (2001). Structure topology and hydrogen bonding in mathozite, $Cu[(UO_2)_3(SeO_3)_2O_2](H_2O)_8$, a comparison with guilleminite, $Ba[(UO_2)_3(SeO_3)_2O_2](H_2O)_3$. *Can. Mineral.* **39**, 797–807.

Cooper, M. A., Hawthorne, F. C., Grice, J. D., and Haynes, P. (2003). Anorthominasragrite, $V^{4+}O(SO_4)(H_2O)_5$, a new mineral species from Temple Mountain, Emery County, Utah, U.S.A.: description, crystal structure and hydrogen bonding. *Can. Mineral.* **41**, 959–979.

Cotton, F. A., Diebold, M. P., Llusar, R., and Roth, W. J. (1986). New chemistry of oxo trinuclear, metal-metal bonded niobium compounds. *J. Chem. Soc. Chem. Commun.* **619**, 1276–1278.

Cotton, F. A., Diebold, M. P., and Roth, W. J. (1988). Further studies of bi-oxo-capped triniobium cluster complexes. *Inorg. Chem.* **27**, 2347–2352.

Cowley, A. R. and Chippindale, A. M. (2000). Ambient-temperature syntheses of layered iron(III) phosphates in silica gels. *Dalton Trans.* **2000**, 3425–3428.

Cudennec, Y. and Riou, A. (1977). Etude structurale des chromates doubles d'aluminium et d'alcalin: $NaAl(CrO_4)_2(H_2O)_2$ et $KAl(CrO_4)_2(H_2O)_2$. *C. R. Hebd. Seanc. Acad. Sci., Ser. C*, **284**, 565–568

Cudennec, Y., Riou, A., and Gerault, Y. (1989). Manganese(II) hydrogenphosphate trihydrate. *Acta Crystallogr.* **C45**, 1411–1412.

Curry, N. A. and Jones, D. W. (1971). Crystal structure of brushite, calcium hydrogen orthophosphate dihydrate: A neutron-diffraction investigation. *J. Chem. Soc. A* **1971**, 3725–3729.

Dadachov, M. S. and Harrison, W. T. A. (1997). Synthesis and crystal structure of $Na_4[(TiO)_4(SiO_4)_3] \cdot 6H_2O$, a rhombohedrally distorted sodium titanium silicate pharmacosiderite analogue. *J. Solid State Chem.* **134**, 409–415.

Dadachov, M. S. and Le Bail, A. (1997). Structure of zeolitic $K_2TiSi_3O_9 \cdot H_2O$ determined ab initio from powder diffraction data. *Eur. J. Solid State Inorg. Chem.* **34**, 381–390.

Dahlman, B. (1952). The crystal structures of kroehnkite, $CuNa_2(SO_4)_2(H_2O)_2$ and brandtite, $MnCa_2(AsO_4)_2(H_2O)_2$. *Arkiv Mineral. Geol.* **1**, 339–366.

Dahm, M. and Adam, A. (2002). Darstellung und Kristallstrukturen der ersten Alkalimetall-hexacarbonatooxotetraberyllate: $K_6[Be_4O(CO_3)_6] \cdot 7H_2O$ und $K_6[Be_4O(CO_3)_6]$. *Z. Anorg. Allg. Chem.* **628**, 1861–1867.

Dai, Z., Shi, Z., Li, G., Chen, X., Li, X., Xu, Y., and Feng, S. (2003). Hydrothermal synthesis and structures of two novel vanadium selenites, $\{[VO(OH)(H_2O)](SeO_3)\}_4 \cdot 2H_2O$ and $(H_3NCH_2CH_2NH_3)[(VO)(SeO_3)_2]$. *J. Solid State Chem.* **172**, 205–211.

Damodaran, K. V. and Rao, K. J. (1988). A molecular dynamics investigation of the structure of $PbO-PbF_2$ glasses. *Chem. Phys. Lett.* **148**, 57–61.

Dan, M., Udayakumar, D., and Rao, C. N. R. (2003). Transformation of a 4-membered ring zinc phosphate SBU to a sodalite-related 3-dimensional structure through a linear chain structure. *Chem. Commun.* **2003**, 2212–2213.

Danis, J. A., Runde, W. H., Scott, B., Fettinger, J., and Eichhorn, B. (2001). Hydrothermal synthesis of the first organically templated open-framework uranium phosphate. *Chem. Commun.* **2001**, 2378–2379.

de Bord, J. R. D., Reiff, W. M., Warren, C. J., Haushalter, R. C., and Zubieta, J. (1997). A 3-D organically templated mixed valence (Fe^{2+}/Fe^{3+}) iron phosphate with oxide-centered $Fe_4O(PO_4)_4$ cubes: hydrothermal synthesis, crystal structure, magnetic susceptibility, and Mössbauer spectroscopy of $[H_3NCH_2CH_2NH_3]_2[Fe_4O(PO_4)_4] \cdot H_2O$. *Chem. Mater.* **9**, 1994–1998.

Deacon, G. B., Gatehouse, B. M., Ward, G. N. (1994). $Cu_3Bi_4V_2O_{12}$, a new compound. *Acta Crystallogr.* **C50**, 1178–1180.

Demartin, F., Diella, V., Donzelli, S., Gramaccioli, C. M., and Pilati, T. (1991). The importance of accurate crystal structure determination of uranium minerals. I. Phosphuranylite $KCa(H_3O)_3(UO_2)_7(PO_4)_4O_4 \cdot 8H_2O$. *Acta Crystallogr.* **B47**, 439–446.

Dent Glasser, L. S., Howie, R. A., and Smart, R. M. (1981). The structure of lead 'orthosilicate', $(PbO)_2SiO_2$. *Acta Crystallogr.* **B37**, 303–306.

Delgado-Friedrichs, O., O'Keeffe, M., and Yaghi, O. M. (2003). Regular and quasiregular nets. *Acta Crystallogr.* **A59**, 22–27.

Dhingra, S. S. and Haushalter, R. C. (1994). Synthesis and crystal structure of the octahedral-tetrahedral framework indium phosphate $Cs(In_2(PO_4)(HPO_4)_2(H_2O)_2)$. *J. Solid State Chem.* **112**, 96–99.

Dick, S. (1997). Die Struktur von $GaAsO_4 \cdot 2(H_2O)$: Ein neues Mitglied der Variscit-Familie. *Z. Naturforsch.* **B52**, 1337–1340.

Dick, S. (1999). Über die Struktur von synthetischem Tinsleyit $K(Al_2(PO_4)_2OH)$ $(H_2O)) \cdot (H_2O)$. *Z. Naturforsch.* **B54**, 1385–1390.

Dick, S., Gossner, U., Grossmann, G., Ohms, G., and Zeiske, T. (1997). Aluminiumphosphate mit azentrischen Schicht- und Raumnetzstrukturen aus topologisch verwandten Motiven: 1. $KAl_2(PO_4)_2(OH) \cdot 4(H_2O)$. *Z. Naturforsch.* **52b**, 1439–1446.

Dickens, B., Brown, W., Kruger, G., and Stewart, J. (1973a). Crystal structure of $Ca_4(PO_4)_2O$. *Acta Crystallogr.* **B29**, 2046–2056.

Dickens, B., Prince, E., Schroeder, L. W., and Brown, W. E. (1973b). $Ca(H_2PO_4)_2$, a crystal structure containing unusual hydrogen bonding. *Acta Crystallogr.* **B29**, 2057–2070.

Dimaras, P. I. (1957). Morphology and structure of anhydrous nickel sulphate. *Acta Crystallogr.* **10**, 313–315.

Dinnebier, R. E., Stephens, P. W., Wies, S., and Eysel, W. (1996). Structures and phase transition of $Bi_2CdO_2[GeO_4]$. *J. Solid State Chem.* **123**, 371–377.

Dobley, A., Zavalii, P. Yu., and Whittingham, M. S. (2000). Sodium trivanadium(III) bis(sulfate) hexahydroxide. *Acta Crystallogr.* **C56**, 1294–1295.

Doran, M. B., Norquist, A. J., and O'Hare, D. (2003a). Poly[[1,4-bis-(3-aminopropyl) piperazinium] [[dioxouranium(VI)]-di-μ2, μ3-sulfato]]. *Acta Crystallogr.* **E59**, m762–m764.

Doran, M. B., Norquist, A. J., and O'Hare, D. (2003b). *catena*-Poly[cyclohexane-1, 4-diammonium [[dioxo(sulfato-κ^2-O,O')uranium(VI)]-μ-sulfato] dihydrate]. *Acta Crystallogr.* **E59**, m765–m767.

Doran, M. B., Norquist, A. J., and O'Hare, D. (2003c). *catena*-Poly[tetramethylammonium [[(nitrato-κ^2-O,O')dioxouranium]-μ_3-sulfato]]. *Acta Crystallogr.* **E59**, m373–m375.

Doran, M. B., Norquist, A. J., and O'Hare, D. (2003d). Exploration of composition space in templated uranium sulfates. *Inorg. Chem.* **42**, 6989–6994.

Doran, M. B., Norquist, A. J., Stuart, C. L., and O'Hare, D. (2004). $(C_8H_{26}N_4)_{0.5}$ $[(UO_2)_2(SO_4)_3(H_2O)] \cdot 2H_2O$, an organically templated uranyl sulfate with a novel layer type. *Acta Crystallogr.* **E60**, m996–m998.

Doran, M. B., Norquist, A. J., and O'Hare, D. (2005a). $(C_3H_{12}N_2)_2[UO_2(H_2O)_2(SO_4)_2]_2 \cdot$ $2H_2O$: an organically templated uranium sulfate with a novel dimer type. *Acta Crystallogr.* **E61**, m881–m884.

Doran, M. B., Cockbain, B. E., and O'Hare, D. (2005b). Structural variation in organically templated uranium sulfate fluorides. *Dalton Trans.* **2005**, 1774–1780.

Dumas, E., Taulelle, F., and Férey, G. (2001). Synthesis and crystal structure of the synthetic analogue of mineral minyulite $K[Al_2F(H_2O)_4(PO_4)_2]$. Structural correlations with $AlPO_4$-CJ2. *Solid State Sci.* **3**, 613–621.

Dunn, P. J., Peacor, D. R., Sturman, D. B., Ramik, R. A., Roberts, W. L., and Nelen, J. A. (1986). Johnwalkite, the Mn-analogue of olmsteadite, from South Dakota. *N. Jb. Mineral. Mh.* **1986**, 115–120.

Dunstetter, F., de Noiffontaine, M.-N., and Courtial, M. (2006). Polymorphism of tricalcium silicate, the major compound of Portland cement clinker. 1. Structural data: review and unified analysis. *Cement Concr. Res.* **36**, 39–53.

Dupre, N., Wallez, G., Gaubicher, J., and Quarton, M. (2004). Phase transition induced by lithium insertion in (alpha I)- and (alpha II)-$(VOPO_4)$. *J. Solid State Chem.* **177**, 28096–2902.

Durif, A. (1995). *Crystal chemistry of condensed phosphates.* Plenum Press, New York and London.

Dzhurinskii, B. F., Tananaev, I. V., Tselebrovskaya, E. G., and Prozorovskii, A. I. (1991). Lanthanoid germanophosphates. *Izv. Akad. Nauk SSSR, Neorg. Mater.* **27**, 334–339.

Eddaoudi, M., Moler, D. B., Li, H., Chen, B., Reineke, T. M., O'Keeffe, M., and Yaghi, O. M. (2001). Modular chemistry: secondary building units as a basis for the design of highly porous and robust metal-organic carboxylate frameworks. *Acc. Chem. Res.* **34**, 319–330.

Effenberger, H. (1985a). $Cu(SeO_2OH)_2$: synthesis and crystal structure. *Z. Kristallogr.* **173**, 267–272.

Effenberger, H. (1985b). $Cu_2O(SO_4)$, dolerophanite: refinement of the crystal structure with a comparison of $OCu(II)_4$ tetrahedra in inorganic compounds. *Monatsh. Chem.* **116**, 927–931.

Effenberger, H. (1986). $PbCu_3(OH)(NO_3)(SeO_3)_3(H_2O)_5$ und $Pb_2Cu_3O_2(NO_3)_2$ $(SeO_3)_2$. Synthese und Kristallstrukturuntersuchung. *Monatsh. Chem.* **117**, 1099–1106.

Effenberger, H. (1987). Crystal structure and chemical formula of schmiederite, $Pb_2Cu_2(OH)_4(SeO_3)(SeO_4)$, with a comparison to linarite $PbCu(OH)_2(SO_4)$. *Mineral. Petrol.* **36**, 3–12.

Effenberger, H. (1992). Structure refinement of $Co_3(OH)_2(PO_3OH)_2$ and $Co(PO_2(OH)_2)_2 \cdot 2H_2O$. *Acta Crystallogr.* **C48**, 2104–2107.

Effenberger, H. (1998). The Bi(III)-selenite $(Bi_2O)Cu(SeO_3)_3 \cdot (H_2O)$. *J. Alloys Compd.* **281**, 152–156.

Effenberger, H. and Langhof, H. (1984). Die Kristallstruktur von Dikalium-tricobalt(II)-dihydroxid trisulfat-dihydrat, $K_2Co_3(OH)_2(SO_4)_3(H_2O)_2$. *Monatsch. Chem.* **115**, 165–177.

Effenberger, H. and Miletich, R. (1995). $Na_2[Bi_2Cu_3O_4(AsO_4)_2] \cdot H_2O$ and $K_2[Bi_2Cu_3O_4$ $(AsO_4)_2] \cdot 2H_2O$: two related crystal structures with topologically identical layers. *Z. Kristallogr.* **210**, 421–426.

Effenberger, H. and Zemann, J. (1984). The crystal structure of caratiite. *Mineral. Mag.* **48**, 541–546.

Effenberger, H., Pertlik, F., and Zemann, J. (1986). Refinement of the crystal structure of krausite: a mineral with an interpolyhedral oxygen–oxygen contact shorter than the hydrogen bond. *Am. Mineral.* **71**, 202–205.

Effenberger, H., Hejny, C., and Pertlik, F. (1996). The crystal structure of $PbFe(AsO_4)$ $(AsO_3(OH))$. *Monatsh. Chem.* **127**, 127–133.

Efremov, V. A., Tyulin, A. V., and Trunov, V. K. (1987). Actual structure of tetragonal $Ln_2O_2MoO_4$ and factors, determining the forming structure of the coordination polyhedra. *Koord. Khim.* **13**, 1276–1282.

Ehrenberg, H., Hartmann, T., Witschek, G., Fuess, H., Morgenroth, W., and Krane, H.-G. (1999). The crystal structure of $Tm_5Re_2O_{12}$. *Acta Crystallogr.* **B55**, 849–852

Eick, H. A. and Kihlborg, L. (1966). The crystal structure of $VOMoO_4$. *Acta Chem. Scand.* **20**, 722–729.

Ekambaram, S. and Sevov, S. C. (2000). Synthesis and characterization of four ethylenediamine-templated iron arsenates. *Inorg. Chem.* **39**, 2405–2410.

Engelen, B., Boldt, K., Unterderweide, K., and Baeumer, U. (1995). Zur Kenntnis der Hydrate $M(HSeO_3)_2 \cdot 4(H_2O)$ (M = Mg,Co,Ni,Zn). Roentgenstrukturanalytische, schwingungsspektroskopische und thermoanalytische Untersuchungen. *Z. Anorg. Allg. Chem.* **621**, 331–339.

Eriksen, K. M., Nielsen, K., and Fehrmann, R. (1996). Crystal structure and spectroscopic characterization of $K_6(VO)_4(SO_4)_8$ containing mixed-valent vanadium(IV)-vanadium(V). *Inorg. Chem.* **35**, 480–484.

Etheredge, K. M. S. and Hwu, S.-J. (1995). A novel copper (I/II) oxophosphate chloride with a quasi-one-dimensional μ_4-oxo-bridged copper(II) chain. Crystal structure and magnetic properties of $[Na_2Cu_3^{II}(PO_4)_2][Cu^IOCl]$. *Inorg. Chem.* **35**, 5278–5282.

Evans, H. T. Jr. (1973). The crystal structures of cavansite and pentagonite. *Am. Mineral.* **58**, 412–424.

Eysel, W. and Breuer, K. H. (1983). Crystal chemistry of compounds $M_3O(TO_4)$. *Z. Kristallogr.* **163**, 1–17.

Fan, J., Slebodnick, C., Troya, D., Angel, R., and Hanson, B. E. (2005). Chiral layered zincophosphate $[d\text{-}Co(en)_3]Zn_3(H_{0.5}PO_4)_2(HPO_4)_2$ assembled about $d\text{-}Co(en)_3^{3+}$ complex cations. *Inorg. Chem.*, **44**, 2719–2727.

Fanfani, L. and Zanazzi, P. F. (1967). Structural similarities of some secondary lead minerals. *Mineral. Mag.* **36**, 522–529.

Fanfani, L., Nunzi, A., and Zanazzi, P. F. (1970a). The crystal structure of fairfieldite. *Acta Crystallogr.* **B26**, 640–645.

Fanfani, L., Nunzi, A., and Zanazzi, P. F. (1970b). The crystal structure of roemerite. *Am. Mineral.* **55**, 78–89.

Fanfani, L., Nunzi, A., and Zanazzi, P. F. (1971). The crystal structure of butlerite. *Am. Mineral.* **56**, 751–757.

Fanfani, L., Nunzi, A., Zanazzi, P. F., and Zanzari, A. R. (1973). The copiapite problem: the crystal structure of a ferrian copiapite. *Am. Mineral.* **58**, 314–322.

Fanfani, L., Nunzi, A., Zanazzi, P. F., and Zanzari, A. R. (1976). Additional data on the crystal structure of montgomeryite. *Am. Mineral.* **61**, 12–14.

Fanfani, L., Tomassini, M., Zanazzi, P. F., and Zanzari, A. R. (1978). The crystal structure of strunzite, a contribution to the crystal chemistry of basic ferric-manganous hydrated phosphates. *Tscherm. Mineral. Petrogr. Mitt.* **25**, 77–87.

Fang, J. H. and Robinson, P. D. (1970a). Crystal structures and mineral chemistry of hydrated ferric sulfates. I. The crystal structure of coquimbite. *Am. Mineral.* **55**, 1534–1540.

Fang, J. H. and Robinson, P. D. (1970b). Crystal structures and mineral chemistry of double-salt hydrates: II. The crystal structure of loeweite. *Am. Mineral.* **55**, 378–386.

Fang, J. H. and Robinson, P. D. (1972). Crystal structures and mineral chemistry of double-salt hydrates: II. The crystal structure of mendozite, $NaAl(SO_4)_2(H_2O)_{11}$. *Am. Mineral.* **57**, 1081–1088.

Fayos, J., Glasser, F. P., Howie, R. A., Lachowski, E. E., and Perez-Mendez, M. (1985). Structure of dodecacalcium potassium fluoride dioxide tetrasilicate bis(sulfate), $KF(Ca_6(SO_4)(SiO_4)_2O)_2$, a fluorine-containing phase encountered in cement clinker. *Acta Crystallogr.* **C41**, 814–816.

Fedoseev, A. M., Budantseva, N. A., Grigoriev, M. S., Bessonov, A. A., Astafurova, L. N., Lapitskaya, T. S., and Krupa, J. C. (1999). Sulfate compounds of hexavalent neptunium and plutonium. *Radiochim. Acta* **86**, 17–22.

Fedoseev, A. M., Budantseva, N. A., Grigor'ev, M. S., Guerman, K. E., and Krupa, J.-C. (2003). Synthesis and properties of neptunium(VI, V) and plutonium(VI) pertechnetates. *Radiochim. Acta* **91**, 147–152.

Fedorov, E. S. (1885). An introduction to the theory of figures. *Zap. Imper. Sank-Peterb. Mineral. Obshch.* **28**, 1–146 (in Russian). English translation: ACA Monograph number 7, Amer. Crystallogr. Assoc., 1971.

Fehrmann, R., Krebs, B., Papatheodorou, G.N., Berg, R.W., Bjerrum, N.J. (1986) Crystal structure and infrared Raman spectra of $KV(SO_4)_2$. *Inorg. Chem.*, **25**, 1571–1577.

Fejdi, P., Poullen, J. F., and Gasperin, M. (1980). Affinement de la structure de la vivianite $Fe_3(PO_4)_2(H_2O)_8$. *Bull. Mineral.* **103**, 135–138.

Fehrmann, R., Boghosian, S., Papatheodorou, G.N., Nielsen, K., Willestofte B.R., Janniksen B.N. (1991) The crystal structure of $NaV(SO_4)_2$. *Acta Chem. Scand.*, **45**, 961–964

Feng, P., Bu, X., and Stucky, G. D. (2000). Control of structural ordering in crystalline lamellar aluminophosphates with periodicity from 51 to 62 Å. *Inorg. Chem.* **39**, 2–3.

Férey, G. (1995). Oxyfluorinated microporous compounds ULM-*n*: chemical parameters, structures and a proposed mechanism for their molecular tectonics. *J. Fluor. Chem.* **72**, 187–193.

Férey, G. (1998). The new microporous compounds and their design. *C. R. Acad. Sci. Ser. IIc* **1**, 1–13.

Férey, G. (2001). Microporous solids: from organically templated inorganic skeletons to hybrid frameworks. Ecumenism in chemistry. *Chem. Mater.* **13**, 3084–3098.

Fernandez, S., Mesa, J., Pizarro, J., Lezama, L., Arriortua, M., and Rojo T. (2003). Hydrothermal synthesis, crystal structures, spectroscopic, and magnetic properties of two new organically templated monodimensional phosphite compounds: $(C_2H_{10}N_2)[M(HPO_3)F_3]$, M = V(III) and Cr(III). *Chem. Mater.* **15**, 1204–1209.

Fernandez-Armas, S., Mesa, J. L., Pizarro, J. L., Lezama, L., Arriortua, M. I., and Rojoa, T. (2004). A new organically templated gallium(III)-doped chromium(III) fluorophosphite, $(C_2H_{10}N_2)[Ga_{0.98}Cr_{0.02}(HPO_3)F_3]$ hydrothermal synthesis, crystal structure and spectroscopic properties. *J. Solid State Chem.* **177**, 765–771.

Ferraris, G. and Abbona, F. (1972). The crystal structure of $Ca_5(HAsO_4)_2(AsO_4)_2(H_2O)_4$ (sainfeldite). *Bull. Soc. Franc. Mineral. Cristallogr.* **95**, 33–41.

Ferraris, G. and Chiari, G. (1970). The crystal structure of $CaHAsO_4$ (weilite). *Acta Crystallogr.* **26**, 403–409.

Ferraris, G. and Ivaldi, G. (1984). X-OH and O-H$\cdots$O bond lengths in protonated oxoanions. *Acta Crystallogr.* **B40**, 1–6.

Ferraris, G., Jones, D. W., and Yerkess, J. (1971). Determination of hydrogen atom positions in the crystal structure of pharmacolite, $CaHAsO_4(H_2O)_2$, by neutron diffraction. *Acta Crystallogr.* **27**, 349–354.

Ferraris, G., Jones, D. W., and Yerkess, J. (1972a). A neutron diffraction study of the crystal structure of calcium bis(dihydrogen arsenate), $Ca(H_2AsO_4)_2$. *Acta Crystallogr.* **B28**, 2430–2437.

Ferraris, G., Jones, D. W., and Yerkess, J. (1972b). A neutron and an X-ray refinement of the crystal structure of $CaHAsO_4(H_2O)$ (haidingerite). *Acta Crystallogr.* **28**, 209–214.

Ferraris, G., Makovicky, E., and Merlino, S. (2004). *Crystallography of modular materials.* Oxford University Press, Oxford.

Figulla-Kroschel, C. and Jansen, M. (2000). Darstellung, Kristallstrukturen und Eigenschaften von $Ln_4Au_2O_9$ (Ln = Nd, Sm, Eu). *Z. Anorg. Allg. Chem.* **626**, 2178–2184.

Filatov, S. K., Semenova, T. F., and Vergasova, L. P. (1992). Types of polymerization of $[OCu_4]^{6+}$ tetrahedra in compounds with 'additional' oxygen atoms. *Dokl. AN SSSR* **322**, 536–539.

Filipenko, O. S., Leonova, L. S., Atovmyan, L. O., and Shilov, G. V. (1998). Crystal structure and protonic conductivity of a new monoclinic modification of $Ce(H_2O)_4(SO_4)_2$. *Dokl. Ross. Akad. Nauk* **360**, 73–76.

Finger, L. W. (1985). Fingerite $Cu_{11}O_2(VO_4)_6$: a new vanadium sublimate from Izalco Volcano, El Salvador: crystal structure. *Am. Mineral.* **70**, 197–199.

Finger, L. W., Hazen, R. M., and Fursenko, B. A. (1995). Refinement of the crystal structure of $BaSi_4O_9$ in the benitoite form. *J. Phys. Chem. Solids* **56**, 1389–1393.

Finney, J. J., Graeber, E. J., Rosenzweig, A., and Hamilton, R. D. (1977). The structure of chloroxiphite, $Pb_3CuO_2(OH)_2Cl_2$. *Mineral. Mag.* **41**, 357–361.

Fischer, K. (1969). Verfeinerung der Kristallstruktur von Benitoit BaTi(Si$_3$O$_9$). *Z. Kristallogr.* **129**, 222–243.

Fischer, W. and Hellner, E. (1964). Über die Struktur des Vanthoffits. *Acta Crystallogr.* **17**, 1613–1613.

Fischer, T., Eisenmann, B., and Kniep, R. (1996a). Crystal structure of dialuminium tris(sulfate) pentahydrate. *Z. Kristallogr.* **211**, 471–472.

Fischer, T., Eisenmann, B., and Kniep, R. (1996b). Crystal structure of oxonium bis(sulfato)aluminate, H$_3$O(Al(SO$_4$)$_2$). *Z. Kristallogr.* **211**, 465–465.

Fischer, T., Eisenmann, B., and Kniep, R. (1996c). Crystal structure of aquaoxonium diaquabis(sulfato)aluminate, H$_5$O$_2$(Al(H$_2$O)$_2$(SO$_4$)$_2$). *Z. Kristallogr.* **211**, 466–466.

Fischer, T., Eisenmann, B., and Kniep, R. (1996d). Crystal structure of dialuminium tris(sulfate) 10.5 hydrate, (Al$_2$(SO$_4$)$_3$(H$_2$O)$_{10.5}$. *Z. Kristallogr.* **211**, 475–476.

Fischer, T., Eisenmann, B., and Kniep, R. (1996e). Crystal structure of trialuminium monohydrogensulfate tetrakis(sulfate) nonahydrate, Al$_3$(HSO$_4$)(SO$_4$)$_4 \cdot$9H$_2$O. *Z. Kristallogr.* **211**, 467–468.

Fischer, T., Eisenmann, B., and Kniep, R. (1996f). Crystal structure of dialuminium tris(sulfate) octahydrate, Al$_2$(SO$_4$)$_3$(H$_2$O)$_8$. *Z. Kristallogr.* **211**, 473–474.

Fischer, T., Kniep, R., and Wunderlich, H. (1996g). Crystal structure of dioxonium bis(tris(sulfato)dialuminate) sulfate docosahydrate, (H$_3$O)$_2 \cdot$(Al$_2$(SO$_4$)$_3$)$_2 \cdot$(SO$_4$)$\cdot$22H$_2$O. *Z. Kristallogr.* **211**, 469–470.

Fleck, M. and Giester, G. (2003). Crystal structures of the new double salts K$_2$Cd$_3$(SO$_4$)$_4 \cdot$5(H$_2$O), Rb$_2$Cu$_3$(SO$_4$)$_3$(OH)$_2$ and Cs$_2$Cu(SeO$_4$)$_2 \cdot$4(H$_2$O). *J. Alloys Compd.* **351**, 77–83.

Fleck, M. and Kolitsch, U. (2003). Natural and synthetic compounds with kröhnkite-type chains. An update. *Z. Kristallogr.* **218**, 553–567.

Fleck, M., Kolitsch, U., Hertweck, B. (2002). Natural and synthetic compounds with kröhnkite-type chains: review and classification. *Z. Kristallogr.* **217**, 435–443.

Fleet, S. G. (1965). The crystal structure of dalyite. *Z. Kristallogr.* **121**, 349–368.

Forbes, T. Z. and Burns, P. C. (2005). Structures and syntheses of four Np^{5+} sulfate chain structures: divergence from U^{6+} crystal chemistry. *J. Solid State Chem.* **178**, 3455–3462.

Francis, R. J., Drewitt, M. J., Halasyamani, P. S., Ranganathachar, C., O'Hare, D., Clegg, W., and Teat, S. J. (1998). Organically templated layered uranium(vi) phosphates: hydrothermal syntheses and structures of [NHEt$_3$][(UO$_2$)$_2$(PO$_4$)(HPO$_4$)] and [NPr$_4$][(UO$_2$)$_3$(PO$_4$)(HPO$_4$)$_2$]. *Chem. Commun.* **1998**, 279–280.

Franke, W. A., Luger, P., Weber, M., and Ivanova, T. I. (1997). Low hydrothermal growth of sincosite Ca(VOPO$_4$)$_2 \cdot$4(H$_2$O). *Zapiski VMO* **126**, 85–86.

Frit, B., Holmberg, B., and Galy, J. (1970). Structure cristalline de l'oxychlorure de barium, Ba$_4$OCl$_6$. *Acta Crystallogr.* **B26**, 16–19.

Fu, Y., Xu, Z., Ren, J., Wu, H., and Yuan R. (2006). Organically directed iron sulfate chains: structural diversity based on hydrogen bonding interactions. *Inorg. Chem.* **46**, 8452–8458.

Fuess, H. and Will, G. (1968). Bestimmung der Kristallstruktur der Selenate $MSeO_4$ (M = Mn, Co, Ni) durch Roentgen- und Neutronenbeugung. *Z. Anorg. Allg. Chem.* **358**, 125–137.

Gaebler, F. and Niewa, R. (2007). Stacking design in inverse perovskites: the system $(Sr_{3-x}Ba_xN)E$, E = Bi, Sb. *Inorg. Chem.* **46**, 859–865.

Gaebler, F., Kirchner, M., Schnelle, W., Schwarz, U., Schmitt, M., Rosner, H., and Niewa, R. (2004). $(Sr_3N)E$ and $(Ba_3N)E$ (E = Sb, Bi): synthesis, crystal structures, and physical properties. *Z. Anorg. Allg. Chem.* **630**, 2292–2298.

Gaebler, F., Kirchner, M., Schnelle, W., Schmitt, M., Rosner, H., and Niewa, R. (2005). $(Sr_3N_x)E$ and $(Ba_3N_x)E$ (E = Sn, Pb): preparation, crystal structures, physical properties and electronic structures. *Z. Anorg. Allg. Chem.* **631**, 397–402.

Gaebler, F., Prots, Yu., and Niewa, R. (2007). First observation of an inverse Ruddlesden-Popper series: $(A_{3n+1}ON_{n-1})Bi_{n+1}$ with A = Sr, Ba and n = 1, 3. *Z. Anorg. Allg. Chem.* **633**, 93–97.

Gaines, R. V., Skinner, H. C. W., Foord, E. E., Mason, B., Rosenzweig, A., King, V. T., and Dowty, E. (1997). *Dana's new mineralogy.* 8th edition. Wiley, New York.

Galliski, M. A., Hawthorne, F. C., and Cooper, M. A. (2002). Refinement of the crystal structure of ushkovite from Nevados de Palermo, Republica Argentina. *Can. Miner.* **40**, 929–937.

Galy, J., Bonnet, J. J., and Andersson, S. (1979). The crystal structure of a new oxide chloride of copper(II) and selenium(IV). $Cu_5Se_2O_8Cl_2$. *Acta Chem. Scand.* **A33**, 383–389.

Gao, Q., Li, B., Chen, J., Li, S., Xu, R., Williams, I. D., Zheng, J., and Barber, D. (1997). Nonaqueous synthesis and characterization of a new 2-dimensional layered aluminophosphate $[Al_3P_4O_{16}]_3[CH_3CH_2NH_3]$. *J. Solid State Chem.*, **129**, 37–44.

Garcia-Moreno, O., Alvarez-Vega, M., Garcia-Alvarado, F., Garcia-Jaca, J., Gallardo Amores, J. M., Sanjuan, M. L., and Amador, U. (2001). Influence of the structure on the electrochemical performance of lithium transition metal phosphates as cathodic materials in rechargeable lithium batteries: a new high-pressure form of $LiMPO_4$ (M = Fe and Ni). *J. Mater. Chem.* **13**, 1570–1576.

Gasanov, Yu. M., Iskhakova, L. D., and Trunov, V.K. (1985). The preparation and crystal structure of $RbNd(SeO_4)_2(H_2O)_3$. *Russ. J. Inorg. Chem.* **30**, 1731–1734.

Gasanov, Yu. M., Proskuryakova, E. V., and Efremov, V. A. (1990). Crystal structure of $Cs_3Nd(CrO_4)_3$. *Kristallogr.* **35**, 1275–1276.

Gasparik, T., Parise, J. B., Eiben, B. A., and Hriljac, J. A. (1995). Stability and structure of a new high pressure silicate, $Na_{1.8}Ca_{1.1}Si_6O_{14}$. *Am. Mineral.* **80**, 1269–1276.

Gastaldi, L., Carre, D., and Pardo, M. P. (1982). Structure de l'oxysulfure d'indium et de lanthane $In_6La_{10}O_6S_{17}$. *Acta Crystallogr.* **B38**, 2365–2367.

Gebert Sherry, E. (1976). The structure of $Pr_2(SO_4)_3(H_2O)_8$ and $La_2(SO_4)_3(H_2O)_9$. *J. Solid State Chem.* **19**, 271–279.

Gebert, W., Medenbach, O., and Floerke, O. W. (1983). Darstellung und Kristallographie von $K_2TiSi_6O_{15}$ isotyp mit Dalyit $K_2ZrSi_6O_{15}$. *Tscherm. Mineral. Petrogr. Mitt.* **31**, 69–79.

Genkina, E. A., Mel'nikov, O. K., Gorbunov, Yu. A., Maksimov, B. A., Kreuer, K. D., and Rabenau, A. (1987). Synthesis and structure of a new Na, Zr phosphate. *Kristallogr.* **32**, 1137–1142.

Geb, J. and Jansen, M. (1996). Bi_2AuO_5 and $Bi_4Au_2O_9$, two novel ternary oxoaurates. *J. Solid State Chem.* **122**, 364–370.

Gesing, T. M. and Rüscher, C. H. (2000). Structure and properties of $UO_2(H_2AsO_4)_2 \cdot H_2O$. *Z. Anorg. Allg. Chem.* **626**, 1414–1420.

Gesing, T. M., Uecker, R., and Buhl, J.-C. (1999). Crystal structure of praseodymium gallate, $Pr_4Ga_2O_9$. *Z. Kristallogr. NCS* **214**, 431.

Ghose, S. and Thakur, P. (1985). The crystal structure of georgechaoite $NaKZr(Si_3O_9)$ $(H_2O)_2$. *Can. Mineral.* **23**, 5–10.

Ghose, S. and Wan, C. (1978). Zektzerite, $NaLiZrSi_6O_{15}$: a silicate with six-tetrahedral-repeat double chains. *Am. Mineral.* **63**, 304–310.

Ghose, S., Wan, C., and Chao, G. Y. (1980). Petarasite, $Na_5Zr_2Si_6O_{18}(Cl,OH) \cdot 2H_2O$, a zeolite-type zirconosilicate. *Can. Mineral.* **18**, 503–509.

Ghose, S., Sen Gupta, P. K., and Campana, C. F. (1987). Symmetry and crystal structure of montregianite $Na_4K_2Y_2Si_{16}O_{38}(H_2O)_{10}$, a double-sheet silicate with zeolitic properties. *Am. Mineral.* **72**, 365–374.

Giacovazzo, C., Menchetti, S., and Scordari, F. (1973). The crystal structure of caledonite, $CuPb_5(SO_4)_3CO_3(OH)_6$. *Acta Crystallogr.* **B29**, 1986–1990.

Giacovazzo, C., Scandale, E., and Scordari, F. (1976a). The crystal structure of chlorothionite $CuK_2Cl_2SO_4$. *Z. Kristallogr.* **144**, 226–237.

Giacovazzo, C., Scordari, F., Todisco, A., and Menchetti, S. (1976b). Crystal structure model for metavoltine from Sierra Gorda. *Tscherm. Mineral. Petrogr. Mitt.* **23**, 155–166.

Giester, G. (1993a). Crystal structure of $KFe(SeO_3)_2$. *Z. Kristallogr.* **207**, 1–7.

Giester, G. (1993b). The crystal structure of $Fe_2(SeO_3)_3 \cdot H_2O$. *J. Solid State Chem.* **103**, 451–457.

Giester, G. (1993c) Crystal structure of the yavapaiite type compound $NaFe(SeO_4)_2$. Mineralogy and Petrology 48, 227–233.

Giester, G. (1994). Crystal structure of $LiFe(SeO_3)_2$. *Monatsh. Chem.* **125**, 535–538.

Giester, G. (1995) Crystal structure of $KMn^{3+}(SeO_4)_2$ – a triclinic distorted member of the yavapaiite family. Mineralogy and Petrology 53, 165–171.

Giester, G. (1996). Synthesis and crystal structures of $Ca_3Fe_2(SeO_3)_6$ and $Sr_3Fe_2(SeO_3)_6$. *Z. Anorg. Allg. Chem.* **622**, 1788–1792.

Giester, G. (2000). Syntheses and crystal structures of the new compounds $BaFe_2(SeO_3)_4$, $AgFe(SeO_3)_2$ and $RbFe(SeO_4)(SeO_3)$. *J. Alloys Compd.* **308**, 71–76.

Giester, G. and Miletich, R. (1995). Crystal structure and thermal decomposition of the coquimbite-type compound $Fe_2(SeO_4)_3 \cdot 9(H_2O)$. *N. Jb. Mineral. Mh.* **1995**, 211–223.

Giester, G. and Rieck, B. (1995). Mereiterite, $K_2Fe(SO_4)_2 \cdot 4(H_2O)$, a new leonite-type mineral from the Lavrion Mining District, Greece. *Eur. J. Mineral.* **7**, 559–566.

Giester, G. and Wildner, M. (1991). Synthesis and crystal structure of monoclinic $Fe_2(SeO_4)_3$. *Monatsh. Chem.* **122**, 617–623.

Giester, G. and Wildner, M. (1992). The crystal structures of kieserite-type compounds. II. Crystal structures of $Me(II)SeO_4 \cdot H_2O$ (Me = Mg, Mn, Co, Ni, Zn). *N. Jb. Mineral. Mh.* **1992**, 135–144.

Giester, G. and Wildner, M. (1996a). Crystal structures and structural relationships of $KFe_2(SeO_2OH)(SeO_3)_3$ and $SrCo_2(SeO_2OH)_2(SeO_3)_2$. *J. Alloys Compd.* **240**, 25–32.

Giester, G. and Wildner, M. (1996b). Crystal structures of the new pseudo-isotypic compounds $NaFe(SeO_3)_2$ and $BaCo(SeO_3)_2$. *J. Alloys Compd.* **239**, 99–102.

Giester, G. and Zemann, J. (1987). The crystal structure of the natrochalcite-type compounds $(Me^+)Cu_2(OH)(ZO_4)_2H_2O$ (Me^+ = Na,K,Rb; Z = S,Se), with special reference to the hydrogen bonds. *Z. Kristallogr.* **179**, 431–442.

Giester, G., Lengauer, C. L., Redhammer, G. J. (1994). Characterization of the $FeSO_4 \cdot (H_2O)$ – $CuSO_4 \cdot (H_2O)$ solid-solution series, and the nature of poitevinite, $(Cu,Fe)SO_4 \cdot (H_2O)$. *Can. Mineral.* **32**, 873–884.

Giester, G., Mikenda, W., and Pertlik, F. (1996). Kleinite from Terlingua, Brewster County, Texas: investigations by single crystal X-ray diffraction, and vibrational spectroscopy. *N. Jb. Mineral. Mh.* **1996**, 49–56.

Ginderow, D. (1988). Structure de l'uranophane alpha, $Ca(UO_2)_2(SiO_3OH)_2(H_2O)_5$. *Acta Crystallogr.* **44**, 421–424.

Ginderow, D. and Cesbron, F. (1983). Structure de la demesmaekerite, $Pb_2Cu_5(SeO_3)_6(UO_2)_2(OH)_6 \cdot 2H_2O$. *Acta Crystallogr.* **C39**, 824–827.

Giuseppetti, G. and Tadini, C. (1973). Riesame della struttura cristallina della nadorite: $PbSbO_2Cl$. *Per. Mineral.* **42**, 335–345

Giuseppetti, G. and Tadini, C. (1980). The crystal structure of osarizawaite. *N. Jb. Mineral. Mh.* **1980**, 401–407.

Giuseppetti, G. and Tadini, C. (1982). The crystal structure of cabrerite $(Ni,Mg)_3(AsO_4)_2(H_2O)_8$ a variety of annabergite. *Bull. Mineral.* **105**, 333–337.

Giusepetti, G. and Tadini, C. (1987). Corkite, $PbFe_3(SO_4)(PO_4)(OH)_6$, its crystal structure and ordered arrangement of the tetrahedral cations. *N. Jb. Mineral. Mh.* **1987**, 71–81.

Giuseppetti, G. and Tadini, C. (1988). The crystal structure of austinite, $CaZn(AsO_4)$ (OH) from Kamareza, Laurion (Greece). *N. Jb. Mineral. Mh.* **1988**, 159–166.

Glaum, R. and Gruehn, R. (1992). Beitraege zum thermischen Verhalten wasserfreier Phosphate: VI. Einkristallstrukturverfeinerung der Metall(III)-orthophosphate $TiPO_4$ und VPO_4. *Z. Kristallogr.* **198**, 41–47.

Gobechiya, E. R., Pekov, I. V., Pushcharovskii, D. Y., Ferraris, G., Gula, A., Zubkova, N. V., and Chukanov, N. V. (2003). New data on vlasovite: Refinement of the crystal structure and the radiation damage of the crystal during the X-ray diffraction experiment. *Crystallogr. Rep.* **48**, 750–754.

Golovastikov, N. I., Matveeva, R. G., and Belov, N. V. (1975). Crystal structure of the tricalcium silicate $(CaOSiO_2)_3 = C_3S$. *Kristallogr.* **20**, 721–729.

Golubev, A. M., Andrianov, V. I., Sigarev, S. E., and Timofeeva, V. A. (1988). Crystal structure of $Na_5TiP_2O_9F$. *Sov. Phys. Crystallogr.* **33**, 653–655.

Gopal, R. and Calvo, C. (1973). Crystal structure of α-$Zn_2V_2O_7$. *Can. J. Chem.* **51**, 1004–1009.

Gorskaya, M. G., Filatov, S. K., Rozhdestvenskaya, I. V., and Vergasova, L. P. (1992). The crystal structure of klyuchevskite, $K_3Cu_3(Fe,Al)O_2(SO_4)_4$, a new mineral from Kamchatka volcanic sublimates. *Mineral. Mag.* **56**, 411–416.

Gougeon, P., Gall, P., and McCarley, R. E. (1990). Structure of $Ho_4Mo_4O_{11}$. *Acta Crystallogr.* **C47**, 1585–1588.

Graeber, E. J. and Rosenzweig, A. (1971). The crystal structures of yavapaiite, $KFe(SO_4)_2$, and goldichite, $KFe(SO_4)_2(H_2O)_4$. *Am. Mineral.* **56**, 1917–1933.

Greaves, C. and Blower, S. K. (1988). Structural relationships between $Bi_2O_2CO_3$ and β-Bi_2O_3. *Mater. Res. Bull.* **23**, 1001–1008.

Grey, I. E., Roth, R. S., Balmer, M. L. (1997). The crystal structure of $Cs_2TiSi_6O_{15}$. *J. Solid State Chem.* **131**, 38–42.

Grice, J. D., Burns, P. C., and Hawthorne, F. C. (1998). Borate minerals II: A hierarchy of structures based upon the borate fundamental building block. *Can. Mineral.* **37**, 731–762.

Grice, J. D., Cooper, M. A., and Hawthorne, F. C. (1999). Crystal-structure of twinned kettnerite. *Can. Mineral.* **37**, 923–927.

Griffen, D. T. and Ribbe, P. H. (1979). Distortions in the tetrahedral oxyanions of crystalline substances. *N. Jb. Mineral. Abh.* **137**, 54–73.

Grigor'ev, M. S., Charushnikova, I. A., Fedoseev, A. M., Budantseva, N. A., Baturin, N. A., and Regel', L. L. (1991a). Synthesis, crystal and molecular structure of double potassium and neptunoile molybdate. *Radiokhim.* **33**, 19–27.

Grigor'ev, M. S., Fedoseev, A. M., Budantseva, N. A., Yanovskii, A. I., Struchkov, Yu. T., Krot, N. N. (1991b). Synthesis, crystal and molecular structure of complex neptunium(V) sulfates $(Co(NH_3)_6)(NpO_2(SO_4)_2) \cdot 2H_2O$ and $(Co(NH_3)_6)$ $H_8O_3(NpO_2(SO_4)_3)$. *Radiokhim.* **33**, 54–60.

Grigor'ev, M. S., Baturin, N. A., Fedoseev, A. M., and Budantseva, N. A. (1991c). Crystal and molecular structure of neptunium(V) complex chromates $CsNpO_2(CrO_4) \cdot H_2O$ and $(NH_4)_4(NpO_2)_2(CrO_4)_3$. *Radiokhim.* **33**, 53–63.

Grigor'ev, M. S., Plotnikova, T. E., Budantseva, N. A., Fedoseev, A. M., Yanovskii, A. I., and Struchkov, Yu. T. (1992a). Synthesis and crystal structure of neptunium(V) complex selenate $(Co(NH_3)_6)NpO_2(SeO_4)_2(H_2O)_3$. *Radiokhim.* **34**, 1–6.

Grigor'ev, M. S., Charushnikova, I. A., Fedoseev, A. M., Budantseva, N. A., Yanovskii, A. I., and Struchkov, Yu. T. (1992b). Crystal and molecular structure of neptunium(V) complex molybdate $K_3NpO_2(MoO_4)_2$. *Radiokhim.* **34**, 7–12.

Grigor'ev, M. S., Fedoseev, A. M., and Budantseva, N. A. (2003). Crystal structure of the mixed-valence neptunium compound $Na_6[(Np^VO_2)_2(Np^{VI}O_2)(MoO_4)_5] \cdot 13H_2O$. *Russ. J. Coord. Chem.*, **29**, 877–879.

Grigor'ev, M. S., Fedoseev, A. M., Budantseva, N. A., Bessonov, A. A., and Krupa, J.-C. (2004). Synthesis and X-ray diffraction study of complex neptunium(VI) potassium chromate $K_2[(NpO_2)_2(CrO_4)_3(H_2O)] \cdot 3H_2O$. *Crystallogr. Rep.* **49**, 598–602.

Groat, L. A. and Hawthorne, F. C. (1986). Structure of ungemachite, $K_3Na_8Fe^{3+}$ $(SO_4)_6(NO_3)_2(H_2O)_6$ a mixed sulfate-nitrate mineral. *Am. Mineral.* **71**, 826–829.

Groat, L. A., Jambor, J. L., and Pemberton, B. C. (2003). The crystal structure of argentojarosite $AgFe_3(SO_4)_2(OH)_6$. *Can. Mineral.* **41**, 921–928.

Guesdon, A., Borel, M. M., Leclaire, A., Grandin, A., and Raveau, B. (1993). A series of mixed-valent molybdenum monophosphates, isotypic with leucophosphite represented by $CsMo_2P_2O_{10}$ and $K_{1.5}Mo_2P_2O_{10} \cdot H_2O$. *Z. Anorg. Allg. Chem.* **619**, 1841–1849.

Guesdon, A., Chardon, J., Provost, J., and Raveau, B. (2002). A copper uranyl monophosphate built up from $[CuO_2]_\infty$ chains: $Cu_2UO_2(PO_4)_2$. *J. Solid State Chem.* **165**, 89–93.

Guesmi, A., Zid, M. F., and Driss, A. (2000). $Na_2Co(H_2PO_4)_4 \cdot 4(H_2O)$. *Acta Crystallogr.* **C56**, 511–512.

Gulya, A. P., Shova, S. G., Rudik, V. F., Biyushkin, V. N., and Antosyak, B.M. (1994). Molecular and crystal structure of cobalt(II) hydroselenite dihydrate. *Koord. Khim.*, **20**, 368–370.

Guo, Y., Shi, Z., Yu, J., Wang, J., Liu, Y., Bai, N., and Pang, W. (2001). Solvothermal synthesis and characterization of a new titanium phosphate with a one-dimensional chiral chain. *Chem. Mater.* **13**, 203–207.

Guse, W., Klaska, K. H., Saalfeld, H., and Adiwidjaja, G. (1985). The crystal structure of iron(II) dihydrogen phosphate dihydrate $Fe(H_2PO_4)_2(H_2O)_2$. *N. Jb. Mineral. Mh.* **1985**, 433–438.

Haddad, S., Awwadi, F., and Willett, R. D. (2003). A planar bibridged $Cu_{10}Br_{22}$ oligomer: dimensional reduction and recombination of the $CuBr_2$ lattice via the N-H$\cdots$Br$^-$ and the C-Br$\cdots$Br$^-$ synthons. *Cryst. Growth Des.* **3**, 501–505.

Hagemann, H., Kubel, F., and Bill, H. (1996). Crystal structure of Sr_4OCl_6. *Eur. J. Solid State Inorg. Chem.* **33**, 1101–1109.

Haile, S. M. and Wuensch, B. J. (2000a). Structure, phase transitions and ionic conductivity of $K_3NdSi_6O_{15}(H_2O)_x$. I. α-$K_3NdSi_6O_{15}(H_2O)_2$ and its polymorphs. *Acta Crystallogr.* **B56**, 335–348.

Haile, S. M. and Wuensch, B. J. (2000b). Structure, phase transitions and ionic conductivity of $K_3NdSi_6O_{15}(H_2O)_x$. II. Structure of β-$K_3NdSi_6O_{15}$. *Acta Crystallogr.* **B56**, 349–362.

Haile, S. M., Wuensch, B. J., Laudise, R. A., and Maier, J. (1997). Structure of $Na_3NdSi_6O_{15} \cdot 2(H_2O)$ – a layered silicate with paths for possible fast-ion conduction. *Acta Crystallogr.* **53**, 7–17.

Halasyamani, P. S., Francis, R. J., Walker, S. M., and O'Hare, D. (1999). New layered uranium(VI) molybdates: syntheses and structures of $(NH_3(CH_2)_3NH_3)$ $(H_3O)_2(UO_2)_3(MoO_4)_5$, $C(NH_2)_3(UO_2)(OH)(MoO_4)$, $(C_4H_{12}N_2)(UO_2)(MoO_4)_2$, and $(C_5H_{14}N_2)(UO_2)(MoO_4)_2 \cdot H_2O$. *Inorg. Chem.* **38**, 271–279.

Haley, M. J., Wallwork, S. C., Duffin, B., Logan, N., and Addison, C. C. (1997). Hexa-μ-nitrato-μ_4-oxo-tetraberillium. *Acta Crystallogr.* **C53**, 829–830.

Harrison, W. T. A. (2000). Synthetic mansfieldite, $AlAsO_4 \cdot 2H_2O$. *Acta Crystallogr.* **C56**, E421.

Harrison, W. T. A. and Buttery, J. H. N. (2000). Synthesis and crystal structure of $Cs(VO_2)_3(TeO_3)_2$, a new layered cesium vanadium(v) tellurite. *Z. Anorg. Allg. Chem.* **626**, 867–870.

Harrison, W. T. A. and Hannooman, L. (1997). Two new tetramethylammonium zinc phosphates: $N(CH_3)_4Zn(HPO_4)(H_2PO_4)$, an open framework phase built up from a low-density 12-ring topology, and $N(CH_3)_4Zn(H_2PO_4)_3$, a molecular cluster. *J. Solid State Chem.* **131**, 363–369.

Harrison, W. T. A., Gier, T. E., Stucky, G. D., and Schultz, A. J. (1990). The crystal structure of the nonlinear optical material thallium titanyl phosphate, $TlTiOPO_4$, above the ferroelectric to paraelectric phase transition. *J. Chem. Soc. Chem. Comm.* **1990**, 540–542.

Harrison, W. T. A., Dussack, L. L., and Jacobson, A. J. (1994a). Syntheses, crystal structures, and properties of new layered molybdenum(VI) selenites: $(NH_4)_2(MoO_3)_3(SeO_3)$ and $Cs_2(MoO_3)_3(SeO_3)$. *Inorg. Chem.* **33**, 6043–6049.

Harrison, W. T. A., Lim, S. C., Vaughey, J. T., Jacobson, A. J., Goshorn, D. P., and Johnson, J. W. (1994b). Synthesis, structure, and magnetism of $Ba_2VO(PO_4)_2 \cdot (H_2O)$, a new barium vanadium(IV) phosphate hydrate. *J. Solid State Chem.* **113**, 444–447.

Harrison, W. T. A., Lim, S. C., Vaughey, J. T., Jacobson, A. J., Goshorn, D. P., and Johnson, J. W. (1994c). Synthesis, structure, and magnetism of $Ba_2VO(PO_4)_2 \cdot H_2O$, a new barium vanadium(IV) phosphate hydrate. *J. Solid State Chem.*, **113**, 444–447.

Harrison, W. T. A., Gier, T. E., Calabrese, J. C., and Stucky, G. D. (1994d). Two new noncentrosymmetric rubidium titanium phosphate phases: $Rb_2Ti_3O_2(PO_4)_2(HPO_4)_2$ and $Rb_3Ti_3O(P_2O_7)(PO_4)_3$. *J. Solid State Chem.* **111**, 257–266.

Harrison, W. T. A., Lim, S. C., Dussack, L. L., Jacobson, A. J., Goshorn, D. P., and Johnson, J. W. (1995a). Synthesis, structure and magnetism of $BaVO(PO_4)$ $(H_2PO_4) \cdot (H_2O)$, a new layered barium vanadium(IV) phosphate hydrate. *J. Solid State Chem.* **118**, 241–246.

Harrison, W. T. A., Vaughey, J. T., Jacobson, A. J., Goshorn, D. P., and Johnson, J. W. (1995b). Two new barium-vanadium(IV) phases: $Ba(VO)_2(SeO_3)_2(HSeO_3)_2$, the first barium vanadium selenite, and $Ba_8(VO)_6(PO_4)_2(HPO_4)_{11} \cdot 3(H_2O)$, a compound built up from two types of one-dimensional chains. *J. Solid State Chem.* **116**, 77–86.

Harrison, W. T. A., Dussack, L. L., and Jacobson, A. J. (1995c). Potassium vanadium selenite, $K(VO_2)_3(SeO_3)_2$. *Acta Crystallogr.*, **C51**, 2473–2476.

Harrison, W. T. A., Gier, T. E., and Stucky, G. D. (1995d). Single-crystal structure of $Cs_3HTi_4O_4(SiO_4)_3 \cdot 4(H_2O)$, a titanosilicate pharmacosiderite analog. *Zeolites* **5**, 408–412.

Harrison, W. T. A., Vaughey, J. T., Jacobson, A. J., Goshorn, D. P., and Johnson, J. W. (1995e). Two new barium-vanadium(IV) phases: $Ba(VO)_2(SeO_3)_2(HSeO_3)_2$, the first barium vanadium selenite, and $Ba_8(VO)_6(PO_4)_2(HPO_4)_{11} \cdot 3(H_2O)$, a compound built up from two types of one-dimensional chains. *J. Solid State Chem.* **116**, 77–86.

Harrison, W. T. A., Dussack, L. L., and Jacobson, A. J. (1996a). Hydrothermal investigation of the barium/molybdenum(VI)/selenium(IV) phase space: single-crystal structures of $BaMoO_3SeO_3$ and $BaMo_2O_5(SeO_3)_2$. *J. Solid State Chem.* **125**, 234–242.

Harrison, W. T. A., Broach, R. W., Bedard, R. A., Gier, T. E., Bu, X. H., Stucky, G. D. (1996b). Synthesis and characterization of a new family of thermally stable open-framework zincophosphate/arsenate phases: $M_3Zn_4O(XO_4)_3 \cdot n(H_2O)$.

Crystal structures of $Rb_3Zn_4O(PO_4)_3 \cdot 3.5(H_2O)$, $K_3Zn_4O(AsO_4)_3 \cdot 4(H_2O)$ and $Na_3Zn_4O(PO_4)_3 \cdot 6(H_2O)$. *Chem. Mater.* **8**, 691–700.

Harrison, W. T. A., Phillips, M. L. F., and Stucky, G. D. (1997). Single-crystal structures and optical properties of the nonlinear optical $KTiOPO_4$ – type materials $K_{0.55}Li_{0.45}TiOPO_4$ and $K_{0.54}Li_{0.46}TiOAsO_4$ prepared by molten salt ion-exchange. *Chem. Mater.* **9**, 1138–1144.

Harrison, W. T. A., Bircsak, Z., Hannooman, L., and Zhang, Z. (1998). Two new 1,3-diammonium-propane zinc hydrogen phosphates: $H_3N(CH_2)_3NH_3 \cdot Zn_2(HPO_4)_2$ $(H_2PO_4)_2$, with 12-ring layers, and $H_3N(CH_2)_3NH_3 \cdot Zn(HPO_4)_2$, with 4-ring ladders. *J. Solid State Chem.* **136**, 93–102.

Harrison, W. T. A., Phillips, M. L. F., Bu, X. H. (2000). Synthesis and single-crystal structure of $Cs_3Zn_4O(AsO_4)_3 \cdot 4(H_2O)$, an open-framework zinc arsenate. *Micropor. Mesopor. Mater.* **39**, 359–365.

Hashimoto, M., Kubata, M., and Yagasaki, A. (2000). $(NH_4)_3(VO_2(SO_4)_2(OH_2)_2) \cdot 1.5$ (H_2O). *Acta Crystallogr.* **C56**, 1411–1412.

Haushalter, R. C. (1987). $(Mo_4O_4)^{6+}$ cubes in $Cs_3Mo_4P_3O_{16}$. *J. Chem. Soc. Chem. Commun.* **1987**, 1566–1568.

Haushalter, R. C. and Mundi, L. A. (1992). Reduced molybdenum phosphates: octahedral-tetrahedral framework solids with tunnels, cages, and micropores. *Chem. Mater.* **4**, 31–48.

Haushalter, R. C., Strohmaier, K. G., and Lai, F. W. (1989). Structure of a three-dimensional microporous molybdenium phosphate with large cavities. *Science* **246**, 1289–1291.

Haushalter, R. C., Wang, Z., Thompson, M. E., Zubieta, J., and O'Connor, C. J. (1993a). A stair step layer structure encapsulating interlayer K^+ cations: hydrothermal synthesis, crystal structure, and magnetic properties of $K_2((V^{IV}O)_2V^{III}(PO_4)_2(HPO_4)$ $(H_2PO_4)(H_2O)_2)$. *Inorg. Chem.* **32**, 3966–3969.

Haushalter, R. C., Wang, Z., Thompson, M. E., and Zubieta, J. (1993b). Octahedral – tetrahedral framework solids of the vanadium phosphate system. Hydrothermal syntheses and crystal structures of $Cs(V^{III}_2(PO_4)(HPO_4)_2(H_2O)_2)$ and $K((V^{IV}O)$ $V^{III}(HPO_4)_3(H_2O)_2)$. *Inorg. Chem.* **32**, 3700–3704.

Haushalter, R. C., Chen, Q., Soghomonian, V., Zubieta, J., and O'Connor, C. J. (1994a). Synthesis, structure, and magnetic properties of $(NH_4)VOPO_4$: a structure with one-dimensional chains of VO_6 octahedra. *J. Solid State Chem.* **108**, 128–133.

Haushalter, R. C., Wang, Z., Thompson, M. E., Zubieta, J., and O'Connor, C. J. (1994b). Hydrothermal synthesis, crystal structure, and magnetic properties of $Cs((V_2O_3)(HPO_4)_2(H_2O))$, a mixed-valence vanadium (IV,V) hydrogen phosphate with a one-dimensional $(-V^{IV}-O-V^V-O-)$ chain of corner-sharing VO_6 octahedra. *J. Solid State Chem.* **109**, 259–264.

Häussermann, U., Svensson, C., and Lidin, S. (1998). Tetrahedral stars as flexible basis clusters in sp-bonded intermetallic frameworks and the compound $BaLi_7Al_6$ with the $NaZn_{13}$ structure. *J. Amer. Chem. Soc.* **120**, 3867–3880.

Hawthorne, F. C. (1976). Hydrogen positions in scorodite. *Acta Crystallogr.* **B32**, 2891–2892.

Hawthorne, F. C. (1979). The crystal structure of morinite. *Can. Mineral.* **17**, 93–102.

Hawthorne, F. C. (1982). The crystal structure of boggildite. *Can. Mineral.* **20**, 263–270.

Hawthorne, F. C. (1983). Graphical enumeration of polyhedral clusters. *Acta Crystallogr.*, **A39**, 724–736.

Hawthorne, F. C. (1984). The crystal structure of mandarinoite, $Fe^{3+}_2Se_3O_9(H_2O)_6$. *Can. Mineral.* **22**, 475–480.

Hawthorne, F. C. (1985). Towards a structural classification of minerals: the $^{vi}M^{iv}T_2O_n$ minerals. *Am. Mineral.*, **70**, 455–473.

Hawthorne, F. C. (1986). Structural hierarchy in $^{iv}M_x^{iii}T_y\varphi_z$ minerals. *Can. Mineral.*, **24**, 625–642.

Hawthorne, F. C. (1987). The crystal chemistry of the benitioite group minerals and structural relations in (Si_3O_9) ring structures. *N. Jb. Mineral. Mh.* **1987**, 16–30.

Hawthorne, F. C. (1988). Sigloite: the oxidation mechanism in $(M(III)_2(PO_4)_2 (OH)_2(H_2O)_2)_2$–structures. *Mineral. Petrol.* **38**, 201–211.

Hawthorne, F. C. (1990). Structural hierarchy in $M^{[6]}T^{[4]}\varphi_n$ minerals. *Z. Kristallogr.*, **192**, 1–52.

Hawthorne, F. C. (1992). The role of OH and H_2O in oxide and oxysalt crystals. *Z. Kristallogr.*, **201**, 183–206.

Hawthorne, F. C. (1994). Structural aspects of oxide and oxysalt crystals. *Acta Crystallogr.*, **B50**, 481–510.

Hawthorne, F. C. (1998). Structure and chemistry of phosphate minerals. *Mineral. Mag.*, **62**, 141–164.

Hawthorne, F. C. and Faggiani, R. (1979). Refinement of the structure of descloizite. *Acta Crystallogr.* **35**, 717–720.

Hawthorne, F. C. and Ferguson, R. B. (1975). Refinement of the crystal structure of kroehnkite. *Acta Crystallogr.* **B31**, 1753–1755.

Hawthorne, F. C. and Ferguson, R. B. (1977). The crystal structure of roselite. *Can. Mineral.* **15**, 36–42.

Hawthorne, F. C. and Huminicki, D. M. C. (2002). The crystal chemistry of beryllium. *Rev. Mineral. Geochem.* **50**, 333–403.

Hawthorne, F. C., Groat, L. A., Raudsepp, M., and Ercit, T. S. (1987a). Kieserite, $MgSO_4(H_2O)$, a titanite-group mineral. *N. Jb. Mineral. Abh.* **157**, 121–132.

Hawthorne, F. C., Groat, L. A., and Ercit, T. S. (1987b). Structure of cobalt deselenite. *Acta Crystallogr.* **C43**, 2042–2044.

Hawthorne, F. C., Krivovichev, S. V., and Burns, P. C. (2000). Crystal chemistry of sulfate minerals. *Rev. Mineral. Geochem.* **40**, 1–112.

Hayden, L. A. and Burns, P. C. (2002a). The sharing of an edge between a uranyl pentagonal bipyramid and sulfate tetrahedron in the structure of $KNa_5((UO_2)(SO_4)_4)(H_2O)$. *Can. Mineral.* **40**, 211–216.

Hayden, L. A. and Burns, P. C. (2002b). A novel uranyl sulfate cluster in the structure of $Na_6(UO_2)(SO_4)_4(H_2O)_2$. *J. Solid State Chem.* **163**, 313–318.

Hector, A. L., Henderson, S., Levason, W., and Webster, M. (2002). Hydrothermal synthesis of scandium, yttrium and lanthanide iodates from the corresponding periodates: X-ray crystal structures of $Sc(IO_3)_3$, $Y(IO_3)_3 \cdot 2(H_2O)$, $La(IO_3)_3 \cdot 1/2(H_2O)$ and $Lu(IO_3)_3 \cdot 2(H_2O)$. *Z. Anorg. Allg. Chem.* **628**, 198–202.

Heeg, M. J., Redman, W., and Frech, R. E. (1986). Structure of hexasodium trans-diaquatetrasulfatozinc(II). *Acta Crystallogr.* **C42**, 949–952.

Hess, H., Keller, P., and Riffel, H. (1988). The crystal structure of chenite, $Pb_4Cu(OH)_6(SO_4)_2$. *N. Jb. Mineral. Mh.* **1988**, 259–264.

Hesse, K. F., Liebau, F., and Merlino, S. (1992). Crystal structure of rhodesite, $HK_{1-x}Na_{x+2y}Ca_{2-y}(lB,3,2^2_\infty)(Si_8O_{19})(6-z)H_2O$, from three localities and its relation to other silicates with dreier double layers. *Z. Kristallogr.* **199**, 25–48.

Hexiong, Y. and Pingqiu, F. (1988). Crystal structure of zincobotryogen. *Kuangwu Xuebao* **8**, 1–12.

Higgins, J. B. and Ribbe, P. H. (1977). The structure of malayaite, $CaSnOSiO_4$, a tin analog of titanite. *Am. Mineral.* **62**, 801–806,

Hill, R.J. (1979). The crystal structure of koettigite. *Am. Mineral.* **64**, 376–382.

Hiltunen, L. and Niinisto, L. (1976a). Ytterbium sulfate octahydrate, $Yb_2(SO_4)_3(H_2O)_8$. *Cryst. Struct. Commun.* **5**, 561–566.

Hiltunen, L. and Niinisto, L. (1976b). Ytterbium selenate octahydrate, $Yb_2(SeO_4)_3$ $(H_2O)_8$. *Cryst. Struct. Commun.* **5**, 567–570.

Hiltunen, L., Leskela, M., Niinisto, L., and Tammenmaa, M. (1985). Crystal structure of copper hydrogenselenite monohydrate. *Acta Chem. Scand.* **A39**, 809–812.

Hinsch, T. R. (1985). The crystal structure of magnesium-dihydrogen phosphate dihydrate $Mg(H_2PO_4)_2(H_2O)_2$. *N. Jb. Mineral. Mh.* **1985**, 439–445.

Hippler, K., Sitta, S., Vogt, P., and Sabrowsky, H. (1990). Structure of Na_3OCl. *Acta Crystallogr.* **C46**, 736–738.

Hizaoui, K., Jouini, N., and Jouini, T. (1999). Synthese et structure cristalline de $Na_3NbO(AsO_4)_2$. *J. Solid State Chem.* **144**, 53–61.

Hopfgarten, F. (1975). Crystal structure of $Bi_6O_7FCl_3$. *Acta Crystallogr.* **B31**, 1087–1092.

Hopfgarten, F. (1976). Crystal structure of $Bi_{12}O_{15}Cl_6$. *Acta Crystallogr.* **B32**, 2570–2573.

Hoppe, R. and Koehler, J. (1988). SCHLEGEL projections and SCHLEGEL diagrams – new ways to describe and discuss solid state compounds. *Z. Kristallogr.* **183**, 77–111.

Hriljac, J. A., Grey, C. P., Cheetham, A. K., VerNooy, P. D., and Torardi, C. C. (1996). Synthesis and structure of $KIn(OH)PO_4$: chains of hydroxide-bridged $InO_4(OH)_2$ octahedra. *J. Solid State Chem.* **123**, 243–248.

Huan, G., Johnson, J. W., Jacobson, A. J., Goshorn, D. P., and Merola, J. S. (1991). Hydrothermal synthesis, single-crystal structure and magnetic properties of $VOSeO_3 \cdot H_2O$. *Chem. Mater.* **3**, 539–541.

Huang, J., Gu, Q., and Sleight, A. W. (1993). Synthesis and characterization of bismuth magnesium phosphate and arsenate: $BiMg_2PO_6$ and $BiMg_2AsO_6$. *J. Solid State Chem.* **105**, 599–606.

Hubert, P. H., Michel, P., and Thozet, A. (1973). Structure du molybdite de neodyme $Nd_5Mo_3O_{16}$. *C. R. Hebd. Seances Acad. Sci., Ser. C* **276**, 1779–1781.

Hughes, J. M. and Brown, M. A. (1989). The crystal structure of ziesite, β-$Cu_2V_2O_7$, a thortveitite-type structure with a non-linear X-O-X inter-tetrahedral bond. *N. Jb. Mineral., Mh.* **1989**, 41–47.

Huntelaar, M. E., Cordfunke, E. H. P., van Vlaanderen, P., and Ijdo, D. J. W. (1994). $SrZr(Si_2O_7)$. *Acta Crystallogr.* **C50**, 988–991.

Hummel, H.-U., Fischer, E., Fischer, T., Joerg, P., and Pezzei, G. (1993). Struktur und thermisches Verhalten von Gadolinium(III)-sulfat-octahydrat $Gd_2(SO_4)_3 \cdot 8H_2O$. *Z. Anorg. Allg. Chem.* **619**, 805–810.

Huminicki, D. M. C. and Hawthorne, F. C. (2001). Refinement of the crystal structure of swedenborgite. *Can. Mineral.* **39**, 153–158.

Huminicki, D. M. C. and Hawthorne, F. C. (2002). The crystal chemistry of phosphate minerals. *Rev. Mineral. Geochem.* **48**, 123–254.

Huvé, M., Colmont, M., and Mentré, O. (2004). HREM: a useful tool to formulate new members of the wide Bi^{3+}/M^{2+} oxide phosphate series. *Chem. Mater.* **16**, 2628–2638.

Huvé, M., Colmont, M., and Mentré, O. (2006). Bi^{3+}/M^{2+} oxyphosphate: a continuous series of polycationic species from the 1D single chain to the 2D planes. Part 1: from HREM images to crystal-structure deduction. *Inorg. Chem.* **45**, 6604–6611.

Iijima, S. (1991). Helical microtubules of graphitic carbon. *Nature* **354**, 56–59.

Ilyukhin, A. B. and Dzhurinskii, B. F. (1993). Crystal structures of double oxoborates $LnCa_4O(BO_3)_3$ (Ln = Gd, Tb, Lu) and $Eu_2CaO(BO_3)_2$. *Zh. Neorg. Khim.* **38**, 917–920.

Ilyushin, G. D., Voronkov, A. A., Nevsky, N. N., Ilyukhin, V. V., and Belov, N. V. (1981a). Crystal structure of hilairite $Na_2ZrSi_3O_9 \cdot 3(H_2O)$. *Sov. Phys. Dokl.* **26**, 916–917.

Ilyushin, G. D., Pudovkina, Z. V., Voronkov, A. A., Khomyakov, A. P., Ilyukhin, V. V., and Pyatenko, Yu. A. (1981b). The crystal structure of the natural modification of $K_2ZrSi_3O_9 \cdot H_2O$. *Sov. Phys. Dokl.* **26**, 257–258.

Ilyushin, G. D. (1993). New data on the crystal structure of umbite $K_2ZrSi_3O_9 \cdot H_2O$. *Izv. Akad. Nauk SSSR Neorg. Mater.* **29**, 971–975.

Ishigami, H., Sumita, M., and Sato, S. (1999). Phase transition in $K_2Fe(SO_4)_2 \cdot 2(H_2O)$. *Ferroelectrics* **229**, 109–114.

Iskhakova, L. D. (1995). Crystal structure of $NH_4Pr(SeO_4)_2(H_2O)_5$. *Kristallogr.* **40**, 631–634.

Iskhakova, L. D., Starikova, Z. A., and Trunov, V. K. (1981). Crystal structure of $RbPr(SO_4)_2(H_2O)_4$. *Koord. Khim.* **7**, 1713–1718.

Jambor, J. L., Owens, D. R., Grice, J. D., and Feinglos, M. N. (1996). Gallobeudantite, $PbGa_3[(AsO_4),(SO_4)](OH)_6$, a new mineral species from Tsumeb, Namibia, and associated new gallium analogues of the alunite-jarosite family. *Can. Mineral.* **34**, 1305–1315.

Jarosch, D. (1985). Kristallstruktur des Leonits; $K_2Mg(SO_4)_2(H_2O)_4$. *Z. Kristallogr.* **173**, 75–79.

Jasty, S., Robinson, P. D., and Malhotra, V. M. (1991). Crystal structure of and low-temperature phase transitions in isostructural $RbSm(SO_4)_2 \cdot 4(H_2O)$ and $(NH_4)_4Sm(SO_4)_2 \cdot 4(H_2O)$. *Phys. Rev.* **B43**, 13215–13227.

Jeffrey, G. A. (1997). *An introduction to hydrogen bonding*. Oxford University Press: New York, Oxford.

Johansson, G. (1961). The crystal structure of $InOHSO_4(H_2O)_2$. *Acta Chem. Scand.* **15**, 1437–1453.

Johnston, M.G. and Harrison, W. T. A. (2004). Two new octahedral/pyramidal frameworks containing both cation channels and lone-pair channels: synthesis and structures of $Ba_2Mn(II)Mn(III)_2(SeO_3)_6$ and $PbFe_2(SeO_3)_4$. *J. Solid State Chem.* **177**, 4680–4686.

Jolibois, B., Laplace, G., Abraham, F., and Novogorocki, G. (1980). The low-temperature forms of some $M(I)_3M(III)(XO_4)_3$ compounds: structure of triammonium indium(III) trisulfate. *Acta Crystallogr.* **B36**, 2517–2519.

Jolibois, B., Laplace, G., Abraham, F., and Novogorocki, G. (1981). Monoclinic-trigonal transition in some $M(I)_3M'(III)(SO_4)_3$ compounds: The high temperature form of $(NH_4)_3In(SO_4)_3$. *J. Solid State Chem.* **40**, 69–74.

Jolicart, G., le Blanc, M., Morel, B., Dehaudt, P., and Dubois, S. (1996). Hydrothermal synthesis and determination of $Cs_2ZrSi_6O_{15}$. *Eur. J. Solid State Inorg. Chem.* **33**, 647–657.

Jolly, J. H. and Foster, H. L. (1967). X-ray diffraction data of aluminocopiapite. *Am. Mineral.* **52**, 1220–1223.

Jones, R. H., Thomas, J. M., Xu, R., Huo, Q., Cheetham, A. K., and Powell, A. V. J. (1991). Synthesis and structure of a novel aluminium phosphate anion – $(Al_3P_4O_{16})^{3-}$. *Chem. Commun.* **1991**, 1266–1268.

Jones, R. H., Chippindale, A. M., Natarajan, S., and Thomas, J. M. (1994). A reactive template in the synthesis of a novel layered aluminum phosphate $(Al_3P_4O_{16})^{3-}$ $[NH_3(CH_2)_5NH_3]^{2+}(C_5H_{10}NH_2)^+$. *Chem.Commun.* **1994**, 565–566.

Johnston, M. G. and Harrison, W. T. A. (2000). Syntheses and structures of two selenite chloride hydrates: $Co(HSeO_3)Cl \cdot 3H_2O$ and $Cu(HSeO_3)Cl \cdot 2H_2O$. *Z. Anorg. Allg. Chem.* **626**, 2487–2490.

Junk, P. C., Kepert, C. J., Skelton, B. W., and White, A. H. (1999). Structural systems of rare earth complexes. XX. (Maximally) hydrated rare earth sulfates and the double sulfates $(NH_4)Ln(SO_4)_2 \cdot 4(H_2O)$. *Austr. J. Chem.* **52**, 601–615.

Kabalov, Y. K., Zubkova, N. V., Pushcharovsky, D. Y., Schneider, J., and Sapozhnikov, A. N. (2000). Powder Rietveld refinement of armstrongite, $CaZr[Si_6O_{15}] \cdot 3H_2O$. *Z. Kristallogr.* **215**, 757–761.

Kadoshnikova, N. V., Shumyatskaya, N. G., and Tananaev, I. V. (1978). The formation and the crystal structure of $In_2(SeO_4)_3(H_2O)_5$. *Dokl. Akad. Nauk SSSR* **243**, 116–118.

Kahlenberg, V., Piotrowski, A., and Giester, G. (2000). Crystal structure of $Na_4[Cu_4O_2(SO_4)_4]$ MeCl (Me: Na, Cu, Box) – the synthetic Na-analogue of piypite (caratiite). *Mineral. Mag.* **64**, 1099–1108.

Kamenar, B., Matkovic-Calogovic, D., and Nagl, A. (1986). Structural study of the system $Hg_2O–N_2O_5–H_2O$; crystal structure of three basic mercury(I) nitrates hydrolysis products of mercury(I) nitrate dihydrate. *Acta Crystallogr.* **C42**, 385–389.

Kampf, A. R. (1977). Minyulite: its atomic arrangement. *Am. Mineral.* **62**, 256–262.

Kampf, A. R. (1991). Grandreefite, $Pb_2F_2SO_4$: crystal structure and relationship to the lanthanide oxide sulfates, $Ln_2O_2SO_4$. *Am. Mineral.* **76**, 278–282.

Kang, Z. C. and Eyring, L. (1997). A compositional and structural rationalization of the higher oxides of Ce, Pr and Tb. *J. Alloys Compd.* **249**, 206–212.

Kang, Z. C. and Eyring, L. (1998). The prediction of the structure of members of the homologous series of the high rare earth oxides. *J. Alloys Compd.* **275**, 30–36.

Kang, H. Y., Lee, W.-C., and Wang, S.-L. (1992a). Hydrothermal synthesis and structural characterization of four layered vanadyl(IV) phosphate hydrates $A(VO)_2(PO_4)_2 \cdot 4(H_2O)$ (A= Co, Ca, Sr, Pb). *Inorg. Chem.* **31**, 4743–4748.

Kang, H. Y., Wang, S.-L., and Lii, K.-H. (1992b). Hydrothermal synthesis and structures of two layered dioxovanadium(V) phosphates $A(VO_2)PO_4$ (A= Ba, Sr). *Acta Crystallogr.* **C48**, 975–978.

Kang, H.-Y., Wang, S.-L., Tsai, P.-P., and Lii, K.-H. (1993). Hydrothermal synthesis. Crystal structure and ionic conductivity of $Ag_2VO_2PO_4$: a new layered phosphate of vanadium(V). *Dalton Trans.* **1993**, 1525–1528.

Karimova, O. V. and Burns, P. C. (2007). Structural units in three uranyl perrhenates. *Inorg. Chem.* **46**, 10106–1013.

Karpov, O. G., Pushcharovskii, D. Yu., Pobedimskaya, E. A., Burshtein, I. F., and Belov, N. V. (1977). The crystal structure of the rare-earth silicate $NaNdSi_6O_{13}(OH)_2 \cdot n(H_2O)$. *Sov. Phys. Dokl.* **22**, 464–466.

Kashaev, A. A. and Sokolova, G. V. (1973). Crystal structure of $CsNbO(SO_4)_2$. *Kristallogr.* **18**, 620–621.

Kashaev, A. A., Postoenko, G. E., and Zel'bst, E. A. (1973). Crystal structures of $RbNbO(SO_4)_2$ and $NH_4NbO(SO_4)_2$. *Kristallogr.* **18**, 1278–1280.

Kato, T. (1977). Further refinement of the woodhouseite structure. *N. Jb. Mineral. Mh.* **1977**, 54–58.

Kato, K. (1979). Die Kristallstruktur von Pentablei(II)-germanat-trioxid. *Acta Crystallogr.* **B35**, 795–797.

Kato, K. (1982). Die Kristallstruktur des Bleisilicats $Pb_{11}Si_3O_{17}$. *Acta Crystallogr.* **B38**, 57–62.

Kato, T. (1990). The crystal structure of florencite. *N. Jb. Mineral. Mh.* **1990**, 227–231.

Kato, T. and Miura, Y. (1977). The crystal structure of jarosite and svanbergite. *Mineral. J.* **8**, 419–430.

Kato, K., Hirota, K., Kanke, Y., Sato, A., Ohsumi, K., Takase, T., Uchida, M., Jarchow, O., Friese, K., and Adiwidjaja, G. (1995). Die Kristallstruktur des Bleigermanats $Pb_{11}Ge_3O_{17}$. *Z. Kristallogr.* **210**, 188–194.

Kawahara, A., Yamakawa, J., Yamada, T., and Kobashi, D. (1995). Synthetic magnesium sodium hydrogen monophosphate: $MgNa_3H(PO_4)_2$. *Acta Crystallogr.* **C51**, 2220–2222.

Keller, H. L. (1982). Darstellung und Kristallstruktur von $TlPb_8O_4Br_9$. *Z. Anorg. Allg. Chem.* **491**, 191–198.

Keller, H. L. (1983). Eine neuartige Blei-Sauerstoff-Baugruppe: $(Pb_8O_4)^{8+}$. *Angew. Chem.* **95**, 318–319.

Keller, H. L. and Langer, R. (1994). $HgPb_2O_2Cl_2$, ein "perforiertes" Blei(II)-oxid. *Z. Anorg. Allg. Chem.* **620**, 977–980.

Kellersohn, T. (1992). Structure of cobalt sulfate tetrahydrate. *Acta Crystallogr.* **C48**, 776–779.

Kemnitz, E., Werner, C., Stiewe, A., Worzala, H., and Troyanov, S. I. (1996). Synthese und Struktur von $Zn(HSO_4)_2(H_2SO_4)_2$ und $Cd(HSO_4)_2$. *Z. Naturforsch.* **51b**, 14–18.

Keramidas, K. G., Voutsas, G. P., and Rentzeperis, P. I. (1993). The crystal structure of BiOCl. *Z. Kristallogr.* **205**, 35–40.

Ketatni, E. M., Mentré, O., Abraham, F., Kzaiber, F., and Mernari, B. (1998). Synthesis and crystal structure of $Bi_{6.67}(PO_4)_4O_4$ oxyphosphate: the $Bi_6M^{2+}(PO_4)_4O_4$ and $Bi_{6.5}A^+{}_{0.5}(PO_4)_4O_4$ series. *J. Solid State Chem.* **139**, 274–280.

Ketatni, E. M., Huvé, M., Abraham, F., and Mentré, O. (2003). Characterization of the new $Bi_{~6.2}Cu_{~6.2}O_8(PO_4)_5$ oxyphosphate; a disordered compound containing 2 and 3 $O(Bi, Cu)_4$ tetrahedra – wide polycationic ribbons. *J. Solid State Chem.* **172**, 327–338.

Ketterer, J. and Kraemer, V. (1985). Structural characterization of the synthetic per-ites $PbBiO_2X$, X=I, Br, Cl. *Mater. Res. Bull.* **20**, 1031–1036.

Ketterer, J. and Kraemer, V. (1986a). Crystal structure of bismuth silicate Bi_2SiO_5. *N. Jb. Mineral., Mh.* **1986**, 13–18.

Ketterer, J. and Kraemer, V. (1986b). Structure refinement of bismuth oxide bromide, BiOBr. *Acta Crystallogr.* **C42**, 1098–1099.

Khamaganova, T. N., Trunov, V. K., and Dzhurinskii, B. F. (1991). Crystal structure of calcium samarium double oxoborate $Sm_2Ca_8O_2(BO_3)_6$. *Zh. Neorg. Khim.* **36**, 855–857.

Khan, A. A. and Baur, W. H. (1972). Salt hydrates. VIII. The crystal structures of sodium ammonium orthochromate dihydrate and magnesium diammonium bis(hydrogen orthophosphate) tetrahydrate and a discussion of the ammonium ion. *Acta Crystallogr.* **B28**, 683–693.

Kharisun, Taylor, M. R., and Bevan, D. J. M. (1988). The crystal chemistry of duftite, $PbCu(AsO_4)(OH)$ and the beta-duftite problem. *Mineral. Mag.* **62**, 121–130.

Kharisun, Taylor, M. R., Bevan, D. J. M., Rae, A. D., and Pring, A. (1997). The crystal structure of mawbyite, $PbFe_2(AsO_4)_2(OH)_2$. *Mineral. Mag.* **61**, 685–691.

Khomyakov, A. P., Kazakova, M. E., and Pushcharovskii, D. Yu. (1980). Nacaphite $Na_2Ca(PO_4)F$ – a new mineral. *Zapiski VMO* **109**, 50–52.

Khosrawan-Sazedj, F. (1982a). On the space group of threadgoldite. *Tscherm. Mineral. Petrogr. Mitt.* **30**, 111–115.

Khosrawan-Sazedj, F. (1982b). The crystal structure of meta-uranocircite II, $Ba(UO_2)_2(PO_4)_2(H_2O)_6$. *Tscherm. Mineral. Petrogr. Mitt.* **29**, 193–204.

Khrustalev, V. N., Andreev, G. B., Antipin, M. Yu., Fedoseev, A. M., Budantseva, N. A., and Shirokova, I. B. (2000). Synthesis and crystal structure of rubidium monohy-drate dimolybdatouranylate, $Rb_2[(UO_2)(MoO_4)_2]\cdot H_2O$. *Russ. J. Inorg. Chem.* **45**, 1845–1847.

Kierkegaard, P. and Longo, J. M. (1970). A refinement of the crystal structure of $MoOPO_4$. *Acta Chem. Scand.* **24**, 427–432.

Kim, Y. H., Lee, K.-S., Kwon, Y.-U., and Han, O. H. (1996). K(VO)(SeO$_3$)$_2$H: a new one-dimensional compound with strong hydrogen bonding. *Inorg. Chem.* **35**, 7394–7398

King, H. E. Jr., Mundi, L. A., Strohmaier, K. G., and Haushalter, R. C. (1991a). A synchrotron single crystal X-ray structure determination of a small crystal: Mo–Mo double bonds in the 3-D microporous molybdenumphosphate NH$_4$(Mo$_2$P$_2$O$_{10}$)·H$_2$O. *J. Solid State Chem.* **92**, 1–7.

King, H. E. Jr., Mundi, L. A., Strohmaier, K. G., and Haushalter, R. C. (1991b). A synchrotron single crystal X-ray structure determination of (NH$_4$)$_3$Mo$_4$P$_3$O$_{16}$: a microporous molybdenum phosphate with Mo$_4$O$_4^{6+}$ cubes. *J. Solid State Chem.* **92**, 154–158.

Kirby, E. C. (1997). Recent work on toroidal and other exotic fullerene structures. In: *From chemical topology to three-dimensional geometry*. Plenum: New York, pp. 263–296.

Kirkby, S. J., Lough, A. J., and Ozin, G. A. (1995). Crystal structure of potassium aluminium fluoride phosphate, KAlFPO$_4$. *Z. Kristallogr.* **210**, 956–956.

Klevtsov, P. V., Kharchenko, L. Yu., and Klevtsova, R. F. (1975). Crystallization and polymorphism of rare earth oxymolybdates with the composition Ln$_2$MoO$_6$. *Kristallogr.* **20**, 571–578.

Klevtsova, R.F., Klevtsov, P.V. (1970) Synthesis and crystal structure of double molybdates KR(MoO$_4$)$_2$ for R(III) = Al, Sc, and Fe and tungstate KSc(WO$_4$)$_2$. *Kristallografiya*, **15**, 953–959.

Kniep, R. and Wilms, A. (1979). Zur Strukturchemie der Phase Al$_2$O$_3$(P$_2$O$_5$)$_2$(H$_2$O)$_9$. *Z. Naturforsch.* **34b**, 750–751.

Kniep, R., Mootz, D., and Vegas, A. (1977). Variscite. *Acta Crystallogr.* **B33**, 263–265.

Kniep, R., Mootz, D., and Wilms, A. (1978). Al(H$_2$PO$_4$)(HPO$_4$)(H$_2$O), ein saures Phosphat mit zweidimensionaler Al-O-P-Vernetzung. *Z. Naturforsch.* **33b**, 1047–1048.

Kokkoros, P. A. and Rentzeperis, P. J. (1958). The crystal structure of the anhydrous sulfates of copper and zinc. *Acta Crystallogr.* **11**, 361–364.

Kolitsch, U. (2001). Refinement of pyrobelonite, PbMn(II)VO$_4$(OH), a member of the descloizite group. *Acta Crystallogr.* **E57**, 119–121.

Kolitsch, U. (2004). (NH$_4$)$_4$Cd(HSe^{IV}O$_3$)$_2$(Se^{VI}O$_4$)$_2$, a new structure type with kröhnkite-like chains. *Acta Crystallogr.* **C60**, i3–i6.

Kolitsch, U. (2006). NaIn(CrO$_4$)$_2$·2H$_2$O, the first indium(III) member of the kröhnkite family. *Acta Crystallogr.* **C62**, i35-i37.

Kolitsch, U. and Tillmans, E. (2003). The crystal structure of anthropogenic Pb$_2$(OH)$_3$(NO$_3$), and a review of Pb-(O,OH) clusters and lead nitrates. *Mineral. Mag.* **67**, 79–93.

Kolitsch, U., Tiekink, E. R. T., Slade, P. G., Taylor, M. R., and Pring, A. (1999a). Hinsdalite and plumbogummite, their atomic arrangements and disordered lead sites. *Eur. J. Mineral.* **11**, 513–520.

Kolitsch, U., Slade, P. G., Tiekink, E. R. T., and Pring, A. (1999b). The structure of antimonian dussertite and the role of antimony in oxysalt minerals. *Mineral. Mag.* **63**, 17–26.

Kolitsch, U., Maczka, J., and Hanuza, J. (2003). $NaAl(MoO_4)_2$: a rare structure type among layered yavapaiite-related $AM(XO_4)_2$ compounds. *Acta Crystallogr.* **E59**, i10-i13.

Kongshaug, K. O., Fjellvåg, H., and Lillerud, K. P. (1999). Layered aluminophosphates II. Crystal structure and thermal behaviour of the layered aluminophosphate UiO-15 and its high temperature variants. *J. Mater. Chem.* **9**, 1591–1598.

Koo, H.-J., Whangbo, M.-H., VerNooy, P. D., Torardi, C. C., and Marshall, W. J. (2002). Flux growth of vanadyl pyrophosphate, $(VO)_2P_2O_7$ and spin dimer analysis of the spin exchange interactions of $(VO)_2P_2O_7$ and vanadyl hydrogen phosphhate, $VO(HPO_4) \cdot 0.5(H_2O)$. *Inorg. Chem.* **41**, 4664–4672.

Koskenlinna, M., and Valkonen, J. (1996). Ammonium uranyl hydrogenselenite selenite. *Acta Crystallogr.* **C52**, 1857–1859.

Koskenlinna, M. and Valkonen, J. (1977). Jahn-Teller distortions in the structure of manganese(III) selenite trihydrate, $Mn_2(SeO_3)_3(H_2O)_3$. *Acta Chem. Scand.* **A31**, 611–614.

Koskenlinna, M., Kansikas, J., and Leskela, T. (1994). Crystal structure and thermal properties of cobalt hydrogenselenitedihydrate, $Co(HSeO_3)_2 \cdot 2(H_2O)$. *Acta Chem. Scand.* **48**, 783–787.

Koskenlinna, M., Mutikainen, I., Leskela, T., and Leskela, M. (1997). Low-temperature crystal structures and thermal decomposition of uranyl hydrogen selenite monohydrate, $UO_2(HSeO_3)_2 \cdot (H_2O)$ and diammonium uranyl selenite hemihydrate, $(NH_4)_2UO_2(SeO_3)_2 \cdot 0.5(H_2O)$. *Acta Chem. Scand.* **51**, 264–269.

Kramer, V. and Post, E. (1985). Preparation and structural characterization of the lead oxide iodide $Pb_3O_2I_2$. *Mater. Res. Bull.* **20**, 407–412.

Krause, M. and Gruehn, R. (1995). Contributions on the thermal behaviour of sulphates XVII. Single crystal structure refinements of $In_2(SO_4)_3$ and $Ga_2(SO_4)_3$. *Z. Kristallogr.* **210**, 427–431.

Krause, W., Effenberger, H., and Brandstaetter, F. (1995). Orthowalpurgite, (UO_2) $Bi_4O_4(AsO_4)_2 \cdot 2(H_2O)$, a new mineral from the Black Forest, Germany. *Eur. J. Mineral.* **7**, 1313–1324.

Krause, W., Bellendorff, K., Bernhardt, H.-J., McCammon, C., Effenberger, H., and Mikenda, W. (1998). Crystal chemistry of tsumcorite-group minerals. New data on ferrilotharmeyerite, tsumcorite, thometzekite, mounanaite, helmutwinklerite, and a redefinition of gartrellite. *Eur. J. Mineral.* **10**, 179–206.

Krause, W., Effenberger, H., Bernhardt, H.-J., and Martin, M. (1999). Cobaltlotharmeyerite, $Ca(Co,Fe,Ni)_2(AsO_4)_2(OH,H_2O)_2$, a new mineral from Schneeberg, Germany. *N. Jb. Mineral. Mh.* **1999**, 505–517.

Krause, W., Bernhardt, H.-J., Effenberger, H., and Martin, M. (2001a). Cobalttsumcorite and nickellotharmeyerite, two new minerals from Schneeberg, Germany. *N. Jb. Mineral. Mh.* **2001**, 558–576.

Krause, W., Blass, G., Bernhardt, H.-J., and Effenberger, H. (2001b). Lukhranite, $CaCuFe^{3+}(AsO_4)_2(H_2O)(OH)$, the calcium analogue of gartrellite. *N. Jb. Mineral. Mh.* **2001**, 481–492.

Krause, W., Bernhardt, H.-J., Effenberger, H., and Witzke, T. (2002). Schneebergite and nickelschneebergite, from Schneeberg, Saxony, Germany: the first Bi-bearing members of the tsumcorite group. *Eur. J. Mineral.* **14**, 115–126.

Krivovichev, S. V. (1997). On the use of Schlegel diagrams for description and classification of mineral crystal structures. *Zap. Vses. Mineral. Obsch. (Proc. Russ. Mineral. Soc.)* **126**(2), 37–46.

Krivovichev, S. V. (1999a). Systematics of fluorite-related structures. I. General principles. *Solid State Sci.* **1**, 211–219.

Krivovichev, S. V. (1999b). Systematics of fluorite-related structures. II. Structural diversity. *Solid State Sci.* **1**, 221–231.

Krivovichev, S. V. (2004a). Topological and geometrical isomerism in minerals and inorganic compounds with laueite-type heteropolyhedral sheets. *N. Jb. Mineral. Mh.* **2004**, 209–220.

Krivovichev, S. V. (2004b). Comparative study of flexibilities of structural units in uranyl sulfates, chromates and molybdates. *Radiochem.* **46**, 401–404.

Krivovichev, S. V. (2006). Crystal chemistry of selenates with mineral-like structures. IV. Crystal structure of $Zn(SeO_4)(H_2O)_2$ – a new compound with a variscite-type heteropolyhedral framework. *Zapiski RMO* **135**(5), 95–101.

Krivovichev, S. V. (2008a). Crystal chemistry of selenates with mineral-like structures. V. Crystal structures of $(H_3O)_2[(UO_2)(SeO_4)_2(H_2O)](H_2O)_2$ and $(H_3O)_2[(UO_2)(SeO_4)_2(H_2O)](H_2O)$ – new compounds with rhomboclase and goldichite topologies. *Zapiski RMO*, **137**(1), 54–61.

Krivovichev, S. V. and Burns, P. C. (2000a). Crystal chemistry of basic lead carbonates. II. Crystal structure of synthetic 'plumbonacrite', $Pb_5O(OH)_2(CO_3)_3$. *Mineral. Mag.* **64**, 1069–1076.

Krivovichev, S. V. and Burns, P. C. (2000b). Crystal chemistry of basic lead carbonates. III. Crystal structures of $Pb_3O_2(CO_3)$ and $NaPb_2(OH)(CO_3)_2$. *Mineral. Mag.* **64**, 1077–1088.

Krivovichev, S. V. and Burns, P.C. (2001a). Crystal chemistry of uranyl molybdates. IV. The structures of $M_2[(UO_2)_6(MoO_4)_7(H_2O)_2]$ (M = Cs, NH_4). *Can. Mineral.*, **39**, 207–214.

Krivovichev, S. V. and Burns, P. C. (2001b). Crystal chemistry of uranyl molybdates. III. New structural themes in the structures of $Na_6[(UO_2)_2O(MoO_4)_4]$, $Na_6[(UO_2)(MoO_4)_4]$ and $K_6[(UO_2)_2O(MoO_4)_4]$. *Can. Mineral.* **39**, 197–206.

Krivovichev, S. V. and Burns, P. C. (2001c). Crystal chemistry of lead oxide chlorides. I. Crystal structures of synthetic mendipite, $Pb_3O_2Cl_2$, and synthetic damaraite, $Pb_3O_2(OH)Cl$. *Eur. J. Mineral.* **13**, 801–809.

Krivovichev, S. V. and Burns, P. C. (2001d). Crystal structure of Br-damaraite, $Pb_3O_2(OH)Br$. *Solid State Sci.* **3**, 455–459.

Krivovichev, S. V. and Burns, P. C. (2002a). Crystal chemistry of rubidium uranyl molybdates: crystal structures of $Rb_6(UO_2)(MoO_4)_4$, $Rb_6(UO_2)_2O(MoO_4)_4$, $Rb_2(UO_2)(MoO_4)_2$, $Rb_2(UO_2)_2(MoO_4)_3$ and $Rb_2(UO_2)_6(MoO_4)_7(H_2O)_2$. *J. Solid State Chem.* **168**, 245–258.

Krivovichev, S. V. and Burns, P. C. (2002b). Crystal chemistry of uranyl molybdates. VI. New uranyl molybdate units in structures of $Cs_4[(UO_2)_3Mo_3O_{14}]$ and $Cs_6[(UO_2)(MoO_4)_4]$. *Can. Mineral.* **40**, 201–209.

Krivovichev, S. V. and Burns, P. C. (2002c). Crystal chemistry of lead oxide chlorides. II. Crystal structure of $Pb_7O_4(OH)_4Cl_2$. *Eur. J. Mineral.* **14**, 135–140.

Krivovichev, S. V. and Burns, P. C. (2002d). Crystal structure of $Pb_{10}O_7(OH)_2F_2(SO_4)$ and crystal chemistry of lead oxysulfate minerals and inorganic compounds. *Z. Kristallogr.* **217**, 451–459.

Krivovichev, S. V. and Burns, P. C. (2003a). Crystal chemistry of K uranyl chromates: crystal structures of $K_8[(UO_2)(CrO_4)_4](NO_3)_2$, $K_5[(UO_2)(CrO_4)_3](NO_3)(H_2O)_3$, $K_4[(UO_2)_3(CrO_4)_5](H_2O)_8$ and $K_2[(UO_2)_2(CrO_4)_3(H_2O)_2](H_2O)_4$. *Z. Kristallogr.* **218**, 725–732.

Krivovichev, S. V. and Burns, P. C. (2003b). Crystal chemistry of uranyl molybdates. VIII. Crystal structures of $Na_3Tl_3[(UO_2)(MoO_4)_4]$, $Na_{13}Tl_3[(UO_2)(MoO_4)_3]_4(H_2O)_5$, $Na_3Tl_5[(UO_2)(MoO_4)_3]_2(H_2O)_3$ and $Na_2[(UO_2)(MoO_4)_2](H_2O)_4$. *Can. Mineral.* **41**, 707–720.

Krivovichev, S. V. and Burns, P. C. (2003c). Combinatorial topology of uranyl molybdate sheets: syntheses and crystal structures of $(C_6H_{14}N_2)_3[(UO_2)_5(MoO_4)_8](H_2O)_4$ and $(C_2H_{10}N_2)[(UO_2)(MoO_4)_2]$. *J. Solid State Chem.* **170**, 106–117.

Krivovichev, S. V. and Burns, P. C. (2003e). Geometrical isomerism in uranyl chromates II. Crystal structures of $Mg_2[(UO_2)_3(CrO_4)_5](H_2O)_{17}$ and $Ca_2[(UO_2)_3(CrO_4)_5](H_2O)_{19}$. *Z. Kristallogr.* **218**, 683–690.

Krivovichev, S. V. and Burns, P. C. (2003f). Synthesis and crystal structure of $Li_2[(UO_2)(MoO_4)_2]$, a uranyl molybdate with chains of corner-sharing uranyl square bipyramids and MoO_4 tetrahedra. *Solid State Sci.* **5**, 481–489.

Krivovichev, S. V. and Burns, P. C. (2003g). Geometrical isomerism in uranyl chromates I. Crystal structures of $(UO_2)(CrO_4)(H_2O)_2$, $[(UO_2)(CrO_4)(H_2O)_2](H_2O)$ and $[(UO_2)(CrO_4)(H_2O)_2]_4(H_2O)_9$. *Z. Kristallogr.* **218**, 568–574.

Krivovichev, S. V. and Burns, P. C. (2003h). First sodium uranyl chromate, $Na_4[(UO_2)(CrO_4)_3]$: synthesis and crystal structure determination. *Z. Anorg. Allg. Chem.* **629**, 1965–1968.

Krivovichev, S. V. and Burns, P. C. (2003i). Crystal chemistry of lead oxide phosphates: crystal structures of $Pb_4O(PO_4)_2$, $Pb_8O_5(PO_4)_2$ and $Pb_{10}(PO_4)_6O$. *Z. Kristallogr.* **218**, 357–365.

Krivovichev, S. V. and Burns, P. C. (2003j). Chains of edge-sharing OPb_4 tetrahedra in the structures of $Pb_4O(VO_4)_2$ and related minerals and inorganic compounds. *Can. Mineral.* **41**, 951–958.

Krivovichev, S. V. and Burns, P. C. (2004). Synthesis and crystal structure of $Cu_2[(UO_2)_3((S,Cr)O_4)_5](H_2O)_{17}$. *Radiochem.* **46**, 408–411.

Krivovichev, S. V. and Burns, P. C. (2005). Crystal chemistry of uranyl molybdates. XI. Crystal structures of $Cs_2[(UO_2)(MoO_4)_2]$ and $Cs_2[(UO_2)(MoO_4)_2](H_2O)$. *Can. Mineral.* **43**, 713–720.

Krivovichev, S. V. and Burns, P. C. (2006). The crystal structure of $Pb_8O_5(OH)_2Cl_4$ – a synthetic analogue of blixite? *Can. Mineral.* **44**, 515–522.

Krivovichev, S. V. and Filatov, S. K. (1998). Conformation of single chains of anion-centred edge-sharing tetrahedra. *Z. Kristallogr.* **213**, 316–318.

Krivovichev, S. V. and Filatov, S. K. (1999a). Structural principles for minerals and inorganic compounds containing anion-centered tetrahedra. *Am. Mineral.* **84**, 1099–1106.

Krivovichev, S. V. and Filatov, S. K. (1999b). Metal arrays in structural units based on anion-centered metal tetrahedra. *Acta Crystallogr.* **B55**, 664–676.

Krivovichev, S. V. and Filatov, S. K. (2001). *Crystal chemistry of minerals and inorganic compounds with complexes of anion-centered tetrahedra*. St. Petersburg University Press, St. Petersburg.

Krivovichev S. V. and Kahlenberg V. (2004). Synthesis and crystal structures of α- and β-$Mg_2[(UO_2)_3(SeO_4)_5](H_2O)_{16}$. *Z. Anorg. Allg. Chem.* **630**, 2736–2742.

Krivovichev, S. V. and Kahlenberg, V. (2005a). Structural diversity of sheets in Rb uranyl selenates: synthesis and crystal structures of $Rb_2[(UO_2)(SeO_4)_2(H_2O)](H_2O)$, $Rb_2[(UO_2)_2(SeO_4)_3(H_2O)_2](H_2O)_4$, $Rb_4[(UO_2)_3(SeO_4)_5(H_2O)]$. *Z. Anorg. Allg. Chem.* **631**, 739–744.

Krivovichev, S. V. and Kahlenberg, V. (2005b). Preparation and crystal structures of $M[(UO_2)(SeO_4)_2(H_2O)](H_2O)_4$ (M = Mg, Zn). *Z. Naturforsch.* **62b**, 538–542.

Krivovichev, S. V. and Kahlenberg, V. (2005c). Low-dimensional structural units in amine-templated uranyl selenates: synthesis and structures of $[C_3H_{12}N_2]$ $[(UO_2)(SeO_4)_2(H_2O)_2](H_2O)$, $[C_5H_{16}N_2]_2[(UO_2)(SeO_4)_2(H_2O)](NO_3)_2$, $[C_4H_{12}N]$ $[(UO_2)(SeO_4)(NO_3)]$, and $[C_4H_{14}N_2][(UO_2)(SeO_4)_2(H_2O)]$. *Z. Anorg. Allg. Chem.* **631**, 2352–2357.

Krivovichev S. V. and Kahlenberg, V. (2005d). Crystal structure of $(H_3O)_2[(UO_2)_2$ $(SeO_4)_3(H_2O)_2](H_2O)_{3.5}$. *Radiochem.* **47**, 412–414.

Krivovichev S. V. and Kahlenberg, V. (2005e). Crystal structure of $(H_3O)_6$ $[(UO_2)_5(SeO_4)_8(H_2O)_5](H_2O)_5$. *Radiochem.* **47**, 415–418.

Krivovichev S. V. and Kahlenberg, V. (2005f). Synthesis and crystal structures of $M_2[(UO_2)_3(SeO_4)_5](H_2O)_{16}$ (M = Co, Zn). *J. Alloys Compd.* **395**, 41–47.

Krivovichev S. V. and Kahlenberg, V. (2005g). Synthesis and crystal structure of $Zn_2[(UO_2)_3(SeO_4)_5](H_2O)_{17}$. *J. Alloys Compd.* **389**, 55–60.

Krivovichev, S. V., Filatov, S. K., and Semenova, T. F. (1997). On the systematics of polyions of linked polyhedra. *Z. Kristallogr.* **212**, 411–417.

Krivovichev, S. V., Filatov, S. K., and Semenova, T. F. (1998a). Types of cationic complexes on the base of oxocentered tetrahedra $[OM_4]$ in crystal structures of inorganic compounds. *Russ. Chem. Rev.* **67**, 137–155.

Krivovichev, S. V., Filatov, S. K., Semenova, T. F., and Rozhdestvenskaya, I. V. (1998b). Crystal chemistry of inorganic compounds based on chains of oxocentered tetrahedra. I. Crystal structure of chloromenite, $Cu_9O_2(SeO_3)_4Cl_6$. *Z. Kristallogr.* **213**, 645–649.

Krivovichev, S. V., Starova, G. L., Filatov, S. K. (1999a). "Face-to-face" relationships between oxocentered tetrahedra and cation-centered tetrahedral oxyanions in crystal structures of minerals and inorganic compounds. *Mineral. Mag.* **63**, 263–266.

Krivovichev, S. V., Shuvalov, R. R., Semenova, T. F., and Filatov, S. K. (1999b). Crystal chemistry of inorganic compounds based on chains of oxocentered tetrahedra. III. The crystal structure of georgbokiite. *Z. Kristallogr.* **214**, 135–138.

Krivovichev, S. V., Li, Y., and Burns, P. C. (2001). Crystal chemistry of lead oxide hydroxide nitrates. III. The crystal structure of $[Pb_3O_2](OH)(NO_3)$. *J. Solid State Chem.* **158**, 78–81.

Krivovichev, S. V., Finch, R., and Burns, P. C. (2002a). Crystal chemistry of uranyl molybdates. V. Topologically different uranyl molybdate sheets in structures of $Na_2[(UO_2)(MoO_4)_2]$ and $K_2[(UO_2)(MoO_4)_2](H_2O)$. *Can. Mineral.* **40**, 193–200.

Krivovichev, S. V., Britvin, S. N., Burns, P. C., and Yakovenchuk, V. N. (2002b). Crystal structure of rimkorolgite, $Ba[Mg_5(H_2O)_7(PO_4)_4](H_2O)$, and its comparison with bakhchisaraitsevite. *Eur. J. Mineral.* **14**, 397–402.

Krivovichev, S. V., Cahill, C. L., and Burns, P. C. (2002c). Syntheses and structures of two topologically related modifications of $Cs_2[(UO_2)_2(MoO_4)_3]$. *Inorg. Chem.* **41**, 34–39.

Krivovichev, S. V., Filatov, S. K., and Burns, P. C. (2002d). Cuprite-like framework of OCu_4 tetrahedra in the crystal structure of synthetic melanothallite, Cu_2OCl_2, and its negative thermal expansion. *Can. Mineral.* **40**, 1185–1190.

Krivovichev, S. V., Yakovenchuk, V. N., Burns, P. C., Pakhomovsky, Ya. A., and Menshikov, Yu. P. (2003a). Cafetite, $Ca[Ti_2O_5](H_2O)$: crystal structure and revision of chemical formula. *Am. Mineral.* **88**, 424–429.

Krivovichev, S. V., Armbruster, T., and Pekov, I. V. (2003b). Cation frameworks in crystal structures of Ba and rare earth fluorocarbonates: crystal structure of kukharenkoite-(La), $Ba_2(La,Ce)(CO_3)_3F$. *Zapiski RMO* **132**(3), 65–72.

Krivovichev, S. V., Cahill, C. L., and Burns P. C. (2003c). New uranyl molybdate open framework in the structure of $(NH_4)_4(UO_2)_5(MoO_4)_7](H_2O)$. *Inorg. Chem.* **42**, 2459–2464.

Krivovichev, S. V., Armbruster, T., and Depmeier, W. (2004a). Crystal structures of $Pb_8O_5(AsO_4)_2$ and $Pb_5O_4(CrO_4)$, and review of PbO-related structural units in inorganic compounds. *J. Solid State Chem.* **177**, 1321–1332.

Krivovichev, S. V., Filatov, S. K., Armbruster, T., and Pankratova, O. Yu. (2004b). Crystal structure of $Cu(I)Cu(II)_4O(SeO_3)Cl_5$, a new heterovalent copper compound. *Dokl. Chem.* **399**, 226–228.

Krivovichev, S. V., Avdontseva, E. Yu., and Burns, P. C. (2004c). Synthesis and crystal structure of $Pb_3O_2(SeO_3)$. *Z. Anorg. Allg. Chem.* **630**, 558–562.

Krivovichev, S. V., Yakovenchuk, V. N., and Pakhomovsky, Ya. A. (2004d). Topology and symmetry of titanosilicate framework in the crystal structure of shcherbakovite, $Na(K,Ba)_2(Ti,Nb)_2O_2[Si_4O_{12}]$. *Zapiski VMO* **133**(3), 55–63.

Krivovichev, S. V., Yakovenchuk, V. N., Armbruster, T., Döbelin, N., Pattison, P., Weber, H.-P., and Depmeier, W. (2004e). Porous titanosilicate nanorods in the structure of yuksporite, $(Sr,Ba)_2K_2(Ca,Na)_{14}(\square,Mn,Fe)\{(Ti,Nb)_4(O,OH)_4[Si_6O_{17}]_2 [Si_2O_7]_3\}(H_2O,OH)_n$, resolved using synchrotron radiation. *Am. Mineral.* **89**, 1561–1565.

Krivovichev, S. V., Kahlenberg, V., Tananaev, I. G., and Myasoedov, B. F. (2005a). Amine-templated uranyl selenates with layered structures. I. Structural diversity of sheets with U:Se = 1:2. *Z. Anorg. Allg. Chem.* **631**, 2358–2364.

Krivovichev, S. V., Tananaev, I. G., Kahlenberg, V., and Myasoedov, B. F. (2005b). $[C_5H_{14}N][(UO_2)(SeO_4)(SeO_2OH)]$ – the first uranyl selenite(IV)–selenate(VI). *Dokl. Phys. Chem.* **403**, 124–127.

Krivovichev, S. V., Locock, A. J., and Burns, P. C. (2005c). Lone electron pair stereoactivity, cation arrangements and distortion of heteropolyhedral sheets in the structure of $Tl_2[(UO_2)(AO_4)_2]$ (A = Cr, Mo). *Z. Kristallogr.* **220**, 10–18.

Krivovichev, S. V., Kahlenberg, V., Kaindl, R., and Mersdorf, E. (2005d). Self-assembly of protonated 1,12-dodecanediamine molecules and strongly undulated uranyl selenate sheets in the structure of amine-templated uranyl selenate: $(H_3O)_2[C_{12}H_{30}N_2]_3[(UO_2)_4(SeO_4)_8](H_2O)_5$. *Eur. J. Inorg. Chem.* **2005**, 1653–1656.

Krivovichev, S. V., Filatov, S. K., Cherepansky, P. N., Armbruster, T., and Pankratova, O. Yu. (2005e). Crystal structure of γ-$Cu_2V_2O_7$ and its comparison to blossite (α-$Cu_2V_2O_7$) and ziesite (β-$Cu_2V_2O_7$). *Can. Mineral.* **43**, 671–678.

Krivovichev, S. V., Kahlenberg, V., Kaindl, R., Mersdorf, E., Tananaev, I. G., and Myasoedov, B. F. (2005f). Nanoscale tubules in uranyl selenates. *Angew. Chem. Int. Ed.* **44**, 1134–1136.

Krivovichev, S. V., Kahlenberg, V., Tananaev, I. G., Kaindl, R., Mersdorf, E., and Myasoedov, B. F. (2005g). Highly porous uranyl selenate nanotubules. *J. Am. Chem. Soc.* **127**, 1072–1073.

Krivovichev, S. V., Cahill, C. L., Nazarchuk, E. V., Burns, P. C., Armbruster, T., and Depmeier W. (2005h). Chiral open-framework uranyl molybdates. 1. Topological diversity: synthesis and crystal structure of $[(C_2H_5)_2NH_2]_2[(UO_2)_4(MoO_4)_5(H_2O)]$ (H_2O). *Micropor. Mesopor. Mater.* **78**, 209–215.

Krivovichev, S. V., Burns, P. C., Armbruster, T., Nazarchuk, E. V., and Depmeier, W. (2005i). Chiral open-framework uranyl molybdates. 2. Flexibility of the U:Mo = 6:7 frameworks: syntheses and crystal structures of $(UO_2)_{0.82}[C_8H_{20}N]_{0.36}[(UO_2)_6$ $(MoO_4)_7(H_2O)_2](H_2O)_n$ and $[C_6H_{14}N_2][(UO_2)_6(MoO_4)_7(H_2O)_2](H_2O)_m$. *Micropor. Mesopor. Mater.* **78**, 217–224.

Krivovichev, S. V., Armbruster, T., Chernyshov, D. Yu., Burns, P. C., Nazarchuk, E. V., and Depmeier W. (2005j). Chiral open-framework uranyl molybdates. 3. Synthesis, structure and the $C222_1 \rightarrow P2_12_12_1$ low-temperature phase transition of $[C_6H_{16}N]_2[(UO_2)_6(MoO_4)_7(H_2O)_2](H_2O)_2$. *Micropor. Mesopor. Mater.* **78**, 225–234.

Krivovichev, S. V., Gurzhii, V. V., Tananaev, I. G., and Myasoedov, B. F. (2006a). Topology of inorganic complexes as a function of amine molecular structure in layered uranyl selenates. *Dokl. Phys. Chem.* **409**, 228–232.

Krivovichev, S. V., Tananaev, I. G., Kahlenberg, V., and Myasoedov, B. F. (2006b). Synthesis and crystal structure of a new uranyl selenite(IV)–selenate(VI), $[C_5H_{14}N]_4$ $[(UO_2)_3(SeO_4)_4(HSeO_3)(H_2O)](H_2SeO_3)(HSeO_4)$. *Radiochem.* **48**, 217–222.

Krivovichev, S. V., Tananaev, I. G., and Myasoedov, B. F. (2006c). Geometrical isomerism of layered complexes in uranyl selenates: synthesis and structure of

$(H_3O)[C_5H_{14}N]_2[(UO_2)_3(SeO_4)_4(HSeO_4)(H_2O)]$ and $(H_3O)[C_5H_{14}N]_2[(UO_2)_3(SeO_4)_4$ $(HSeO_4)(H_2O)](H_2O)$. *Radiochem.* **48**, 552–560.

Krivovichev, S. V., Siidra, O. I., Nazarchuk, E. V., Burns, P. C., and Depmeier, W. (2006d). Particular topological complexity of lead oxide blocks in $Pb_{31}O_{22}X_{18}$ (X = Br,Cl). *Inorg. Chem.* **45**, 3846–3848.

Krivovichev, S. V., Filatov, S. K., Burns, P. C., and Vergasova, L. P. (2006e). The crystal structure of allochalcoselite, $Cu^+Cu^{2+}_5PbO_2(SeO_3)_2Cl_5$, a mineral with well-defined Cu^+ and Cu^{2+} positions. *Can. Mineral.* **44**, 507–514.

Krivovichev, S. V., Gurzhiy, V.V., Tananaev, I. G., and Myasoedov, B. F. (2007a). Microscopic model of crystallogenesis from aqueous solutions of uranyl selenate. In: Krivovichev, S. V. (ed.). Crystallogenesis and mineralogy. spec. issue. Zapiski RMO, pp. 91–114.

Krivovichev, S. V., Vergasova, L. P., Britvin, S. N., Filatov, S. K., Kahlenberg, V., and Ananiev, V. V. (2007b). Pauflerite, β-$VO(SO_4)$, a new mineral from the Tolbachik volcano, Kamchatka peninsula, Russia. *Can. Mineral.* **45**, 921–927.

Krivovichev, S. V., Pakhomovsky, Ya. A., Ivanyuk, G. Yu., Mikhailova, J. A., Men'shikov, Yu. P., Armbruster, T., Selivanova, E. A., and Meisser, N. (2007c). Yakovenchukite-(Y) $K_3NaCaY_2[Si_{12}O_{30}](H_2O)_4$, a new mineral from the Khibiny massif, Kola Peninsula, Russia: A novel type of octahedral–tetrahedral open-framework structure. *Am. Mineral.* **92**, 1525–1530.

Krivovichev, S. V., Filatov, S. K., Burns, P. C., and Vergasova, L. P. (2007d). The crystal structure of parageorgbokiite. *Can. Mineral.* **45**, 929–934.

Krivovichev, S. V., Yakovenchuk, V. N., Ivanyuk, G. Yu., Pakhomovsky, Ya. A., Armbruster, T., and Selivanova, E. A. (2007e). The crystal structure of nacaphite, $Na_2Ca(PO_4)F$: A re-investigation. *Can. Mineral.* **45**, 915–920.

Krivovichev, S. V., Burns, P. C., and Tananaev, I. G., eds. (2007f). *Structural chemistry of inorganic actinide compounds.* Elsevier, Amsterdam.

Krivovichev, S. V., Burns, P. C., Tananaev, I. G., and Myasoedov, B. F. (2007g). Nanostructured actinide compounds. *J. Alloys Comp.* **444–445**, 457–463.

Krogh Andersen, A.M. and Norby, P. (2000). Ab initio structure determination and Rietveld refinement of a high-temperature phase of zirconium hydrogen phosphate and a new polymorph of zirconium pyrophosphate from in-situ temperature-resolved powder diffraction data. *Acta Crystallogr.* **B56**, 618–625.

Krogh Andersen, A. M., Norby, P., and Vogt, T. (1998). Determination of formation regions of titanium phosphates; determination of the crystal structure of beta-titanium phosphate, $Ti(PO_4)(H_2PO_4)$ from neutron powder data. *J. Solid State Chem.* **140**, 266–271

Kuznetsov, B. Ya., Rogachev, D. L., Porai-Koshits, M. A., and Dikareva, L. M. (1974). Crystal structure of rubidium oxodisulfatoniobate. *Izv. Akad. Nauk SSSR, Ser. Khim.* **1974**, 2167–2170.

Kuznicki, S. M., Bell, V. A., Nair, S., Hillhouse, H. W., Jacubinas, R. M., Braunbarth, C. M., Toby, B. H., and Tsapatis, M. (2001). A titanosilicate molecular sieve with adjustable pores for size-selective adsorption of molecules. *Nature* **412**, 720–724.

Lafront, A.-M., Trombe, J. C., and Bonvoisin, J. (1995). Layered hydrogenselenites. II. Synthesis, structure studies and magnetic properties of a novel series of bimetallic hydrogenselenites: $(Cu(HSeO_3)_2MCl_2(H_2O)_4)$, M(II) = Mn, Co, Ni, Cu, Zn. *Inorg. Chim. Acta* **238**, 15–22.

Lagercrantz, A. and Sillén, L. G. (1948). On the crystal structure of $Bi_2O_2CO_3$ (bismutite) and $CaBi_2O_2(CO_3)_2$ (beyerite). *Ark. Kemi, Miner. Geol.* **A25**, 1–21.

Langecker, C. and Keller, H. L. (1994). $Ag_2Pb_8O_7Cl_4$, ein neues Blei(II)-oxidhalogenid mit Silber. *Z. Anorg. Allg. Chem.* **620**, 1229–1233.

Latrach, A., Mentzen, B. F., and Bouix, J. (1985a). Polymorphisme du dioxotriplomb(II) tetraoxosulfate(VI). *Mater. Res. Bull.* **20**, 853–861.

Latrach, A., Mentzen, B. F., and Bouix, J. (1985b). Polymorphisme du dioxotri-plomb (II) tetraoxosulfate(VI) $Pb_3O_2(SO_4)$ a basse temperature. II Etude structurale de la phase ferroelastique gamma- $Pb_3O_2(SO_4)$ par diffraction neutonique sur poudres a 2 et 45 K. *Mater. Res. Bull.* **20**, 1081–1090.

le Fur, E. and Pivan, J. Y. (1999). Synthesis, crystal structure and magnetic properties of the novel stacking variant lamellar lead oxovanadium phosphate hydrate β-$(Pb(VOPO_4)_2 \cdot 4(H_2O))$. Crystal chemical relationships with related intercalated $M(VOPO_4)_2 \cdot y(H_2O)$. *J. Mater. Chem.* **9**, 2589–2594.

le Fur, E. and Pivan, J. Y. (2001). The new mercury vanadium phosphate $Hg_4VO(PO_4)_2$ containing Hg_2^{2+} dumbbells: crystal structure and thermal and magnetic properties. *J. Solid State Chem.* **158**, 94–99.

le Fur, E., Pena, O., and Pivan, J. Y. (1999). Magnetic and thermal properties of vanadium phosphates hydrates $M(II)(VOPO_4)_2 \cdot 4(H_2O)$ $(M(II) = Ca^{2+}, Ba^{2+}$ and $Cd^{2+})$. *J. Alloys Compd.* **285**, 89–97.

le Marouille, J. Y., Bars, O., and Grandjean, D. (1980). Etude de chromates, molybdates et tungstates hydrates. III. Etude structurale du molybdate de zinc dihydrate. *Acta Crystallogr.* **B36**, 2558–2560.

le Meins, J.-M., Hemon-Ribaud, A., Laligant, Y., and Courbion, G. (1997). A new fluorophosphate with a laueite-type structural unit: synthesis, TEM study and crystal structure of $SrFePO_4F_2$. *Eur. J. Solid State Inorg. Chem.* **34**, 391–404.

le Meins, J.-M., Hemon-Ribaud, A., and Courbion, G. (1998). Synthesis and crystal structure of two fluorophosphated compounds with different infinite sheets: $Sr_2Ga(HPO_4)(PO_4)F_2$ and $Sr_2Fe_2(HPO_4)(PO_4)_2F_2$. *Eur. J. Solid State Inorg. Chem.* **35**, 117–132.

Leavens, P. B. and Rheingold, A. L. (1988). Crystal structures of gordonite, $MgAl_2(PO_4)_2(OH)_2(H_2O)_6(H_2O)_2$, and its Mn analog. *N. Jb. Mineral. Mh.* **1988**, 265–270.

Leblanc, M. and Férey, G. (1990). Room-temperature structures of oxocopper(II) vanadate(V) hydrates, $Cu_3V_2O_8(H_2O)$ and $CuV_2O_6(H_2O)_2$. *Acta Crystallogr.* **C46**, 15–18.

Leclaire, A., Borel, M. M., Chardon, J., Grandin, A., and Raveau, B. (1992). A niobium silicophosphate belonging to the niobium phosphate bronze series: $K_4Nb_8P_4SiO_{34}$. *Acta Crystallogr.* **C48**, 1744–1747.

Leclaire, A., Borel, M. M., Grandin, A., Raveau, B. (1994). The mixed valent molybdenum monophosphate $RbMo_2P_2O_{10} \cdot (1-x)H_2O$: an intersecting tunnel structure isotypic with leucophosphite. *J. Solid State Chem.* **108**, 177–183.

Lee, K.-S. and Kwon, Y.-U. (1996). Crystal structure of $V_2Se_2O_9$ and phase relations in the $V_2O_5 - SeO_2$ system. *J. Kor. Chem. Soc.* **40**, 379–383.

Leech, M. A., Cowley, A. R., Prout, K., and Chippindale, A. M. (1998). Ambient-temperature synthesis of new layered AlPOs and GaPOs in silica gels. *Chem. Mater.* **10**, 451–456.

Lefebre, J. J. and Gasparrini, C. (1980). Florensite, an occurence in the Zairian copperbelt. *Can. Mineral.* **18**, 301–311.

Lehmann, M. S., Christensen, A. N., and Fjellvag, H. (1987). Structure determination by use of pattern decomposition and the Rietveld method on synchroton X-ray and neutron powder data; the structures of $Al_2Y_4O_9$ and I_2O_4. *J. Appl. Crystallogr.* **20**, 123–129.

Lengauer, C. L., Giester, G., and Irran, E. (1994). $KCr_3(SO_4)_2(OH)_6$: synthesis, characterization, powder diffraction data, and structure refinement by the Rietveld technique and a compilation of alunite-type compounds. *Powder Diffr.* **9**, 265–271.

Leonowicz, M. E., Johnson, J. W., Brody, J. F., Shannon, H. F., and Newsam, J. M. (1985). Vanadyl hydrogenphosphate hydrates: $VO(HPO_4)(H_2O)_4$ and $VO(HPO_4)$ $(H_2O)_{.5}$. *J. Solid State Chem.* **56**, 370–378.

Liebau, F. (1985). *Structural chemistry of silicates. Structure, bonding and classification.* Springer, Berlin.

Liebau, F. (2003). Ordered microporous and mesoporous materials with inorganic hosts: definitions of terms, formula notation, and systematic classification. *Micropor. Mesopor. Mater.* **58**, 15–72.

Li, Y. and Burns, P. C. (2001). The structures of two sodium uranyl compounds relevant to nuclear waste disposal. *J. Nucl. Mater.* **299**, 219–226.

Li, N. and Xiang, S. (2002). Hydrothermal synthesis and crystal structure of two novel aluminophosphites containing infinite Al-O-Al chains. *J. Mater. Chem.* **12**, 1397–1400.

Li, J.-J., Zhou, J.-L., and Dong, W. (1990). The structure of amarillite. *Chinese Sci. Bull.* **35**, 2073–2075.

Li, G., Peacor, D. R., Essene, E. J., Brosnahan, D. R., and Beane, R. E. (1992). Walthierite, $Ba_{0.5}\square_{0.5}[Al_3(OH)_6(SO_4)_2]$, and huangite, $Ca_{0.5}\square_{0.5}[Al_3(OH)_6(SO_4)_2]$, two new minerals of the alunite group from the Coquimbo region, Chile. *Am. Mineral.* **77**, 1275–1284.

Li, H., Eddaoudi, M., O'Keeffe, M., and Yaghi, O. M. (1999). Design and synthesis of an exceptionally stable and highly porous metal-organic framework. *Nature* **402**, 276–279.

Li, Y., Krivovichev, S. V., and Burns, P. C. (2000). Crystal chemistry of lead oxide hydroxide nitrates. I. The crystal structure of $[Pb_6O_4](OH)(NO_3)(CO_3)$. *J. Solid State Chem.* **153**, 365–370.

Li, Y., Krivovichev, S. V., and Burns, P. C. (2001). Crystal chemistry of lead oxide hydroxide nitrates. II. The crystal structure of $[Pb_{13}O_8](OH)_6(NO_3)_4$. *J. Solid State Chem.* **158**, 74–77.

Liao, Y.-C., Luo, S.-H., Wang, S.-L., Kao, H.-M., and Lii, K.-H. (2000). Synthesis and characterization of organically templated metal arsenates with a layer structure: $[NH_3(CH_2)_3NH_3]_{0.5}[M(OH)AsO_4]$ (M = Ga, Fe). *J. Solid State Chem.* **155**, 37–41.

Liebau, F. (1985). *Structural chemistry of silicates. Structure, bonding and classification.* Springer, Berlin.

Lightfoot, P., Cheetham, A. K., and Sleight, A. W. (1988). Synthesis and crystal structure of $KMn_2O(PO_4)(HPO_4)$. *J. Solid State Chem.* **73**, 325–329.

Lii, K.-H. (1996a). Hydrothermal synthesis and crystal structure of $Na_3In(PO_4)_2$. *Eur. J. Solid State Inorg. Chem.* **33**, 519–526.

Lii, K.-H. (1996b). $RbIn(OH)PO_4$: an indium(III) phosphate containing spirals of corner-sharing InO_6 octahedra. *J. Chem. Soc.,Dalton Trans.* **1996**, 815–818.

Lii, K.-H. and Huang, Y.-F. (1997a). $[HN(CH_2CH_2)_3NH]_3[Fe_8(HPO_4)_{12}(PO_4)_2(H_2O)_6]$: an organically templated iron phosphate with a pillared layer structure. *Dalton Trans.* **1997**, 2221–2225.

Lii, K.-H. and Huang, Y. F. (1997b). Large tunnels in the hydrothermally synthesized open-framework iron phosphate $[H_3N(CH_2)_3NH_3]_2[Fe_4(OH)_3(HPO_4)_2(PO_4)_3] \cdot xH_2O$. *Chem. Commun.* **1997**, 839–840.

Lii, K.-H. and Mao, L. F. (1992). Hydrothermal synthesis and structural characterization of a nickel(II) vanadyl(IV) phosphate hydrate: $Ni_{0.5}VOPO_4 \cdot 1.5H_2O$. *J. Solid State Chem.* **96**, 436–441.

Lii, K.-H. and Tsai, H.-J. (1991a). Hydrothermal synthesis and crystal structure of the vanadyl(IV) hydrogen phosphate $K_2(VO)_2P_3O_9(OH)_3 \cdot 1.125H_2O$. *Inorg. Chem.* **30**, 446–448.

Lii, K.-H. and Tsai, H.-J. (1991b). Hydrothermal synthesis and characterization of vanadyl(IV) hydrogenphosphate, $K_2(VO)_3(HPO_4)_4$. *J. Solid State Chem.* **91**, 331–338.

Lii, K.-H. and Wang, S.-L. (1997). $Na_3M(OH)(HPO_4)(PO_4)$, M = Al, Ga: two phosphates with a chain structure. *J. Solid State Chem.* **128**, 21–26.

Lii, K.-H., Wang, Y. P., Cheng, C.-Y., Wang, S.-L., and Ku, H. C. (1990). Crystal structure and magnetic properties of a vanadium(IV) pyrophosphate: $Rb_2V_3P_4O_{17}$. *J. Chin. Chem. Soc.* **37**, 141–149.

Lii, K.-H., Li, C.-H., Cheng, C.-Y., and Wang, S.-L. (1991a). Hydrothermal synthesis, structure and magnetic properties of a new polymorph of lithium vanadyl(IV) orthophosphate : β-LiVOPO$_4$. *J. Solid State Chem.* **95**, 352–359.

Lii, K.-H., Li, C.-H., Chen, T.-M., and Wang, S.-L. (1991b). Synthesis and structural characterization of sodium vanadyl(IV) orthophosphate $NaVOPO_4$. *Z. Kristallogr.* **197**, 67–73.

Lii, K.-H., Lee, T.-C., Liu, S.-N., and Wang, S.-L. (1993a). Hydrothermal synthesis and crystal structures of $Sr_2V(PO_4)_2(H_2PO_4)$ and $Sr_2Fe(PO_4)_2(H_2PO_4)$. *Dalton Trans.* **1993**, 1051–1054.

Lii, K., Wu, L., and Gau, H. (1993b). Crystal structure of $Ni_{0.5}VOPO_4 \cdot 2H_2O$, a layered vanadyl(IV) phosphate hydrate with tetragonal symmetry. *Inorg. Chem.* **32**, 4153–4154.

Lima-de-Faria, J., Hellner, E., Liebau, F., Makovicky, E., and Parthé, E. (1990). Nomenclature of inorganic structure types: report of the IUCr comission on crystallographic nomenclature subcommittee on the nomenclature of inorganic structure types. *Acta Crystallogr.* **A46**, 1–11.

Lin, H.-M. and Lii, K.-H. (1998). Synthesis and structure of [(1*R*,2*R*)-C$_6$H$_{10}$(NH$_3$)$_2$] [Ga(OH)(HPO$_4$)$_2$] · H$_2$O, the first metal phosphate containing a chiral amine. *Inorg. Chem.* **37**, 4220–4222.

Lin, Z., Rocha, J., Brandao, P., Ferreira, A., Esculcas, A. P., Pedrosa de Jesus, J. D., Philippou, A., and Anderson, M. W. (1997). Synthesis and structural characterization of microporous umbite, penkvilksite, and other titanosilicates. *J. Phys. Chem.* **B101**, 7114–7120.

Linde, S. A., Gorbunova, Yu. E., Lavrov, A. V., Kuznetsov, V. G. (1979). The crystal structure of the vanadyl acid orthophospate VO(H$_2$PO$_4$)$_2$. *Dokl. Akad. Nauk SSSR* **244**, 1411–1414.

Lindgren, O. (1977a). The crystal structure of cerium(IV) sulfate tetrahydrate, Ce(SO$_4$)$_2$(H$_2$O)$_4$. *Acta Chem. Scand.* **A31**, 453–456.

Lindgren, O. (1977b). The crystal structure of cerium(IV) dichromate dihydrate, Ce(CrO$_4$)$_2$(H$_2$O)$_2$. *Acta Chem. Scand.* **A31**, 167–170.

Liu, Y., Na, L., Zhu, G., Xiao, F.S., Pang, W., and Xu, R. (2000). Synthesis and structure of a novel layered zinc–cobalt phosphate, Zn$_{(2-x)}$Co$_x$(HPO$_4$)$_3$ · C$_3$N$_2$H$_{12}$, *x*~0.05, with 12 Rings. *J. Solid State Chem.* **149**, 107–112.

Liu, Z. C., Weng, L. H., Zhou, Y. M., Chen, Z. X., and Zhao, D. Y. (2003a). Synthesis of a new organically templated zeolite-like zirconogermanate (C$_4$N$_2$H$_{12}$)[ZrGe$_4$O$_{10}$F$_2$] with cavansite topology. *J. Mater. Chem.* **13**, 308–311.

Livage, C., Millange, F., Walton, R.I., Loiseau, T., Simon, N., O'Hare, D., and Férey, G. (2001). Ambient temperature crystallisation of a lamellar gallium fluorophosphate from the synthetic solution of microporous ULM-5. *Chem. Commun.* **2001**, 994–995.

Locock, A. J. (2007). Crystal chemistry of actinide phosphates and arsenates. In: Krivovichev, S. V., Burns, P. C., and Tananaev, I. G. *Structural chemistry of inorganic actinide compounds*, Elsevier, Amsterdam, pp. 217–278.

Locock, A. J. and Burns, P. C. (2002a). The crystal structure if triuranyl diphosphate tetrahydrate. *J. Solid State Chem.* **163**, 275–280.

Locock, A. J. and Burns, P. C. (2002b). Crystal structures of three framework alkali metal uranyl phosphate hydrates. *J. Solid State Chem.* **167**, 226–236.

Locock, A. J. and Burns, P. C. (2003a). Crystal structures and synthesis of the copper-dominant members of the autunite and meta-autunite groups: torbernite, zeunerite, metatorbernite and metazeunerite. *Can. Mineral.* **41**, 489–502.

Locock, A. J. and Burns, P. C. (2003b). The crystal structure of synthetic autunite, Ca[(UO$_2$)(PO$_4$)]$_2$(H$_2$O)$_{11}$. *Am. Mineral.* **88**, 240–244.

Locock, A. J. and Burns, P. C. (2003c). The structure of hugelite, an arsenate of the phosphuranylite group, and its relationship to dumontite. *Mineral. Mag.* **67**, 1109–1120.

Locock, A. J. and Burns, P. C. (2003d). The crystal structure of bergenite, a new geometrical isomer of the phosphuranylite group. *Can. Mineral.* **41**, 91–101.

Locock, A. J. and Burns, P. C. (2003e). Structures and syntheses of framework triuranyl diarsenate hydrates. *J. Solid State Chem.* **176**, 18–26.

Locock, A. J., Burns, P. C., and Flynn, T. M. (2005). The role of water in the structures of synthetic hallimondite, $Pb_2[(UO_2)(AsO_4)_2](H_2O)_n$ and synthetic parsonsite, $Pb_2[(UO_2)(PO_4)_2](H_2O)_n$, $0 < n < 0.5$. *Am. Mineral.* **90**, 240–246.

Loiseau, T. and Férey, G. (1994). Crystal structure of $(NH_4)(Ga_2(PO_4)_2(OH)(H_2O))\cdot(H_2O)$, isotypic with ALPO4–15. *Eur. J. Solid State Inorg. Chem.* **31**, 575–581.

Loiseau, T., Calage, Y., Lacorre, P., and Férey, G. (1994). NH_4FePO_4F: structural study and magnetic properties. *J. Solid State Chem.* **111**, 390–396.

Loiseau, T., Paulet, C., and Férey G. (1998). Crystal structure determination of the hydrated gallium phosphate $GaPO_4 \cdot 2(H_2O)$, analog of variscite. *C.R. Acad. Sci. Paris. Ser. II*, **1**, 667–674.

Loiseau, T., Paulet, C., Simon, N., Munch, V., Taulelle, F., and Férey, G. (2000). Hydrothermal synthesis and structural characterization of $(NH_4)GaPO_4F$, KTP-type and $(NH_4)_2Ga_2(PO_4)(HPO_4)F_3$, pseudo-KTP-type materials. *Chem. Mater.* **12**, 1393–1399.

Loiseau, T., Beitone, L., Millange, F., Taulelle, F., O'Hare, D., and Férey, G. (2004). Observation and reactivity of the chainlike species $([Al(PO_4)_2]^{3-})_n$ during the X-ray diffraction investigation of the hydrothermal synthesis of the super-sodalite sodium aluminophosphate MIL-74 $(Na_2Al_7(PO_4)_{12} \cdot 4trenH_3 \cdot Na(H_2O)_{16})$. *J. Phys. Chem.* **B108**, 20020–20029.

Long, J. R., McCarty, L. S., and Holm, R. H. (1996). A solid state route to molecular clusters: access to the solution chemistry of $[Re_6Q_8]^{2+}$ ($Q = S$, Se) core-containing clusters via dimensional reduction. *J. Am. Chem. Soc.* **118**, 4603–4616.

Longo, J. M. and Arnott, R. J. (1970). Structure and magnetic properties of $VOSO_4$. *J. Solid State Chem.* **1**, 394–398.

Longo, J. M., Pierce, J. W., and Kafalas, J. A. (1971). The tetragonal high-pressure form of $TaOPO_4$. *Mater. Res. Bull.* **6**, 1157–1166.

Losilla, E. R., Aranda, M. A. G., and Bruque, S. (1996). Structural features of the reactive sites in α-$M(DPO_4)_2 \cdot (D_2O)$ (M = Ti, Zr, Pb): hydrogen-bond network and framework. *J. Solid State Chem.* **125**, 261–269

Losilla, E. R., Salvado, M. A., Aranda, M. A. G., Cabeza, A., Pertierra, P., Garcia-Granda, S., and Bruque, S. (1998). Layered acid arsenates α-$M(HAsO_4)_2 \cdot (H_2O)$ (M = Ti, Sn, Pb): synthesis optimization and crystal structures. *J. Mol. Struct.* **470**, 93–104.

Louisnathan, S. J., Hill, R. J., and Gibbs, G. V. (1977). Tetrahedral bond length variations in sulfates. *Phys. Chem. Miner.* **1**, 53–69.

Lundgren, G., Wernfors, G., and Yamaguchi, T. (1982). The structure of ditin(II) oxide sulfate. *Acta Crystallogr.* **B38**, 2357–2361.

Lyakhov, A. S., Selevich, A. F., and Verenich, A. I. (1993). Crystal structure of rubidium titanilphosphate alpha $RbTiOPO_4$. *Zh. Neorg. Khim.* **38**, 1121–1124.

Maciček, J., Gradinarov, S., Bontchev, R., and Balarew, C. (1994). A short dynamically symmetrical hydrogen bond in the structure of $K(Mg(H_{0.5}SO_4)_2(H_2O)_2)$. *Acta Crystallogr.* **C50**, 1185–1188.

Magarill, S. A., Romanenko, G. V., Pervukhina, N. V., Borisov, S. V., and Palchik, N. A. (2000). Oxocentered polycationic complexes – an alternative approach to the description of crystal chemistry of natural and synthetic mercury oxysalts. *Zh. Strukt. Khim.* **41**, 116–126.

Mahesh, S., Green, M. A., and Natarajan, S. (2002). Synthesis of open-framework iron phosphates, $[C_6N_2H_{14}][Fe^{III}_2F_2(HPO_4)_2(H_2PO_4)_2] \cdot 2H_2O$ and $[C_6N_2H_{14}][Fe^{III}_3(OH)F_3(PO_4)(HPO_4)_2]_2 \cdot H_2O$, with one- and three-dimensional structures. *J. Solid State Chem.* **165**, 334–344.

Majzlan, J., Stevens, R., Boerio-Goates, J., Woodfield, B. F., Navrotsky, A., Burns, P. C., Crawford, M. K., and Amos, T. G. (2004). Thermodynamic properties, low-temperature heat-capacity anomalies, and single-crystal X-ray refinement of hydronium jarosite, $(H_3O)Fe_3(SO_4)_2(OH)_6$. *Phys. Chem. Miner.* **31**, 518–531.

Maksimov, B. A., Kharitonov, Yu. A., Gorbunov, Yu. A., and Belov, N. V. (1974). Crystal structure of scandium oxyorthogermanate Sc_2GeO_5. *Kristallogr.* **19**, 1081–1083.

Maksimov, B. A., Muradyan, L. A., Genkina, E. A., and Simonov, V. I. (1986). Crystal structure of the monoclinic modification of $Li_3Fe_2(PO_4)_3$. *Dokl. Akad. Nauk SSSR* **288**, 634–638.

Mandal, S., Natarajan, S., Klein, W., Panthöfer, M., and Jansen, M. (2003). Synthesis, structure and magnetic characterization of a one-dimensional iron phosphate, $[NH_3CH_2CH_2CH(NH_3)CH_2CH_3]^{2+}_{\infty}[FeF(HPO_4)_2]$. *J. Solid State Chem.* **173**, 367–373.

Manoli, J. M., Herpin, P., and Dereigne, A. (1972). Etude cristallochimique d'hydrates intermediaires de la serie des aluns. Cas du dihydrate $CsTl(III)(SO_4)_2(H_2O)_2$. *Acta Crystallogr.* **B28**, 806–810.

Marichal, C., Chézeau, J.M., Roux, M., Patarin, J., Jordá, J.L., McCusker, L.B., Baerlocher, C., and Pattison, P. (2006). Synthesis and structure of Mu-33, a new layered aluminophosphate $|((CH_3)_3CNH_3^+)_{16}(H_2O)_4|[Al_{16}P_{24}O_{88}(OH)_8]$. *Micropor. Mesopor. Mater.*, **90**, 5–15.

Martin, F.-D. and Müller-Buschbaum, H. (1994). Ein neues Alkalimetall-Kupfer-Oxovanadat: $KCu_5V_3O_{13}$. *Z. Naturforsch.* **49b**, 1137–1140.

Martinez, M. L., Rodriguez, A., Mestres, L., Solans, X., and Bocanegra, E. H. (1990). Synthesis, crystal structure and thermal studies of $(NH_4)_2Cd_2(SeO_4)_3 \cdot 3H_2O$. *J. Solid State Chem.* **89**, 88–93.

Mathew, M., Brown, W. E., Schroeder, L. W., and Dickens, B. (1988). Crystal structure of octacalcium bis(hydrogenphosphate) tetrakis(phosphate) pentahydrate, $Ca_8(HPO_4)_2(PO_4)_4(H_2O)_5$. *J. Crystallogr. Spectr. Res.* **18**, 235–250.

Mattfeld, H. and Meyer, G. (1994). $Na_2(Pr_4O_2)Cl_9$, the first reduced quaternary Praseodymium Chloride with Anti-SiS_2 analogous Pr_4O_2-chains. *Z. Anorg. Allg. Chem.* **620**, 85–89.

McNear, E., Vincent, M. G., and Parthé, E. (1976). The crystal structure of vuagnatite, $CaAl(OH)SiO_4$. *Am. Mineral.* **61**, 831–838.

Medrano, M.D., Evans, H.T.Jr., Wenk, H.R., and Piper, D.Z. (1998). Phosphovanadylite: a new vanadium phosphate mineral with a zeolite-type structure. *Am. Mineral.* **83**, 889–895.

Mercurio-Lavaud, D. and Frit, M. B. (1973). Structure cristalline de la variété basse température du pyrovanadate de cuivre: $Cu_2V_2O_7\alpha$. *Acta Crystallogr.* **B29**, 2737–2741.

Medrish, I. V., Vologzhanina, A. V., Starikova, Z. A., Antipin, M. Yu., Serezhkina, L. B., and Serezhkin, V. N. (2005). Crystal structure of $[CN_4H_7]_2[UO_2(SO_4)_2(H_2O)]$. *Russ. J. Inorg. Chem.* **50**, 360–364.

Menchetti, S. and Sabelli, C. (1974). Alunogen. Its structure and twinning.Tscherm. *Mineral. Petrogr. Mitt.* **21**, 164–178.

Menchetti, S. and Sabelli, C. (1976a). Crystal chemistry of the alunite series: crystal structure refinement of alunite and synthetic jarosite. *N. Jb. Mineral. Mh.* **1976**, 406–417.

Menchetti, S. and Sabelli, C. (1976b). The halotrichite group: the crystal structure of apjohnite. *Mineral. Mag.* **40**, 599–608.

Mentré, O., Ketatni, E. M., Colmont, M., Huvé, M., Abraham, F., and Petricek, V. (2006). Structural features of the modulated $BiCu_2(P_{1-x}V_x)O_6$ solid solution; 4-D treatment of $x = 0.87$ compound and magnetic spin-gap to gapless transition in new Cu^{2+} two-leg ladder systems. *J. Amer. Chem. Soc.* **128**, 10857–10867.

Mentzen, B. F., Latrach, A., Bouix, J., and Hewat, A. W. (1984a). The crystal structures of $PbOPbXO_4$ (X = S, Cr, Mo) at 5 K by neutron powder profile refinement. *Mater. Res. Bull.* **19**, 549–554.

Mentzen, B. F., Latrach, A., Bouix, J., Boher, P., and Garnier, P. (1984b). Structure cristalline de la phase haute temperature de $PbSO_4 \cdot 2PbO$ a 973K. *Mater. Res. Bull.* **19**, 925–934.

Mercier, N. and Leblanc, M. (1993). Crystal growth and structures of rare earth fluorocarbonates: I. Structures of $BaSm(CO_3)_2F$ and $Ba_3La_2(CO_3)_5F_2$: revision of the corresponding huanghoite and cebaite type structures. *Eur. J. Solid State Inorg. Chem.* **30**, 195–205.

Mercier, R., Pham, T. M., and Colomban, P. (1985). Structure, vibrational study and conductivity of the trihydrated uranyl bis(dihydrogenophosphate): $UO_2(H_2PO_4)_2$ $(H_2O)_3$. *Solid State Ion.* **15**, 113–126.

Mereiter, K. (1974). Die Kristallstruktur von Rhomboklas $(H_5O_2)^+(Fe(SO_4)_2(H_2O)_2)$. *Tscherm. Mineral Petrogr Mitt.* **21**, 216–232.

Mereiter, K. (1976). Die Kristallstruktur des Ferrinatrits, $Na_3Fe(SO_4)_3(H_2O)_3$. *Tshcherm. Mineral. Petrogrogr. Mitt.* **23**, 317–327.

Mereiter, K. (1979). Die Kristallstruktur von Mangan(III)-hydroxid-sulfat-dihydrate, $Mn(OH)SO_4(H_2O)_2$. *Acta Crystallogr.* **B35**, 579–585.

Mereiter, K. (1980). The structure of potassium thallium triaqua-μ_3-oxo-hexa-μ-sulfato-triferrate(III) dihydrate, $K_{2.64}Tl_{2.36}Fe_3O(SO_4)_6(H_2O)_5$. *Acta Crystallogr.* **B36**, 1283–1288.

Mereiter, K. (1982a). The crystal structure of walpurgite, $(UO_2)Bi_4O_4(AsO_4)_2 \cdot 2(H_2O)$. *Tshcherm. Mineral. Petrogrogr. Mitt.* **30**, 129–139.

Mereiter, K. (1982b). Die kristallstruktur des johannits, $Cu(UO_2)_2(OH)_2(SO_4)_2 \cdot 8H_2O$. *Tscherm. Mineral. Petrogr. Mitt.* **30**, 47–57.

Mereiter, K. (1990). Structure of a triclinic rubidium ammonium triaqua-triaqua-μ_3-oxo-hexa-μ-sulfato-triferrate(III) tetrahydrate, $Rb_{2.74}(NH_4)_{2.26}Fe_3O((SO_4)_6) \cdot 7H_2O$. *Acta Crystallogr.* **C46**, 972–976.

Mereiter, K. and Voellenkle, H. (1978). Die Kristallstruktur von β-Pentakalium-μ_3-oxo-hexa-μ-sulfato-triaquaeisen(III))-Heptahydrat, eine monokline Modifikation des Mausschen Salzes. *Acta Crystallogr.* **B34**, 378–384

Mereiter, K. and Voellenkle, H. (1980). The structure of pentarubidium triaqua-μ_3-oxo-hexa-μ-sulfato-triferrate(III) dihydrate. *Acta Crystallogr.* **B36**, 1278–1283.

Mer'kov, A. N., Bussen, I. V., Goiko, E. A., Kul'chitskaya, E. A., Men'shikov, Yu. P., and Nedorezova, A. P. (1973). Raite and zorite, new minerals from the Lovozero tundra. *Zapiski VMO* **102**, 54–62.

Merlino, S., ed. (1997). *Modular aspects of minerals.* EMU Notes in Mineralogy, Vol. 1. Eötvös University Press, Budapest.

Mestres, L., Martines, M. L., Rodriguez, A., Solans, X., and Font-Altaba, M. (1985). Study on the system $CoSeO_4$–$NiSeO_4$–H_2O at 30°C. Crystal structure of $CoSeO_4(H_2O)_5$. *Z. Anorg. Allg. Chem.* **528**, 183–190.

Meyer, H.-J., Meyer, G., and Simon, M. (1991). Über ein Oxidchlorid des Calcium: Ca_4OCl_6. *Z. Anorg. Allg. Chem.* **596**, 89–92.

Mgaidi, A., Boughzala, H., Driss, A., Clerac, R., and Coulon, C. (1999). Structure et propriétés magnétiques du composé $NH_4Fe_3(H_2PO_4)_6(HPO_4)_2.4H_2O$. *J. Solid State Chem.* **144**, 163–168.

Mi, J.-X., Wang, C.-X., Wei, Z.-B., Chen, F.-J., Xua, C.-Y., and Mao, S.-Y. (2005). $K_2Fe[H(HPO_4)_2]F_2$. *Acta Crystallogr.* **E61**, i143–i145.

Micka, Z., Nemec, I., Vojtisek, P., Ondracek, J., and Hoelsae, J. (1994). Crystal structure, thermal behavior, and infrared absorption spectrum of cobalt(II) hydrogen selenite dihydrate $Co(HSeO_3)_2 \cdot 2H_2O$. *J. Solid State Chem.* **112**, 237–242.

Micka, Z., Nemec, I., Vojtisek, P., and Ondracek, J. (1996). Crystal structure and infrared absorption spectra of magnesium(II) hydrogen selenite tetrahydrate, $Mg(HSeO_3)_2 \cdot 4(H_2O)$. *J. Solid State Chem.* **122**, 338–342.

Mikhailov, Yu. N., Kokh, L. A., Kuznetsov, V. G., Grevtseva, T. G., Sokol, S. K., and Ellert, G. V. (1977). Synthesis and crystal structure of potassium trisulfatouranylate $K_4(UO_2(SO_4)_3)$. *Koord. Khim.* **3**, 508–513.

Mikhailov, Yu. N., Mistryukov, V. E., Serezhkina, L. B., Demchenko, E. A., Gorbunova, Yu. E., and Serezhkin, V. N. (1995). Crystal structure of $[UO_2SO_4 \cdot 2H_2O] \cdot CH_2ClCONH_2$. *Russ. J. Inorg. Chem.* **40**, 1238–1240.

Mikhailov, Yu. N., Gorbunova, Yu. E., Serezhkina, L. B., Demchenko, E. A., and Serezhkin, V. N. (1997a). Crystal structure of $(NH_4)_2UO_2(SeO_4)_2 \cdot 3H_2O$. *Zh. Neorg. Khim.* **42**, 1413–1417.

Mikhailov, Yu. N., Gorbunova, Yu. E., Serezhkina, L. B., and Serezhkin, V. N. (1997b). Crystal structure of $(NH_4)_2(UO_2)_2(CrO_4)_3 \cdot 6H_2O$. *Zh. Neorg. Khim.* **42**, 734–738.

Mikhailov, Yu. N., Gorbunova, Yu. E., Demchenko, E. A., Serezhkina, L. B., and Serezhkin, V. N. (2000). Crystal structure of $[C_2H_4(NH_3)_2][UO_2(SO_4)_2 \cdot H_2O]$. *Zh. Neorg. Khim.* **45**, 1711–1713.

Mikhailov, Yu. N., Gorbunova, Yu. E., Shishkina, O. V., Serezhkina, L. B., and Serezhkin, V. N. (2001a). Crystal structure of $Cs_2[(UO_2)(SeO_4)_2 \cdot H_2O]H_2O$. *Russ. J. Inorg. Chem.* **46**, 1661–1665.

Mikhailov, Yu. N., Gorbunova, Yu. E., Baeva, E. E., Serezhkina, L. B., and Serezhkin, V. N. (2001b). Crystal structure of $Na_2[UO_2(SeO_4)_2] \cdot 4H_2O$. *Zh. Neorg. Khim.* **46**, 2017–2021.

Mikhailov, Yu. N., Gorbunova, Yu. E., Mit'kovskaya, E. V., Serezhkina, L. B., and Serezhkin, V. N. (2002). Crystal structure of $Rb(UO_2(SO_4)F)$. *Radiochem.* **44**, 315–318.

Millange, F., Walton R. I., Guillou, N., Loiseau, T., O'Hare, D., and Férey, G. (2002a). Two chain gallium fluorodiphosphates: synthesis, structure solution, and their transient presence during the hydrothermal crystallisation of a microporous gallium fluorophosphate. *Chem. Commun.* **2002**, 826–827.

Millange, F., Walton, R. I., Guillou, N., Loiseau, T., O'Hare, D., Férey, G. (2002b). Synthesis and structure of low-dimensional gallium fluorodiphosphates seen during the crystallization of the three-dimensional microporous gallium fluorophosphate ULM-3. *Chem. Mater.* **14**, 4448–4459.

Miller, S. A. and Taylor, J. C. (1986). The crystal structure of saleeite, $Mg(UO_2PO_4)_2 \cdot 10H_2O$. *Z. Kristallogr.* **177**, 247–253.

Miller, S. R., Slawin, A. M. Z., Wormald, P., and Wright, P. A. (2005). Hydrothermal synthesis and structure of organically templated chain, layered and framework scandium phosphates. *J. Solid State Chem.* **178**, 1737–1751.

Milton, D. J. and Bastron, H. (1971). Churchite and florensite (Nd) from Sausalito, California. *Mineral. Rec.* **2**, 166–168.

Ming, Y., Chern, F. J., Disalvo, J. B., Parise, J., and Goldstone, J. A. (1992). The structural distortion of the anti-perovskite nitride Ca_3AsN. *J. Solid State Chem.* **96**, 426–435.

Mirceva, A. and Golic, L. (1995). Hydroxylammonium scandium sulfate sesquihydrate, $Sc(NH_3OH)(SO_4)_2 \cdot 1.5(H_2O)$. *Acta Crystallogr.* **C51**, 175–177.

Mistryukov, V. E. and Mikhailov, Yu. N. (1983). The characteristic properties of the structural function of the selenitogroup in the uranyl complex with neutral ligands. *Koord. Khim.* **9**, 97–102.

Mitchell, R. H. (2002). *Perovskites. Modern and ancient.* Almaz Press, Thunder Bay.

Mit'kovskaya, E. V., Mikhailov, Yu. N., Gorbunova, Yu. E., Serezhkina, L. B., and Serezhkin, V. N. (2003). Crystal structure of $(H_3O)_3[(UO_2)_3O(OH)_3(SeO_4)_2]$. *Russ. J. Inorg. Chem.* **48**, 666–670.

Miyake, M., Matsuda, N., Sato, S., Matsuda, M., Nakagawa, S., and Ochiai, J. (1998). $Mg(H_2PO_4)_2 \cdot 4(H_2O)$. *Acta Crystallogr.* **C54**, 702–703.

Mizrahi, A., Wignacourt, J.-P., and Steinfink, H. (1997). $Pb_2BiO_2PO_4$, a new oxyphosphate. *J. Solid State Chem.* **133**, 516–521.

Montgomery, H., Chastain, R. V., Natt, J. J., Witkowska, A. M., and Lingafelter, E. C. (1967). The crystal structure of Tutton's salts. VI. vanadium(II), iron(II) and cobalt(II) ammonium sulfate hexahydrates. *Acta Crystallogr.* **22**, 775–780.

Mooney, R. C. L. (1956). Crystal structure of anhydrous indium phosphate and thallic phosphate by X-ray diffraction. *Acta Crystallogr.* **9**, 113–117.

Mooney-Slater, R. C. L. (1966). The crystal structure of hydrated gallium phosphate of composition $GaPO_4(H_2O)_2$. *Acta Crystallogr.* **20**, 526–534.

Moore, P. B. (1965). The crystal structure of laueite, $MnFe_2(OH)_2(PO_4)_2(H_2O)_6(H_2O)_2$. *Am. Mineral.* **50**, 1884–1892.

Moore, P. B. (1970a). Structural hierarchies among minerals containing octahedrally coordinating oxygen. I. Stereoisomerism among corner-sharing octahedral and tetrahedral chains. *N. Jb. Mineral. Mh.* **1970**, 163–173.

Moore, P. B. (1970b). Crystal chemistry of the basic iron phosphates. *Am. Mineral.* **55**, 135–169.

Moore, P. B. (1972). Octahedral tetramer in the crystal structure of leucophosphite, $K_2[Fe^{3+}_4(OH)_2(H_2O)_2(PO_4)_4](H_2O)_2$. *Am. Mineral.* **57**, 397–410.

Moore, P. B. (1975). Laueite, pseudolaueite, stewartite and metavauxite: a study in combinatorial polymorphism. *N. Jb. Mineral. Abh.* **1975**, 148–159.

Moore, P. B. and Araki, T. (1974a). Stewartite, $Mn(II)Fe(III)_2(OH)_2(H_2O)_6(PO_4)_2$ $(H_2O)_2$. Its atomic arrangement. *Am. Mineral.* **59**, 1272–1276.

Moore, P. B. and Araki, T. (1974b). Montgomeryite, $Ca_4Mg(H_2O)_{12}[Al_4(OH)_4(PO_4)_6]$. Its crystal structure and relation to Vauxite, $Fe_2(H_2O)_4(Al_4(OH)_4(H_2O)_4(PO_4)_4)$ $(H_2O)_4$. *Am. Mineral.* **59**, 843–850.

Moore, E. P., Chen, H. Y., Brixner, L. H., and Foris, C. M. (1982). The crystal structure of $Pb_8Bi_2(PO_4)_6O_2$. *Mater. Res. Bull.* **17**, 653–660.

Moore, P. B., Araki, T., Kampf, A. R., and Steele, I. M. (1976). Olmsteadite, $K_2(Fe^{2+})_2((Fe^{2+})_2((Nb^{6+})(Ta^{6+}))_2O_4(H_2O)_4(PO_4)_4)$, a new species, its crystal structure and relation to vauxite and montgomeryite. *Am. Mineral.* **61**, 5–11

Moore, P. B., Kampf, A. R., and Sen Gupta, P. K. (2000). The crystal structure of phi-lolithite, a trellis-like open framework based on cubic closest-packing of anions. *Am. Mineral.* **85**, 810–816

Morgan, K. R., Gainsford, G. J., and Milestone, N. B. (1997). A new type of layered aluminium phosphate $[NH_4]_3[Co(NH_3)_6]_3[Al_2(PO_4)_4]_2$. *Chem. Commun.* **1997**, 61–62.

Mori, H. and Ito, T. (1950). The structure of vivianite and symplesite. *Acta Crystallogr.* **3**, 1–6.

Morita, S. and Toda, K. (1984). Determination of the crystal structure of Pb_2CrO_5. *J. Appl. Phys.* **55**, 2733–2737.

Moriyoshi, C. and Itoh, K. (1996). Structural study of phase transition mechanism of langbeinite-type $K_2Zn_2(SO_4)_3$ crystals. *J. Phys. Soc. Jpn.* **65**, 3537–3543.

Morosov, I., Troyanov, S.I., Stiewe, A., Kemnitz, E. (1998) Synthese und Kristallstruktur von Hydrogenselenaten zweiwertiger Metalle – $M(HSeO_4)_2$ (M = Mg, Mn, Zn) und $M(HSeO_4)_2 \cdot (H_2O)$ (M = Mn, Cd). *Z. Anorg. Allg. Chem.*, **624**, 135–140.

Morozov, V. A., Lazoryak, B. I., Malakho, A. P., Pokholok, K. V., Polyakov, S. N., and Terekhina, T. P. (2001). The glaserite-like structure of double sodium and iron phosphate $Na_3Fe(PO_4)_2$. *J. Solid State Chem.* **160**, 377–381.

Morris, R. E., Harrison, W. T. A., Stucky, G. D., and Cheetham, A. K. (1991). The syntheses and crystal structures of two novel aluminium selenites, $Al_2(SeO_3)_3 \cdot 6H_2O$ and $AlH(SeO_3)_2 \cdot 2H_2O$. *J. Solid State Chem.* **94**, 227–235.

Morris, R. E., Harrison, W. T. A., Stucky, G. D., and Cheetham, A. K. (1992). On the structure of $Al_2(SeO_3)_3 \cdot 6H_2O$. *J. Solid State Chem.* **99**, 200.

Moser, P. and Jung, W. (2000). $TlCu_5O(VO_4)_3$ mit $KCu_5O(VO_4)_3$-Struktur – ein Thallium-kupfer(II)-oxidvanadat als Oxidationsprodukt einer Tl/Cu/V-Legierung. *Z. Anorg. Allg. Chem.* **626**, 1421–1425.

Muilu, H. and Valkonen, J. (1987). Crystal structures, thermal behaviour and IR spectra of iron(III) diselenite hydrogenselenite and iron(III) tris(hydrogenselenite). *Acta Chem. Scand.* **A41**, 183–189.

Mukhtarova, N. N., Rastsvetaeva, R. K., Ilyukhin, V. V., and Belov, N. V. (1977). The crystal structure of $NaIn(SeO_4)_2(H_2O)_6$. *Dokl. Akad. Nauk SSSR* **235**, 575–577

Mukhtarova, N. N., Rastsvetaeva, R. K., Ilyukhin, V. V., and Belov, N. V. (1978). About the crystal structure of $NH_4In(SO_4)_2(H_2O)_4$. *Dokl. Akad. Nauk SSSR* **239**, 322–325.

Mukhtarova, N. N., Rastsvetaeva, R. K., Ilyukhin, V. V., and Belov, N. V. (1979a). Crystal structure of $KIn(SO_4)_2(H_2O)_4$. *Dokl. Akad. Nauk SSSR* **245**, 589–593.

Mukhtarova, N. N., Rastsvetaeva, R. K., Ilyukhin, V. V., and Belov, N. V. (1979b). Refinement of the crystal structure of $RbIn(SO_4)_2(H_2O)_4$. *Dokl. Akad. Nauk SSSR* **247**, 600–603.

Mukhtarova, N. N., Rastsvetaeva, R. K., Ilyukhin, V. V., and Belov, N. V. (1979c). Crystal structure of $Na_3In(SO_4)_3(H_2O)_3$. *Dokl. Akad. Nauk SSSR* **244**, 602–606.

Muller, O., White, W. B., and Roy, R. (1969). X-ray diffraction study of the chromates of nickel, magnesium and cadmium. *Z. Kristallogr.* **130**, 112–120.

Müller-Buschbaum, H. and Mertens, B. (1997). Zu $BaMg_2Cu_8V_6O_{26}$ isotype Alkalimetall-Cadmium-Kupferoxovanadate $KCd_{0.67}Cu_{4.33}V_3O_{13}, RbCd_{0.5}Cu_{4.5}V_3O_{13}$ sowie $TlCd_{0.5}Cu_{4.5}V_3O_{13}$. *Z. Naturforsch.* **52b**, 639–642.

Müller-Buschbaum, H. and Wadewitz, C. (1996). Strukturelle Unterschiede zwischen $Sr_2(VO)(AsO_4)_2$ und $Ba_2(VO)(PO_4)_2$. *Z. Naturforsch.* **51b**, 1290–1294.

Mumme, W. G. (1995). Crystal structure of tricalcium silicate from a Portland cement clinker and its application to quantitative XRD analysis. *N. Jb. Mineral. Mh.* **1995**, 145–160.

Nagornyi, P. G., Kapshuk, A. A., Kornienko, Z. I., Mitkevich, V. V., and Tret'yak, S.-M. (1990). Synthesis and structure of new fluorophosphate $Na_5Cr(PO_4)_2F_2$. *Zh. Neorg. Khim.* **35**, 839–842.

Nair, S., Jeong, H. K., Chandrasekaran, A., Braunbarth, C., Tsapatis, M., and Kuznicki, S. M. (2001). Synthesis and structure determination of ETS-4 single crystals. *Chem. Mater.* **13**, 4247–4254.

Natarajan, S., Attfield, M. P., and Cheetham, A. K. (1997). Synthesis and characterization of a new zinc phosphate, $[NH_3(CH_2)_4NH_3]^{2+}[Zn_2P_3O_9(OH)_3]^{2-}$, containing alternating inorganic–organic layers. *J. Solid State Chem.* **132**, 229–234.

Natarajan, S., van Wullen, L., Klein, W., and Jansen, M. (2003). Synthesis of a single four-ring (S4R) molecular zinc phosphate and its assembly to an extended polymeric structure: a single-crystal and in-situ MAS NMR investigation. *Inorg. Chem.* **42**, 6265–6273.

Nazarchuk, E. V., Krivovichev, S. V., and Filatov, S.K. (2004). Phase transitions and high-temperature crystal chemistry of the $Cs_2(UO_2)_2(MoO_4)_3$ polymorphs. *Radiochem.* **46**, 405–407.

Neronova, N. N. and Belov, N. V. (1963). Crystal structure of elpidite Na_2Zr $[Si_6O_{15}] \cdot 3H_2O$, and dimorphism of $[Si_6O_{15}]$ dimetasilicate radicals. *Dokl. Akad. Sci. USSR, Earth Sci.* **150**, 115–118.

Neeraj, S., Natarajan, S., and Rao, C. N. R. (2000). Isolation of a zinc phosphate primary building unit $[C_6N_2H_{18}]^{2+}[Zn(HPO_4)(H_2PO_4)_2]^{2-}$ and its transformation to open-framework phosphate $[C_6N_2H_{18}]^{2+}[Zn_3(H_2O)_4(HPO_4)_4]^{2-}$. *J. Solid State Chem.* **150**, 417–422.

Neronova, N. N. and Belov, N. V. (1964). Crystal structure of elpidite, $Na_2Zr(Si_6O_{15}) \cdot 3H_2O$. *Sov. Phys. Crystallogr.* **9**, 700–705.

Nielsen, K., Fehrmann, R., and Eriksen, K. M. (1993). Crystal structure of $Cs_4(VO)_2O(SO_4)_4$. *Inorg. Chem.* **32**, 4825–4828.

Nielsen, K., Boghosian, S., Fehrmann, R., and Willestofte Berg, R. (1999). Crystal structure and spectroscopic characterization of a green V(IV) compound, $Na_8(VO)_2(SO_4)_6$. *Acta Chem. Scand.* **53**, 15–23.

Niinisto, L., Toivonen, J., and Valkonen, J. (1978). Uranyl(VI) compounds. I. The crystal structure of ammonium uranyl sulfate dihydrate, $(NH_4)_2UO_2(SO_4)_2(H_2O)_2$. *Acta Chem. Scand.* **A32**, 647–651.

Niinisto, L., Toivonen, J., and Valkonen, J. (1979). Uranyl(VI) compounds. II. The crystal structure of potassium uranyl sulfate dihydrate $K_2UO_2(SO_4)_2(H_2O)_2$. *Acta Chem. Scand.* **A33**, 621–624.

Nikitin, A. V. and Belov, N. V. (1962). Crystal structure of batisite $Na_2BaTi_2Si_4O_{14} = Na_2BaTi_2O_2[Si_4O_{12}]$. *Dokl. Akad. Sci. USSR, Earth. Sci.* **146**, 142–143.

Nishi, F. and Takeuchi, Y. (1986). Structures of two modifications of Ca_3GeO_5. *Z. Kristallogr.* **176**, 303–317.

Nishi, F., Takeuchi, Y., and Maki, I. (1985). Tricalcium silicate $Ca_3O(SiO_4)$: The monoclinic superstructure. *Z. Kristallogr.* **172**, 297–314.

Noerbygaard, T., Berg, R. W., and Nielsen, K. (1998). The reaction between MoO_3 and molten $K_2S_2O_7$ forming $K_2MoO_2(SO_4)_2$, studied by Raman and IR spectroscopy and X-ray crystal structure determination. *Electrochem. Soc. Proc.* **98**, 553–573.

Norby, P. (1997). Synchrotron powder diffraction using imaging plates: crystal structure determination and Rietveld refinement. *J. Appl. Crystallogr.* **30**, 21–30.

Norlund Christensen, A., Krogh Andersen, E., Krogh Andersen, I. G., Alberti, G., Nielsen, M., and Lehmann, M.S. (1990). X-ray powder diffraction study of layer compounds. The crystal structure of α-$Ti(HPO_4)_2 \cdot H_2O$ and a proposed structure for γ-$Ti(H_2PO_4)(PO_4) \cdot 2H_2O$. *Acta Chem. Scand.* **44**, 865–872.

Norrestam, R., Nygren, M., and Bovin, J. O. (1992). Structural investigations of new calcium-rare earth (R) oxyborates with the composition $Ca_4RO(BO_3)_3$. *Chem. Mater.* **4**, 737–743.

Norquist, A. J. and O'Hare, D. (2004). Kinetic and mechanistic investigations of hydrothermal transformations in zinc phosphates. *J. Am. Chem. Soc.* **126**, 6673–6679.

Norquist, A. J., Thomas, P. M., Doran, M. B., and O'Hare, D. (2002). Synthesis and cyclical diamine templated uranium sulfates. *Chem. Mater.* **14**, 5179–5184.

Norquist, A. J., Doran, M. B., Thomas, P. M., and O'Hare, D. (2003a). Structural diversity in organically templated sulfates. *Dalton Trans.* **2003**, 1168–1175.

Norquist, A. J., Doran, M. B., Thomas, P. M., and O'Hare, D. (2003b). Controlled structural variations in templated uranyl sulfates. *Inorg. Chem.* **42**, 5949–5953.

Norquist, A. J., Doran, M. B., and O'Hare, D. (2003c). The effects of linear diamine chain length in uranium sulfates. *Solid State Sci.* **5**, 1149–1158.

Norquist, A. J., Doran, M. B., and O'Hare D. (2005a). The role of amine sulfates in hydrothermal uranium chemistry. *Inorg. Chem.* **44**, 3837–3843.

Norquist, A. J., Doran, M. B., and O'Hare D. (2005b). $(C_7H_{20}N_2)[(UO_2)_2(SO_4)_3(H_2O)]$: an organically templated uranium sulfate with a novel layer topology. *Acta Crystallogr.* **E61**, m807–m810.

Nowotny, H. and Wittmann, A. (1954). Zeolithische Alkaligermanate. *Monatsh. Chem.* **85**, 558–574.

Nyfeler, D. and Armbruster, T. (1998). Silanol groups in minerals and inorganic compounds. *Am. Mineral.* **83**, 119–125.

Nyfeler, D., Hoffmann, C., Armbruster, T., Kunz, M., and Libowitzky, E. (1997). Orthorhombic Jahn-Teller distortion and Si-OH in mozartite, $CaMn^{3+}O(SiO_3OH)$: a single-crystal X-ray, FTIR, and structure modelling study. *Am. Mineral.* **82**, 841–848.

Nyman, H. and Andersson, S. (1979). The stella quadrangula as a structure building unit. *Acta Crystallogr.* **A35**, 934–936.

Nyman, M., Bonhomme, F., Teter, D. M., Maxwell, R. S., Gu, B. X., Wang, L. M., Ewing, R. C., and Nenoff, T. M. (2000). Integrated experimental and computational methods for structure determination and characterization of a new, highly stable cesium silicotitanate phase, $Cs_2TiSi_6O_{15}$ (SNL-A). *Chem. Mater.* **12**, 3449–3458.

Nyman, M., Bonhomme, F., Maxwell, R. S., and Nenoff, T. M. (2001). First Rb silicotitanate phase and its K-structural analogue: new members of the SNL-A family $(Cc-A_2TiSi_6O_{15}; A = K, Rb, Cs)$. *Chem. Mater.* **13**, 4603–4611.

Ok, K. M. and Halasyamani, P. S. (2006). Synthesis, structure, and characterization of a new one-dimensional tellurite phosphate, $Ba_2TeO(PO_4)_2$. *J. Solid State Chem.* **179**, 1317–1322.

Ok, K. M., Baek, J., Halasyamani, P. S., and O'Hare, D. (2006). New layered uranium phosphate fluorides: syntheses, structures, characterizations, and ion-exchange properties of $A(UO_2)F(HPO_4) \cdot xH_2O$ (A = Cs^+, Rb^+, K^+; x = 0–1). *Inorg. Chem.* **45**, 10207–10214.

Oka, Y., Yao, T., Yamamoto, N., Ueda, Y., Kawasaki, S., Azuma, M., and Takano, M. (1996). Hydrothermal synthesis, crystal structure, and magnetic properties of $FeVO_4$-II. *J. Solid State Chem.* **123**, 54–59.

Okada, K., Hirabayashi, J., and Ossaka, J. (1982) Crystal structure of natroalunite and crystal chemistry of the alunite group. *N. Jb. Mineral. Mh.* **1982**, 534–540.

O'Keeffe, M. and Hyde, B. G. (1985). An alternative approach to non-molecular crystal structures with emphasis on the arrangements of cations. *Struct. Bond.* **61**, 77–144.

O'Keeffe, M., Eddaoudi, M., Li, H., Reineke, T., and Yaghi O. M. (2000). Frameworks for extended solids: geometrical design principles. *J. Solid State Chem.* **152**, 3–20.

Oliver, S., Kuperman, A., Lough, A., and Ozin, G. A. (1996a). The synthesis and structure of two novel layered aluminophosphates containing interlamellar cyclohexylammonium. *Chem. Commun.* **1996**, 1761–1762.

Oliver, S., Kuperman, A., Lough, A., and Ozin, G. A. (1996b). Aluminophosphate chain-to-layer transformation. *Chem. Mater.* **8**, 2391–2398.

Oliver, S., Kuperman, A., Lough, A., and Ozin G. A. (1996c). Synthesis and crystal structures of two novel anionic aluminophosphates: a one-dimensional chain, UT-7 ([$Al_3P_5O_{20}H$]$^{5-}$[$C_7H_{13}NH_3{}^+$]$_5$), and a layer containing two cyclic amines, UT-8 ([$Al_3P_4O_{16}$]$^{3-}$[$C_4H_7NH_3{}^+$]$_2$[$C_5H_{10}NH_2{}^+$]). *Inorg. Chem.* **35**, 6373–6380.

Oliver, S., Kuperman, A., and Ozin, G. A. (1998). A new model for aluminophosphate formation: transformation of a linear chain aluminophosphate to chain, layer, and framework structures. *Angew. Chem. Int. Ed.* **37**, 46–62.

Ossaka, J., Hirabayashi, J., Okada, K., Kobayashi, R., and Hayashi, T. (1982). Crystal structure of minamiite, a new mineral of the alunite group. *Am. Mineral.* **67**, 114–119.

Ovanisyan, S. M., Iskhakova, L. D., and Trunov, V. K. (1987a). Crystal structure of RbCe(SeO_4)$_2 \cdot 5H_2O$. *Kristallogr.* **32**, 1148–1152.

Ovanisyan, S. M., Iskhakova, L. D., and Trunov, V. K. (1987b). The preparation and crystal structure of the compound CsLn(SeO_4)$_2$(H_2O)$_4$. *Russ. J. Inorg. Chem.*, **32**, 501–503.

Ovanisyan, S. M., Iskhakova, L. D., and Trunov, V. K. (1988). Structure of a new modification of Tr_2(SeO_4)$_3$(H_2O)$_8$. *Sov. Phys. Crystallogr.* **33**, 37–40.

Pabst, A. (1934). The crystal structure of sulphohalite. *Z. Kristallogr.* **89**, 514–517.

Palazzi, M. and Jaulmes, S. (1981). Structure d'un oxysulfure a deux cations d'un type nouveau: La_4O_3(AsS_3)$_2$. *Acta Crystallogr.* **B37**, 1340–1342.

Palkina, K. K., Kuz'mina, N. E., Lysanova, G. V., Dzhurinskii, B. F., and Komova, M. G. (1994). Structure and synthesis of lutetium germanate $Lu_6Ge_4O_{17}$. *Zh. Neorg. Khim.* **39**, 184–187.

Palkina, K. K., Kuz'mina, N. E., Dzhurinskii, B. F., and Tselebrovskaya, E. G. (1995). Neodymium oxyphosphate Nd_3PO_7. *Dokl. Akad. Nauk* **341**, 644–648.

Palmer, D. J., Wong, R. Y., and Lee, K.-S. (1972). The crystal structure of ferric ammonium sulfate trihydrate, $FeNH_4$(SO_4)$_2$(H_2O)$_3$. *Acta Crystallogr.* **B28**, 236–241.

Pasha, I., Choudhury, A., and Rao, C. N. R. (2003a). An organically templated open-framework cadmium selenite. *Solid State Sci.* **5**, 257–262.

Pasha, I., Choudhury, A., and Rao, C. N. R. (2003b). Organically templated vanadyl selenites with layered structures. *Inorg. Chem.* **42**, 409–415.

Pasha, I., Choudnury, A., and Rao, C. N. R. (2003c). The first organically templated linear metal selenate. *J. Solid State Chem.* **174**, 386–391.

Patarin, J., Marler, B., and Huve, L. (1994). Synthesis and structure of Zn(H_2PO_4)$_2$(HPO_4) $\cdot$ $H_2N_2C_6H_{12}$. *Eur. J. Solid State Inorg. Chem.* **31**, 909–920.

Paul, G., Choudhury, A., and Rao, C. N. R. (2002a). Organically templated linear and layered cadmium sulfates. *Dalton Trans.* **2002**, 3859–3867.

Paul, G., Choudhury, A., Sampathkumaran, E. V., and Rao, C. N. R. (2002b). Organically templated mixed-valent iron sulfates possessing kagomé and other types of layered networks. *Angew. Chem.* **114**, 4473–4476.

Paul, G., Choudhury, A., and Rao, C. N. R. (2003). Organically templated linear and layered iron sulfates. *Chem. Mater.* **15**, 1174–1180.

Peacor, D. R., Rouse, R. C., Coskren, T. D., and Essene, E. J. (1999). Destinezite ("diadochite"), $Fe_2(PO_4)(SO_4)(OH) \cdot 6H_2O$: its crystal structure and role as a soil mineral at Alum Cave Bluff, Tennessee. *Clay and Clay Miner.* **47**, 1–11.

Pelloquin, D., Louer, M., and Louer, D. (1994). Powder diffraction studies in the $YONO_3$-Y_2O_3 system. *J. Solid State Chem.* **112**, 182–188.

Pernet, M., Quezel, G., Coing-Boyat, J., and Bertaut, E. F. (1969). Structures magnetiques des chromates de cobalt et de nickel. *Bull. Soc. Franc. Mineral. Crist.* **92**, 264–273.

Pertlik, F. (1987). The structure of freedite, $Pb_8Cu(AsO_3)_2O_3Cl_5$. *Mineral. Petrol.* **36**, 85–92.

Pertlik, F. (1989). The crystal structure of cechite, $Pb(Fe,Mn)(VO_4)(OH)$ with $Fe > Mn$. A mineral of the descloizite group. *N. Jb. Mineral. Mh.* **1989**, 34–40.

Pertlik, F. and Zemann, J. (1982). Neubestimmung der Kristallstruktur des Koechlinits. *Fortschr. Mineral.* **60**, 162–163.

Peterson, R. C., Roeder P. L., and Zhang, Z. (2003). The atomic structure of siderotil, $(Fe,Cu)SO_4 \cdot 5H_2O$. *Can. Mineral.* **41**, 671–676.

Petrosyants, S. P., Ilyukhin A. B., and Sukhorukov A. Yu. (2005). Coordination polymers of scandium sulfate. Crystal structures of $(H_2Bipy)[Sc(H_2O)(SO_4)_2]_2 \cdot 2H_2O$ and $(H_2Bipy)[HSO_4]_2$. *Russ. J. Coord. Chem.* **31**, 545–551.

Peytavin, S., Philippot, E., and Lindqvist, O. (1973). Potassium cadmium selenate dihydrate, $K_2Cd(SeO_4)_2(H_2O)_2$. *Cryst. Struct. Commun.* **2**, 163–165.

Peytavin, S., Philippot, E., and Lindqvist, O. (1974). Etude structurale du sel double dihydrate $K_2Cd(SeO_4)_2(H_2O)_2$. *Rev. Chim. Mineral.* **11**, 37–47.

Philippou, A. and Anderson, M. W. (1996). Structural investigation of ETS-4. *Zeolites* **16**, 98–107.

Phillips, M. L. F., Harrison, W. T. A., Gier, T. E., Stucky, G. D., Kulkarni, G. V., and Burdett, J. K. (1990). Electronic effects of substitution chemistry in the $KTiOPO_4$ structure field: structure and optical properties of potassium vanadyl phosphate. *Inorg. Chem.* **29**, 2158–2163.

Phillips, M. L. F., Nenoff, T. M., Thompson, C. T., and Harrison, W. T. A. (2002). Variations on the (3,4)-net motif in organo-zincophosphite chemistry: syntheses and structures of $(CN_3H_6)_2 \cdot Zn_3(HPO_3)_4 \cdot H_2O$ and $H_3N(CH_2)_3NH_3 \cdot H_2O$. *J. Solid State Chem.* **167**, 337–343.

Piret, P. and Declercq, J.-P. (1983). Structure cristalline de l'upalite $Al[(UO_2)_3 O(OH)(PO_4)_2] \cdot 7H_2O$. Un exemple de macle mimétique. *Bull. Minéral.* **106**, 383–389.

Piret, P. and Deliens, M. (1982). La vanmeersscheite $U(UO_2)_3(PO_4)_2(OH)_6 \cdot 4H_2O$ et la méta-vanmeersscheite $U(UO_2)_3(PO_4)_2(OH)_6 \cdot 2H_2O$, nouveaux minéraux. *Bull. Minéral.* **105**, 125–128.

Piret, P. and Deliens, M. (1987). Les phosphates d'uranyle et d'aluminium de Kobokobo. IX. L'althupite, $AlTh(UO_2)[(UO_2)_3O(OH)(PO_4)_2]_2(OH)_3 \cdot 15H_2O$, nouveau minéral; propriétés et structure cristalline. *Bull. Minéral.* **110**, 65–72.

Piret, P. and Piret-Meunier, J. (1988). Nouvelle détermination de la structure cristalline de la dumontite $Pb_2[(UO_2)_3O_2(PO_4)_2] \cdot 5H_2O$. *Bull. Minéral.* **111**, 439–442.

Piret, P. and Piret-Meunier, J. (1994). Structure de la seelite de Rabejac (France). *Eur. J. Mineral.* **6**, 673–677.

Piret, P. and Piret-Meunier, J. (1991). Composition chimique et structure cristalline de la phosphuranylite $Ca(UO_2)[(UO_2)_3(OH)_2(PO_4)_2]_2 \cdot 12H_2O$. *Eur. J. Mineral.* **3**, 69–77.

Piret, P., Piret Meunier, J., and Declercq, J.-P. (1979a). Structure of phuralumite. *Acta Crystallogr.* **B35**, 1880–1882.

Piret, P., Declercq, J. P., and Wauters-Stoop, D. (1979b). Structure of threadgoldite. *Acta Crystallogr.* **B35**, 3017–3020.

Piret, P., Deliens, M., and Piret-Meunier, J. (1988). La françoisite-(Nd), nouveau phosphate d'uranyle et de terres rares; propriétés et structure cristalline. *Bull. Minéral.* **111**, 443–449.

Piret, P., Piret-Meunier, J., and Deliens, M. (1990). Composition chimique et structure cristalline de la dewindtite $Pb_3[H(UO_2)_3O_2(PO_4)_2]_2 \cdot 12H_2O$. *Eur. J. Mineral.* **2**, 399–405.

Plaisier, J. R., Huntelaar, M. E., de Graaff, R. A. G., and Ijdo, D. J. W. (1994). A contribution to the understanding of phase equilibria (structure of $Sr_7ZrSi_6O_{21}$). *Mater. Res. Bull.* **29**, 701–707.

Plevert, J., Sanchez-Smith, R., Gentz, T. M., Li, H., Groy, T. L., Yaghi, O. M., and O'Keeffe, M. (2003). Synthesis and characterization of zirconogermanates. *Inorg. Chem.* **42**, 5954–5959.

Pluth, J. J. and Smith, J. V. (1984). Structure of $NH_4Al_2(OH)(H_2O)(PO_4)_2(H_2O)$, the ammonium aluminium analog of $GaPO_4(H_2O)_2$ and leucophosphite. *Acta Crystallogr.* **C40**, 2008–2011.

Podberezskaya, N. V. and Borisov, S. V. (1976). Refinement of the crystal structure of $Sm_2(SO_4)_3(H_2O)_8$. *Zh. Strukt. Khim.* **17**, 186–188.

Poljak, R. J. (1958). On the structure of anhydrous nickel sulfate. *Acta Crystallogr.* **11**, 306–306.

Polyanskaya, T. M., Borisov, S. V., and Belov, N. V. (1970). A new form of the scheelite structural type: crystal structure of Nd_2WO_6. *Dokl. Akad. Nauk SSSR* **193**, 83–86.

Poojary, D. M. and Clearfield, A. (1994). Crystal structure of sodium zirconium phosphate, $Zr_2(NaPO_4)_4 \cdot 6(H_2O)$, from X-ray powder diffraction data. *Inorg. Chem.* **33**, 3685–3688.

Poojary, D. M., Cahill, R. A., and Clearfield, A. (1994). Synthesis, crystal structures and ion-exchange properties of a novel porous titanosilicate. *Chem. Mater.* **6**, 2364–2368.

Poojary, D. M., Bortun, A. I., Bortun, L. N., and Clearfield, A. (1997a). Syntheses and X-ray powder structures of $K_2(ZrSi_3O_9) \cdot (H_2O)$ and its ion-exchanged phases with Na and Cs. *Inorg. Chem.* **36**, 3072–3079.

Poojary, D. M., Bortun, A. I., Bortun, L. N., Clearfield, A. (1997b). Synthesis and X-ray powder structures of three novel titanium phosphate compounds. *J. Solid State Chem.* **132**, 213–223.

Pring, A., Gatehouse, B. M., and Birch, W. D. (1990). Francisite, $Cu_3Bi(SeO_3)_2O_2Cl$, a new mineral from Iron Monarch, South Australia: description and crystal structure. *Am. Mineral.* **75**, 1421–1425.

Protas, J. and Gindt, R. (1976). Structure cristalline de la brassite, $MgHAsO_4$ $(H_2O)_4$ produit de deshydratation de la roesslerite. *Acta Crystallogr.* **B32**, 1460–1466.

Pudovkina, Z. V. and Chernitsova, N. M. (1991). Crystal structure of terskite $Na_4Zr[H_4Si_6O_{18}]$. *Sov. Phys. Dokl.* **36**, 201–203.

Pushcharovskii, D. Yu., Lima-de-Faria, J., and Rastsvetaeva, R. K. (1998). Main structural subdivisions and structural formulas of sulphate minerals. *Z. Kristallogr.* **213**, 141–150.

Pushcharovskii, D. Yu., Rastsvetaeva, R. K., and Sarp, H. (1996). Crystal structure of deloryite, $Cu_4(UO_2)(Mo_2O_8)(OH)_6$. *J. Alloys Compd.* **239**, 23–26.

Pushcharovskii, D. Y., Pekov, I. V., Pasero, M., Gobechiya, E. R., Merlino, S., and Zubkova, N. V. (2002). Crystal structure of cation-deficient calciohilairite and possible mechanisms of decationization in mixed-framework minerals. *Crystallogr. Rep.* **47**, 748–752.

Qurashi, M. M. and Barnes, W. H. (1963). The structures of the minerals of the descloizite and adelite group. IV. Descloizite and conichalcite. *Can. Mineral.* **7**, 561–577.

Radoslavljevic, I., Evans, J. S. O., and Sleight, A. W. (1998). Synthesis and structure of $BiCa_2VO_6$. *J. Solid State Chem.* **137**, 143–147.

Radoslovich, E. W. (1982). Refinement of gorceixite structure in *Cm*. *N. Jb. Mineral. Mh.* **1982**, 446–464.

Radoslovich, E. W. and Slade, P. G. (1980). Pseudo-trigonal symmetry and the structure of gorceixite. *N. Jb. Mineral. Mh.* **1980**, 157–170.

Rae-Smith, A. R., Cheetham, A. K., and Fuess, H. (1984). Preparation and crystal structure of La_3ReO_8. *Z. Anorg. Allg. Chem.* **510**, 46–50.

Ralle, M. and Jansen, M. (1994). Darstellung und Kristallstruktur des neuen Lanthanaurates $La_4Au_2O_9$. *J. Alloys Compd.* **203**, 7–13.

Rao, C. N. R., Natarajan S., Choudhury A., Neeraj S., and Ayi A. A. (2001a). Aufbau principle of complex open-framework structures of metal phosphates with different dimensionalities. *Acc. Chem. Res.* **34**, 80–87.

Rao, C. N. R., Natarajan, S., Choudhury, A., Neeraj, S. and Vaidhyanathan, R. (2001b). Synthons and design in metal phosphates and oxalates with open architectures. *Acta Crystallogr.* **B57**, 1–12.

Rao, C. N. R., Sampathkumaran, E. V., Nagarajan, R., Paul, G., Behera, J. N., and Choudhury, A. (2004). Synthesis, structure, and the unusual magnetic properties of an amine-templated iron(II) sulfate possessing the Kagomé lattice. *Chem. Mater.*, **16**, 1441–1446.

Rastsvetaeva, R. K. and Khomyakov, A. P. (1992). Crystal structure of a rare earth analog of hilairite. *Sov. Phys. Crystallogr.* **37**, 845–847.

Rastsvetaeva, R. K. and Khomyakov, A. P. (1996). Pyatenkoite-(Y) $Na_5YTiSi_6O_{18} \cdot 6H_2O$ a new mineral of the hilairite group: Crystal structure. *Dokl. Chem.* **351**, 283–286.

Rastsvetaeva, R. K. and Mukhtarova, N. N. (1987). Role of univalent cations and hydrogen bonds in stabilizing the structures of goldichites. *Kristallogr.* **32**, 1419–1427.

Rastsvetaeva, R. K. and Pushcharovskii, D. Yu. (1989). Crystal chemistry of sulfates. *Itogi Nauki I Tekhniki, ser Kristallokhimiya,* Vol. 23 (= Advances in Science and Technology, ser Crystal Chemistry).

Rastsvetaeva, R. K., Andrianov, V. I., and Volodina, A. N. (1984). The crystal structure of $CsIn_3H_2(SeO_3)_6(H_2O)_2$. *Dokl. Akad. Nauk SSSR* **277**, 871–875.

Rastsvetaeva, R. K., Maksimov, B. A., and Timofeeva, V. A. (1996). Crystal structure of the new phosphate $Na_5Fe(PO_4)_2F_2$. *Dokl. Ross. Akad. Nauk* **350**, 499–502.

Rastsvetaeva, R. K., Pushcharovskii, D. Yu., Konev, A. A., and Evsyunin, V. G. (1997). Crystal structure of K-containing batisite. *Crystallogr. Rep.* **42**, 770–773.

Rastsvetaeva, R. K., Barinova, A. V., Fedoseev, A. M., Budantseva, N. A., and Nikolaev, Yu. V. (1999). Synthesis and crystal structure of a new hydrated cesium uranyl molybdate $Cs_2UO_2(MoO_4)_2 \cdot H_2O$. *Dokl. Ross. Akad. Nauk* **365**, 68–71.

Rauch, P. E. and Simon, A. (1992). Das neue Subnitrid $NaBa_3N$ – eine Erweiterung der Alkalimetallsuboxid-Chemie. *Angew. Chem.* **104**, 1505–1506.

Reckeweg, O. and DiSalvo, F. J. (2006). Crystal structure of distrontium oxide diiodide, Sr_2OI_2. *Z. Kristallogr. NCS* **221**, 271–272.

Reckeweg, O. and Meyer, H.-J. (1997). Crystal structure of tetrastrontium oxide hexachloride, Sr_4OCl_6. *Z. Kristallogr.* **212**, 235.

Rentzeperis, P. J. (1958). Die Kristallstruktur der beiden Modifikationen von wasserfreiem $CoSO_4$. *N. Jb. Mineral. Mh.* **1958**, 226–233.

Rentzeperis, P. J. and Soldatos, C. T. (1958). The crystal structure of the anhydrous magnesium sulfate. *Acta Crystallogr.* **11**, 686–688.

Richter, K. L. and Mattes, R. (1991). Hydrated phases in the V_2O_5–K_2O–SO_3–H_2O system. Preparation and structures of $K(VO_2(SO_4)(H_2O))$ and $K(VO_2(SO_4)(H_2O)_2)$ H_2O. *Inorg. Chem.* **30**, 4367–4369.

Richter, K. L. and Mattes, R. (1992). Darstellung, Ramanspektren und Kristallstrukturen von $V_2O_3(SO_4)_2$, und $NH_4(VO(SO_4)_2)$. *Z. Anorg. Allg. Chem.* **611**, 158–164.

Riebe, H.-J. and Keller, H. L. (1988). Die Kristallstruktur von $AgPb_4O_4Cl$, eine kuriose Variante des Blei(II)-oxid-Strukturtyps. *Z. Anorg. Allg. Chem.* **566**, 62–70.

Riebe, H.-J. and Keller, H. L. (1989a). Darstellung und Kristallstruktur von HgPb$_2$O(OH)Br$_3$. *Z. Anorg. Allg. Chem.* **574**, 191–198.

Riebe, H.-J. and Keller, H. L. (1989b). Pb$_{13}$O$_{10}$Br$_6$, ein neuer Vertreter der Blei(II)-oxidhalodenide. *Z. Anorg. Allg. Chem.* **571**, 139–147.

Riebe, H.-J. and Keller, H. L. (1991). AgPbOBr – ein neuer Sillen-Typ? Darstellung und Kristallstruktur. *Z. Anorg. Allg. Chem.* **597**, 151–161.

Rinaldi, R., Pluth, J. J., and Smith, J. V. (1975). Crystal structure of cavansite dehydrated at 220°C. *Acta Crystallogr.* **B31**, 1598–1602.

Riou-Cavellec, M., Riou, D., and Férey, G. (1999). Magnetic iron phosphates with an open framework. *Inorg. Chim. Acta* **291**, 317–325.

Roberts, M. A. and Fitch, A. N. (1996). The crystal structures of hydrated and partially dehydrated M$_3$HGe$_7$O$_{16}$·n(H$_2$O) (M = K, Rb, Cs) determined from powder diffraction data using synchrotron radiation. *Z. Kristallogr.* **211**, 378–387.

Roberts, M. A., Fitch, A. N., and Chadwick, A. V. (1995). The crystal structures of (NH$_4$)$_3$HGe$_7$O$_{16}$·n(H$_2$O) and Li$_{4-x}$H$_x$Ge$_7$O$_{16}$·n(H$_2$O) determined from powder diffraction data using synchrotron radiation. *J. Phys. Chem. Solids* **56**, 1353–1358.

Robinson, P. D. and Fang, J. H. (1969). Crystal structures and mineral chemistry of double-salt hydrates: I. Direct determination of the crystal structure of tamarugite. *Am. Mineral.* **54**, 19–30.

Robinson, P. D. and Fang, J. H. (1971). Crystal structures and mineral chemistry of hydrated ferric sulphates: II. The crystal structure of paracoquimbite. *Am. Mineral.* **56**, 1567–1571.

Robinson, P. D. and Fang, J. H. (1973). Crystal structures and mineral chemistry of hydrated ferric sulphates. III. The crystal structure of kornelite. *Am. Mineral.* **58**, 535–539.

Roca, M., Marcos, M. D., Amoros, P., Alamo, J., Beltran-Porter, A., and Beltran-Porter, D. (1997). Synthesis and crystal structure of a novel lamellar barium derivative: Ba(VOPO$_4$)$_2$·4(H$_2$O). Synthetic pathways for layered oxovanadium phosphate hydrates M(VOPO$_4$)$_2$·n(H$_2$O). *Inorg. Chem.* **36**, 3414–3421.

Rocha, J., Ferreira, P., Lin, Z., Brandaõ, P., Ferreira, A., and Pedrosa de Jesus, J. D. (1997). The first synthetic microporous yttrium silicate containing framework sodium atoms. *Chem. Commun.* **1997**, 2103–2104.

Rocha, J., Ferreira, P., Carlos, L. D., and Ferreira, A. (2000). The first microporous framework cerium silicate. *Angew. Chem. Int. Ed.* **39**, 3276–3279.

Rodriguez, J., Suarez, M., Luz Rodriguez, M., Llavona, R., Arce, M. J., Salvado, M. A., Pertierra, P., and Garcia-Granda, S. (1999). Crystal structure and intercalation properties of γ-Zr(AsO$_4$)(H$_2$AsO$_4$)·2(H$_2$O). *Eur. J. Inorg. Chem.* **1999**, 61–65

Roelofsen-Ahl, J. N. and Peterson, R. C. (1989). Gittinsite: a modification of the thortveitite structure. *Can. Mineral.* **27**, 703–708.

Rogers, R. D., Bond, A. H., Hipple, W. G., Rollins, A. N., and Henry, R. F. (1991). Synthesis and structural elucidation of novel uranyl-crown ether compounds isolated from nitric, hydrochloric, sulfuric, and acetic acids. *Inorg. Chem.* **30**, 2671–2679.

Rosenzweig, A. and Ryan, R. R. (1975). Refinement of the crystal structure of cuprosklodowskite, $Cu(UO_2)_2(SiO_3OH)_2(H_2O)_6$. *Am. Mineral.* **60**, 448–453.

Rosenzweig, A. and Ryan, R. R. (1977). Kasolite, $Pb(UO_2)(SiO_4)(H_2O)$. *Cryst. Struct. Commun.* **6**, 617–621.

Ross, M. and Evans, H.T., Jr. (1960). The crystal structure of cesium biuranyl trisulfate. *J. Inorg. Nucl. Chem.* **15**, 338–351.

Ross, M. and Evans, H.T., Jr. (1964). Studies of the torbernite minerals (I): The crystal structure of abernathyite and the structurally related compounds $NH_4UO_2AsO_4 \cdot (H_2O)_3$ and $K(H_3O)(UO_2AsO_4)_2 \cdot (H_2O)_6$. *Am. Mineral.* **49**, 1578–1602.

Rouse, R. C. and Dunn, P. J. (1985). The structure of thorisokite, a naturally occurring member of the bismuth oxyhalide group. *J. Solid State Chem.* **57**, 389–395.

Rouse, R. C., Peacor, D. R., Dunn, P. J., Criddle, A. J., Stanley, C. J., and Innes, J. (1988). Asisite, a silicon bearing lead oxychloride from the Kombat mine, South West Africa (Namibia). *Am. Mineral.* **73**, 643–650.

Rudolf, P. R. and Clearfield, A. (1985a). Time of flight neutron powder Rietveld refinement of the $ZrKH(PO_4)_2$ structure. *Inorg. Chem.* **24**, 3714–3715.

Rudolf, P. R. and Clearfield, A. (1985b). The solution of unknown crystal structure from X-ray powder diffraction data. Technique and an example, $ZnNaH(PO_4)_2$. *Acta Crystallogr.* **B41**, 418–425.

Rudolf, P. R. and Clearfield, A. (1989). X-ray powder structure and Rietveld refinement of the monosodium-exchanged monohydrate of alpha-zirconium phosphate, $Zr(NaPO_4)(HPO_4) \cdot H_2O$. *Inorg. Chem.* **28**, 1706–1710.

Ruckman, J. C., Morrison, R. T. W., and Buck, R. H. (1972). Structure and lattice parameters of dilead(II) pentaoxochromate(VI). *J. Chem. Soc., Dalton Trans.* **1972**, 426–427

Rumanova, I. M. and Malitskaya, G. I. (1959). Revision of the structure of astrakhanite by weighted phase projection methods. *Kristallogr.* **4**, 510–525.

Rumanova, I. M. and Volodina, G. F. (1958). The crystal structure of natrochalcite $NaCu_2(OH)(SO_4)_2(H_2O) = Na(SO_4)_2(Cu_2,OH,H_2O)$. *Sov. Phys. Dokl.* **3**, 1093–1096.

Rumanova, I. M. and Znamenskaya, M. N. (1960). The crystal structure of anapaite. *Kristallogr.* **5**, 681–688.

Runde, W., Bean, A. C., Albrecht-Schmitt, T. E., and Scott, B. L. (2003). Structural characterization of the first hydrothermally synthesized plutonium compound, $PuO_2(IO_3)_2 \cdot H_2O$. *Chem. Commun.* **2003**, 478–479.

Saadi, M., Dion, C., and Abraham, F. (2000). Synthesis and crystal structure of the pentahydrated uranyl orthovanadate $(UO_2)_3(VO_4)_2 \cdot 5(H_2O)$, precursor for the new $(UO_2)_3(VO_4)_2$ uranyl-vanadate. *J. Solid State Chem.* **150**, 72–80.

Sabelli, C. (1985). Uklonskovite, $NaMg(SO_4)F(H_2O)_2$: new mineralogical data and structure refinement. *Bull. Mineral.* **108**, 133–138.

Sabelli, C. and Trosti-Ferroni, R. (1985). A structural classification of sulfate minerals. *Per. Mineral.* **54**, 1–46.

Yakubovich, O. V., Kabalob, Yu. K., Gavrilenko, P. G., Liferovich, R. P., and Massa, W. (2003a). Strontium in collinsite structure: Rietveld refinement. *Crystallogr. Rep.* **48**, 226–232.

Yakubovich, O. V., Massa, W., Liferovich, R. P., Gavrilenko, P. G., Bogdanova, A. N., and Tuisku, P. (2003b). Hillite, a new member of the fairfieldite group: its description and crystal structure. *Can. Mineral.* **41**, 981–988.

Yamane, H., Sakamoto, T., Kubota, S., and Shimada, M. (1999). Gd_3GaO_6 by X-ray powder diffraction. *Acta Crystallogr.* **C55**, 479–481.

Yan, W., Yu, J., Li, Y., Shi, Z., and Xu, R. (2002). Synthesis and characterization of a new layered aluminophosphate $[Al_3P_4O_{16}][(CH_3)_2NHCH_2CH_2NH(CH_3)_2][H_3O]$. *J. Solid State Chem.* **167**, 282–288.

Yang, Z.-M. (1995). Structure redetermination of natural cebaite-(Ce), $Ba_3Ce_2(CO_3)_5F_2$. *N. Jb. Mineral. Mh.* **1995**, 56–64.

Yang Z.-M. and Pertlik, F. (1993). Huanghoite-(Ce), $BaCe(CO_3)_2F$, from Khibina, Kola peninsula, Russia: redetermination of the crystal structure with a discussion on space group symmetry. *N. Jb. Mineral. Mh.* **1993**, 163–171.

Yang, S., Li, G., Blake, A. J., Sun, J., Xiong, M., Liao, F., and Lin, J. (2008). Oxyfluorotitanophosphate cluster $[Ti_{10}P_4O_{16}F_{44}]^{16-}$: synthesis and characterization of $K_{16}[Ti_{10}P_4O_{16}F_{44}]$. *Inorg. Chem.* **47**, 1414–1416.

Yao, Y., Natarajan, S., Chen, J., and Pang, W. (1999). Synthesis and structural characterization of a new layered aluminophosphate intercalated with triply-protonated triethylenetetramine $[C_6H_{21}N_4][Al_3P_4O_{16}]$. *J. Solid State Chem.* **146**, 458–463.

Yeom, Y. H., Kim, Y., and Seff, K. (1997). Crystal structure of zeolite X exchanged with Pb(II) at pH 6.0 and dehydrated: $(Pb^{4+})_{14}(Pb^{2+})_{18}(Pb_4O_4)_8Si_{100}Al_{92}O_{384}$. *J. Phys. Chem.* **B101**, 5314–5318.

Yeom, Y. H., Kim, Y., and Seff, K. (1999). Crystal structure of $Pb^{2+}_{44}Pb^{4+}_5Tl^+_{18}O^{2-}_{17-}Si_{100}Al_{92}O_{384}$, zeolite X exchanged with Pb^{2+} and Tl^+ and dehydrated, containing $Pb_4O_4(Pb^{2+},Pb^{4+}$ mixed$)_4$ clusters. *Micropor. Mesopor. Mater.* **28**, 103–112.

Yu, J. and Xu, R. (2003). Rich structure chemistry in the aluminophosphate family. *Acc. Chem. Res.* **36**, 481–490.

Yu, J. and Williams, I. D. (1998). Two unusual layer aluminophosphates templated by imidazolium ions; $[N_2C_3H_5]$ $[AlP_2O_8H_2 \cdot 2H_2O]$ and $2[N_2C_3H_5][Al_3P_4O_{16}H]$. *J. Solid State Chem.* **136**, 141–144.

Yu, J., Sugiyama, K., Hiraga, K., Togashi, N.; Terasaki, O., Tanaka, Y., Nakata, S., Qiu, S., and Xu, R. (1998). Synthesis and characterization of a new two-dimensional aluminophosphate layer and structural diversity in anionic aluminophosphates with $Al_2P_3O_{12}^{3-}$ stoichiometry. *Chem. Mater.* **10**, 3636–3642.

Yu, J., Li, J., Sugiyama, K., Togashi, N., Terasaki, O., Hiraga, K., Zhou, B., Qiu, S., and Xu, R. (1999a). Formation of a new layered aluminophosphate $[Al_3P_4O_{16}]$ $[C_5N_2H_9]_2[NH_4]$. *Chem. Mater.* **11**, 1727–1732.

Yu, J., Sung, H. H.-Y., and Williams, I. D. (1999b). A two-dimensional mixed-metal phosphate templated by ethylene diamine, $[enH_2][CoIn(PO_4)_2H(OH_2)_2F_2]$. *J. Solid State Chem.* **142**, 241–246.

Yuan, G and Xue, D. (2007). Crystal chemistry of borates: the classification and algebraic description by topological type of fundamental building blocks. *Acta Crystallogr.* **B63**, 353–362

Yuan, H. M., Zhu, G. S., Chen, J. S., Chen, W., Yang, G. D., and Xu, R. (2000). Dual function of racemic isopropanolamine as solvent and as template for the synthesis of a new layered aluminophosphate: $[NH_3CH_2CH(OH)CH_3]_3 \cdot Al_3P_4O_{16}$. *J. Solid State Chem.* **151**, 145–149.

Zachariasen, W. H. (1930a). The crystal structure of titanite. *Z. Kristallogr.* **73**, 7–16.

Zachariasen, W. H. (1930b). The crystal structure of benitoite $BaTiSi_3O_9$. *Z. Kristallogr.* **74**, 139–146.

Zahrobsky, R. F. and Baur, W. H. (1968). On the crystal chemistry of salt hydrates. V. The determination of the crystal structure of $CuSO_4(H_2O)_3$ (bonattite). *Acta Crystallogr.* **B24**, 508–513.

Zaitsev, A. N., Yakovenchuk, V. N., Chao, G. Y., Gault, R. A., Subbotin, V. V., Pakhomovsky, Y. A., and Bogdanova, A. N. (1996). Kukharenkoite-(Ce), $Ba_2Ce(CO_3)_3F$, a new mineral from Kola Peninsula, Russia and Québec, Canada. *Eur. J. Mineral.* **8**, 1327–1336.

Zakharova, B. S., Ilyukhin, A. B., and Chudinova, N. N. (1994). Synthesis and crystal structure of vanadyl phosphite $(VO(HPO_3)(H_2O)_2)_n \cdot 3n(H_2O)$. *Zh. Neorg. Khim.* **39**, 1443–1445.

Zalkin, A., Ruben, H., and Templeton, D. H. (1978). Structure of a new uranyl sulfate hydrate, α-$(UO_2SO_4)_2(H_2O)_7$. *Inorg. Chem.* **17**, 3701–3702.

Zemann, J. (1948). Formel und Strukturtyp des Pharmakosiderits. *Tscherm. Mineral. Petrogr. Mitt.* **1**, 1–13.

Zemann, J. (1956). Die Kristallstruktur von Koechlinit, Bi_2MoO_6. *Beitr. Mineral. Petrogr.* **5**, 139–145.

Zemann, J. (1959). Isotypie zwischen Pharmakosiderit und zeolithischen Germanaten. *Acta Crystallogr.* **12**, 252.

Zhang, P. and Tao, K. (1981). Zhonghuacerite $Ba_2Ce(CO_3)_3F$ – a new mineral. *Scientia Geol. Sinica.* **4**, 195–196.

Zhou, J., Li, J., and Dong, W. (1988). The crystal structure of xitieshanite. *Kexue Tongbao* **33**, 502–505.

Zhukov, S., Yatsenko, A., Chernyshev, V., Trunov, V., Tserkovnaya, E., Antson, O., Hoelsae, J., Baules, P., and Schenk, H. (1997). Structural study of lanthanum oxysulfate $(LaO)_2SO_4$. *Mater. Res. Bull.* **32**, 43–50

Zima, V. and Lii, K.-H. (2003). Synthesis and crystal structures of $(NH_3CH_2CH_2NH_3)_{1.5}$ $[(VO)_2(HPO_4)_2(PO_4)]$ and $(C_4H_{12}N_2)[V_4O_6(HPO_4)_2(PO_4)_2]$, two layered vanadium phosphates templated with organic diamines. *J. Solid State Chem.* **172**, 424–430.

Zoltai, T. (1960). Classification of silicates and other minerals with tetrahedral structures. *Am. Mineral.* **45**, 960–973.

Zou, X. and Dadachov, M. S. (1999). A new mixed framework compound with corrugated $[Si_6O_{15}]_{\infty\infty}$ layers: $K_2TiSi_6O_{15}$. *J. Solid State Chem.* **156**, 135–142.

Index

abernathyite 38, 287
allochalcoselite 174, 181, 267
althupite 169, 283
alumoklyuchevskite 174
alunite 59, 62, 64, 66, 230, 257, 270,
    274, 281, 282
amarantite 74, 294
amarillite 222, 270
ammonioalunite 63
ammoniojarosite 63
anapaite 86–7, 238, 288
anion topology 163–72
annabergite 71, 250, 301
anorthominasragrite 87, 240
antiperovskite structures 203–9
apjohnite 87, 274
aplowite 87
arctite 205, 207–8
argentojarosite 63, 251
armstrongite 98, 148–9, 258
asisite 195, 287
augmentation of nets 95–6
austinite 132, 239, 250
autunite 35, 37–40, 42, 44, 49, 65, 86, 120, 121,
    124, 126, 127, 142, 145, 272,
averievite 174, 187, 189–90, 293

baricite 71, 302
basic graph 7, 8, 40, 42, 43–46, 53
batisite 98, 108, 110–1, 113, 280, 286, 290
beaudantite 63
beaverite 63, 235
belkovite 98,
benauite 63, 299
benitoite 98, 135–138, 239, 246, 303
bergenite 169, 272
beyerite 195, 268
bismoclite 195
bismuthite 195
blödite 86–7
blossite 133–6, 267
bobierrite 71, 295
bobjonesite 87, 289
boggildite 78
boltwoodite 168, 236
botryogen 73, 294
brandtite 75, 241
brassite 73, 284
brianite 11
brushite 220, 241

burnsite 174, 176–7, 236
butlerite 75, 124, 126–8, 245, 290

cabalzarite 71, 235
calciovolborthite 132
caledonite 74, 249
cassidyite 75
cation arrays 209–214
cavansite 143, 146, 149, 272, 286, 293
cebaite-(Ce) 210–4, 302
cechite 132, 283
chalcanthite 73, 225
chalcocyanite 132, 301
chenite 87, 256
chloromenite 174, 179, 181–2, 265
chlorothionite 74, 249
chloroxiphite 183, 246
cobaltlotharmeyerite 71, 262
collinsite 75, 235, 302
combinatorial polymorphism 2, 160, 277
composite building block 98
connectedness 6, 8, 223, 225
connectivity diagrams 34, 37, 79, 80, 84, 177,
    179, 180
conichalcite 132, 285
coordination sequence 17–8
coparsite 174
copiapite 78, 231, 245, 294
coquimbite 87, 104–5, 222, 245, 249
corkite 63, 250
crandallite 62, 63, 233, 297, 299
cuprosklodowskite 168, 287
curetonite 60, 240

dalyite 148–9, 247, 291
damaraite 183, 263
davanite 148–9
decoration of nets 95–6
delhayelite 149, 237
deloryite 75, 285
demesmaekerite 78, 250
descloizite 132, 230, 240, 255, 261, 283,
    285, 301
destinezite 74, 282
dewindtite 169, 284
dimensional reduction 215
dimensionality, definition of 96
dolerophanite 174, 190, 243
dorallcharite 63, 230
duftite 132, 260, 293

dumontite 169, 272
dussertite 63, 261
dwornikite 125

elpidite 98, 153–4, 156, 237, 279, 289
erythrite 71, 301
euchlorine 176, 290
Euler equation 106
eylettersite 63, 297
expansion of nets 95–7

face symbol 106–10
fairfieldite 75
fedotovite 174, 176, 293
ferrarisite 219–21, 237–8
ferrilotharmeyerite 71, 262
ferrinatrite 78, 222, 290
fibroferrite 73, 290
fingerite 176, 246
florensite 63, 269
francisite 190–2, 233, 284
francoisite-(Nd) 169, 284
freedite 189, 283
fundamental building block 97–99, 101, 103,
    200, 236, 251, 303
fundamental chain 97, 113–116, 123–4,
    127–8, 130–4

galeite 205–7
gallobeudantite 63, 257
gartrellite 71, 262
gaidonnayite 98
georgbokiite 174, 184–6, 189, 266
gittinsite 154, 156, 287
gorceixite 63, 285
goldichite 10, 13, 222, 251, 263, 285
gordonite 60, 269
gottlobite 132, 302
grandreefite 195
guerinite 219–21, 237–8
guilleminite 169, 240
gunningite 125

haidingerite 220, 246
hannayite 23, 237
hatrurite 205, 207
helmutwinklerite 71, 262
hidalgoite 63
hilairite 98, 137–8, 140, 257, 285
hilgenstockite 176
hillite 75, 302
hinsdalite 63, 261
hohmannite 74, 290
huanghoite-(Ce) 210–4
huangite 63, 270
hügelite 169
humberstonite 88, 236
hydronium ion, H_3O^+ 4, 5, 63, 273

ilinskite 174
isomerism
    cis-trans 12, 34–5, 79, 80, 84, 86
    geometrical 2, 10, 18, 21, 24, 27, 28, 32, 34,
        35, 50, 51, 54, 57, 58, 60, 61, 72, 74, 77, 79,
        82, 84, 144, 147–8, 150–2, 169, 170, 263–4,
        267, 272
    structural 16, 131, 255
    topological 15, 56, 60, 86

jarosite 5, 63, 230, 259, 274
johannite 168
jokokuite 73

kaatialaite 78
kamchatkite 174, 181, 298
kasolite 168, 287
kastningite 60, 227
keldyshite 98,
kemmlitzite 63
kettnerite 195, 251
kieserite 124–6, 250, 255, 301
kingite 60–2, 64, 66, 68, 299
kleinite 199
klyuchevskite 174, 183, 251
koechlinite 195
koettigite 71, 256
kogarkoite 205–7
komarovite 98
kombatite 195, 240
komkovite 138
kornelite 20, 287
kostylevite 98
krausite 72, 76, 118, 222, 244
krettnichite 71, 235
kroehnkite 70, 115, 117–119, 121, 124, 241, 255
kukharenkoite-(Ce) 210–1, 214, 304

labuntsovite 98, 1079, 113, 229, 239
lanarkite 173, 179, 183, 204
langbeinite 103–4, 278
laueite 54, 56–58, 60, 61, 64, 65, 231, 263,
    269, 277
lemoynite 98
leonite 87, 225, 249
leucophosphite 98–9, 252, 269, 277, 284, 302
linarite 74, 243
lone pair of electrons 23, 84, 116, 258, 302
lotharmeyerite 71, 235
lukranite 71

macdonaldite 149, 237
malayaite 125, 230, 256
mangangordonite 60
manganlotharmeyerite 71, 235
mansfeldite 142
marecottite 168, 235
marthozite 169

maxwellite 125, 240
mawbyite 71, 260
mendozite 222, 245
mereiterite 87, 249
melanothallite 174, 199, 266
mendipite 183, 263
metasideronatrite 76, 290
metastructures 95, 289
metatorbernite 38, 272
metauranocircite 38
metavauxite 54, 57, 59, 60, 231, 277
metavoltine 87, 249
metazeunerite 38, 272
minamiite 63, 282
minasragrite 87
minyulite 54, 56, 57, 67, 243, 259
mitryaevaite 60–2, 64–5, 68, 237
mixed anionic radicals, theory of 2, 289
modular approach 42, 210, 293
mohrite 225
monetite 220, 238
montgomeryite 64–5, 67–8, 245, 278
montregianite 148–9, 249
morinite 87, 255
mosesite 199
mottramite 132, 240
mounanaite 71, 235, 262
mozartite 132, 230, 281

nacaphite 204–9, 260, 268, 293
nadorite 195, 250
nanotube 86, 89, 136
nanotubules 88, 89, 91, 267
Nasicon 103–5
natroalunite 63, 281
natrochalcite 71, 250, 288
natrojarosite 63
nets, regular 94–5
newberyite 20, 227, 230, 294
nickellotharmeyerite 71, 262
nickelschneebergite 71, 263

olmsteadite 12, 34, 37, 243, 278
orientation matrix 24, 26, 28, 34, 51, 57, 74, 79,
    148, 169, 171
orthowalpurgite 75, 262
osarizawaite 63, 250

parabutlerite 73, 234, 290
paracoquimbite 87, 287
parageorgbokiite 174, 184–6, 189, 268
parallelohedra 106
parasymplesite 71
paravauxite 60, 231
parkinsonite 195, 295
parsonsite 172, 236, 272
pauflerite 124–5
penkvilksite 98, 271

pentagonite 149
pentahydrite 73
perite 195, 260
petarasite 98, 112–3, 249
pharmacolite 220, 246
pharmacosiderite 102–3, 232, 236, 241, 253
phaunouxite 220, 237
philolithite 183, 278
phoenicochroite 187, 301
phosphogartrellite 71
phosphovanadylite 103, 274
phosphuranylite 115, 165, 169, 242, 272, 283
    anion topology 165–171
phuralumite 169, 284
phurcalite 169, 229
piypite 174, 183, 258
plumbogummite 63, 261
plumbojarosite 63, 295
"plumbonacrite" 176,
poitevnite 125, 225
polyhalite 88
polyhedral units 97, 105–112, 135–6
polyphite 205, 208–9
ponomarevite 174
pre-nucleation building blocks 82
protonation of tetrahedral oxyanions 3, 4
pseudolaueite 54, 56, 57, 59, 60, 62, 66, 231, 277
pyatenkoite-(Y) 138
pyrobelonite 132, 261

quenstedtite 86–7, 296

rauenthalite 220, 237
rhodesite 149, 256
rhomboclase 10, 11, 263
ring symbol 7, 11–14, 16, 20, 23, 27, 143, 144,
    146, 148, 150
roemerite 86–7, 244
roselite 75, 255
rozenite 87
russelite 195

sahlinite 195, 234
sainfeldite 220, 246
saleeite 38
sazhinite-(Ce) 148–9, 291
sazykinaite-(Y) 138
schairerite 205–7
schertelite 87, 237
schlossmacherite 63
schmiederite 74, 243
schneebergite 71, 263
scorodite 142, 254
secondary building block 42
seelite 168, 283
shcherbakovite 98, 108, 110–1, 113, 266, 297
sideronatrite 76, 290
siderotil 73, 283

sidpietersite 181, 240
sigloite 60, 255
sincosite 38, 247
sitinakite 103
sklodowskite 168,
slavikite 56, 62
spheniscidite 99
starkeyite 87
stella quadrangula 177–8, 200, 281
stereohedra 106
stewartite 54, 56–58, 60, 277
stoiberite 179, 181
strengite 142
structural hierarchy 1, 98, 236, 255
structural subunit 1, 98
strunzite 60, 245
sulphohalite 204–6, 282
svanbergite 63, 259
swedenborgite 176, 257
symesite 195, 301
szmikite 125
szomolnokite 125

talmessite 75, 237
tamarugite 222, 287
terskite 98, 285
thometzekite 71, 262
thorikosite 195
threadgoldite 38, 260, 284
tilasite 125, 233
tilings 106–9, 137, 164
tinsleyite 99
titanite 125–6, 229, 230, 255–6, 296, 303
torbernite 38, 272, 287
tsumcorite 67, 71–2, 235, 262–3, 296
tsumebite 77
tubular units 97, 135–140
twin boundary 46

uklonskovite 73, 288
umbite 98, 150, 152, 155, 257, 271
ungemachite 88, 251

upalite 169, 283
uranophane, alpha 168, 250
uranophane, beta 168, 298
ushkovite 60, 248

vanadomalayaite 125, 230
vanmeersscheite 169, 283
vanthoffite 42
variscite 141–2, 162, 261, 263, 272, 296
vergasovaite 181, 233
vuagnatite 132, 274
vertex symbol 106–7, 109–10
vivianite 71, 230, 245, 278, 301
vladimirite 220, 237
vlasovite 98, 137–9, 250, 299
vlodavetsite 124, 127, 293

wadeite 98
walpurgite 75, 183, 275
walthierite 63, 270
waylandite 63, 239
weilerite 63
weilite 220, 246
wherryite 77, 240
wilhelmkleinite 125, 227
woodhouseite 63, 259

xitieshanite 87, 304

yakovenchukite 148, 153, 268
yanomamite 142
yavapaite 11, 222, 229, 249, 251, 262
yuksporite 88, 91–93, 268

zairite 63, 297
zavaritskite 195
zektzerite 98, 110, 112, 249
zeunerite 38, 272
ziesite 133–6, 257, 267
zincobotryogen 73, 256
zincosite 132, 301
zorite 98, 134–5, 137, 275, 288